T0330547

Introduction to
Statistical
Process Control

CHAPMAN & HALL/CRC
Texts in Statistical Science Series

Series Editors

Francesca Dominici, *Harvard School of Public Health, USA*
Julian J. Faraway, *University of Bath, UK*
Martin Tanner, *Northwestern University, USA*
Jim Zidek, *University of British Columbia, Canada*

Texts in Statistical Science

Introduction to
Statistical
Process Control

Peihua Qiu

Department of Biostatistics
University of Florida, USA

CRC Press
Taylor & Francis Group
Boca Raton London New York

CRC Press is an imprint of the
Taylor & Francis Group an **informa** business

A CHAPMAN & HALL BOOK

CRC Press
Taylor & Francis Group
6000 Broken Sound Parkway NW, Suite 300
Boca Raton, FL 33487-2742

© 2014 by Taylor & Francis Group, LLC
CRC Press is an imprint of Taylor & Francis Group, an Informa business

Printed on acid-free paper
Version Date: 20130703

International Standard Book Number-13: 978-1-4398-4799-2 (Hardback)

Library of Congress Cataloging-in-Publication Data

Qiu, Peihua, 1965-
    Introduction to statistical process control / Peihua Qiu.
        pages cm. -- (Chapman & Hall/CRC texts in statistical science)
    Includes bibliographical references and index.
    ISBN 978-1-4398-4799-2 (hardback : alk. paper)
    1. Process control--Statistical methods. I. Title.

TS156.8.Q48 2013
658.5--dc23   2013023425

Visit the Taylor & Francis Web site at
http://www.taylorandfrancis.com

and the CRC Press Web site at
http://www.crcpress.com

To the memory of my mother

Contents

List of Figures

List of Tables

Preface

Statistical process control (SPC) is for monitoring sequential processes (e.g., production lines, Internet traffic, medical systems, social or economic status) to make sure that they work stably and satisfactorily. It is a major tool for quality control and management. In the past 10–20 years, SPC has made great progress. Many new SPC methods have been developed for improving traditional SPC methods and for handling new SPC applications. This book aims to make a systematic description of both the traditional and recent SPC methods.

I started doing my research on SPC around 1998 after I joined the faculty of the School of Statistics at the University of Minnesota. At that time, I was doing research on jump regression analysis, which is about regression modeling when the underlying regression function has jumps. My senior colleague, Professor Doug Hawkins, published a book on CUSUM charts that year, and he kindly gave me a free copy of his book. Before I read the book, I had heard of CUSUM charts, but never had a chance to learn that subject systematically. The book makes a thorough description about the existing CUSUM charts. It has 10 chapters. Nine of them are on cases when a univariate quality characteristic variable is of interest for process monitoring, and only one of them is on multivariate SPC, which is about cases with multiple quality characteristic variables being monitored. All multivariate SPC methods described in that chapter are based on the assumption that the multiple quality characteristic variables follow a joint normal distribution. In my opinion, there are two major limitations in the SPC research of that time. First, in most applications, the quality of a product is affected by multiple characteristics of the product (cf., Section 7.1 for a more detailed explanation). Therefore, the SPC research should focus on multivariate cases, instead of univariate cases. Second, my extensive consulting experience tells me that multivariate data would hardly follow a joint normal distribution. Therefore, we should address the multivariate SPC problem without the normality assumption. With encouragement from Doug, I started my own research on multivariate SPC. After finding the fact that traditional multivariate SPC charts would be unreliable in cases when the normality assumption was invalid (cf., Figures 9.1 and 9.2), I started to study existing statistical methods for describing multivariate non-normal data and for transforming multivariate non-normal data to multivariate normal data. After more than one year's study, I realized that existing statistical tools for describing and handling multivariate non-normal data were limited, and the multivariate SPC problem should be handled using certain statistical methods that were not based on the normality assumption. Since then, my co-authors and I have proposed antirank-based multivariate SPC charts that are appropriate to use in most applications and a general framework to handle the multivariate SPC problem using the log-linear model (cf., Sections 9.2 and 9.3). Besides our research on multivariate SPC, my co-authors and I have also made some contributions to univariate SPC, SPC based on change-point detection, and profile monitoring.

This book has 10 chapters. Chapter 1 describes briefly the concept of quality, the early history of research on quality improvement, some basic concepts on quality management, the role of SPC and other statistical methods in quality control and management, and the overall scope of the book. Chapter 2 describes some basic statistical concepts and methods that are useful in SPC. Chapters 3–5 make a systematic description of some traditional SPC charts, including the Shewhart, CUSUM, and EWMA charts. Some more recent control charts based on change-point detection are described in Chapter 6. Some fundamental multivariate SPC charts under the normality assumption are described in Chapter 7. Then, Chapters 8 and 9 introduce some recent univariate and multivariate control charts designed for cases when the normality assumption is invalid. Control charts for profile monitoring are discussed in Chapter 10. At the end of each chapter, some exercises are provided for readers to practice the methods described in the chapter. Computations involved in all examples in the book are accomplished using the statistical software package R. Some basic R functions are introduced in Appendix A, along with some R packages developed specifically for SPC analysis and all R functions written by the author for the book. A list of all datasets used in the book is given in Appendix B.

The mathematical and statistical levels required are intentionally low. Readers with some background in basic linear algebra, calculus through integration and differentiation, and an introductory level of statistics can understand most parts of the book without much difficulty. For a given topic, some major methods and procedures are introduced in detail, and some more advanced or more technical material is briefly discussed in the section titled "Some Discussions" of each chapter. For some important methods, pseudo computer codes are given in the book. All R functions and datasets used in the book are posted on the author's web page for free download.

This book can be used as a primary textbook for a one-semester course on statistical process control. This course should be appropriate for both undergraduate and graduate students from statistics, industrial engineering, systems engineering, management sciences, and other related disciplines that are concerned about process quality control. It can also be used as a supplemental textbook for courses on quality improvement and system management. SPC researchers from both universities and industries should find this book useful because it includes many of the most recent research results in various SPC research areas, including univariate and multivariate nonparametric SPC, SPC based on change-point detection, and profile monitoring. Quality control practitioners in almost all industries should find this book useful as well, because many state-of-the-art SPC techniques are described in the book, their major advantages and limitations are discussed, and some practical guidelines about their implementations are provided.

I am grateful to Doug Hawkins for introducing the SPC topic to me, for his guidance on my early SPC research, for his constant encouragement and help, and for his numerous comments and suggestions during the course of my SPC research.

I thank all my co-authors of SPC research, including Singdhansu Chatterjee, Doug Hawkins, Chang Wook Kang, Zhonghua Li, Zhaojun Wang, Jingnan Zhang, Jiujun Zhang, and Changliang Zou, for their patience, stimulating discussions, and helpful comments and suggestions. I am fortunate to have had Josh Wiltsie read the entire manuscript. He provided a great amount of constructive comments and suggestions. Both Giovanna Capizzi and Arthur Yeh provided detailed review reports about the book manuscript that greatly improved the quality of the book. Part of the book manuscript was used as lecture notes in my recent advanced topic course offered at the School of Statistics of University of Minnesota in Fall 2012. Students from that class, especially Mr. Yicheng Kang, corrected a number of typos and mistakes in the manuscript. Dr. Changliang Zou kindly helped me with the computation of the nonparametric EWMA chart described in Subsection 8.2.3.

This book project took more than 3 years to finish. During that period of time, my wife, Yan, gave me a great amount of support and help, by taking care of our two sons and much household work as well. My two sons, Andrew and Alan, helped me in their own way by not interrupting me during my writing of the book manuscript at home and by keeping my home office quiet. I thank all of them for their love and constant support.

PEIHUA QIU
Gainesville, Florida
August 2013

Chapter 1

Introduction

In our daily life, we often talk about the *quality* of a product. The product can be small items, such as the clothes, shoes, watches, food, and so forth, that we could not live without. It can also be larger items, such as bikes, cars, televisions, and so forth, that our families routinely use. It can even be the network systems that manage our communities or even the entire society, such as the banking system, health care system, internet system, and so forth. In cases when the quality of any of these products has a problem, our life could become miserable. Besides the products mentioned above that affect everyone's life, we are actually concerned about the quality of any products that we produce, although some of them may not affect everyone's life, or they may not affect some people's lives directly (e.g., certain satellites, luxury products). This raises an important question. How can we control or assure the quality of products? To answer the question, many management philosophies and statistical methodologies have been developed in the literature. A central part of these methodologies is the so-called *statistical process control (SPC)*. This book describes some fundamental SPC methodologies.

Although SPC provides major statistical methodologies for quality control, there are some other scientific methods that are helpful for improving the quality of products. In this chapter, we give a brief overview of these methods, and introduce some basic concepts and terminologies that are related to quality, quality improvement, and SPC. From the introduction, we hope to provide a big picture about quality control and the major role of SPC in quality control.

1.1 Quality and the Early History of Quality Improvement

Most people have their own conceptual understanding of quality. Intuitively, a product with good quality should meet the requirements of its users. This is the so-called "fitness-for-use" criterion that is commonly used in the literature for defining quality. Garvin (1987) gives eight dimensions to the definition of quality, which are briefly summarized in the box below.

1

Eight Dimensions of Quality

Performance concerns how well a product performs certain specific functions.
Features refer to the functions that the product performs, especially to those functions that its competitors do not have.
Reliability concerns how often the product fails.
Conformance concerns whether the product meets its designed standard.
Durability refers to the effective service life of the product.
Serviceability concerns the maintenance of the product, and the product is good in this dimension if its maintenance is easy and convenient.
Aesthetics is related to the visual appeal of the product.
Perceived quality refers to the reputation of the product.

Therefore, quality is a multifaceted concept. It is also a dynamic concept in the sense that it keeps changing over time. For instance, a good quality personal computer bought ten years ago would not be good any more in today's standards.

In the history of human beings, improving the quality of life might be one major motivation for the entire society to keep making progress. For instance, in order to improve communication, our ancestors created numbers and languages. To further improve communication and computation, computers and the internet were developed not so long ago. Nowadays, more and more modern communication techniques are developed on a daily basis, making our communication even more convenient. On the other hand, the quality of our life is reflected in the quality of many different products and systems that we use. It is difficult to imagine a good life with bad services of our banks and medical systems and with our cars, computers, and other daily used products frequently broken. So, quality and its improvement is not new to us. It actually exists in our daily life and in the entire history of mankind. For instance, the ancient Chinese people invented the movable type system of printing about one thousand years ago (Needham, 1986), and we have been trying to improve the quality of printing since then.

Although quality and quality improvement are an indispensable component of our society, we did not have systematic theory and methods about them until about a century ago. At that time, several western companies, including the AT&T Company, realized the importance of quality assurance of their products and services. The Western Electric Company was reorganized by AT&T in 1907 for inspection and testing of manufactured or installed products and purchased materials of the AT&T Company (Wadsworth et al., 2002). The Inspection Department within the Western Electric Company had more than 5000 members by 1924. One prominent figure in that department was Joseph M. Juran who later became a well-known international consultant in quality control and management. In 1925, a new Inspection Department was created in the newly formed Bell Telephone Laboratories. This department had several pioneers of modern quality control and management, including Donald A. Quarles, Walter A. Shewhart, Harold F. Dodge, and George D. Edwards. Many important statistical concepts and terminologies, including the Shewhart control chart, risks of type I and type II errors in statistical hypothesis testing, were created by them around that time.

During World War II, because of the US involvement in the war, a large amount of war materials were needed, which resulted in the rapid expansion of the US manufacturing industry. To assure the quality of the manufactured goods, many skilled people in quality inspection and testing were needed. Consequently, some training programs in quality inspection and testing were established by individual companies and by government organizations as well, such as the War Department. One important figure working with the War Department was Dr. W. Edwards Deming, who later became a leading quality control consultant. In the 1940s, several research groups were established across the country to perform research in quality control. The famous Hotelling's T^2 statistic was proposed by Harold Hotelling in 1947. (Hotelling was a member of the Statistical Research Group at Columbia University.) In 1946, the American Society of Quality Control was established, with George Edwards of the Bell Telephone Laboratories as the first president of the new society. For many other early developments in quality control and management, readers are encouraged to read an overview given in Chapter 1 of Wadsworth et al. (2002).

1.2   Quality Management

A *production process* turns materials, workers' labor and skills, manufacturing facilities (e.g., machines), and other necessary inputs into desired outputs, in the form of products. Then, the products are properly evaluated and monitored. Any problems found in the evaluation and monitoring will be investigated, and the process is further improved. See Figure 1.1 for a demonstration. The products of the production process can be soft-drink cans, shoes, and other physically visible items. They can also be software packages, services, information, and other things that may not be physically visible but still meet the needs of certain people. In this book, if there is no further specification, products refer to both types, although the first type is mentioned more often in our description for simplicity of perception.

The quality of a production process is determined by the quality of its products. The production process is of high quality only if its products are overall of high quality. The quality characteristics of a product can usually be measured numerically, and thus be denoted by numerical variables. For instance, if the weight of a soft-drink can is an important quality characteristic, then that characteristic can be denoted by a numerical variable "weight." However, certain quality characteristics cannot be conveniently measured numerically. For instance, a soft-drink can should be classified as a defective product if it is dented. In this case, the shape of the soft-drink can is categorical, and it has two categories "dented" and "not-dented." In the literature, such binary categorical quality characteristics are often called *attributes*. As a distinction, the numerical quality characteristics are often called variables. To make the related terminologies consistent with those in other areas of statistics, in this book, we call all quality characteristics variables, and they are classified as *continuous numerical variables, discrete numerical variables,* and *categorical variables* (cf., Peck and Devore, 2012). The traditional attributes are just binary categorical

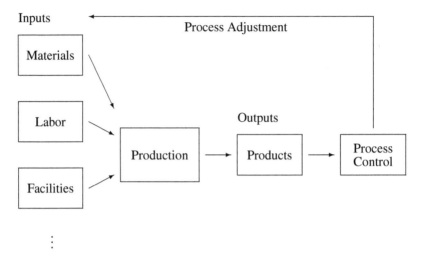

Figure 1.1: *Demonstration of a production process.*

variables with two categories "defective" and "non-defective," or "conforming" and "non-conforming." One benefit with the proposed classification system of the quality characteristics is that all possible quality characteristics can be properly classified in this system. In the traditional classification system using variables and attributes, categorical quality characteristics with more than two categories would be difficult to classify, and such categorical quality characteristics are popular in applications (e.g., color, taste, customers' feedback on certain services in categories of "satisfactory," "neutral," "dissatisfactory," and so forth).

As described in Section 1.1, quality is a multifaceted concept. So, to have a complete evaluation of the quality of a production process, we should consider different quality characteristics that can be used for evaluating different dimensions of the quality. For instance, for the soft-drink can mentioned above, besides the amounts of major ingredients in the drink (for evaluating its performance, conformance, and features), we may also want to consider the decoration on the can surface (for evaluating its aesthetics), its lifetime from production to the time that the drink inside the can turns bad (for evaluating its durability), and so forth.

The values of the quality characteristics of a product are obviously related to the input variables of the related production process (cf., Figure 1.1). Because of the fact that it is impossible to exhaustively list all possible input variables that have an impact on the quality characteristics of the product and that some listed input variables are uncontrollable, the relationship between the quality characteristics and the controllable input variables has a certain randomness involved (cf., Qiu, 2005, Section 2.1). When we design a product or a production process, we usually set some specific requirements on the randomness. For instance, for a specific type of soft drink, we may require the average value of sugar content in a can to be at a given level, with a given upper bound on a spread measure (e.g., the standard deviation) of the sugar content readings of different cans. The quality of the products that meet such designed requirements is referred to as the *quality of design*. Intuitively, the soft drink would have a good quality of design if the designed spreads of the quality characteristics are small, the designed average values of the healthy ingredients are high, and the designed average values of the unhealthy ingredients are low. It is natural to require a good quality of design on all products. But, in practice, a better quality of design usually implies a higher cost and a lower productivity. Therefore, we need to make compromises among these related considerations.

After the designed requirements on a product have been specified, the manufactured products of the related production process may not be able to meet all these designed requirements, due to various reasons, including defects in materials, inadequate skills of the workforce, poorly planned manufacturing, inappropriate inspection and tests, and so forth. Quality of a product related to its conformance to the designed requirements is called the *quality of conformance*.

To assure the quality of a production process, it is clear that the quality of all components and/or stages of the process should be assured, including the quality of the raw materials, education and skills of the workforce, manufacturing planning and operation, tests and monitoring of the products, and so forth. To this end, some quality management systems have been developed and implemented in industries, among which the *total quality management (TQM)* and *six-sigma* systems are especially popular. TQM places emphasis on identifying customers' needs and requirements, and on satisfying all these needs and requirements by continuously improving all aspects of the production process. Six-sigma takes a project-by-project approach. Each project focuses on improving certain substantial components of the production process, using a specific five-step problem-solving approach: define, measure, analyze, improve, and control (DMAIC). The DMAIC approach uses many statistical tools, including control charts, experimental design, regression analysis, and so forth.

In quality management, statistics turn out to play an important role. As demonstrated by Figure 1.1, quality of a production process depends on various input variables, some of which are uncontrollable or even difficult for us to recognize as existing. As a consequence, the relationship between the quality characteristics of a product and the controllable input variables is random. To find and quantitatively describe such a random relationship, *design of experiments (DOE)*, *analysis of variance (ANOVA)*, and *regression analysis* provide major statistical tools. More specifically, DOE provides a structured and organized method for determining the relationship between the controllable input variables and the output variables (i.e., the quality characteristics in the current setting). It usually involves designing a set of experiments, in which all relevant input variables are varied systematically. When the results of these experiments are obtained (i.e., the output variables are observed in all experiments) and analyzed by ANOVA or regression, those input variables that most influence the output variables can be identified, along with the input variables that do not influence the output variables significantly. To investigate whether manufactured products meet all their designed requirements, control charts are especially useful. By a control chart, manufactured products of a production process are sampled sequentially, and their quality characteristics are monitored over time. Any deviations of the quality characteristics from their designed requirements would be signaled as soon as possible, and the root causes of such deviations would be identified and properly deleted, which is usually achieved by adjusting the controllable input variables. Besides the statistical methods mentioned above, there are some other methods that are commonly used in quality management. For instance, when inspecting raw materials or manufactured products, *acceptance sampling* plans provide appropriate sampling methodologies for inspection and make decisions on whether the raw materials or manufactured products should be accepted or rejected after the sampled items are inspected.

1.3 Statistical Process Control

From the brief discussion about quality control and management in the above two sections, it can be seen that it is important to check whether the manufactured products conform to their designed requirements. If the answer is negative, then the production process should be stopped as soon as possible. Otherwise, much time and money is wasted because the products would not be able to meet consumers' needs in such cases. SPC is a major statistical tool for checking the conformance of the products to their designed requirements, and it is the focus of this book. For those who are interested in reading a broader description about quality control and management, or about other statistical tools that are useful in quality control, we recommend books by Montgomery (2009), Ryan (2000), Wadsworth et al. (2002), among many others.

As mentioned in Section 1.2, even in cases when the production process works stably, the quality characteristics of interest would still have randomness or variability involved. This type of variability is mainly caused by certain uncontrollable (or difficult/expensive to control) input variables, and is referred to as *common cause variation*. Common cause variation is considered to be an inherent part of the production process and cannot be changed without changing the process itself. However, with the progress in science and technology, certain uncontrollable (or difficult/expensive to control) input variables in the past might now be partially or completely controllable. Consequently, certain common cause variation can be removed, and the quality of design of the production process can be improved. In cases when only common cause variation is present in the production process, the process is considered to be *stable*, *in statistical control*, or simply *in-control (IC)*. When some component(s) of the production process become(s) out of order, certain corresponding quality characteristics would have a relatively large variability or a systematic variation from the designed requirements, and consequently many manufactured products would not be able to meet the designed requirements. This type of variation is referred to as *special (or assignable) cause variation*, and the component(s) of the process that cause(s) the variation is the special cause. Examples of special causes of variation include defective raw materials, improper operation of the workers, improperly adjusted machines, and so forth. When a production process has special cause variation present, it is considered to be *unstable*, *out of statistical control*, or simply *out-of-control (OC)*. The major goal of SPC is to distinguish special cause variation from common cause variation, and give a signal as soon as special cause variation occurs.

SPC is often divided into two different phases. In phase I, we try to properly set up a production process and make it run stably. Because of the fact that we do not know much about the process at the beginning of this phase, statistical analysis involved in this phase has an exploratory nature. Usually, the numerical relationship between the quality characteristics of the manufactured products and certain controllable input variables is first studied, using DOE, ANOVA, regression, and other related statistical methods. Then, the controllable input variables are set at some specific levels such that the quality characteristics would roughly meet their designed requirements by the established relationship between the controllable input variables and the quality characteristics. Under this condition, a set of process data (i.e., observations of the quality characteristics) is gathered and analyzed by a SPC chart. Any unusual "patterns" in the data lead to adjustments and fine tuning of the controllable input variables of the process. Then, a new set of process data is collected under the adjusted condition, and analyzed by a SPC chart again. This control-and-adjustment step is repeated several times, until all special causes are believed to be removed and the process works stably. Then, we are left with a clean set of data, gathered under stable operating conditions and illustrative of the actual process performance. This set is then used for estimating the IC distribution of the quality characteristics. In phase II, the process is believed to be IC at the beginning, and our major goal is to monitor the process online to make sure that it keeps running stably. To this end, manufactured products are sampled sequentially over time, and observations of the quality characteristics of the sampled products are monitored using a SPC chart, and the chart gives a signal once it detects a significant special cause deviation of the quality characteristics. After a signal of a special cause deviation is delivered, the process should be stopped immediately, and the special causes should be figured out and removed. From the above description, we can see some substantial differences between phase I and phase II SPC problems. First, the size of the observed process data is usually fixed in the phase I problem, while the observed data in the phase II problem increase sequentially. Second, the distribution of the quality characteristics is often unknown in the phase I problem, while this distribution is routinely assumed to be known or it can be estimated from an IC data in the phase II problem. Because of these substantial differences between phase I and phase II problems, SPC methodologies designed for solving the two problems are also quite different.

After creation of the Shewhart control chart by Walter A. Shewhart more than 80 years ago (Shewhart, 1931), SPC has made tremendous progress. Noticeable progress in SPC includes the creation of the cumulative sum (CUSUM) control chart by Page (1954) and the exponentially weighted moving average (EWMA) control chart by Roberts (1959). One fundamental difference between a Shewhart chart and a CUSUM or EWMA chart is that the former uses only the data observed at the current time point for detecting variation caused by special causes and it ignores all data that are observed at earlier time points, while the latter uses all available data that are observed at both the current and earlier time points. In recent years, many new SPC methodologies have been developed for improving traditional SPC methods and for handling new SPC applications, which include control charts based on change-point detection, nonparametric control charts designed for cases when the traditional normality assumption is invalid, and control charts for monitoring profiles. This book will systematically describe both the traditional and some recent SPC methods.

1.4 Organization of the Book

This book has 10 chapters. Chapter 2 introduces some basic statistical concepts and methods that are useful in constructing and understanding SPC charts. The remaining eight chapters describe both the traditional and some recent SPC charts that are designed for handling different scenarios. Chapters 3–5 describe three traditional families of control charts that are commonly used in practice. They are the Shewhart charts, the CUSUM charts, and the EWMA charts, respectively. Chapter 6 introduces an alternative family of control charts that were proposed recently based on change-point detection. Compared to the Shewhart, CUSUM, or EWMA charts, control charts based on change-point detection have the advantage that the occurrence time of a special cause deviation can be estimated simultaneously when a signal of the special cause deviation is delivered. Control charts described in Chapters 3–6 are mainly for situations when a single quality characteristic is involved and its IC and OC distributions are normal. Such control charts are routinely called univariate SPC charts in the literature. When multiple quality characteristics are involved, the corresponding SPC problem is often referred to as the multivariate SPC problem. Some fundamental multivariate SPC charts under the normality assumption are described in Chapter 7. Then, Chapters 8 and 9 introduce some recent control charts designed for various cases when the normality assumption is invalid: Those for univariate SPC are introduced in Chapter 8 and those for multivariate SPC are introduced in Chapter 9. In certain applications, instead of monitoring one or more quality characteristics, we are interested in monitoring the functional relationship between a response variable and some predictors. This is the so-called *profile monitoring* problem in the literature. Some fundamental control charts for profile monitoring are discussed in Chapter 10. The book also has two appendices at the end. The first appendix introduces some basic functions of the statistical software package R, and some R packages and functions developed specifically for SPC analysis. The second appendix gives a list of all datasets used in the book.

This book is written in such a way that readers with some background in basic linear algebra, calculus through integration and differentiation, and an introductory level of statistics can easily understand most parts of the book. For a given topic, some major methods and procedures will be introduced in detail, and some more advanced or more technical material will be briefly discussed in the section titled "Some Discussions" of each chapter. For some important methods, pseudo computer codes will be given in the book, and all the datasets and source codes in R that are used in the examples, figures, tables, and exercises of the book will be posted on the book web page for free download. At the end of each chapter, some exercises are provided for readers to practice the methods described in the chapter.

1.5 Exercises

1.1 Choose a product that you are familiar with, and discuss the eight dimensions of its quality.

1.2 For the production process of soft-drink cans of a given brand in a factory, list all possible input variables, and answer the following questions:

(i) Among the variables listed, which ones are controllable?

(ii) Among the variables listed, which ones are uncontrollable or controllable but difficult/expensive to control?

(iii) Are there any other factors or variables that might affect the quality of the products but are ignored in your list?

The variables considered in parts (ii) and (iii) are the major source of common cause variation in the quality characteristics of products. Discovery of them can potentially improve the quality of design of products.

1.3 Assume that you are interested in buying a car of a specific brand and model, and would like to know the designed requirements of certain quality characteristics. Can you list at least 10 such quality characteristics? Which ones are continuous numerical, discrete numerical, or categorical variables?

1.4 Using the soft-drink example in exercise 1.2, discuss the difference between quality of design and quality of conformance. To improve the quality of design of the products, what can we potentially do? What kind of actions or adjustments can potentially improve the quality of conformance?

1.5 Both quality of design and quality of conformance are related to costs and productivity. Using the soft-drink example in exercise 1.2, discuss this relationship in detail.

1.6 SPC charts try to distinguish special cause variation from common cause variation. Using a quality characteristic found in the car example in exercise 1.3, describe the two types of variation and their major causes, and then answer the following questions:

(i) In which situations is the related production process considered IC?

(ii) In which situations is the related production process considered OC?

(iii) If an OC signal is given by a control chart, what appropriate actions should we take?

1.7 Using the soft-drink example discussed in exercise 1.2, describe the phase I and phase II SPC problems.

Chapter 2

Basic Statistical Concepts and Methods

2.1 Introduction

As described in Chapter 1, statistical process control (SPC) is a major statistical tool for monitoring a production process to make sure that it works stably. The stability of the production process is reflected by the conformance of the quality characteristics of its products to their designed requirements. Because the quality characteristics are affected by both controllable and uncontrollable input variables of the production process (cf., Figure 1.1), observations of the quality characteristics would have random *noise* involved, which represents the common cause variation in the quality characteristics (cf., the related discussion in Section 1.3). To describe and analyze such data with random noise, the use of various statistical concepts and methods is necessary, and is briefly introduced in this chapter.

This chapter is written for the convenience of those readers who do not know or do not remember some of the basic statistical concepts and methods well. It can be skipped by readers with a background in introductory-level statistics. The introduction here is kept to a minimum. For a more complete discussion about statistical theory and inference, see, for example, Lehmann and Casella (1998), Lehmann and Romano (2005), and Casella and Berger (2002). For a more complete introduction about commonly used statistical methods, see. for example, Devore (2011).

2.2 Population and Population Distribution

In statistics, any statement and/or conclusion is only applicable to a specific *population*, which is the entire collection of members or subjects about which information is desired. For instance, if we are interested in knowing the working status of a production process of soft-drink cans of a given brand, then the collection of all manufactured soft-drink cans of that production process is our population.

In a particular application, usually we are only interested in one or several specific characteristics of the members in a population, e.g., indices of certain ingredients in a soft-drink can in the above example. These characteristics are often called *variables* because their values can change among different members in the population. As another example, suppose that we are interested in knowing the numerical relationship between the height and weight of all current college students in this country. Then, the collection of all current college students in the country is our population, and height and weight are two variables of interest. By the nature of their values, all

Table 2.1: *Probability distribution function of the grades of an introductory statistics class.*

Grade	A	B	C	D	F
Probability	0.09	0.35	0.38	0.15	0.03

variables are classified into three categories: *categorical variables, discrete numerical variables*, and *continuous numerical variables*. A variable is called a categorical variable if all its values are categories. Examples of categorical variables include gender, race, color, conformance status of a product, and so forth. A variable is discrete numerical if all its values are isolated numbers on the number line. It is continuous numerical if the set of all its values is an interval on the number line. Height and weight mentioned above are examples of continuous numerical variables, while the number of accidents on a given road segment during a given time period is an example of a discrete numerical variable.

Let us first focus on single-variable cases. Suppose that a member is randomly selected from a population. After the selection, the variable value of the selected member becomes known. But before the selection, the variable value of the member to be selected could be any possible value. In that sense, the variable is random and is therefore called a *random variable*. In the case when several variables are involved, the random variable is multivariate, or it is a random vector.

For a given population, it is often of interest to know how all the values in the population are distributed, or equivalently, how all the possible values of the related random variable are distributed. This distribution is called the *population distribution*. By the connection between a population and a random variable, the population distribution is the same as the distribution of the related random variable.

We now consider how to describe the population distribution. If the related random variable is categorical in the sense that all its possible values belong to several categories, then a table listing all the categories and the corresponding proportions of the population members in the categories is sufficient for describing the population distribution. This table is often called the *probability distribution function* or the *probability mass function* (see Table 2.1).

Example 2.1 *An introductory statistics class has 100 students. Assume that the course grading system has 5 levels: A, B, C, D, and F. At the end of the semester, 9 students in the class get the grade of A, 35 students get the grade of B, 38 students get the grade of C, 15 students get the grade of D, and 3 students get the grade of F. If we are interested in the distribution of students' grades in this class, then all students in the class constitute a population. Let X denote the grade of a randomly selected student in the class. Then, X is the random variable of our interest, and its probability distribution function, which is also the probability distribution function of the population, is given in Table 2.1.*

If a random variable X is univariate numerical, then its distribution can be described by the following *cumulative distribution function* (cdf):

$$F(x) = P(X \leq x), \qquad \text{for } x \in R, \tag{2.1}$$

where $P(X \leq x)$ denotes the probability of the event that X is less than or equal to a given value x on the number line R. From (2.1), $F(x)$ is obviously a non-decreasing and right-continuous function on R with its values in $[0,1]$. When x gets larger, $F(x)$ is closer to 1; it is closer to 0 when x gets smaller.

In the case when the cdf $F(x)$ is absolutely continuous, in the sense that there is a nonnegative, real-valued, measurable function f on R such that

$$F(x) = \int_{-\infty}^{x} f(u)\, du, \qquad \text{for } x \in R, \qquad (2.2)$$

the random variable X is said to be *absolutely continuous*, and f is called its *probability density function* (pdf). The corresponding curve of a pdf is called a (probability) density curve. Thus, for an absolutely continuous random variable X, its pdf f can be computed easily from its cdf F by the relationship

$$f(x) = F'(x), \qquad \text{for } x \in R. \qquad (2.3)$$

From the definition (2.2), a pdf must be a nonnegative integrable function, and the area under its entire curve is one. These are also the sufficient conditions for a measurable function on R to be a pdf. If X has a pdf f, then the area under the density curve and above an interval $[a,b]$ equals $P(a \leq X \leq b)$, for any $-\infty \leq a \leq b \leq \infty$. That is, areas underneath the density curve give probabilities for the random variable X.

If a random variable X is discrete, in the sense that all its possible values are isolated on the number line R, then the center of its possible values can be measured by

$$\mu_X = \sum_{j=1}^{N} x_j p_j, \qquad (2.4)$$

where $\{x_j,\ j = 1, 2, \ldots, N\}$ are all the values of X, and $\{p_j,\ j = 1, 2, \ldots, N\}$ are the corresponding probabilities. Obviously, μ_X is a weighted average of $\{x_j,\ j = 1, 2, \ldots, N\}$ with the probabilities $\{p_j,\ j = 1, 2, \ldots, N\}$ being the weights. In the case when the probabilities are all the same, μ_X is just the simple average of $\{x_j,\ j = 1, 2, \ldots, N\}$. In the literature, μ_X is often called the *mean* of X or the *expected value* of X. Another commonly used notation for μ_X is $\mathrm{E}(X)$, where E is the first letter of "expected value." The spread of all possible values of X can be measured by

$$\sigma_X^2 = \sum_{j=1}^{N} (x_j - \mu_X)^2 p_j, \qquad (2.5)$$

where σ_X^2 is called the *variance* of X. Sometimes we also write σ_X^2 as $\mathrm{Var}(X)$, where Var denotes "variance." Its square root σ_X is called the *standard deviation* of X. Because σ_X has the same unit as X, it is often more convenient for measuring the spread; but, mathematically, σ_X^2 is easier to handle. From (2.4) and (2.5), it is obvious that

$$\sigma_X^2 = \mathrm{E}(X - \mu_X)^2 = \mathrm{E}\left(X^2\right) - \mu_X^2.$$

Namely, σ_X^2 is the mean squared distance from X to its center μ_X, and it can also be computed from $E(X^2)$ and μ_X.

In cases when X is an absolutely continuous random variable with a pdf f, its mean is defined by

$$\mu_X = \int_{-\infty}^{\infty} u f(u)\, du, \tag{2.6}$$

and its variance is defined by

$$\sigma_X^2 = \int_{-\infty}^{\infty} (u - \mu_X)^2 f(u)\, du = E(X - \mu_X)^2. \tag{2.7}$$

In cases when $p > 1$ characteristics of the members in a population are of interest (e.g., p quality characteristics of the products of a production process), a p-dimensional random vector $\mathbf{X} = (X_1, X_2, \ldots, X_p)'$ can be used for denoting the p characteristics of a randomly selected member from the population. The (joint) cdf of $\mathbf{X}$ can be defined similarly to (2.1) by

$$F(\mathbf{x}) = P(X_1 \le x_1, X_2 \le x_2, \ldots, X_p \le x_p), \qquad \text{for } \mathbf{x} = (x_1, x_2, \ldots, x_p)' \in R^p, \tag{2.8}$$

where R^p denotes the p-dimensional Euclidean space. If there is a nonnegative, real-valued, p-dimensional, measurable function f on R^p such that

$$F(\mathbf{x}) = \int_{-\infty}^{x_1} \cdots \int_{-\infty}^{x_p} f(\mathbf{u})\, d\mathbf{u}, \qquad \text{for } \mathbf{x} \in R^p, \tag{2.9}$$

then $\mathbf{X}$ is said to be absolutely continuous, and f is its (joint) pdf.

In cases when the cdf of $\mathbf{X}$ has the property that

$$F(\mathbf{x}) = P(X_1 \le x_1) P(X_2 \le x_2) \cdots P(X_p \le x_p), \qquad \text{for any } \mathbf{x} \in R^p, \tag{2.10}$$

the p random variables $X_1, X_2, \ldots, X_p$ are said to be *independent*. Intuitively speaking, a sequence of random variables are independent of each other if the values of any subset of the sequence provide no information about the values of the remaining random variables in the sequence. Based on (2.1)–(2.3) and (2.8)–(2.10), if $\mathbf{X}$ has a pdf f, then each of $X_1, X_2, \ldots, X_p$ must have a pdf, and $X_1, X_2, \ldots, X_p$ are independent if and only if

$$f(\mathbf{x}) = f_{X_1}(x_1) f_{X_2}(x_2) \cdots f_{X_p}(x_p), \qquad \text{for any } \mathbf{x} \in R^p, \tag{2.11}$$

where $f_{X_j}(x_j)$ denotes the pdf of X_j, for $j = 1, 2, \ldots, p$.

2.3　Important Continuous Distributions

In statistics, a number of parametric distribution families are frequently used in developing various statistical methods and theories. Some important distribution families of absolutely continuous numerical random variables are introduced in this section. Some important distribution families of discrete numerical random variables are introduced in the next section. For a more complete discussion on parametric distribution families, see Johnson et al. (1992), Johnson et al. (1994), and Johnson et al. (1995).

2.3.1 Normal distribution

A very important family of parametric distributions is the *normal* (or *Gaussian*) distribution family. When an absolutely continuous random variable X has the following pdf:

$$f(x) = \frac{1}{\sqrt{2\pi}\sigma} \exp\left[-\frac{(x-\mu)^2}{2\sigma^2}\right], \qquad \text{for } x \in R, \qquad (2.12)$$

where μ and σ are two parameters, then its distribution is called a normal distribution. It can be easily checked, using (2.6) and (2.7), that if X has a normal distribution with parameters μ and σ as defined in equation (2.12), then $\mu_X = \mu$ and $\sigma_X^2 = \sigma^2$. Therefore, a normal distribution is uniquely determined by its mean and variance. We use the conventional notation $X \sim N(\mu, \sigma^2)$ to denote "X has a normal distribution with mean μ and variance σ^2." The normal distribution is important in statistics for two major reasons. One is that distributions of many continuous variables in practice, such as the height or weight of all people in this country, can be described reasonably well by normal distributions. The second major reason is that much statistical theory is developed under the assumption of normality.

Some properties of a normal distribution are as follows. Its density curve is bell-shaped, symmetric about the mean μ, and its spread is controlled by the standard deviation σ. Therefore, μ is a *location parameter* and σ is a *scale parameter*. If $X \sim N(\mu, \sigma^2)$, then the random variable defined by

$$Z = \frac{X - \mu}{\sigma}$$

has a normal distribution with mean zero and variance one. This specific normal distribution with mean zero and variance one is called the *standard normal distribution*. Its cdf $\Phi(x)$ and pdf $\phi(x)$ are displayed in Figure 2.1. From Figure 2.1(b), the density curve of Z indeed looks bell-shaped. By the way, in statistics, it is a convention to use Z to denote a random variable that has the standard normal distribution.

2.3.2 Chi-square distribution

If $X_1, X_2, \ldots, X_k$ are independent normal random variables with means $\mu_1, \mu_2, \ldots, \mu_k$, respectively, and with a common variance one, then the distribution of

$$Q = X_1^2 + X_2^2 + \cdots + X_k^2$$

is called the non-central *chi-square distribution* with k degrees of freedom (df) and with a non-central parameter $\delta = \mu_1^2 + \mu_2^2 + \ldots + \mu_k^2$. By notation, it is denoted as $Q \sim \chi_k^2(\delta)$. If $\delta = 0$ (i.e., all $\mu_1, \mu_2, \ldots, \mu_k$ are 0), then the corresponding distribution is called the (central) chi-square distribution with df equal to k, which is simply denoted as χ_k^2. The pdf of the χ_k^2 distribution has the expression

$$f(x) = \frac{1}{2^{k/2}\Gamma(k/2)}x^{k/2-1}e^{-x/2}, \qquad \text{for } x \geq 0,$$

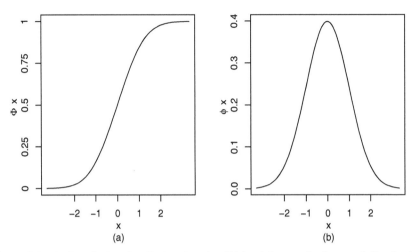

Figure 2.1 *(a) Cumulative distribution function $\Phi(x)$ of the standard normal distribution. (b) Probability density function $\phi(x)$ of the standard normal distribution.*

where $\Gamma(x) = \int_0^\infty u^{x-1}e^{-u}\, du$ is the gamma function. If $Q \sim \chi_k^2$, then it can be checked using (2.6) and (2.7) that

$$\mu_Q = k, \qquad \sigma_Q^2 = 2k.$$

When $k = 1, 2, 3$, and 4, the density curves of χ_k^2 are shown in Figure 2.2(a), from which it can be seen that the curves tend to be more and more symmetric when k increases. One important property of the chi-square distribution is that, if Q_1 and Q_2 are two independent random variables, $Q_1 \sim \chi_{k_1}^2$, and $Q_2 \sim \chi_{k_2}^2$, then $Q_1 + Q_2 \sim \chi_{k_1+k_2}^2$.

2.3.3   t distribution

Another important continuous distribution is the *t* distribution defined as follows. Assume that Z and V are two independent random variables, Z has the standard normal distribution, and V has a chi-square distribution with df equal to k. Then, the distribution of

$$T = \frac{Z}{\sqrt{V/k}}$$

is called the *t distribution* with k degrees of freedom. From its definition, the *t* distribution is uniquely determined by its only parameter k. For that reason, the distribution is conventionally denoted as t_k. Its pdf is

$$f(x) = \frac{\Gamma((k+1)/2)}{\sqrt{k\pi}\Gamma(k/2)} \left(1 + \frac{x^2}{k}\right)^{-(k+1)/2}, \qquad \text{for } x \in R.$$

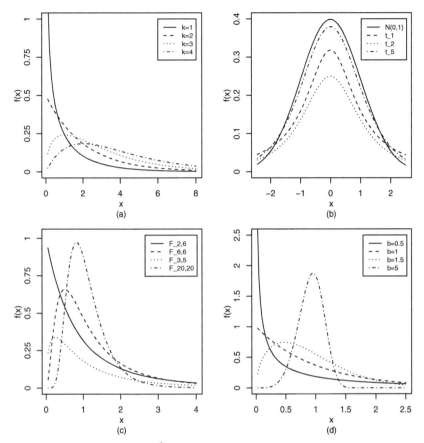

Figure 2.2 *(a) Density curves of* χ_k^2 *in cases when* $k = 1, 2, 3,$ *and 4. (b) Density curves of* t_k *in cases when* $k = 1, 2,$ *and 5. The solid curve in the plot is the density curve of the standard normal distribution. (c) Density curves of* F_{k_1, k_2} *in cases when* $(k_1, k_2) = (2, 6), (6, 6), (3, 5),$ *and* $(20, 20)$. *(d) Density curves of Weibull*$(1, b)$ *in cases when* $b = 0.5, 1, 1.5,$ *and 5.*

By (2.6) and (2.7), it can be checked that, when $T \sim t_k$, μ_T exists only when $k > 1$ and σ_T^2 exists and is finite only when $k > 2$. In such cases,

$$\mu_T = 0, \text{ when } k > 1, \qquad \sigma_T^2 = \frac{k}{k-2}, \text{ when } k > 2.$$

When $k = 1, 2,$ and 5, the density curves of t_k are shown in Figure 2.2(b) where the solid curve is the density curve of the standard normal distribution. From the plot, we can have the following conclusions:

(i) For a given value of k, the density curve of t_k is bell-shaped and symmetric about 0;

(ii) When k increases, the density curve of t_k gets closer to the density curve of the standard normal distribution; and

(iii) For any constant $a > 0$ and any finite positive integer k, it is always true that

$$P(T \geq a) > P(Z \geq a),$$

where $T \sim t_k$ and $Z \sim N(0,1)$. Note that the probabilities $P(T \geq a)$ and $P(Z \geq a)$ equal the areas of the t_k and $N(0,1)$ density curves, respectively, in the right-tail region of $[a, \infty)$. Therefore, this conclusion implies that the density curve of a t distribution has *heavier tails*, compared to the density curve of the standard normal distribution.

2.3.4 F distribution

If X_1 and X_2 are two independent random variables, $X_1 \sim \chi^2_{k_1}$, and $X_2 \sim \chi^2_{k_2}$, then the distribution of

$$F = \frac{X_1/k_1}{X_2/k_2}$$

is called the *F distribution* with numerator degrees of freedom k_1 and denominator degrees of freedom k_2. It is denoted as $F \sim F_{k_1,k_2}$. The pdf of the F_{k_1,k_2} distribution is

$$f(x) = \frac{1}{xB(k_1/2, k_2/2)} \sqrt{\frac{(k_1 x)^{k_1} k_2^{k_2}}{(k_1 x + k_2)^{k_1+k_2}}}, \qquad \text{for } x \geq 0,$$

where $B(x,y) = \int_0^1 u^{x-1}(1-u)^{y-1}\, du$ is the beta function. Its mean exists and is finite only when $k_2 > 2$, and its variance exists and is finite only when $k_2 > 4$. In such cases,

$$\mu_F = \frac{k_2}{k_2 - 2}, \text{ when } k_2 > 2, \qquad \sigma_F^2 = \frac{2k_2^2(k_1 + k_2 - 2)}{k_1(k_2 - 2)^2(k_2 - 4)}, \text{ when } k_2 > 4.$$

In cases when $(k_1, k_2) = (2,6), (6,6), (3,5)$, and $(20,20)$, the density curves of F_{k_1,k_2} are shown in Figure 2.2(c), from which it can be seen that the density curve of the F distribution is quite flexible and can have many different shapes. From the definitions of the t and F distributions, it is obvious that, if $T \sim t_k$, then $T^2 \sim F_{1,k}$.

2.3.5 Weibull distribution and exponential distribution

In life science and reliability analysis, the *Weibull distribution* family is widely used for describing the distribution of the life times of products. Its pdf is

$$f(x) = \frac{b}{a}\left(\frac{x}{a}\right)^{b-1} e^{-(x/a)^b}, \qquad \text{for } x \geq 0,$$

where $a > 0$ is a scale parameter, and $b > 0$ is a shape parameter. If a random variable W has a Weibull distribution with scale parameter a and shape parameter b, denoted as $W \sim Weibull(a,b)$, then its mean and variance are

$$\mu_W = a\Gamma(1 + 1/b), \qquad \sigma_W^2 = a^2\left[\Gamma(1 + 2/b) - \Gamma^2(1 + 1/b)\right].$$

When $a = 1$ and $b = 0.5, 1, 1.5$, and 5, the density curves of $Weibull(a,b)$ are shown in Figure 2.2(d). In the special case when $b = 1$, the Weibull distribution becomes the *exponential distribution*, which is another continuous distribution that is widely used in applications.

2.4 Important Discrete Distributions

2.4.1 Binary variable and Bernoulli distribution

In practice, many variables are *binary* in the sense that they only take two possible values. Examples of binary variables include gender (either male or female), result of an inspection of a product (pass or fail the inspection), status of a cancer patient after a five-year period (alive or dead), and so forth. For convenience, in statistics, we usually label the value of a binary variable that we are interested in studying as success (S), and the other value as failure (F). If a random variable X is binary, then its probability distribution function is determined by

$$\pi = P(X = S), \qquad 1 - \pi = P(X = F),$$

which is called the *Bernoulli distribution.* Without losing any information, we can assign the number 1 to S, and the number 0 to F. After this assignment, X becomes a binary numerical random variable, and it is obvious that

$$\mu_X = \pi, \qquad \sigma_X^2 = \pi(1 - \pi).$$

2.4.2 Binomial and multinomial distributions

Now, let us consider the experiment of flipping a coin n times. This experiment satisfies the following four requirements:

(i) The experiment consists of n trials with n fixed.

(ii) The n trials are independent of each other, in the sense that the result of one trial does not affect the result of any other trials.

(iii) Each trial has only two possible outcomes: S and F. Such a trial is often called a Bernoulli trial.

(iv) The probability of S is the same from trial to trial.

An experiment meeting the above four requirements is called a *binomial experiment.* So, flipping a coin n times is an example of a binomial experiment. Now, let X denote the number of S's in a binomial experiment with n trials, and π denote the probability of S. Then, X is a discrete numerical random variable taking the values of $0, 1, 2, \ldots, n$. Its probability distribution can be described by the following formula:

$$P(X = x) = \binom{n}{x} \pi^x (1 - \pi)^{n-x}, \qquad \text{for } x = 0, 1, 2, \ldots, n, \qquad (2.13)$$

where

$$\binom{n}{x} = \frac{n!}{x!(n-x)!}$$

is called the binomial coefficient, which denotes the number of combinations when choosing x subjects from a total of n subjects. The distribution described by (2.13) is called the *binomial distribution*, denoted as $Binomial(n, \pi)$. If $X \sim Binomial(n, \pi)$, then X is often called a binomial random variable, and its mean and variance are

$$\mu_X = n\pi, \qquad \sigma_X^2 = n\pi(1 - \pi). \qquad (2.14)$$

The binomial distribution can be generalized as follows. Assume that, in an experiment of n trials, each trial has k possible outcomes with $k \geq 2$, the probabilities of the k outcomes are $\{\pi_1, \pi_2, \ldots, \pi_k\}$ in each trial, and the trials are independent of each other. Let X_j be the number of trials having the j-th outcome, for $j = 1, 2, \ldots, k$. Then, the distribution of $(X_1, X_2, \ldots, X_k)$ is

$$P(X_1 = x_1, X_2 = x_2, \ldots, X_k = x_k) = \frac{n!}{x_1! x_2! \cdots x_k!} \pi_1^{x_1} \pi_2^{x_2} \cdots \pi_k^{x_k},$$

for any non-negative integers $(x_1, x_2, \ldots, x_k)$ that satisfy $\sum_{j=1}^{k} x_j = n$. This distribution is called the *multinomial distribution*, denoted as $Multinomial(n, \pi_1, \pi_2, \ldots, \pi_k)$. Obviously, when $k = 2$, the multinomial distribution is equivalent to a binomial distribution. Further, in cases when $k > 2$, the distribution of a single X_j, for any j, is $Binomial(n, \pi_j)$. Therefore, by (2.14), for $j = 1, 2, \ldots, k$,

$$\mu_{X_j} = n\pi_j, \qquad \sigma_{X_j}^2 = n\pi_j(1 - \pi_j).$$

2.4.3 Geometric distribution

Let X be the number of Bernoulli trials needed to get the first S. Then, X is a discrete numerical random variable taking the values of $\{1, 2, \ldots\}$. Its probability distribution can be described by

$$P(X = x) = (1 - \pi)^{x-1} \pi, \qquad \text{for } x = 1, 2, \ldots$$

This distribution is called the *geometric distribution*, denoted as $Geom(\pi)$. If $X \sim Geom(\pi)$, then it is easy to check that

$$\mu_X = \frac{1}{\pi}, \qquad \sigma_X^2 = \frac{1 - \pi}{\pi^2}.$$

The geometric distribution plays an important role in statistical process control (SPC), because in SPC we are mainly concerned about the first time when a control chart gives a signal that a production process is out-of-control. See Section 3.2 in Chapter 3 for a related discussion.

2.4.4 Hypergeometric distribution

In statistics, *ball models* are often used for introducing certain important discrete distributions. To introduce the hypergeometric distribution, a related ball model can be described as follows. Assume that an urn contains a total of N balls, among which there are M red balls and $N - M$ black balls, where $N > 0$ and $0 \leq M \leq N$ are two integers. We select n balls from the urn without replacement. Let X be the number of selected red balls. Then, X is a discrete random variable taking the values of $\{0, 1, \ldots, \min(n, M)\}$. Its distribution can be described by

$$P(X = x) = \frac{\binom{M}{x} \binom{N-M}{n-x}}{\binom{N}{n}}, \qquad \text{for } x = 0, 1, \ldots, \min(n, M).$$

This distribution is called the *hypergeometric distribution*, and its mean and variance are

$$\mu_X = \frac{nM}{N}; \qquad \sigma_X^2 = \frac{nM(N-n)(N-M)}{N^2(N-1)}, \qquad \text{when } N > 1.$$

2.4.5 Poisson distribution

When describing the distribution of a discrete random variable X whose value is a count (e.g., the number of traffic accidents on a specific segment of a road in a given time period), the *Poisson distribution* is often useful. By definition, X has a Poisson distribution if it takes count values in $\{0, 1, 2, \ldots\}$ and

$$P(X = x) = \frac{\lambda^x e^{-\lambda}}{x!}, \qquad \text{for } x = 0, 1, \ldots, \tag{2.15}$$

where $\lambda > 0$ is a parameter. If X has a Poisson distribution with parameter λ, denoted as $X \sim Poisson(\lambda)$, then it can be checked using (2.4), (2.5), and (2.15) that

$$\mu_X = \lambda, \qquad \sigma_X^2 = \lambda. \tag{2.16}$$

From (2.16), we can notice an important property of the Poisson distribution; its mean and variance are the same. In practice, people often use this property to verify whether a dataset follows a Poisson distribution. Another important property of the Poisson distribution is that, if X_1 and X_2 are two independent random variables, $X_1 \sim Poisson(\lambda_1)$, and $X_2 \sim Poisson(\lambda_2)$, then $X_1 + X_2 \sim Poisson(\lambda_1 + \lambda_2)$. In quality control, the Poisson distribution is often used for describing the distribution of the number of defects in an inspection unit. See related discussion in Subsection 3.3.2.

2.5 Data and Data Description

In applications, it is often of interest to know the population distribution. For instance, before a presidential election, people are interested in knowing the proportion of approvals of a specific candidate in the entire population of legitimate voters. Because individuals of the population in this example take only two possible values, either approval or disapproval, the population distribution is uniquely determined by the approval rate in the population, or the *population proportion* of approvals. More generally, when a population distribution has a parametric form, such as those discussed in Sections 2.3 and 2.4, the population parameters appearing in the parametric form uniquely determine the entire population distribution. Therefore, it suffices to know the values of these parameters in order to know the population distribution. In some instances, our major interest is in one or more population parameters instead of the entire population distribution. As an example, it is often sufficient to know the average exam score of the students in a class in order to have a rough idea about the overall performance of these students in the exam. However, to know the population distribution or its parameters, we need to know variable values for all members in the population. In many applications, the related population is large. It is therefore time-consuming or even impossible to observe the variable value for each member in the population. To overcome this difficulty, in statistics, we often use the idea of sampling the population, described below.

A *sample* of the population is a subset of the population, selected in some pre-scribed manner for study. After a sample is obtained, the population distribution or its parameters can be estimated based on the sample. In practice, most people call the observed sample selected properly from a population *data*, although the word "data" is also used for some other purposes in our daily life (e.g., to represent certain generic information that may not necessarily be a sample from a population).

To have an accurate estimation of the population distribution or its parameters, the sample should represent the population well. In the literature, many different sampling techniques have been proposed to handle different cases. Interested readers can read Cochran (1977) or other textbooks on statistical sampling for a systematic discussion about this topic. In this book, if there is no further specification, we as-sume that all samples are *simple random samples*. A simple random sample of size n is a sample that consists of n selected members, which is generated in a way that the result of one selection has nothing to do with the result of any other selection, and every member in the population has the same chance to be selected to the sam-ple. By this definition, each observation in the sample can be thought of as a random variable whose distribution is just the population distribution, because the value of the observation can be any member in the population (i.e., the observation value is random) and each member in the population has the same chance to be selected to the sample. Of course, after the sample is physically obtained, observations in the sample are uniquely determined and non-random. To make the distinction, we con-ventionally use capital letters to denote observations in a sample that are treated as random variables, and little letters to denote observations in a physically obtained sample.

Simple random samples have some nice statistical properties. One property can be described as follows. Assume that the characteristic of interest in each member of a population is univariate numerical, the cdf of the population distribution is F, and $\{X_1, X_2, \ldots, X_n\}$ is a simple random sample from the population. Then, by the definition of the simple random sample, $\{X_1, X_2, \ldots, X_n\}$ is a sequence of independent and identically distributed (i.i.d.) random variables with a common cdf F. The joint cdf of $\{X_1, X_2, \ldots, X_n\}$ is

$$P(X_1 \leq x_1, X_2 \leq x_2, \ldots, X_n \leq x_n) = \Pi_{i=1}^n P(X_i \leq x_i) = \Pi_{i=1}^n F(x_i),$$

for any $(x_1, x_2, \ldots, x_n)' \in R^n$. If the population distribution has a pdf f, then the joint distribution of $\{X_1, X_2, \ldots, X_n\}$ also has a pdf, and the pdf is

$$\Pi_{i=1}^n f(x_i), \qquad \text{for } (x_1, x_2, \ldots, x_n)' \in R^n.$$

See (2.11) and the related discussion in Section 2.2.

For a sample $\{X_1, X_2, \ldots, X_n\}$, there are several ways to describe its *center*. One natural way is to use the *sample mean*

$$\overline{X} = \frac{1}{n} \sum_{i=1}^n X_i,$$

which is a simple average of all n observations. By using the sample mean, if there are some extremely large or extremely small values in the data, which are often called *outliers*, then they would affect the sample mean quite dramatically, which is demonstrated by the example below.

Example 2.2 *A state government wants to know the salary information of faculty members in high education institutes in that state. For that purpose, 10 faculty members are randomly chosen and their most recent salaries (in thousands) are listed below.*

$$67, 84, 56, 210, 79, 85, 73, 64, 88, 93$$

The value of sample mean is $\bar{x} = 89.9$. But, obviously this is not a good measure of the data center, because there are only two observations larger than $\bar{x}$ and the remaining 8 observations are all below $\bar{x}$. This phenomenon is due to the outlier "210" which pulls the mean salary up quite dramatically.

From Example 2.2, in cases when outliers are present, the sample mean may not be a good measure of the data center. In such cases, to remove the impact of outliers, people often use the *sample median* for measuring the data center. To compute the sample median, the observations in the sample $\{X_1, X_2, \ldots, X_n\}$ are first ordered from the smallest to the largest as follows:

$$X_{(1)} \leq X_{(2)} \leq \cdots \leq X_{(n)}. \tag{2.17}$$

Then, $X_{(1)}$ is the first *order statistic*, $X_{(2)}$ is the second order statistic, and so on. Roughly speaking, the sample median is defined by the observation at the middle position of the ordered data. So, about half of the observations in the data are smaller than the sample median, and the other half are larger than the sample median. More specifically, the sample median, denoted as $\widetilde{X}$, is defined by

$$\widetilde{X} = \begin{cases} X_{((n+1)/2)}, & \text{if } n \text{ is odd} \\ \frac{X_{(n/2)} + X_{(n/2+1)}}{2}, & \text{if } n \text{ is even.} \end{cases}$$

By this definition, obviously, the sample median is not affected by a small amount of outliers in the data.

An alternative method to remove the impact of outliers is to use the so-called *trimmed sample mean*, computed as follows. First, we need to choose a trimming percentage $q\%$, where $q \in [0, 50)$. Then, the data are ordered, as in (2.17), and the $q\%$ smallest observations and the $q\%$ largest observations are removed. Finally, the $q\%$ trimmed sample mean is computed as the simple average of the remaining observations.

Example 2.2 (continued) *For the data discussed in Example 2.2, the ordered data are*

$$56, 64, 67, 73, 79, 84, 85, 88, 93, 210.$$

So, the sample median is $\widetilde{x} = (79 + 84)/2 = 81.5$, which is smaller than the sample mean $\bar{x} = 89.9$ because the former is not affected by the outlier 210. The 10%

trimmed sample mean is $(64 + 67 + 73 + 79 + 84 + 85 + 88 + 93)/8 = 79.125$, *after the* 10% *(i.e., one in this example) smallest observations and the* 10% *largest observations are removed.*

Besides the center of a dataset. another important feature of the dataset is its spread. Consider the following three datasets, each of which has five observations:

Dataset 1: 1,   3,   5,   7,   9

Dataset 2: 1,   4.5,   5,   5.5,   9

Dataset 3: 4,   4.5,   5,   5.5,   6

Obviously, the sample means of the three datasets are all the same with a value of 5. But, their distributions have quite different spreads. The first dataset seems to have the largest spread, while the spread of the third dataset seems to be the smallest. To describe the spread of a dataset of size n, denoted by $\{X_1, X_2, \ldots, X_n\}$, a conventional measure is the *sample variance*, defined by

$$s^2 = \frac{1}{n-1} \sum_{i=1}^{n} (X_i - \overline{X})^2.$$

The sample variance s^2 is basically the average of the squares of the n deviations $\{X_i - \overline{X}, i = 1, 2, \ldots, n\}$. Therefore, if its value is large, then overall the individual observations are far away from the sample mean, implying that the data spread is large. The square root of the sample variance, s, is another spread measure, which is called the *sample standard deviation*. Regarding the two measures, the sample standard deviation is easier to interpret as a measure of the data spread because it has the same unit as the original observations, but the sample variance is easier to handle mathematically.

Similar to the sample mean, the sample variance or the sample standard deviation would be affected in a substantial way by possible outliers in the data. To eliminate such effect, we can once again consider the ordered observations in (2.17). Then, the sample first quartile, denoted as Q_1, is defined by the median of the first half of the ordered data. More specifically,

$$Q_1 = \begin{cases} \text{median of } \{X_{(1)}, \ldots, X_{((n-1)/2)}\}, & \text{if } n \text{ is odd} \\ \text{median of } \{X_{(1)}, \ldots, X_{(n/2)}\}, & \text{if } n \text{ is even.} \end{cases}$$

The sample third quartile, denoted as Q_3, is defined similarly, by the median of the second half of the ordered data. Then, in cases when outliers are present, we can use the sample *inter-quartile range (IQR)*, defined by $Q_3 - Q_1$, as a measure of the data spread. Obviously, IQR would not be affected by a small number of outliers in the data.

It should be pointed out that all the measures of the data center and data spread discussed in this section can also be applied to a population of numerical members to measure the center and spread of all members in the population, although some of these measures of the population distribution are not explicitly defined in the book. For instance, we can define population median in the same way as the sample median,

Table 2.2: *Frequency table of the party status of 1,000 randomly chosen voters.*

Party	Democrat	Republican	Green Party	Other
Frequency	437	486	53	24

except that all population members should be used in defining the population median. If a sample represents a population well, then the sample version of a given measure should provide a good estimate of the corresponding population version. See Section 2.7 for a related discussion.

2.6   Tabular and Graphical Methods for Describing Data

2.6.1   *Frequency table, pie chart, and bar chart*

To discuss tabular and graphical methods for describing categorical data, let us consider the following example.

Example 2.3 *Before a presidential election, a TV network made a survey in the nation, and they randomly selected 1,000 legitimate voters to ask for their favorite candidates. Each voter was also asked to tell the TV network his or her party status. The data about the party status are summarized in Table 2.2.*

In Example 2.3, the data about party status is categorical with four categories. Table 2.2 lists all the categories of the observations and the corresponding counts, or frequencies. This table is called the *frequency table*. In the table, sometimes it is more convenient to list relative frequencies (or proportions) of the categories. The relative frequency of a category is defined by

$$\text{relative frequency} = \frac{\text{frequency}}{\text{sample size}}.$$

For the data summarized in Table 2.2, the relative frequencies of the categories Democrat, Republican, Green Party, and Other are 0.437, 0.486, 0.053, and 0.024, respectively. In the table, if relative frequencies, instead of frequencies, are listed, then the table is often called the *relative frequency table*. Sometimes, in a frequency table, both frequencies and relative frequencies are listed for convenience of its applications.

There are two commonly used graphical methods to describe categorical data. One is the *pie chart*, by which a circle is divided into slices, each slice denotes a category, and the slice size is proportional to the relative frequency of the category. For the data in Table 2.2, the corresponding pie chart is shown in Figure 2.3(a). The other popular graphical method for describing categorical data is the *bar chart*, by which all the categories are listed in the *x*-axis, a bar is drawn above the corresponding category label, and the height of the bar equals the corresponding frequency (or relative frequency). For the data in Table 2.2, its bar chart is shown in Figure 2.3(b), in the case when relative frequencies are used. By the way, because the positions of the category labels can be switched on the *x*-axis, the shape of a bar chart usually does not provide any helpful information about the related data.

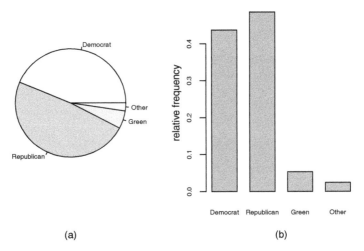

(a)   (b)

Figure 2.3: *(a) Pie chart of the data shown in Table 2.2. (b) The corresponding bar chart.*

2.6.2   Dot plot, stem-and-leaf plot, and box plot

In this and the next subsections, we will introduce several graphical methods for describing numerical data. To introduce these methods, let us start with the following example.

Example 2.4 *To estimate the age distribution of all members of a recreation center, 25 members of the center were randomly selected. Their ages were recorded as follows.*

$$18, 27, 56, 19, 33, 24, 19, 48, 37, 25, 20, 22, 31,$$
$$29, 65, 41, 22, 39, 37, 22, 45, 22, 43, 61, 53$$

For this data, it can be computed by the formulas in Section 2.5 that the sample mean is $\bar{x} = 34.32$, the sample median is $\tilde{x} = 31$, the sample variance is $s^2 = 198.56$, and the sample standard deviation is $s = 14.091$.

For the data in Example 2.4, the *dot plot* is shown in Figure 2.4, from which we can see that a dot plot can be constructed as follows. First, draw a horizontal line and mark it properly with a measurement scale. Then, locate each observation in the data along the measurement scale, and represent it by a dot above the scaled line. If there are two or more observations with the same value, stack the dots vertically. Usually, a dot plot is appropriate for displaying a relatively small dataset, because it lists all observations in the plot and the plot would look messy if the dataset is large. By the dot plot, the distribution of the observed data and any unusual observations in the data can be visually displayed.

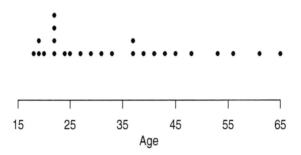

Figure 2.4: *A dot plot of the data in Example 2.4.*

The *stem-and-leaf plot* provides a different way to display a numerical dataset by listing all its observations in a plot. It is constructed as follows. First, select one or more leading digits as stems, and the remaining digits as leaves. Second, list all stem values in a vertical column. Third, draw a vertical line on the right-hand side of the stem values, and list the leaf of each observation beside the corresponding stem value on the right side of the vertical line. Finally, indicate the units of the stems and leaves somewhere in the plot (e.g., at the lower-right corner). For the data in Example 2.4, the tens digit of each observation can be chosen as the stem, and the ones digit as its leaf. The resulting stem-and-leaf plot is shown in Figure 2.5.

```
1 | 899
2 | 022224579
3 | 13779
4 | 1358
5 | 36              Stem:  tens digit
6 | 15              Leaf:  ones digit
```

Figure 2.5: *A stem-and-leaf plot of the data in Example 2.4.*

Instead of displaying all observations of a dataset, the *box plot* only displays the following five summary values of the dataset: the minimum $X_{(1)}$, the first quartile Q_1, the median $\tilde{X}$, the third quartile Q_3, and the maximum $X_{(n)}$. It is constructed as follows. First, draw a horizontal line and scale it properly. Second, construct a rectangular box with its left edge at the first quartile and its right edge at the third quartile. Third, draw a vertical line segment inside the box at the location of the median. Fourth, extend horizontal line segments from each end of the box to the smallest and the largest observations of the data. For the data in Example 2.4, the box plot is shown in Figure 2.6.

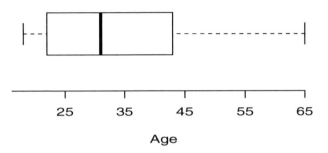

Figure 2.6: *The box plot of the data in Example 2.4.*

Table 2.3: *A frequency table of the data in Example 2.4.*

Class Interval	Frequency	Relative Frequency
$[10, 20)$	3	0.12
$[20, 30)$	9	0.36
$[30, 40)$	5	0.20
$[40, 50)$	4	0.16
$[50, 60)$	2	0.08
$[60, 70)$	2	0.08

2.6.3   Frequency histogram and density histogram

If observations in a dataset are discrete numerical and the number of their different values is small, then a frequency table can be constructed in the same way as that described in Subsection 2.6.1, with each different value as a category. In cases when the observations in a dataset are discrete but the number of different values is quite large, or when the observations are continuous numerical, it does not make much sense to use each different value as a category when constructing a frequency table, because the resulting table would be too large to convey any helpful information about the data pattern. In such cases, an alternative method is to combine certain observation values to form *class intervals*, and then construct a frequency table using the class intervals.

As a demonstration, let us use the dataset in Example 2.4. For this dataset, let us consider the class intervals: $[10, 20), [20, 30), [30, 40), [40, 50), [50, 60), [60, 70)$. The corresponding frequency table is shown in Table 2.3.

A *frequency histogram* turns a frequency table into a graph in the following way. First, draw a horizontal line, and mark the boundaries of the class intervals on it. Second, draw a vertical line and scale it properly to represent frequencies. Third, draw a rectangular bar for each class interval above the corresponding interval on the x-axis, with its height equal to the corresponding frequency. The frequency histogram for Table 2.3 is shown in Figure 2.7. From the histogram, it can be seen that the data are not symmetrically distributed. Instead, the mode (i.e., the peak) of the histogram is closer to the left end, compared to its distance to the right end; or, the right tail of the histogram is longer than the left tail, implying the existence of a number of relatively large observations in the data. By the way, from the definition of relative frequencies, a histogram using frequencies and a histogram using the corresponding relative frequencies would have exactly the same shape. Their only difference would be in the scale of the heights of the rectangular bars. The latter histogram is sometimes called a relative frequency histogram.

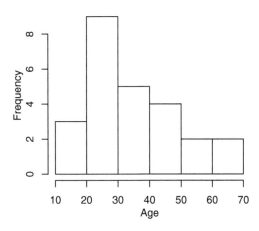

Figure 2.7: *A frequency histogram of the data in Example 2.4.*

Example 2.5 *To monitor an aluminum smelter, 189 observations of the content of SiO_2 in its products are obtained. The data are summarized in Table 2.4, which is a frequency table using 15 class intervals with the same length, and the corresponding frequency histogram is shown in Figure 2.8(a). From the histogram, it can be seen that there are only a small number of observations larger than or equal to 1. So, we combine the last 11 intervals into a larger interval $[1, 3.75)$. The modified class intervals and the corresponding frequencies and relative frequencies are shown in Table 2.5, and the frequency histogram using the modified intervals is shown in Figure 2.8(b). By comparing the two histograms in Figure 2.8, we may have two conflicting impressions about the distribution of the same dataset. By Figure 2.8(a), it seems that the chance to have an observation of SiO_2 larger than 1 is really small. However, from Figure 2.8(b), we may have a different impression that the chance to have such an observation is not that small. The main reason for the latter impression is that the last class interval in Figure 2.8(b) is much longer than the other class intervals, and the height of the bar above that interval is determined by the frequency of all observations in that interval, which does not take the length of the interval into account.*

Table 2.4 *Frequency table of the aluminum smelter data discussed in Example 2.5, using class intervals with the same length.*

Class Interval	Frequency	Relative Frequency
$[0, 0.25)$	22	0.116
$[0.25, 0.5)$	70	0.370
$[0.5, 0.75)$	51	0.270
$[0.75, 1)$	23	0.122
$[1, 1.25)$	8	0.042
$[1.25, 1.5)$	5	0.026
$[1.5, 1.75)$	2	0.011
$[1.75, 2)$	3	0.016
$[2, 2.25)$	1	0.005
$[2.25, 2.5)$	1	0.005
$[2.5, 2.75)$	2	0.011
$[2.75, 3)$	0	0.000
$[3, 3.25)$	0	0.000
$[3.25, 3.5)$	0	0.000
$[3.5, 3.75)$	1	0.005

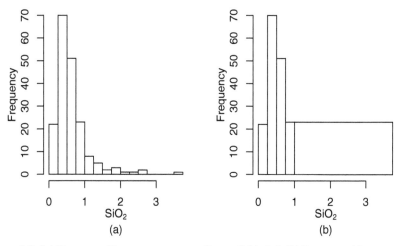

Figure 2.8 *(a) Frequency histogram corresponding to Table 2.4. (b) Frequency histogram corresponding to Table 2.5 using class intervals with different lengths.*

From Figure 2.8, it can be seen that the frequency or relative frequency histogram works well only when the class intervals have the same length. Cases when the class intervals have different lengths may give people a wrong impression about the data distribution. To overcome this limitation, we define the *density* of a class interval by

$$\text{density} = \frac{\text{relative frequency}}{\text{interval length}}.$$

Then, a *density histogram* can be constructed in the same way as that of a frequency

Table 2.5 *Frequency table of the aluminum smelter data discussed in Example 2.5 using class intervals with different lengths.*

Class Interval	Frequency	Relative Frequency	Density
$[0, 0.25)$	22	0.116	0.464
$[0.25, 0.5)$	70	0.370	1.480
$[0.5, 0.75)$	51	0.270	1.080
$[0.75, 1)$	23	0.122	0.488
$[1, 3.75)$	23	0.122	0.044

or relative frequency histogram, except that the height of each rectangular bar in a density histogram is determined by the density of the corresponding class interval, instead of by the frequency or relative frequency. Obviously, the density histogram has the property that summation of the areas of all its rectangular bars is 1, which is in parallel to the property of the density curve of a population distribution that the area under the curve is 1.

The densities of the class intervals in Table 2.5 have been computed and presented in the last column of that table. The corresponding density histogram is presented in Figure 2.9, in which a solid curve has been added to sketch its shape. The solid curve is often called a smoothed histogram, which provides an estimate of the population density curve. From Figure 2.9, it can be seen that the chance to have an observation of SiO_2 that is larger than 1 is indeed small, even in cases when we use unequal class intervals.

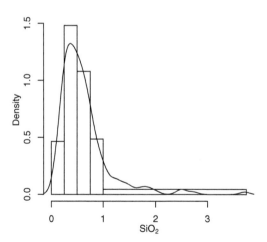

Figure 2.9: *A density histogram corresponding to Table 2.5 with unequal class intervals.*

From the above discussion, in cases when the class intervals have the same length, a frequency histogram, the corresponding relative frequency histogram, and the corresponding density histogram would all have the same shape and their only difference is in the scale of the heights of the rectangular bars. However, in cases when the class intervals have different lengths, as the above example demonstrated, the frequency or relative frequency histogram should be avoided for possible confusion, and the density histogram is the one that is appropriate to use.

Figure 2.10 presents several smoothed histograms with different shapes. In plots (a), (c), and (d), each of the smoothed histograms has a single peak (or mode). Such histograms are called unimodal histograms. The smoothed histogram in plot (b) has two modes, and it is called a bimodal histogram. If a smoothed histogram has more than two modes, then it is called a multimodal histogram. The smoothed histograms in plots (a) and (b) are symmetric, and the ones in plots (c) and (d) are unimodal and skewed. If a unimodal histogram has a longer right tail, compared to its left tail, then we say that it is skewed to the right, or positively skewed. If its left tail is longer than its right tail, then we say that it is skewed to the left, or negatively skewed. So, the one in plot (c) is skewed to the right and the one in plot (d) is skewed to the left.

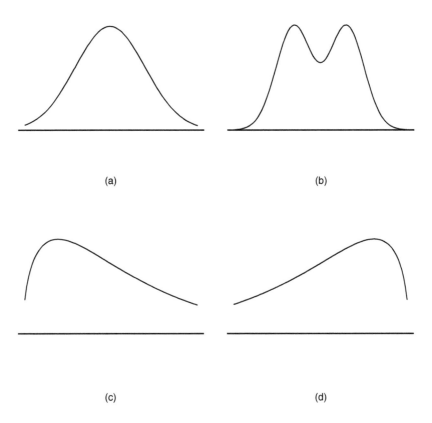

(a)   (b)

(c)   (d)

Figure 2.10 *Smoothed histograms with different shapes. (a) Symmetric and unimodal. (b) Symmetric and bimodal. (c) Skewed to the right and unimodal. (d) Skewed to the left and unimodal.*

2.7 Parametric Statistical Inferences

As discussed in the previous two sections, a sample from a population of interest carries useful information about the population. After a sample is obtained, the next question is how to properly estimate the population distribution or its parameters based on the sample. After an estimator is obtained, we also need to evaluate its performance, so that the best one can be chosen if multiple estimators are available. These are the major goals of statistical inference. In this section, we briefly introduce some basic concepts and methodologies of statistical inference in cases when the population distribution has a parametric form.

2.7.1 Point estimation and sampling distribution

In cases when a population distribution has a parametric form with one or more parameters, the population distribution itself and all of its summary measures (e.g., mean and variance) are uniquely determined by the population distribution parameters. So, in such cases, estimation of the population distribution parameters is important. Based on a sample $\{X_1, X_2, \ldots, X_n\}$, some commonly used estimators are listed below.

- In cases when the population distribution is $N(\mu, \sigma^2)$, the sample mean $\overline{X}$ is a good estimator of the population mean μ, the sample variance s^2 is a good estimator of the population variance σ^2, and the sample standard deviation s is a reasonable estimator of the population standard deviation σ.

- In cases when the population distribution is Bernoulli with π being the probability of success (S) (cf., Subsection 2.4.1), a good estimator of π is the sample proportion of S, defined by

$$p = \frac{\text{the number of S's in the sample}}{n}.$$

The parameter π is also called the population proportion of S, as mentioned in Section 2.5. Clearly, if we use 1 to denote S and 0 to denote F, then p is just $\overline{X}$.

All estimators mentioned above are calculated from the sample $\{X_1, X_2, \ldots, X_n\}$. In statistics, any quantity that is calculated from the sample and uniquely determined by the sample is called a *statistic*. Therefore, $\overline{X}$, s^2, s, and p are all statistics.

To estimate a specific population parameter θ (e.g., θ is the population mean or standard deviation), an appropriate statistic should be chosen as an estimator. It has become a convention in the literature to put a hat above the parameter to denote its estimator. So, $\widehat{\theta}$ is an estimator of θ. Although $\widehat{\theta}$ is a function of the sample, this is often not explicit in notation, for simplicity. To estimate θ by $\widehat{\theta}$, we use a single-valued statistic calculated from the sample (i.e., $\widehat{\theta}$) for estimating a single-valued parameter of the population (i.e., θ). This parameter estimation method is therefore called *point estimation* in the literature, and the estimator is called a *point estimator*. For instance, $\overline{X}$ is a point estimator of μ, and s^2 is a point estimator of σ^2.

Because the sample is random, any statistic computed from the sample, including all point estimators mentioned above, is also random. Thus, a point estimator has its own distribution, which is called the *sampling distribution*. To assess the accuracy of a point estimator for estimating a parameter, we need to study its sampling distribution, especially the mean and variance of the sampling distribution.

In cases when the population distribution is $N(\mu, \sigma^2)$, it can be proved that the sampling distributions of the sample mean and sample variance have the properties summarized in the box below.

Properties of the Sampling Distributions of $\overline{X}$ and s^2

Assume that the population distribution is $N(\mu, \sigma^2)$, and $\overline{X}$ and s^2 are the sample mean and sample variance of a simple random sample $\{X_1, X_2, \ldots, X_n\}$ from the population. Then, their sampling distributions have the following properties:
(i) $\overline{X} \sim N(\mu, \sigma^2/n)$,
(ii) $(n-1)s^2/\sigma^2 \sim \chi_{n-1}^2$, and
(iii) $\overline{X}$ and s^2 are independent of each other.

By combining all these three properties and by the definition of a t distribution (cf., Subsection 2.3.3), we have

$$T = \frac{\overline{X} - \mu}{s/\sqrt{n}} \sim t_{n-1}. \tag{2.18}$$

In cases when the population distribution is unknown, the sampling distribution of $\overline{X}$ would be asymptotically normal. Namely, it is asymptotically true that

$$\frac{\overline{X} - \mu}{\sigma/\sqrt{n}} \sim N(0,1),$$

where "asymptotically true" means that the distribution of $(\overline{X} - \mu)/(\sigma/\sqrt{n})$ is closer and closer to $N(0,1)$ when the sample size n increases. This result plays an important role in statistics, because the population mean is often the population characteristic that we are interested in estimating and this result gives a general conclusion about its point estimator $\overline{X}$ that it is always asymptotically normal no matter what is the real population distribution. Because of its importance, this result is called the *Central Limit Theorem (CLT)* in the statistical literature, and it is formally stated in the box below.

Central Limit Theorem

Assume that $\{X_1, X_2, \ldots, X_n\}$ is a simple random sample from a population, and $\overline{X}$ is its sample mean. Then the sampling distribution of $(\overline{X} - \mu)/(\sigma/\sqrt{n})$ converges to $N(0,1)$ when n increases.

The CLT provides an intuitive explanation about the phenomenon that many variables in our daily life, including height, weight, blood pressure readings, and so forth, would roughly follow normal distributions. For instance, our height is affected by our parents' heights, grandparents' heights, our food intake and other nutritional factors, environmental factors, and many other factors. Therefore, it is a weighted average of many different factors, similar to $\overline{X}$, which is an average of n observations. By the CLT, our height would roughly have a normal distribution. The CLT also explains why the normal distribution family is so popular in statistics. As pointed out in the previous paragraph, many statistical inferences involve the sample mean $\overline{X}$ for estimating the population mean μ. By the CLT, the sampling distribution of $\overline{X}$ is close to normal in all cases, as long as the sample size is large.

In practice, we need to determine a threshold value for the sample size n, so that n can be regarded as "large" and consequently the sampling distribution of $\overline{X}$ can be treated as a normal distribution. Of course, the asymptotic behavior of the sampling distribution of $\overline{X}$ depends on the true population distribution. Intuitively, if the true population distribution is very skewed, then the sampling distribution of $\overline{X}$ would be slow in converging to a normal distribution. On the other hand, if the true population distribution is already quite close to a normal distribution, then the sampling distribution of $\overline{X}$ would be fast in converging to a normal distribution. However, the true population distribution is usually unknown in practice, making the problem of choosing the threshold value complicated. Based on much research, a commonly used threshold value for n is 30. By this threshold value, the sample size can be regarded as "large" if $n \geq 30$.

Example 2.6 *Assume that $\{X_1, X_2, \ldots, X_n\}$ is a simple random sample from a population with mean 2 and variance 9, and $n = 36$. Then, $\overline{X}$ is asymptotically distributed as $N(2, 1/4)$ by the CLT. So, we can compute different probabilities related to $\overline{X}$. For instance,*

$$P(\overline{X} > 3) = P\left(\frac{\overline{X} - 2}{1/2} > \frac{3 - 2}{1/2}\right)$$
$$\approx P(Z > 2) = 0.0228,$$

where Z denotes a random variable with the standard normal distribution, and "$\approx$" denotes the asymptotic equality. As a comparison, we usually cannot easily compute probabilities related to a distribution if we do not know its mathematical expression. Therefore, the CLT is helpful in such cases.

In cases when the population distribution is Bernoulli with π being the population proportion of S, $\{X_1, X_2, \ldots, X_n\}$ is a simple random sample from the population, and S and F are represented by 1 and 0, respectively. The sampling distribution of $\sum_{i=1}^{n} X_i$ is $Binomial(n, \pi)$, and the sampling distribution of the sample proportion $p = \overline{X} = \sum_{i=1}^{n} X_i/n$ can be determined accordingly. When the sample size is large in the sense that both $n\pi$ and $n(1 - \pi)$ are large, by the CLT, the distribution of p would be asymptotically normal. In other words, when the sample size is large, the distribution of p can be regarded as $N(\pi, \pi(1 - \pi)/n)$. In practice, the sample size can be regarded as "large" if $n\pi \geq 10$ and $n(1 - \pi) \geq 10$.

For a given population parameter, there could be multiple point estimators. Let us revisit Example 2.2, in which a state government wants to estimate the population mean salary μ of faculty members of high education institutes in that state. In that problem, different people can come up with different point estimators of μ. For instance, assume that the following three point estimators have all been proposed for estimating μ:

- one person prefers to use the first observation X_1 in the sample to estimate μ,
- another person thinks that $X_1 + 5$ would be a more reasonable estimator of μ, and
- the third person has some statistical knowledge and wants to use the sample mean $\overline{X}$ to estimate μ.

To choose among multiple point estimators, or to convince people why one point estimator is better than another point estimator, we need a criterion for evaluating the performance of a point estimator. For estimating a population parameter θ, if a point estimator $\widehat{\theta}$ satisfies the condition that the mean of its sampling distribution, denoted as $\mu_{\widehat{\theta}}$, equals θ, i.e.,

$$\mu_{\widehat{\theta}} = \theta$$

for all values of θ, then $\widehat{\theta}$ is called an *unbiased estimator* of θ. Otherwise, the estimator is biased and the bias is defined by

$$\text{Bias}\left(\widehat{\theta}, \theta\right) = \mu_{\widehat{\theta}} - \theta.$$

By the above definition, on average, an unbiased estimator equals the parameter to estimate. Therefore, in practice, people often require a good point estimator to be unbiased. In the faculty salary example mentioned in the previous paragraph, both the first and the third estimators are unbiased estimators, and the second estimator is biased. So, by the criterion of unbiasedness, the second estimator should be avoided.

To compare two unbiased estimators, the one with a smaller variance is obviously a better estimator, because its sampling distribution would have a smaller spread and consequently that estimator is generally closer to the true value of the population parameter. Therefore, among all unbiased estimators of θ, the one with the smallest variance should be the best unbiased estimator. This estimator is often called the *minimum variance unbiased estimator* (MVUE). It can be checked that, if the population distribution is normal, then $\overline{X}$ is the MVUE for estimating the population mean μ. Therefore, in the faculty salary example, if it is reasonable to assume that the population distribution of faculty members' salary is normal, then the estimator $\overline{X}$ should be the best unbiased estimator among all unbiased estimators.

Another commonly used criterion for choosing a good point estimator is the following *mean squared error* (MSE):

$$
\begin{aligned}
\text{MSE}\left(\widehat{\theta}, \theta\right) &= \text{E}\left(\widehat{\theta} - \theta\right)^2 \\
&= \text{E}\left(\widehat{\theta} - \mu_{\widehat{\theta}} + \mu_{\widehat{\theta}} - \theta\right)^2 \\
&= \sigma_{\widehat{\theta}}^2 + \text{Bias}^2\left(\widehat{\theta}, \theta\right).
\end{aligned}
$$

The MSE criterion measures the averaged, squared distance between the point estimator $\widehat{\theta}$ and the parameter θ. By this criterion, the best point estimator has the smallest MSE value among all possible point estimators. Obviously, the MSE criterion makes a trade-off between bias and variance of a point estimator. The best point estimator by this criterion may not be unbiased, and the MVUE estimator may not be the best point estimator either, because the MVUE estimator has the smallest MSE value among all unbiased point estimators only, instead of among all point estimators.

Although it is not explicit in notation, both $\text{Bias}(\widehat{\theta}, \theta)$ and $\text{MSE}(\widehat{\theta}, \theta)$ depend on the sample size n. Generally speaking, when n is larger, the sample carries more information about the population, so it is natural to expect that both $\text{Bias}(\widehat{\theta}, \theta)$ and $\text{MSE}(\widehat{\theta}, \theta)$ would be smaller. If an estimator $\widehat{\theta}$ of θ is biased but the bias converges to zero as the sample size increases, then it is called an *asymptotically unbiased estimator*. If $\widehat{\theta}$ satisfies the condition that

$$\lim_{n \to \infty} \text{MSE}\left(\widehat{\theta}, \theta\right) = 0,$$

then we say that $\widehat{\theta}$ is L_2 *consistent*.

"Consistency" is a kind of large-sample property of a point estimator. There are several different versions of consistency in the literature. If the cdf of $\widehat{\theta}$ converges to the cdf of the constant θ at all continuity points of the latter cdf, then $\widehat{\theta}$ is said to be *consistent in distribution*. If for any constant $\rho > 0$,

$$\lim_{n \to \infty} P\left(|\widehat{\theta} - \theta| > \rho\right) = 0,$$

then $\widehat{\theta}$ is said to be *consistent in probability*. Another commonly used consistency is defined as follows: if

$$P\left(\lim_{n \to \infty} \widehat{\theta} = \theta\right) = 1,$$

then $\widehat{\theta}$ is said to be *almost surely (a.s.) consistent*. Based on some routine mathematical manipulations, it is easy to check the following relations among the four types of consistency defined above.

- If $\widehat{\theta}$ is consistent in probability, then it must also be consistent in distribution.

- If $\widehat{\theta}$ is L_2 consistent or a.s. consistent, then it must also be consistent in probability.

- There are L_2 consistent estimators that are not a.s. consistent, and there are a.s. consistent estimators that are not L_2 consistent.

For each type of consistency mentioned above, there is a convergence rate associated with it, which tells us how fast the related convergence is. For example, if $n^{\nu} \text{MSE}(\widehat{\theta}, \theta) = O(1)$ for some positive constant ν, then we say that $\widehat{\theta}$ is L_2 consistent with the convergence rate $O(n^{-\nu})$. Here, the big O notation $a_n = O(b_n)$ has been used, where $\{a_n\}$ and $\{b_n\}$ are two sequences of nonnegative numbers. Its formal definition is that there exist two positive constants A and B such that $A \leq a_n/b_n \leq B$. Thus, if $a_n = O(b_n)$ and b_n converges to 0 as n tends to infinity, then a_n also converges to 0 with the *same* convergence rate. Sometimes, the small o notation will also be used. By definition, the expression $a_n = o(b_n)$ means that $\lim_{n \to \infty} a_n/b_n = 0$. So, if $a_n = o(b_n)$ and b_n converges to 0 as n tends to infinity, then a_n also converges to

0 with a *faster* rate. For other types of consistency, the convergence rate can be discussed similarly. More systematic discussion about large-sample properties of point estimators can be found in textbooks such as Ash (1972) and Chung (2001).

2.7.2 Maximum likelihood estimation and least squares estimation

In statistics, *maximum likelihood estimation* and *least squares estimation* provide two general methods for deriving point estimators of population parameters, which are briefly introduced in this subsection.

Assume that a population distribution has a pdf with the parametric form $f(x; \theta_1, \ldots, \theta_r)$, where $\theta_1, \ldots, \theta_r$ are r unknown population parameters. To estimate these population parameters based on a simple random sample $\{X_1, X_2, \ldots, X_n\}$, the maximum likelihood estimation procedure is based on the following *likelihood function*:

$$L(\theta_1, \ldots, \theta_r; X_1, X_2, \ldots, X_n) = \Pi_{i=1}^{n} f(X_i; \theta_1, \ldots, \theta_r). \tag{2.19}$$

The likelihood function $L(\theta_1, \ldots, \theta_r; X_1, X_2, \ldots, X_n)$ is treated as a function of the unknown parameters $\theta_1, \ldots, \theta_r$ only. The sample $\{X_1, X_2, \ldots, X_n\}$ is assumed to be given. From the discussion about the pdf in Section 2.2, the value of $f(x; \theta_1, \ldots, \theta_r) \Delta x$ is roughly equal to $P(X \in [x, x + \Delta x])$, where Δx is a small positive number. So the likelihood function is proportional to the likelihood that the observations in the sample take values around the observed sample $\{X_1, X_2, \ldots, X_n\}$.

The *maximum likelihood estimators* (MLEs) $\widehat{\theta}_1, \ldots, \widehat{\theta}_r$ of $\theta_1, \ldots, \theta_r$ are defined by the maximizers of the likelihood function $L(\theta_1, \ldots, \theta_r; X_1, X_2, \ldots, X_n)$. So, the likelihood that the observations in the sample take values around $\{X_1, X_2, \ldots, X_n\}$ reaches the maximum when the parameters equal their MLEs, which is reasonable because the sample $\{X_1, X_2, \ldots, X_n\}$ is assumed to have been observed before parameter estimation.

In practice, it is often more convenient to work with the logarithm of the likelihood function because the likelihood function has an exponential form in many cases. Then, the MLEs of $\theta_1, \ldots, \theta_r$ are the solutions to $\tilde{\theta}_1, \ldots, \tilde{\theta}_r$ of the following maximization problem:

$$\max_{\tilde{\theta}_1, \ldots, \tilde{\theta}_r \in R} \sum_{i=1}^{n} \log \left[f(X_i; \tilde{\theta}_1, \ldots, \tilde{\theta}_r) \right]. \tag{2.20}$$

In cases when the population has a normal distribution $N(\mu, \sigma^2)$, the likelihood function is

$$
\begin{aligned}
L(\mu, \sigma^2; X_1, X_2, \ldots, X_n) &= \Pi_{i=1}^{n} \left[\frac{1}{\sqrt{2\pi}\sigma} \exp\left(-\frac{(X_i - \mu)^2}{2\sigma^2} \right) \right] \\
&= \left(\frac{1}{\sqrt{2\pi}\sigma} \right)^n \exp\left(-\sum_{i=1}^{n} \frac{(X_i - \mu)^2}{2\sigma^2} \right).
\end{aligned}
$$

The log-likelihood function is

$$\log\left(L(\mu, \sigma^2; X_1, X_2, \ldots, X_n) \right) = -n\log\left(\sqrt{2\pi}\sigma \right) - \sum_{i=1}^{n} \frac{(X_i - \mu)^2}{2\sigma^2}.$$

It is easy to check that the maximization problem (2.20) with this log-likelihood function gives the MLEs

$$\widehat{\mu} = \overline{X}, \qquad \widehat{\sigma}^2 = \frac{1}{n} \sum_{i=1}^n (X_i - \overline{X})^2.$$

So, $\overline{X}$ is both the MVUE and the MLE of μ. The sample variance s^2 is slightly different from the above MLE of σ^2 in that $s^2 = \frac{n}{n-1}\widehat{\sigma}^2$. Because s^2 is an unbiased estimator of σ^2, $\widehat{\sigma}^2$ is biased, although the bias is $-\sigma^2/n$, which tends to 0 when n increases. For this reason, in practice, most people prefer to use the unbiased estimator s^2 for estimating σ^2, instead of its MLE $\widehat{\sigma}^2$.

In cases when the population distribution is discrete, the likelihood function can still be defined by (2.19), except that the pdf needs to be replaced by the probability distribution function. For instance, when the probability distribution is Bernoulli with π being the probability of S and we use 1 to denote S and 0 to denote F, the likelihood function is defined by

$$\begin{aligned} L(\pi; X_1, X_2, \ldots, X_n) &= \Pi_{i=1}^n \left[\pi^{X_i} (1-\pi)^{1-X_i} \right] \\ &= \pi^{\sum_{i=1}^n X_i} (1-\pi)^{n - \sum_{i=1}^n X_i}. \end{aligned}$$

Then, it is easy to check that the MLE of π is $\widehat{\pi} = \overline{X}$, which is also the sample proportion of S.

Besides MLE, least squares (LS) estimation is another general methodology for estimating population parameters. Usually, LS estimation is used for parametric regression modeling. Assume that there are two variables X and Y, and we are interested in building a numerical relationship between them. Between X and Y, assume that Y is a variable to predict based on the built relationship, and X is a variable to predict from. Then, Y is often called a *response variable*, and X is often called an *explanatory variable* or *predictor*. We further assume that X and Y follow a linear regression model

$$Y = \beta_0 + \beta_1 x + \varepsilon, \qquad (2.21)$$

where $\beta_0 + \beta_1 x$ is the *linear regression function*, which can be written as $\mathrm{E}(Y|X = x) = \beta_0 + \beta_1 x$, denoting the assumption that the mean value of Y when X is given at x is assumed to be a linear function of x, β_0 and β_1 are regression coefficients, and ε is a random error term. Now, assume that we have n observations of (X, Y), denoted as $\{(x_i, Y_i), i = 1, 2, \ldots, n\}$, and they are all generated from the linear regression model (2.21). Namely,

$$Y_i = \beta_0 + \beta_1 x_i + \varepsilon_i, \ i = 1, 2, \ldots, n,$$

where $\{\varepsilon_i, i = 1, 2, \ldots, n\}$ are random errors at the *design points* $\{x_i, i = 1, 2, \ldots, n\}$. For the error terms, we conventionally assume that they are i.i.d. and normally distributed, so that $\varepsilon_i \sim N(0, \sigma^2)$, for all $i = 1, 2, \cdots, n$, and the common variance σ^2 is usually unknown. All these conventional assumptions can be summarized by the four letters in LINE, where L denotes the assumption that the regression function is linear,

I denotes the assumption that the error terms are independent, N denotes the assumption that all error terms are normally distributed, and E denotes the assumption that all error terms have equal variance σ^2.

A widely used criterion for measuring the goodness-of-fit of a candidate estimator $b_0 + b_1 x$ of the true linear regression function $\beta_0 + \beta_1 x$ is the *residual sum of squares* (RSS), defined by

$$\text{RSS}(b_0, b_1) = \sum_{i=1}^{n} [Y_i - (b_0 + b_1 x_i)]^2 .$$

The LS estimators of β_0 and β_1 are defined to be the minimizers of $\text{RSS}(b_0, b_1)$. They can be calculated by the following formulas:

$$\begin{aligned}
\widehat{\beta_1} &= \frac{\sum_{i=1}^{n} (x_i - \bar{x})(Y_i - \bar{Y})}{\sum_{i=1}^{n} (x_i - \bar{x})^2} \\
\widehat{\beta_0} &= \bar{Y} - \widehat{\beta_1} \bar{x},
\end{aligned} \tag{2.22}$$

where $\bar{x}$ and $\bar{Y}$ are the sample means of x and Y values, respectively. Then, the estimated regression model is

$$\widehat{Y} = \widehat{\beta_0} + \widehat{\beta_1} x.$$

Under the four conventional assumptions (i.e., LINE), it can be checked that

$$\begin{pmatrix} \widehat{\beta_0} \\ \widehat{\beta_1} \end{pmatrix} \sim N\left(\begin{pmatrix} \beta_0 \\ \beta_1 \end{pmatrix}, \sigma^2 \begin{pmatrix} \frac{1}{n} + \frac{\bar{x}^2}{sxx}, & -\frac{\bar{x}}{sxx} \\ -\frac{\bar{x}}{sxx}, & \frac{1}{sxx} \end{pmatrix} \right), \tag{2.23}$$

where $sxx = \sum_{i=1}^{n} (x_i - \bar{x})^2$. For σ^2, it is often estimated by

$$\widehat{\sigma}^2 = \frac{1}{n-2} \sum_{i=1}^{n} \left[Y_i - \left(\widehat{\beta_0} + \widehat{\beta_1} x_i \right) \right]^2, \tag{2.24}$$

which is the so-called *residual mean squares (RMS)* of the estimated regression model. For the variance estimator in (2.24), by (2.23), we have

$$\frac{(n-2)\widehat{\sigma}^2}{\sigma^2} \sim \chi^2_{n-2}. \tag{2.25}$$

From the above description, it can be seen that the LS estimation does not use any information about the distribution of Y, which is an advantage, compared to the maximum likelihood estimation. Under the conventional assumptions LINE, $Y_i \sim N(\beta_0 + \beta_1 x_i, \sigma^2)$, for $i = 1, 2, \ldots, n$. Therefore, β_0, β_1 and σ^2 can also be estimated by their MLEs. As a matter of fact, it can be checked that the LS estimators of β_0 and β_1 given in (2.22) are the same as their MLEs, and the MLE of σ^2 is

$$\frac{n-2}{n} \widehat{\sigma}^2,$$

where $\widehat{\sigma}^2$ is the estimator defined in (2.24). By the way, $\widehat{\sigma}^2$ is an unbiased estimator of σ^2, and the MLE of σ^2 is biased.

2.7.3 Confidence intervals and hypothesis testing

Besides point estimation discussed in the previous subsection, there are two other methods of statistical inference about a population parameter. The first method uses confidence intervals, by which a population parameter θ is estimated by an interval. Suppose that a point estimator $\widehat{\theta}$ of θ has a normal distribution with mean θ and variance $\sigma_{\widehat{\theta}}^2$. Then,

$$Z = \frac{\widehat{\theta} - \theta}{\sigma_{\widehat{\theta}}} \sim N(0,1)$$

and

$$P\left(\widehat{\theta} - Z_{1-\alpha/2}\sigma_{\widehat{\theta}} < \theta < \widehat{\theta} + Z_{1-\alpha/2}\sigma_{\widehat{\theta}}\right) = 1 - \alpha,$$

where α is a given number between 0 and 1, and $Z_{1-\alpha/2}$ is the $(1-\alpha/2)$-th *quantile* of the standard normal distribution (cf. Figure 2.1(b)), defined by $P(Z \leq Z_{1-\alpha/2}) = 1 - \alpha/2$. Therefore, the random interval

$$\left(\widehat{\theta} - Z_{1-\alpha/2}\sigma_{\widehat{\theta}}, \ \widehat{\theta} + Z_{1-\alpha/2}\sigma_{\widehat{\theta}}\right)$$

has a $100(1-\alpha)\%$ chance to cover the true value of θ. This interval is called a $100(1-\alpha)\%$ *confidence interval (CI)* for θ, and the number $100(1-\alpha)\%$ is called the *confidence level*. For simplicity, the above CI is often written as

$$\widehat{\theta} \pm Z_{1-\alpha/2}\sigma_{\widehat{\theta}}.$$

Usually, the standard deviation $\sigma_{\widehat{\theta}}$ has some unknown population parameters, such as the population standard deviation σ, involved. So the above CI formula can be used only after these parameters are replaced by their point estimators. In other words, the standard deviation $\sigma_{\widehat{\theta}}$ needs to be estimated by the *standard error* $\widehat{\sigma}_{\widehat{\theta}}$ in which the unknown parameters have been replaced by their point estimators.

In cases when a population distribution is $N(\mu, \sigma^2)$ and $\{X_1, X_2, \ldots, X_n\}$ is a simple random sample from this population, by the discussion in Subsection 2.7.1, $\overline{X} \sim N(\mu, \sigma^2/n)$. Therefore, in cases when σ is known,

$$Z = \frac{\overline{X} - \mu}{\sigma/\sqrt{n}} \sim N(0,1), \tag{2.26}$$

and the $100(1-\alpha)\%$ CI for μ is

$$\overline{X} \pm Z_{1-\alpha/2}\frac{\sigma}{\sqrt{n}}.$$

In cases when σ is unknown, it should be replaced by the sample standard deviation s in (2.26). Because of the extra randomness added by this replacement, the resulting statistic has a t distribution with df equal to $n-1$. Namely,

$$T = \frac{\overline{X} - \mu}{s/\sqrt{n}} \sim t_{n-1}. \tag{2.27}$$

See (2.18) and the related discussion in Subsection 2.7.1. So, based on (2.27), the $100(1-\alpha)\%$ CI for μ is

$$\overline{X} \pm t_{1-\alpha/2}(n-1)\frac{s}{\sqrt{n}}, \tag{2.28}$$

where $t_{1-\alpha/2}(n-1)$ is the $(1-\alpha/2)$-th quantile of the t_{n-1} distribution.

In applications, the exact distribution of $\widehat{\theta}$ is often unknown for a fixed sample size. If its *asymptotic distribution*, which is the limit distribution of $\widehat{\theta}$ when n tends to infinity, can be derived, then the CI for θ can be constructed based on this asymptotic distribution. Of course, in such cases it is only asymptotically true that the CI has $100(1-\alpha)\%$ chance to cover the true value of θ. In cases when we are interested in estimating the population mean μ (i.e., $\theta = \mu$), if the population distribution is unknown, or it is known but non-normal, then by the CLT, (2.26) is asymptotically true. Namely, when the sample size n is large (i.e., $n \geq 30$), the CI formula (2.28) can still be used, in which we can use either $t_{1-\alpha/2}(n-1)$ or $Z_{1-\alpha/2}$ because the two quantities should be almost the same in such cases. See a related discussion in Subsection 2.3.3.

In cases when the population distribution is Bernoulli with π being the probability of success and when the sample size is large (i.e., $n\widehat{\pi} \geq 10$ and $n(1-\widehat{\pi}) \geq 10$), by similar arguments to those in the previous two paragraphs, the large-sample $100(1-\alpha)\%$ CI for π is

$$\widehat{\pi} \pm Z_{1-\alpha/2}\sqrt{\frac{\widehat{\pi}(1-\widehat{\pi})}{n}}, \tag{2.29}$$

where $\widehat{\pi}$ is the sample proportion of success.

Another method of statistical inference about a population parameter involves testing hypotheses. In our daily life, we often make a statement or *hypothesis* about a population parameter. For example, some reports claim that smoking would increase the chance of lung cancer under a general circumstance. In this example, all smokers constitute the population. Assume that the prevalence rate of lung cancer in this population is an unknown parameter π, and that the prevalence rate of the same disease among all non-smokers is known to be π_0. Then, the above statement basically says that $\pi > \pi_0$. A major goal of many research projects is to collect data for verifying such a hypothesis. If the hypothesis is supported by the data obtained from one or more repeated experiments, then it becomes a new theory. In statistics, the hypothesis that we want to validate is called an *alternative hypothesis*. Usually, an alternative hypothesis represents a potential new theory, new method, new discovery, and so forth. It is a competitor to a so-called *null hypothesis*, which often represents existing theory, existing method, existing knowledge, and so forth. Because the null hypothesis is usually verified in the past, it is initially assumed true. It is rejected only in cases when the observed data from the population in question provide convincing evidence against it.

By convention, the null hypothesis is denoted by H_0, and it usually takes the equality form

$$H_0 : \theta = \theta_0,$$

where θ_0 is the hypothesized value of the population parameter θ. The alternative hypothesis is denoted by H_1 or H_a, and it can take one of the following three forms:

$$H_1 : \theta > \theta_0$$
$$H_1 : \theta < \theta_0$$
$$H_1 : \theta \neq \theta_0.$$

The first form is called right-sided or right-tailed, the second one is called left-sided or left-tailed, and the last one is called two-sided or two-tailed.

To test whether H_0 should be rejected in favor of H_1, we need a criterion constructed from the observed data. To this end, assume that $\{X_1, X_2, \ldots, X_n\}$ is a simple random sample from the population of interest and that $\widehat{\theta}$ is a point estimator of θ. To describe how to construct a hypothesis testing procedure, let us assume that $\widehat{\theta} \sim N(\theta, \sigma_{\widehat{\theta}}^2)$ and that $\sigma_{\widehat{\theta}}^2$ does not include any unknown parameters. Then, in cases when H_0 is true, we have

$$Z = \frac{\widehat{\theta} - \theta_0}{\sigma_{\widehat{\theta}}} \sim N(0, 1).$$

Let us assume that the alternative hypothesis of interest is right-sided, and the observed value of Z is denoted by Z^*. Then, the probability

$$P_{H_0}(Z \geq Z^*)$$

would tell us the likelihood that we can observe Z^* or values of Z that are more inconsistent with H_0 when H_0 is assumed true, where P_{H_0} denotes the probability under H_0. The probability $P_{H_0}(Z \geq Z^*)$ is called the p-value and the statistic Z is called a *test statistic*. The p-value is defined by $P_{H_0}(Z \leq Z^*)$ when H_1 is left-sided, and by $P_{H_0}(|Z| \geq |Z^*|)$ when H_1 is two-sided.

From the definition of p-value, the data provide more evidence against H_0 if the p-value is smaller. Then, the question becomes: how small is small? To answer this question, we usually compare the calculated p-value with a pre-specified *significance level*, denoted conventionally as α. If the p-value is smaller than or equal to α, then we reject H_0 and conclude that the data have provided significant evidence to support H_1. Otherwise, we fail to reject H_0 and conclude that there is no significant evidence in the data to support H_1. This process of decision making is called *hypothesis testing*.

For a given significance level α, an alternative approach to perform hypothesis testing is to compare the observed value Z^* of the test statistic Z with its α-*critical value*. In the case considered above when $Z = \frac{\widehat{\theta} - \theta_0}{\sigma_{\widehat{\theta}}} \sim N(0, 1)$, its α-critical value is defined by $Z_{1-\alpha}$ if the alternative hypothesis H_1 is right-sided, by Z_α if H_1 is left-sided, and by $Z_{1-\alpha/2}$ if H_1 is two-sided. Then, the null hypothesis H_0 is rejected when $Z^* > Z_{1-\alpha}$, $Z^* < Z_\alpha$, and $|Z^*| > Z_{1-\alpha/2}$, respectively, for the three types of H_1 considered above. It is easy to check that decisions made by the two approaches are actually equivalent to each other. In applications, the p-value approach is often preferred because it provides us decisions regarding whether H_0 should be rejected at a given significance level and gives us a quantitative measure of the strength of evidence in the observed data that is against H_0 as well.

In our decision making using any hypothesis-testing procedure, mistakes are inevitable. There are two types of mistakes, or, more conventionally, two types of errors that we can make. *Type I error* refers to the case in which H_0 is rejected when it is actually true. *Type II error* refers to the case in which H_0 fails to be rejected when it is actually false. The probabilities of type I and type II errors are denoted by α and β, respectively. Note that α is used to denote both the significance level and the probability of type I error because these two quantities are the same in most situations.

Intuitively, an ideal hypothesis-testing procedure should have small α and small β. However, in reality, if α is kept small, then β would be large, and vice versa. To handle this situation, a conventional strategy is to control α at some fixed level, and let β be as small as possible. By this strategy, H_0 is protected to a certain degree because the chance to reject it when it is actually true cannot exceed the fixed level of α. This is reasonable in the sense that H_0 often represents existing methods or knowledge, and it has been justified in the past. Selection of α usually depends on the consequence of the type I error. If the consequence is serious, then a small α should be chosen. Otherwise, a relatively large value could be chosen. Commonly used α values include 0.1, 0.05, 0.01, 0.005, and 0.001. A default α value adopted by most scientific communities is 0.05.

From the above discussion, it can be seen that a major step for solving a hypothesis-testing problem is to find an appropriate test statistic, which should have the property that its distribution under H_0, or its *null distribution*, is known or can be computed. For a given hypothesis-testing problem, different testing procedures are possible. In statistics, we usually only consider the procedures whose type I error probabilities are all below a given level (i.e., α). Among those procedures, the one with the smallest type II error probability β, or equivalently the largest *power*, defined to be $1 - \beta$, is the best.

A general methodology for deriving a testing procedure for a hypothesis-testing problem is the *likelihood ratio test (LRT)* described below. Let $L(\theta; X_1, X_2, \ldots, X_n)$ be the likelihood function, Θ_0 be the set of θ values under H_0, and Θ_1 be the set of θ values under H_1. Consider the following ratio of two maximum likelihoods:

$$\Lambda(X_1, X_2, \ldots, X_n) = \frac{\max_{\theta \in \Theta_0} L(\theta; X_1, X_2, \ldots, X_n)}{\max_{\theta \in \Theta_0 \cup \Theta_1} L(\theta; X_1, X_2, \ldots, X_n)}. \tag{2.30}$$

Obviously, it is always true that $0 \leq \Lambda(X_1, X_2, \ldots, X_n) \leq 1$. In cases when H_0 is true, the two maximum likelihoods in (2.30) should be close to each other; consequently, $\Lambda(X_1, X_2, \ldots, X_n)$ is close to 1. Otherwise, $\Lambda(X_1, X_2, \ldots, X_n)$ should be small. Therefore, we can make decisions about the hypotheses H_0 and H_1 using the *LRT statistic* $\Lambda(X_1, X_2, \ldots, X_n)$, and reject H_1 in cases when $\Lambda(X_1, X_2, \ldots, X_n)$ is too small. Wilks (1938) showed that, under some regularity conditions, it was asymptotically true that

$$-2\log(\Lambda(X_1, X_2, \ldots, X_n)) \overset{H_0}{\sim} \chi^2_{df}, \tag{2.31}$$

where $\overset{H_0}{\sim} \chi^2_{df}$ means "has the χ^2_{df} distribution under H_0," and df equals the difference in dimensionality of Θ_0 and Θ_1. Then, in large sample cases, H_1 can be rejected at the significance level of α if the observed value of $-2\log(\Lambda(X_1, X_2, \ldots, X_n))$ is larger than the $(1-\alpha)$-th quantile of the χ_{df}^2 distribution. In cases when multiple parameters are involved in the testing problem, the LRT test can be described in a similar way. It has been demonstrated in the literature that LRT tests have certain desirable statistical properties. For a related discussion, see Casella and Berger (2002) and Lehmann and Romano (2005).

When the population distribution is $N(\mu, \sigma^2)$, the null hypothesis is $H_0 : \mu = \mu_0$, and σ is known, a commonly used test statistic is

$$Z = \frac{\overline{X} - \mu_0}{\sigma/\sqrt{n}} \overset{H_0}{\sim} N(0,1).$$

In such cases, it can be checked that

$$
\begin{aligned}
& -2\log(\Lambda(X_1, X_2, \ldots, X_n)) \\
= \; & \frac{\sum_{i=1}^n (x_i - \mu_0)^2 - \sum_{i=1}^n (x_i - \overline{X})^2}{\sigma^2} \\
= \; & n\left(\frac{\overline{X} - \mu_0}{\sigma}\right)^2 \\
= \; & Z^2.
\end{aligned}
$$

Therefore, the test based on Z and the LRT test are exactly the same when the alternative hypothesis H_1 is two-sided. Also, the result (2.31) is exact in such cases. In cases when σ is unknown, it should be replaced by s when defining the test statistic Z. The resulting test statistic and its null distribution are

$$T = \frac{\overline{X} - \mu_0}{s/\sqrt{n}} \overset{H_0}{\sim} t_{n-1}. \qquad (2.32)$$

In cases when the population distribution is non-normal, the test statistic in (2.32) can still be used, as long as the sample size n is large (i.e., $n \geq 30$), because of the central limit theorem discussed in Subsection 2.7.1. In such cases, the asymptotic null distribution of T is $N(0,1)$, to which the t_{n-1} distribution is close. Therefore, either distribution can be used when computing the p-value.

In cases when the population distribution is Bernoulli with π being the probability of success, the null hypothesis is $H_0 : \pi = \pi_0$, and the sample size is large (i.e., $n\pi_0 \geq 10$ and $n(1 - \pi_0) \geq 10$), a commonly used test statistic is

$$Z = \frac{\widehat{\pi} - \pi_0}{\sqrt{\pi_0(1 - \pi_0)/n}} \overset{H_0}{\sim} N(0,1), \qquad (2.33)$$

where π_0 is the hypothesized value of π, and $\widehat{\pi}$ is the sample proportion of success. Of course, the null distribution specified in (2.33) is only asymptotically true in such cases.

In certain applications, we need to compare two populations with regard to a specific population characteristic (e.g., the population mean). To this end, assume that $\{X_{11}, X_{12}, \ldots, X_{1n_1}\}$ and $\{X_{21}, X_{22}, \ldots, X_{2n_2}\}$ are two samples from the two populations, respectively, and that the two samples are independent of each other. The sample means and the sample variances of the two samples are denoted as $\overline{X}_1, \overline{X}_2, s_1^2$, and s_2^2. If we are interested in comparing the two population means μ_1 and μ_2, then the null and alternative hypotheses would be

$$H_0 : \mu_1 - \mu_2 = \delta_0 \qquad \text{versus} \qquad H_1 : \mu_1 - \mu_2 \neq \delta_0 (> \delta_0, < \delta_0), \qquad (2.34)$$

where δ_0 is a given number, and $\mu_1 - \mu_2 \neq \delta_0 (> \delta_0, < \delta_0)$ denotes one of the three forms: $\mu_1 - \mu_2 \neq \delta_0$, $\mu_1 - \mu_2 > \delta_0$, and $\mu_1 - \mu_2 < \delta_0$. In most cases, we are interested in the case when $\delta_0 = 0$. In such cases, H_0 in (2.34) says that the two population means are the same. To test the above hypotheses, it is natural to use $\overline{X}_1 - \overline{X}_2$, which is a good point estimator of $\mu_1 - \mu_2$, according to the related discussion about the sample mean in one-population cases (cf., Subsections 2.7.1 and 2.7.2).

In cases when the two population distributions are $N(\mu_1, \sigma_1^2)$ and $N(\mu_2, \sigma_2^2)$, it is easy to check that

$$Z = \frac{(\overline{X}_1 - \overline{X}_2) - (\mu_1 - \mu_2)}{\sqrt{\sigma_1^2/n_1 + \sigma_2^2/n_2}} \sim N(0, 1).$$

In such cases, if both σ_1^2 and σ_2^2 are known, then a $100(1 - \alpha)\%$ CI for $\mu_1 - \mu_2$ is

$$(\overline{X}_1 - \overline{X}_2) \pm Z_{1-\alpha/2} \sqrt{\sigma_1^2/n_1 + \sigma_2^2/n_2},$$

and a good test statistic for the hypotheses in (2.34) is

$$Z = \frac{(\overline{X}_1 - \overline{X}_2) - \delta_0}{\sqrt{\sigma_1^2/n_1 + \sigma_2^2/n_2}} \overset{H_0}{\sim} N(0, 1).$$

Of course, in practice, the two population variances σ_1^2 and σ_2^2 are often unknown. If the two population distributions are still $N(\mu_1, \sigma_1^2)$ and $N(\mu_2, \sigma_2^2)$ and both σ_1^2 and σ_2^2 are unknown, then we have

$$T = \frac{(\overline{X}_1 - \overline{X}_2) - (\mu_1 - \mu_2)}{\sqrt{s_1^2/n_1 + s_2^2/n_2}} \sim t_{df}, \qquad (2.35)$$

where

$$df = \frac{(V_1 + V_2)^2}{V_1^2/(n_1 - 1) + V_2^2/(n_2 - 1)}, \qquad (2.36)$$

$V_1 = s_1^2/n_1$, and $V_2 = s_2^2/n_2$. Based on the statistic T in (2.35), the $100(1 - \alpha)\%$ CI for $\mu_1 - \mu_2$ is

$$(\overline{X}_1 - \overline{X}_2) \pm t_{1-\alpha/2}(df) \sqrt{s_1^2/n_1 + s_2^2/n_2},$$

where $t_{1-\alpha/2}(df)$ is the $(1 - \alpha/2)$-th quantile of the t_{df} distribution with df defined by (2.36). For testing hypotheses in (2.34), a good test statistic is

$$T = \frac{(\overline{X}_1 - \overline{X}_2) - \delta_0}{\sqrt{s_1^2/n_1 + s_2^2/n_2}} \overset{H_0}{\sim} t_{df}. \tag{2.37}$$

The test using the test statistic in (2.37) is often called the *two independent sample t-test*. In certain cases, it is reasonable to assume that $\sigma_1^2 = \sigma_2^2 = \sigma^2$. In such cases, we can estimate σ^2 by the following *pooled sample variance*:

$$s_p^2 = \frac{(n_1 - 1)s_1^2 + (n_2 - 1)s_2^2}{n_1 + n_2 - 2}.$$

Then, the statistic in (2.35) can be replaced by

$$T = \frac{(\overline{X}_1 - \overline{X}_2) - (\mu_1 - \mu_2)}{s_p \sqrt{1/n_1 + 1/n_2}} \sim t_{n_1 + n_2 - 2}. \tag{2.38}$$

The CI formula and the test statistic for testing the hypotheses in (2.34) can be modified accordingly, using the statistic in (2.38).

In cases when the two population distributions are unknown, but both n_1 and n_2 are large (i.e., $n_1 \geq 30$ and $n_2 \geq 30$), it is asymptotically true that

$$Z = \frac{(\overline{X}_1 - \overline{X}_2) - (\mu_1 - \mu_2)}{\sqrt{s_1^2/n_1 + s_2^2/n_2}} \sim N(0, 1). \tag{2.39}$$

In such cases, the large sample CI for $\mu_1 - \mu_2$ and the large sample testing procedure for testing the hypotheses in (2.34) can be constructed using the statistic in (2.39). For instance, for testing the hypotheses in (2.34), the test statistic would be the same as that in (2.37), except that its null distribution could be approximated well by either t_{df} or $N(0, 1)$.

If we are interested in comparing two population proportions π_1 and π_2 using two independent samples from the two populations, then, in large sample cases,

$$Z = \frac{(\widehat{\pi}_1 - \widehat{\pi}_2) - (\pi_1 - \pi_2)}{\sqrt{\widehat{\pi}_1(1 - \widehat{\pi}_1)/n_1 + \widehat{\pi}_2(1 - \widehat{\pi}_2)/n_2}} \sim N(0, 1),$$

where $\widehat{\pi}_1$ and $\widehat{\pi}_2$ are the two sample proportions. Therefore, a $100(1 - \alpha)\%$ CI for $\pi_1 - \pi_2$ would be

$$(\widehat{\pi}_1 - \widehat{\pi}_2) \pm Z_{1-\alpha/2} \sqrt{\widehat{\pi}_1(1 - \widehat{\pi}_1)/n_1 + \widehat{\pi}_2(1 - \widehat{\pi}_2)/n_2}.$$

The large sample condition is satisfied if $n_1\widehat{\pi}_1 \geq 10$, $n_1(1 - \widehat{\pi}_1) \geq 10$, $n_2\widehat{\pi}_2 \geq 10$, and $n_2(1 - \widehat{\pi}_2) \geq 10$. To test hypotheses

$$H_0 : \pi_1 - \pi_2 = 0 \qquad \text{versus} \qquad H_1 : \pi_1 - \pi_2 \neq 0 (> 0, < 0),$$

a good test statistic is

$$Z = \frac{\widehat{\pi}_1 - \widehat{\pi}_2}{\sqrt{\widehat{\pi}_p(1 - \widehat{\pi}_p)}\sqrt{1/n_1 + 1/n_2}} \overset{H_0}{\sim} N(0,1),$$

where

$$\widehat{\pi}_p = \frac{n_1 \widehat{\pi}_1 + n_2 \widehat{\pi}_2}{n_1 + n_2}$$

is the pooled sample proportion.

When we compare two populations, sometimes certain variables that may affect the variable of interest should be taken into account. For instance, if we are interested in studying whether physical exercise can effectively control our weight, then one possible approach is as follows. First, we specify two populations. For example, one population could be all adults aged 20 and above who go to the gym regularly (e.g., at least two times a week), and the other population includes all adults aged 20 and above who do not go to the gym regularly. Second, we randomly select a group of people from each population and record the weights for all people selected. Finally, we can use the two independent sample t-test to compare the mean weights of the two populations. This approach seems appropriate. But, it can happen that, in the two groups of people selected, one group includes more males than the other group. So, the significant difference between the two population means of weight found by the t-test could be due to the gender difference in the two samples, instead of the difference in physical exercise. Therefore, to make a more appropriate comparison, we should take into account the variables that are not our major interest but may affect the result, such as gender, age, profession, and so forth. These variables are often called the *confounding risk factors*. To avoid the impact of the confounding risk factors, we can collect data in an alternative way as follows. First, we list the major confounding risk factors that we can think of. Second, for each randomly selected member from the first population, we randomly select a member from all members in the second population that match the selected member in the first population by all the confounding risk factors. In that way, the two resulting samples from the two populations are *paired*. Let the two samples be $\{X_{11}, X_{12}, \ldots, X_{1n}\}$ and $\{X_{21}, X_{22}, \ldots, X_{2n}\}$. Then, X_{11} is paired with X_{21}, X_{12} is paired with X_{22}, and so forth. So, the sample sizes of the two samples must be the same. To test the hypotheses in (2.34), we can use the difference between the two samples, by defining $D_i = X_{1i} - X_{2i}$, for $i = 1, 2, \ldots, n$. Then, $\{D_1, D_2, \ldots, D_n\}$ can be regarded as a simple random sample from a "difference" population with the mean $\mu_d = \mu_1 - \mu_2$. Therefore, the original two-population problem becomes a one-population problem with the hypotheses

$$H_0 : \mu_d = \delta_0 \qquad \text{versus} \qquad H_1 : \mu_d \neq \delta_0 (> \delta_0, < \delta_0).$$

And, most statistical inference methods for the one-population problem can be used here. For instance, if it is reasonable to assume that the "difference" population has a normal distribution with an unknown variance, then a good test statistic would be

$$T = \frac{\overline{D} - \delta_0}{s_d/\sqrt{n}} \overset{H_0}{\sim} t_{n-1}, \qquad\qquad (2.40)$$

where $\overline{D}$ and s_d are the sample mean and sample standard deviation of the "difference" sample $\{D_1, D_2, \ldots, D_n\}$. The test based on (2.40) is often called the *two paired sample t-test*. By the way, a $100(1-\alpha)\%$ CI formula for μ_d can be derived accordingly, to be

$$\overline{D} \pm t_{1-\alpha/2}(n-1)\frac{s_d}{\sqrt{n}}.$$

2.7.4 The delta method and the bootstrap method

In the previous subsection, we discussed how to construct a CI or perform a hypothesis test for a population parameter θ using the exact or asymptotic distribution of its point estimator $\widehat{\theta}$. In some cases, we are interested in statistical inferences about a function of θ, denoted as $g(\theta)$. In such cases, it is natural to estimate $g(\theta)$ by $g(\widehat{\theta})$. If the exact distribution of $g(\widehat{\theta})$ is available, then the related inferences can be made accordingly. Otherwise, the method described below can be considered for deriving the asymptotic distribution of $g(\widehat{\theta})$.

Assume that the function g has the second-order derivative at θ, and $\sqrt{n}(\widehat{\theta} - \theta)$ converges in distribution to $N(0, \sigma^2)$, written in notation as

$$\sqrt{n}\left(\widehat{\theta} - \theta\right) \overset{D}{\to} N(0, \sigma^2). \tag{2.41}$$

Then, by the Taylor's expansion, we have

$$g(\widehat{\theta}) = g(\theta) + g'(\theta)\left(\widehat{\theta} - \theta\right) + O\left(\left(\widehat{\theta} - \theta\right)^2\right).$$

So,

$$\sqrt{n}\left(g(\widehat{\theta}) - g(\theta)\right) = g'(\theta)\sqrt{n}\left(\widehat{\theta} - \theta\right) + O\left(\sqrt{n}\left(\widehat{\theta} - \theta\right)^2\right).$$

The second term on the right-hand side of the above expression would converge to 0 in distribution, and by (2.41) the first term would converge to a normal distribution. Therefore, we have

$$\sqrt{n}\left(g(\widehat{\theta}) - g(\theta)\right) \overset{D}{\to} N\left(0, \sigma^2\,[g'(\theta)]^2\right). \tag{2.42}$$

Then, statistical inferences about $g(\theta)$ can proceed using the result in (2.42). This general approach of statistical inferences about $g(\theta)$ based on the Taylor's expansion is often called the *delta method*.

Example 2.7 *Assume that $\{X_1, X_2, \ldots, X_n\}$ is a simple random sample from a population of interest, and that we are interested in estimating $g(\mu) = \mu(\mu+1)$, where μ is the population mean. Then, a reasonable point estimator of $g(\mu)$ is $g(\overline{X})$, where $\overline{X}$ is the sample mean. By the CLT, we have*

$$\sqrt{n}\left(\overline{X} - \mu\right) \overset{D}{\to} N(0, \sigma^2),$$

where σ^2 is the population variance. By (2.42), we have

$$\sqrt{n}\left(g(\overline{X}) - g(\mu)\right) \xrightarrow{D} N\left(0, \sigma^2 \left[g'(\mu)\right]^2\right).$$

So, a large-sample $100(1-\alpha)\%$ CI for $g(\mu)$ would be

$$g(\overline{X}) \pm Z_{1-\alpha/2} g'(\overline{X}) s/\sqrt{n},$$

where $g'(\overline{X}) = 2\overline{X} + 1$ and s is the sample standard deviation.

In some cases, when making statistical inferences about a population parameter θ, the exact distribution of its point estimator $\widehat{\theta}$ is difficult to know. Furthermore, the asymptotic distribution of $\widehat{\theta}$ is also difficult to derive, or it is inappropriate to use the asymptotic distribution because the sample size n in a given application is not large enough. In all such cases, the *bootstrap* approach (cf., Efron (1979), Efron and Tibshirani (1993)) is often helpful. A typical bootstrap algorithm works in several steps as follows.

Step 1 Draw a random sample of size n from the observed data $\{X_1, X_2, \ldots, X_n\}$ with replacement. The new sample, which is called the *bootstrap sample*, is denoted as $\{X_1^*, X_2^*, \ldots, X_n^*\}$. Compute the estimate of θ from the bootstrap sample, and the estimate is denoted as $\widehat{\theta}^*$.

Step 2 Repeat step 1 B times, and the estimates of θ from the B bootstrap samples are denoted as $\{\widehat{\theta}_j^*, \ j = 1, 2, \ldots, B\}$, where B is often called the *bootstrap sample size*.

Step 3 The empirical distribution of $\{\widehat{\theta}_j^*, \ j = 1, 2, \ldots, B\}$ (cf., the related discussion in Subsection 2.8.1 below), denoted as $\widehat{F}_{\widehat{\theta}}$, is used as an estimate of the true distribution of $\widehat{\theta}$, denoted as $F_{\widehat{\theta}}$, for statistical inferences.

For instance, in the case when $B = 10,000$, we can use the interval $(\widehat{\theta}_{(251)}^*, \widehat{\theta}_{(9750)}^*)$ as the 95% CI for θ, where $\widehat{\theta}_{(251)}^*$ and $\widehat{\theta}_{(9750)}^*$ are the 251th and 9750th order statistics of $\{\widehat{\theta}_j^*, \ j = 1, 2, \ldots, 10,000\}$.

2.8 Nonparametric Statistical Inferences

Statistical methods introduced in the previous section are appropriate to use only when the parametric model assumption imposed on the population distribution (e.g., the normal distribution) is valid. In practice, however, the assumed parametric model of the population distribution is often invalid. Statistical inferences without a parametric model of the population distribution are often called *nonparametric statistical inferences*. In this section, we briefly introduce some major methodologies in this area. More complete discussions about this topic can be found in text books, such as Gibbons and Chakraborti (2003), Hollander and Wolfe (1999), and Kvam and Vidakovic (2007).

2.8.1 Order statistics and their properties

When a population characteristic of interest is numeric and a parametric form of the population distribution is unavailable, statistical inferences about the population distribution are often based on the ordering (or *ranking*) information of the observations in a sample. So, in such cases, the order statistics play an important role. In this subsection, we briefly discuss some basic properties of the order statistics.

Let $\{X_1, X_2, \ldots, X_n\}$ be a simple random sample from a population with the cdf F, and $X_{(1)} \leq X_{(2)} \leq \cdots \leq X_{(n)}$ be the order statistics. Then, $\{X_{(1)}, X_{(2)}, \ldots, X_{(n)}\}$ is an ordered version of $\{X_1, X_2, \ldots, X_n\}$. If X_i is the R_i-th order statistic, for $i = 1, 2, \ldots, n$, then R_i takes its value in $\{1, 2, \ldots, n\}$ and $\{R_1, R_2, \ldots, R_n\}$ is a permutation of $\{1, 2, \ldots, n\}$. In the literature, R_i is often called the *rank* of X_i. On the other hand, if $X_{(i)}$ is the A_i-th observation in the sample, for $i = 1, 2, \ldots, n$, then A_i is often called the i-th *inverse rank*, or the i-th *antirank*. Obviously, $\{A_1, A_2, \ldots, A_n\}$ is also a permutation of $\{1, 2, \ldots, n\}$. Regarding the ranks and antiranks, they are all random variables because they are uniquely determined by the random sample. The sampling distribution of $\{R_1, R_2, \ldots, R_n\}$ is that it takes any permutation of $\{1, 2, \ldots, n\}$ with the probability of $1/n!$, and the sampling distribution of $\{A_1, A_2, \ldots, A_n\}$ is the same.

In the above discussion, we have assumed that there are no ties in the observed data, which is usually true in cases when the population distribution is absolutely continuous (i.e., it has a pdf). When there are one or more ties in the observed data, the ranks and antiranks can also be defined properly, which is demonstrated by the following example.

Example 2.8 *Assume that the observed sample consists of the following 10 numbers*

$$2.3, 1.5, 3.4, 1.5, 0.9, 1.7, 1.5, 1.7, 2.1, 0.7.$$

In this data, the number 1.7 is observed twice, and the number 1.5 is observed three times. The ordered observations are

$$0.7, 0.9, 1.5, 1.5, 1.5, 1.7, 1.7, 2.1, 2.3, 3.4.$$

So, $R_1 = 9, R_3 = 10, R_5 = 2, R_9 = 7$, and $R_{10} = 1$. However, each of R_2, R_4, and R_7 can be defined to be 3, 4, or 5, because (X_2, X_4, X_7) is a tie. In such a case, we can define $R_2 = R_4 = R_7 = (3 + 4 + 5)/3 = 4$. Similarly, we can define $R_6 = R_8 = 6.5$. The antiranks can be handled similarly. Of course, other solutions for handling the ties are possible. For instance, to break the tie of (X_6, X_8) when defining R_6 and R_8, we can draw a random number from the uniform distribution on $[0, 1]$, which has the same chance to be any number in the interval $[0, 1]$. If the random number is smaller than or equal to 0.5, then we define $R_6 = 6$ and $R_8 = 7$. Otherwise, we define $R_6 = 7$ and $R_8 = 6$.

For the last order statistic $X_{(n)}$, its cdf $F_{X_{(n)}}$ can be derived easily as follows:

$$
\begin{aligned}
F_{X_{(n)}}(x) &= P(X_{(n)} \leq x) \\
&= P(X_1 \leq x, X_2 \leq x, \ldots, X_n \leq x) \\
&= P(X_1 \leq x)P(X_2 \leq x) \cdots P(X_n \leq x) \\
&= F^n(x).
\end{aligned}
$$

Similarly, the cdf of the first order statistic $X_{(1)}$, denoted as $F_{X_{(1)}}$, can be derived to be

$$F_{X_{(1)}}(x) = 1 - [1 - F(x)]^n.$$

For a general $1 \leq i \leq n$, the cdf of the i-th order statistic $X_{(i)}$ is

$$F_{X_{(i)}}(x) = \sum_{j=i}^{n} \binom{n}{j} F^j(x)[1 - F(x)]^{n-j}.$$

In cases when the population distribution has a pdf f, the pdf of the i-th order statistic $X_{(i)}$, denoted as $f_{X_{(i)}}$, is

$$f_{X_{(i)}}(x) = i \binom{n}{i} F^{i-1}(x)[1 - F(x)]^{n-i} f(x),$$

for $1 \leq i \leq n$. When $i = n$, the pdf of $X_{(n)}$ is

$$f_{X_{(n)}}(x) = nF^{n-1}(x)f(x).$$

When $i = 1$, the pdf of $X_{(1)}$ is

$$f_{X_{(1)}}(x) = n[1 - F(x)]^{n-1} f(x).$$

For a pair of order statistics $(X_{(i_1)}, X_{(i_2)})$ with $1 \leq i_1 < i_2 \leq n$, it can be checked that their joint cdf is

$$F_{X_{(i_1)},X_{(i_2)}}(x,y) =$$
$$\begin{cases} F_{X_{(i_2)}}(y), & \text{if } x > y \\ \sum \frac{n!}{r!s!(n-r-s)!} F^r(x)[F(y) - F(x)]^s[1 - F(y)]^{n-r-s}, & \text{if } x \leq y, \end{cases}$$

where the summation is over the range $\{(r,s) : i_1 \leq r \leq n, i_2 \leq r + s \leq n\}$. If the population distribution has a pdf f, then the joint pdf of $(X_{(i_1)}, X_{(i_2)})$ is

$$f_{X_{(i_1)},X_{(i_2)}}(x,y) =$$
$$\begin{cases} 0, & \text{if } x > y \\ \frac{n!}{(i_1-1)!(i_2-i_1-1)!(n-i_2)!} F^{i_1-1}(x)[F(y) - F(x)]^{i_2-i_1-1}[1 - F(y)]^{n-i_2}, & \text{if } x \leq y. \end{cases}$$

One major application of the order statistics is to construct the *empirical cumulative distribution function (ecdf)*

$$F_n(x) = \begin{cases} 0, & \text{if } x < X_{(1)} \\ 1/n, & \text{if } x \in [X_{(1)}, X_{(2)}) \\ \vdots & \vdots \\ (n-1)/n, & \text{if } x \in [X_{(n-1)}, X_{(n)}) \\ 1, & \text{if } x \geq X_{(n)}. \end{cases} \tag{2.43}$$

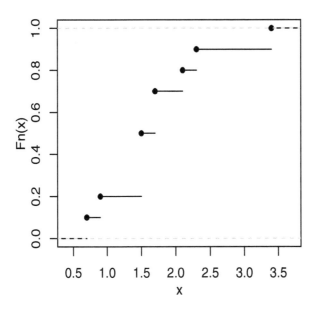

Figure 2.11 *The ecdf constructed from the data in Example 2.8. The dark point at the begin-*
ning of each horizontal line segment denotes that the value of the ecdf at each jump position
equals the height of the dark point.

Example 2.8 (continued) *For the data in Example 2.8, the ecdf is shown in Figure*
2.11. From the plot, it can be seen that the ecdf is a step function with the jumps at
the observed values.

Clearly, the ecdf defined in (2.43) has the property that

$$F_n(x) = \frac{1}{n} \sum_{i=1}^{n} I(X_i \le x),$$

where $I(a)$ is the *indicator function* of a, and it equals 1 when a is "true" and 0
otherwise. So, $F_n(x)$ is the proportion of observations in the sample that are smaller
than or equal to x; it is the cdf of the data in the observed sample and is commonly
used for estimating the population cdf F. As long as the sample $\{X_1, X_2, \ldots, X_n\}$ is a
simple random sample, we have the following result:

$$P\left(\lim_{n \to \infty} \max_{x \in R} |F_n(x) - F(x)| = 0 \right) = 1.$$

This result says that $F_n(x)$ converges to $F(x)$ almost surely (cf., Subsection 2.7.1)
and uniformly over $x \in R$.

2.8.2 Goodness-of-fit tests

In statistical process control, one major goal is to check whether a quality characteristic of a product follows a given distribution specified by some design requirements
(cf., the related discussion in Section 1.2). This and many other applications are related to the testing of the hypotheses:

$$H_0 : F(x) = F_0(x), \qquad \text{for all } x \in R$$
$$\text{versus} \quad H_1 : F(x) \neq F_0(x), \qquad \text{for some } x \in R, \tag{2.44}$$

where F denotes the true cdf of a population, and F_0 is a given cdf that is assumed
completely known. If F is assumed to have a parametric form, then the testing problem of (2.44) can be addressed by the parametric methods discussed in Subsection
2.7.3. However, in many applications, such a parametric form is unavailable. Therefore, nonparametric testing procedures are needed. Because the hypotheses in (2.44)
mainly concern how well the actual population distribution F can be described by
the assumed distribution F_0, tests for (2.44) are often called *goodness-of-fit tests*.

To test the hypotheses in (2.44), one approach is to divide the number line R into
k intervals

$$(-\infty, a_1), [a_1, a_2), \dots, [a_{k-1}, \infty),$$

where $a_1 < a_2 < \cdots < a_{k-1}$ are given cut points. For a simple random sample
$\{X_1, X_2, \dots, X_n\}$, let O_j be the number of observations in the j-th interval, for
$j = 1, 2, \dots, k$. If H_0 is true, then, on average, there should be

$$E_j = n[F_0(a_j) - F_0(a_{j-1})], \qquad \text{for } j = 1, 2, \dots, k,$$

observations in the sample that fall into the j-th interval. Therefore, to test the hypotheses in (2.44), we can compare the *observed counts* $\{O_j, j = 1, 2, \dots, k\}$ with
the *expected counts* $\{E_j, j = 1, 2, \dots, k\}$. If H_0 is true, then their difference should
be small. Otherwise, it is an indication that H_0 is false. Based on this intuition, we
define a test statistic

$$X^2 = \sum_{j=1}^{k} \frac{(O_j - E_j)^2}{E_j}, \tag{2.45}$$

which provides a measure of the difference between $\{O_j, j = 1, 2, \dots, k\}$ and
$\{E_j, j = 1, 2, \dots, k\}$. The test based on X^2 in (2.45) was first suggested by Pearson (1900), and is called the *Pearson's chi-square test* in the literature. Under H_0, the
asymptotic null distribution of X^2 is proved to be χ^2_{k-1}.

After the number line R is divided into k intervals by the cut points $a_1 < a_2 <
\cdots < a_{k-1}$, the LRT testing procedure (cf., (2.30) and (2.31)) can also be used
for testing the hypotheses in (2.44). Let π_j be the probability that a randomly selected member from the population in question is included in the j-th interval, for
$j = 1, 2, \dots, k$. Then, the observed counts $\{O_j, j = 1, 2, \dots, k\}$ of the simple random
sample $\{X_1, X_2, \dots, X_n\}$ have a multinomial distribution (cf., Subsection 2.4.2), and
the likelihood function of the sample is

$$L(\pi_1, \pi_2, \dots, \pi_k; X_1, X_2, \dots, X_n) = \frac{n!}{O_1! O_2! \cdots O_k!} \pi_1^{O_1} \pi_2^{O_2} \cdots \pi_k^{O_k}.$$

Let

$$\Lambda = \frac{\max_{H_0} L(\pi_1, \pi_2, \ldots, \pi_k; X_1, X_2, \ldots, X_n)}{\max_{H_0 \cup H_1} L(\pi_1, \pi_2, \ldots, \pi_k; X_1, X_2, \ldots, X_n)}$$

be the ratio of two maximized likelihoods, where $\max_{H_0}$ denotes "maximization under H_0" and $\max_{H_0 \cup H_1}$ denotes "maximization under $H_0 \cup H_1$." Then, it is easy to check that

$$\max_{H_0} L(\pi_1, \pi_2, \ldots, \pi_k; X_1, X_2, \ldots, X_n) = \frac{n!}{O_1! O_2! \cdots O_k!} \pi_{01}^{O_1} \pi_{02}^{O_2} \cdots \pi_{0k}^{O_k}$$

and

$$\max_{H_0 \cup H_1} L(\pi_1, \pi_2, \ldots, \pi_k; X_1, X_2, \ldots, X_n) = \frac{n!}{O_1! O_2! \cdots O_k!} \pi_{11}^{O_1} \pi_{12}^{O_2} \cdots \pi_{1k}^{O_k},$$

where

$$\begin{aligned} \pi_{0j} &= F_0(a_j) - F_0(a_{j-1}) \\ \pi_{1j} &= O_j/n, \quad \text{for } j = 1, 2, \ldots, k. \end{aligned}$$

Therefore

$$G^2 = -2\log(\Lambda) = 2 \sum_{j=1}^{k} O_j \log(O_j/E_j), \tag{2.46}$$

where $\{E_j\}$ are the expected counts defined before. Similar to X^2, G^2 in (2.46) has an asymptotic null distribution of χ_{k-1}^2.

The X^2 and G^2 tests would give similar results when the sample size n is large. In the literature, the sample size is considered large if each expected count is at least 5 (i.e., $E_j \geq 5$, for each j). In cases when most E_j's are at least 5 and there are a small number of E_j's as small as 1, the X^2 test often gives a more reliable test, compared to the G^2 test. To best satisfy the large sample condition, in applications, the cut points $a_1 < a_2 < \cdots < a_{k-1}$ should be chosen such that the expected counts $\{E_j, j = 1, 2, \ldots, k\}$ are roughly the same, or equivalently, $\{\pi_{0j}, j = 1, 2, \ldots, k\}$ are roughly the same. For detailed discussion on this topic, see Koehler (1986) and Koehler and Larntz (1980).

To test hypotheses in (2.44), it is natural to use the ecdf F_n defined in (2.43). If the difference between F_n and F_0 is large, then we should reject H_0, because F_n is a good estimator of the true population cdf F. To this end, let us define

$$D_n = \max_{x \in R} |F_n(x) - F_0(x)|. \tag{2.47}$$

The exact null distribution of D_n is available, although its expression is quite complicated, based on which we can compute the critical value $D_{n,\alpha}$, defined to be the smallest value satisfying $P(D_n \geq D_{n,\alpha}) \leq \alpha$, for a given significance level α. See, e.g., Dunstan et al. (1979) for extensive tables of $D_{n,\alpha}$. In cases when the sample size is large, Kolmogorov (1933) and Smirnov (1939) proved that, under H_0 and the condition that F_0 is absolutely continuous (i.e., it has a pdf),

$$\lim_{n \to \infty} P\left(\sqrt{n} D_n \leq d\right) = L(d),$$

Table 2.6: *Critical values L_α for several commonly used values of α.*

α	0.20	0.15	0.10	0.05	0.01
L_α	1.07	1.14	1.22	1.36	1.63

where

$$L(d) = 1 - 2 \sum_{s=1}^{\infty} (-1)^{s-1} \exp\left(-2s^2 d^2\right).$$

Therefore, $D_{n,\alpha}$ can also be approximated by $\widetilde{D}_{n,\alpha} = L_\alpha/\sqrt{n}$, where L_α satisfies $L(L_\alpha) = 1 - \alpha$. For several commonly used values of α, the corresponding values of L_α are listed in Table 2.6. In cases when $n \geq 50$ and $\alpha = 0.01$ or 0.05, the ratio $\widetilde{D}_{n,\alpha}/D_{n,\alpha}$ would be between 1 and 1.02. Therefore, $\widetilde{D}_{n,\alpha}$ should be good enough for practical use. The test using D_n in (2.47) is often called the *Kolmogorov-Smirnov test*.

2.8.3 Rank tests

The hypothesis testing procedures described in Subsection 2.7.3 are based on the assumption that the related population distribution has a parametric form (e.g., the normal distribution). In certain applications, the parametric form of the population distribution is unavailable, but we still want to test hypotheses about a population parameter. In such cases, an appropriate nonparametric testing procedure should be considered. This subsection introduces several commonly used nonparametric testing procedures.

Let us first focus on hypothesis testing of a population location parameter. In such cases, the population median, denoted as $\widetilde{\mu}$, is more convenient to use, compared to the population mean μ. Assume that we have a simple random sample $\{X_1, X_2, \ldots, X_n\}$ from a population with the population median $\widetilde{\mu}$, and we are interested in testing

$$H_0 : \widetilde{\mu} = \widetilde{\mu}_0 \qquad \text{versus} \qquad H_1 : \widetilde{\mu} \neq \widetilde{\mu}_0 (> \widetilde{\mu}_0, < \widetilde{\mu}_0), \tag{2.48}$$

where $\widetilde{\mu}_0$ is a hypothesized value of $\widetilde{\mu}$. Define a test statistic Y by

$$Y = \text{the number of observations in the sample that exceeds } \widetilde{\mu}_0.$$

Then, under H_0, $Y \sim Binomial(n, 0.5)$. Therefore, the hypotheses in (2.48) can be tested using Y, with its critical value or the related p-value determined by the above binomial distribution. This testing procedure is called the *sign test*. When n is large in the sense that $n \geq 20$, the null distribution of Y is approximately normal, and consequently we can use the test statistic

$$Z = \frac{Y - n/2}{\sqrt{n}/2} \overset{H_0}{\sim} N(0, 1).$$

The sign test described above only uses the number of observations in the sample that exceeds $\tilde{\mu}_0$ for testing the hypotheses in (2.48). It does not make use of the ordering information among all observations in the sample. In cases when the population distribution is symmetric, this ordering information can be used properly in the following two steps:

Step 1 Order all values of $\{|X_i - \tilde{\mu}_0|, i = 1, 2, \ldots, n\}$ from the smallest to the largest, and then obtain the ranks of all observations from this ordering.

Step 2 Define S_+ to be the summation of the ranks obtained in Step 1 that correspond to the nonnegative values of $\{X_i - \tilde{\mu}_0, i = 1, 2, \ldots, n\}$, and S_- to be the summation of the ranks obtained in Step 1 that correspond to the negative values of $\{X_i - \tilde{\mu}_0, i = 1, 2, \ldots, n\}$.

If H_0 is true (i.e., $\tilde{\mu}_0$ is the true median of the population distribution), then the distributions of S_+ and S_- are approximately the same. Since $S_+ + S_- = 1 + 2 + \cdots + n = n(n+1)/2$, the tests based on S_+, S_- or $S_+ - S_-$ are all asymptotically equivalent. For simplicity, we focus on the test using S_+ here. Obviously, if H_0 is true, then S_+ and S_- should be close to each other. If the value of S_+ is too large or too small, then it indicates that H_0 might be invalid. The null distribution of S_+ is given by Wilcoxon et al. (1972) in cases when $n \leq 50$. Table 2.7 gives some tail probabilities of this distribution when $5 \leq n \leq 20$. In cases when n is large (e.g., $n > 20$), the null distribution of S_+ can be regarded as a normal distribution with

$$\mu_{S_+} = n(n+1)/4, \qquad \sigma_{S_+}^2 = n(n+1)(2n+1)/24.$$

The test described above, using either S_+, or S_-, or $S_+ - S_-$, is called the *Wilcoxon signed-rank test*.

Example 2.9 *Assume that we obtain the following 20 observations from a population with a symmetric distribution:*

$$8.7, 19.5, 12.7, 10.4, 12.6, 6.1, 11.1, 0.7, 2.2, 12.3,$$
$$11.5, 16.2, 16.0, 13.1, -0.8, 18.1, 8.1, 12.8, 11.6, 12.7.$$

We are interested in testing

$$H_0 : \tilde{\mu} = 10 \qquad versus \qquad H_1 : \tilde{\mu} > 10.$$

To perform the Wilcoxon signed-rank test, we need to find the ranks of $\{|X_i - \tilde{\mu}_0|, i = 1, 2, \ldots, n\}$, which are listed in Table 2.8. From Table 2.8,

$$S_+ = 19 + 9.5 + 1 + 8 + 2 + 7 + 4 + 15 + 14 + 12 + 17 + 11 + 5 + 9.5 = 134.$$

By Table 2.7, the p-value is $P_{H_0}(S_+ \geq 134) > P_{H_0}(S_+ \geq 140) = 0.101$. Therefore, we fail to reject H_0 at the significance level of 0.05.

To compare two population means μ_1 and μ_2, if we have two paired samples from the two populations, then the Wilcoxon signed-rank test can still be used, after the difference between the two paired samples is computed and the two-sample problem

Table 2.7 *Some upper-tail probabilities* $P_{H_0}(S_+ \geq s_+)$ *of the null distribution of the Wilcoxon signed-rank test statistic* S_+.

n	s_+	$P_{H_0}(S_+ \geq s_+)$	n	s_+	$P_{H_0}(S_+ \geq s_+)$	n	s_+	$P_{H_0}(S_+ \geq s_+)$
5	13	0.094	11	48	0.103	16	93	0.106
	14	0.062		52	0.051		94	0.096
	15	0.031		55	0.027		100	0.052
6	17	0.109		59	0.009		106	0.025
	20	0.031	12	56	0.102		112	0.011
	21	0.016		60	0.055		113	0.009
7	22	0.109		61	0.046		116	0.005
	24	0.055		64	0.026	17	104	0.103
	26	0.023		68	0.010		105	0.095
	28	0.008		71	0.005		112	0.049
8	28	0.098	13	64	0.108		118	0.025
	30	0.055		65	0.095		125	0.010
	32	0.027		69	0.055		129	0.005
	34	0.012		70	0.047	18	116	0.098
	35	0.008		74	0.024		124	0.049
	36	0.004		78	0.011		131	0.024
9	34	0.102		79	0.009		138	0.010
	37	0.049		81	0.005		143	0.005
	39	0.027	14	73	0.108	19	128	0.098
	42	0.010		74	0.097		136	0.052
	44	0.004		79	0.052		137	0.048
10	41	0.097		84	0.025		144	0.025
	44	0.053		89	0.010		152	0.010
	47	0.024		92	0.005		157	0.005
	50	0.010	15	83	0.104	20	140	0.101
	52	0.005		84	0.094		150	0.049
				89	0.053		158	0.024
				90	0.047		167	0.010
				95	0.024		172	0.005
				100	0.011			
				101	0.009			
				104	0.005			

becomes the one-sample problem, as discussed at the end of Subsection 2.7.3, as long as it is reasonable to assume that the "difference" population has a symmetric distribution. In cases when we have two independent samples from the two populations, there are two commonly used nonparametric tests in the literature, which are described below.

Assume that (i) $\{X_{11}, X_{12}, \ldots, X_{1n_1}\}$ and $\{X_{21}, X_{22}, \ldots, X_{2n_2}\}$ are two samples from the two populations, respectively, (ii) the two samples are independent of each other, and (iii) the two population distributions have exactly the same shape and spread and their only difference is in their means. In such cases, we are interested in testing the hypotheses in (2.34). If H_0 is true, then the values of

Table 2.8: *Ranks of* $\{|X_i - \tilde{\mu}_0|, i = 1, 2, \ldots, n\}$ *for the observed data in Example 2.9.*

X_i	$X_i - \tilde{\mu}_0$	$|X_i - \tilde{\mu}_0|$	Ranks
8.7	-1.3	1.3	3
19.5	9.5	9.5	19
12.7	2.7	2.7	9.5
10.4	0.4	0.4	1
12.6	2.6	2.6	8
6.1	-3.9	3.9	13
11.1	1.1	1.1	2
0.7	-9.3	9.3	18
2.2	-7.8	7.8	16
12.3	2.3	2.3	7
11.5	1.5	1.5	4
16.2	6.2	6.2	15
16.0	6.0	6.0	14
13.1	3.1	3.1	12
-0.8	-10.8	10.8	20
18.1	8.1	8.1	17
8.1	-1.9	1.9	6
12.8	2.8	2.8	11
11.6	1.6	1.6	5
12.7	2.7	2.7	9.5

$\{X_{11} - \delta_0, X_{12} - \delta_0, \ldots, X_{1n_1} - \delta_0\}$ and the values of $\{X_{21}, X_{22}, \ldots, X_{2n_2}\}$ should be similar to each other, and they can be regarded as two independent samples from the same population. Otherwise, their values will be quite different. Based on this intuition, to test the hypotheses in (2.34), we can consider the combined sample

$$\{X_{11} - \delta_0, X_{12} - \delta_0, \ldots, X_{1n_1} - \delta_0, X_{21}, X_{22}, \ldots, X_{2n_2}\}.$$

Let W be the sum of the ranks of $\{X_{11} - \delta_0, X_{12} - \delta_0, \ldots, X_{1n_1} - \delta_0\}$ in the combined sample. Then, if the value of W is too large or too small, it implies that the values of $\{X_{11} - \delta_0, X_{12} - \delta_0, \ldots, X_{1n_1} - \delta_0\}$ are relatively large or small, compared to the values of $\{X_{21}, X_{22}, \ldots, X_{2n_2}\}$. Consequently, the observed data provide us evidence against H_0. Again, the null distribution of W has been tabulated by many authors, including Dixon and Massey (1969). Some of its tail probabilities are listed in Table 2.9 in cases when $4 \leq n_1 \leq n_2 \leq 8$. When both n_1 and n_2 are larger than 8, its null distribution can be well approximated by the normal distribution with

$$\mu_W = n_1(n_1 + n_2 + 1)/2, \qquad \sigma_W^2 = n_1 n_2 (n_1 + n_2 + 1)/12.$$

The test using W as its test statistic is called the *Wilcoxon rank-sum test.*

Under the conditions described above for the Wilcoxon rank-sum test, an alternative approach to test the hypotheses in (2.34) is to use the test statistic

$$U = \sum_{i=1}^{n_1} \sum_{j=1}^{n_2} I(X_{2j} < X_{1i} - \delta_0),$$

Table 2.9 *Some upper-tail probabilities $P_{H_0}(W \geq w)$ of the null distribution of the Wilcoxon rank-sum test statistic W.*

n_1	n_2	w	$P_{H_0}(W \geq w)$	n_1	n_2	w	$P_{H_0}(W \geq w)$
4	4	24	0.057	5	8	47	0.047
		25	0.029			49	0.023
		26	0.014			51	0.009
	5	27	0.056			52	0.005
		28	0.032	6	6	50	0.047
		29	0.016			52	0.021
		30	0.008			54	0.008
	6	30	0.057			55	0.004
		32	0.019		7	54	0.051
		33	0.010			56	0.026
		34	0.005			58	0.011
	7	33	0.055			60	0.004
		35	0.021		8	58	0.054
		36	0.012			61	0.021
		37	0.006			63	0.010
	8	36	0.055			65	0.004
		38	0.024	7	7	66	0.049
		40	0.008			68	0.027
		41	0.004			71	0.009
5	5	36	0.048			72	0.006
		37	0.028		8	71	0.047
		39	0.008			73	0.027
		40	0.004			76	0.010
	6	40	0.041			78	0.005
		41	0.026	8	8	84	0.052
		43	0.009			87	0.025
		44	0.004			90	0.010
	7	43	0.053			92	0.005
		45	0.024				
		47	0.009				
		48	0.005				

where $I(a)$ is the indicator function of a that equals 1 if $a =$ "True" and 0 otherwise. Similar to W, the statistic U makes use of the ordering information between the two samples. If the value of U is too large or too small, then it is an indication that H_0 might be invalid. This test is often called the *Mann-Whitney test*. As a matter of fact, it can be checked that

$$U = W - n_1(n_1 + 1)/2.$$

Therefore, the Wilcoxon rank-sum test and the Mann-Whitney test are actually equivalent.

2.8.4 Nonparametric density estimation

Assume that a population distribution has a pdf f, and we are interested in estimating f from a simple random sample $\{X_1, X_2, \ldots, X_n\}$. If the pdf f has a parametric form, then it can be estimated by parametric methods, such as the maximum likelihood estimation method discussed in Subsection 2.7.2. In this subsection, we discuss estimation of f when its parametric form is not available.

In cases when f does not have any parametric form, one natural approach for estimating f is to use the density histogram that is discussed in Subsection 2.6.3 (see also Figure 2.9). However, the pdf f is usually a continuous function; but, the function represented by the density histogram is stepwise. To make the approximation to f by the density histogram better, we can use smaller intervals when constructing the density histogram. By this approach, the resulting histogram would have a larger variability, which is not good either. To overcome these limitations, the approach called *kernel density estimation* has been proposed in the literature (cf. Parzen, 1962; Rosenblatt, 1956; Wand and Jones, 1995), which is described below.

For any $x \in R$, the *kernel density estimator* of $f(x)$ is defined by

$$\widehat{f}(x) = \frac{1}{nh_n} \sum_{i=1}^{n} K\left(\frac{x - X_i}{h_n}\right), \tag{2.49}$$

where $h_n > 0$ is a *bandwidth* and K is a *kernel function*. The kernel function K is often chosen to satisfy the following conditions:

(i) $K(x) \geq 0$, for any $x \in R$,

(ii) $\int_{-\infty}^{\infty} K(x)\, dx = 1$,

(iii) $K(x)$ is a decreasing function of x when $x > 0$, and

(iv) K is symmetric about 0 (i.e., $K(x) = K(-x)$, for any $x > 0$).

Therefore, the kernel function K itself is a probability density function. From (2.49), it can be seen that the kernel density estimator $\widehat{f}(x)$ is a weighted average of the kernel densities $K((x - X_i)/h_n)$ at the individual observations, and the weights are controlled by the kernel function K and the bandwidth h_n. For the aluminum smelter data discussed in Example 2.5 of Subsection 2.6.3, the kernel density estimates of the population pdf f are shown by the solid and dashed curves, respectively, in cases when $h_n = 0.05$ and 0.25 and the kernel function K is chosen to be the pdf ϕ of the standard normal distribution. It can be seen that when h_n is chosen small, the kernel density estimate is quite noisy, and it is smooth when h_n is chosen large. For this reason, h_n is often called a *smoothing parameter*.

Usually, the kernel function $K(x)$ is chosen to be a smooth density function that is symmetric about 0 and non-decreasing on $(-\infty, 0]$. Commonly used kernel functions include the uniform kernel function $K(u) = I(-1/2 \leq u \leq 1/2)$, the Epanechnikov kernel function $K(u) = \frac{12}{11}(1 - u^2)I(-1/2 \leq u \leq 1/2)$, and the Gaussian kernel function $K(u) = \frac{1}{\sqrt{2\pi}}\exp(-u^2/2)$ or its truncated version $K(u) = 0.6171\exp(-u^2/2)I(-1/2 \leq u \leq 1/2)$, where $I(a)$ is an indicator function defined by $I(a) = 1$ if a is "true" and 0 otherwise. By using the Epanechnikov kernel function or the Gaussian kernel function, observations closer to x receive more weight in

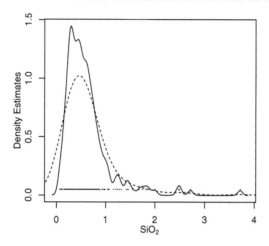

Figure 2.12 *Kernel density estimates of the aluminum smelter data discussed in Example 2.5 of Subsection 2.6.3. The kernel function K is chosen to be the pdf ϕ of the standard normal distribution. The solid curve denotes the kernel density estimate when $h_n = 0.05$, and the dashed curve denotes the kernel density estimate when $h_n = 0.25$.*

defining $\widehat{f}(x)$. Selection of the bandwidth h_n can be discussed in a similar way to that discussed in the next subsection.

2.8.5 *Nonparametric regression*

In the regression model (2.21) discussed in Subsection 2.7.2, the regression function is assumed to be a linear function. In many applications, the true regression function may not be well described by a parametric function. In such cases, *nonparametric regression analysis* would be more appropriate to use. Assume that bivariate observations $\{(x_i, Y_i), i = 1, 2, \ldots, n\}$ follow the regression model

$$Y_i = f(x_i) + \varepsilon_i, \qquad \text{for } i = 1, 2, \ldots, n, \tag{2.50}$$

where $\{x_i, i = 1, 2, \ldots, n\}$ are the design points, $\{Y_i, i = 1, 2, \ldots, n\}$ are observations of the response variable Y, $\{\varepsilon_i, i = 1, 2, \ldots, n\}$ are i.i.d. random errors, and f is the true regression function. For simplicity of discussion, let us further assume that the design interval of f is $[0, 1]$. Then, the major goal of nonparametric regression analysis is to estimate f from the observed data $\{(x_i, Y_i), i = 1, 2, \ldots, n\}$.

In cases when f is continuous in the entire design interval, intuitively, observations farther away from x provide less information about $f(x)$. Therefore, one natural idea is to simply average observations in a neighborhood $[x - h_n/2, x + h_n/2]$ of a given point x and then to use this average as an estimator of $f(x)$, where h_n is the bandwidth. The bandwidth is usually chosen relatively small, especially when the sample size n is large. Based on this idea, $f(x)$ can be estimated by the following simple average (SA):

$$\widehat{f}_{SA}(x) = \frac{1}{N_n(x)} \sum_{x_i \in [x - h_n/2, x + h_n/2]} Y_i,$$

where $N_n(x)$ denotes the number of design points in $[x - h_n/2, x + h_n/2]$. Based on (2.50), we have

$$\widehat{f}_{SA}(x) = \frac{1}{N_n(x)} \sum_{x_i \in [x - h_n/2, x + h_n/2]} f(x_i) + \frac{1}{N_n(x)} \sum_{x_i \in [x - h_n/2, x + h_n/2]} \varepsilon_i. \tag{2.51}$$

The estimator $\widehat{f}_{SA}(x)$ should estimate $f(x)$ well because of the following two facts:

- All values of $f(t)$ for $t \in [x - h_n/2, x + h_n/2]$ should be close to $f(x)$ since f is continuous in the design interval and h_n is small. Consequently, the first term on the right-hand side of (2.51) is close to $f(x)$.

- The second term on the right-hand side of (2.51) is close to zero as long as there are enough terms in the summation, which is guaranteed by the central limit theorem discussed in the previous section. Intuitively, positive errors and negative errors would be canceled out in this term, and thus, on average, the summation of many i.i.d. error terms is close to zero.

The above function estimation procedure that results in the estimator $\widehat{f}_{SA}(x)$ is a simple example of data smoothing. Almost all data smoothing procedures in the literature involve averaging observations for estimating $f(x)$. For any given x in the design interval, procedures that average all observations in the design interval when estimating $f(x)$ are referred to as *global smoothing* procedures in the literature. The linear regression analysis discussed in Subsection 2.7.2 is an example of global smoothing. Other procedures only average observations in a neighborhood of x, and are referred to as *local smoothing* procedures. So, $\widehat{f}_{SA}(x)$ is an example of local smoothing. The two facts mentioned above about the two terms on the right-hand side of (2.51) are commonly used when studying the properties of the local smoothing procedures.

By using $\widehat{f}_{SA}(x)$ for estimating $f(x)$, all observations outside the neighborhood $[x - h_n/2, x + h_n/2]$ are ignored completely, and all observations inside the neighborhood are treated equally. A natural generalization of $\widehat{f}_{SA}(x)$, which does not treat all observations in the neighborhood equally, is

$$\widehat{f}_{NW}(x) = \frac{\sum_{i=1}^{n} Y_i K \left(\frac{x_i - x}{h_n} \right)}{\sum_{i=1}^{n} K \left(\frac{x_i - x}{h_n} \right)}, \tag{2.52}$$

where K is a kernel function that has the support of $[-1/2, 1/2]$. The estimator $\widehat{f}_{NW}(x)$ is simply a weighted average of the observations in the neighborhood $[x - h_n/2, x + h_n/2]$, with weights controlled by the kernel function. As in kernel density estimation, the kernel function K is often chosen to be a smooth density function that is symmetric about 0 and non-decreasing in $[-1/2, 0]$, such as those listed at the end of Subsection 2.8.4. Obviously, the estimator $\widehat{f}_{SA}(x)$ is a special case of $\widehat{f}_{NW}(x)$ when the uniform kernel function is used. The kernel estimator $\widehat{f}_{NW}(x)$ was first suggested by Nadaraya (1964) and Watson (1964). So it is often called a *Nadaraya-Watson (NW) kernel estimator* in the literature, as identified by the subscript of $\widehat{f}_{NW}(x)$.

It can be easily checked that the Nadaraya-Watson kernel estimator $\widehat{f}_{NW}(x)$ is the solution to a of the following minimization problem:

$$\min_{a \in R} \sum_{i=1}^{n} (Y_i - a)^2 K \left(\frac{x_i - x}{h_n} \right). \tag{2.53}$$

Therefore, $\widehat{f}_{NW}(x)$ has the property that, among all constants, its weighted distance to the observations in the neighborhood $[x - h_n/2, x + h_n/2]$ is the smallest. This is illustrated by Figure 2.13(a), in which the solid curve at the bottom denotes the weights $K((x_i - x)/h_n)$ and the dashed horizontal line denotes $\widehat{f}_{NW}(x)$.

A natural generalization of (2.53) is the following minimization problem:

$$\min_{a, b_1, \dots, b_m \in R} \sum_{i=1}^{n} [Y_i - (a + b_1(x_i - x) + \dots + b_m(x_i - x)^m)]^2 K \left(\frac{x_i - x}{h_n} \right), \tag{2.54}$$

where m is a positive integer. Equation (2.54) is used to search for a polynomial function of order m whose weighted distance to the observed data in the neighborhood $[x - h_n/2, x + h_n/2]$ reaches the minimum. Then, the solution to a in (2.54) can be defined as an estimator of $f(x)$, and is called the *m-th order local polynomial kernel estimator*. As a by-product, the solution to b_j in (2.54) can be used as an estimator of $j! f^{(j)}(x)$, for $j = 1, 2, \dots, m$, where $f^{(j)}(x)$ is the j-th order derivative of f at x. Obviously, (2.53) is a special case of (2.54) when $m = 0$. That is, the Nadaraya-Watson kernel estimator is the zeroth order local polynomial kernel estimator, or the *local constant kernel estimator*.

In applications, the most commonly used local polynomial kernel estimator is the *local linear kernel (LK) estimator*, which is the solution to a in (2.54) when $m = 1$, and is denoted by $\widehat{f}_{LK}(x)$. This estimator is illustrated in Figure 2.13(b) by the dashed line. By some routine algebra, when $m = 1$, the solutions to a and b_1 of (2.54) have the following expressions:

$$
\begin{aligned}
\widehat{a}(x) &= \sum_{i=1}^{n} \frac{w_2 - w_1(x_i - x)}{w_0 w_2 - w_1^2} Y_i K \left(\frac{x_i - x}{h_n} \right) \\
\widehat{b}_1(x) &= \sum_{i=1}^{n} \frac{w_0(x_i - x) - w_1}{w_0 w_2 - w_1^2} Y_i K \left(\frac{x_i - x}{h_n} \right),
\end{aligned}
\tag{2.55}
$$

where $w_j = \sum_{i=1}^{n} (x_i - x)^j K(\frac{x_i - x}{h_n})$ for $j = 0, 1$ and 2. Therefore, $\widehat{f}_{LK}(x) = \widehat{a}(x)$ and $\widehat{f}'(x) = \widehat{b}_1(x)$.

Compared to the local constant kernel estimator, the local linear kernel estimator defined in (2.55) has the benefit that the first-order derivative (i.e., the slope) of f would have little impact on its performance; by the definition, most of the slope

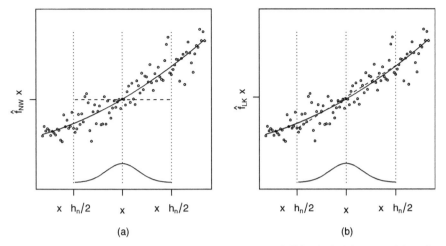

$$\text{x } h_n/2 \qquad \text{x} \qquad \text{x } h_n/2 \qquad\qquad \text{x } h_n/2 \qquad \text{x} \qquad \text{x } h_n/2$$

(a)   (b)

Figure 2.13 (a) The Nadaraya-Watson kernel estimator of $f(x)$ (dashed horizontal line) has the property that, among all constants, its weighted distance to observations in the neighborhood $[x - h_n/2, x + h_n/2]$ is the smallest. (b) The local linear kernel estimator equals the value of the dashed line at x. The dashed line has the property that its weighted distance to observations in the neighborhood $[x - h_n/2, x + h_n/2]$ is the smallest among all possible lines. The little circles in each plot denote the observations. The solid curve going through the data denotes the true regression function. The solid curve at the bottom denotes the weights that are controlled by the kernel function.

effect has been accommodated in the estimation by fitting a local linear function. This benefit makes the local linear kernel estimator $\widehat{f}_{LK}(x)$ have the properties that (i) its bias is not substantially larger when x is in the boundary regions $[0, h_n/2)$ and $(1 - h_n/2, 1]$, compared to its bias when x is in the interior region $[h_n/2, 1 - h_n/2]$, and (ii) its bias is not substantially larger when the design points in $[x - h_n/2, x + h_n/2]$ are unequally distributed, compared to its bias when the design points in $[x - h_n/2, x + h_n/2]$ are equally distributed. As a comparison, the local constant kernel estimator $\widehat{f}_{NW}(x)$ does not have these properties.

Regarding the choice of the power m of the polynomial for local polynomial kernel estimation, under some regularity conditions, it can be shown that: (i) the variance of a $(2k + 1)$-th order local polynomial kernel estimator is asymptotically the same as the variance of a $(2k)$-th order local polynomial kernel estimator, (ii) the variance of the $(2k + 1)$-th order local polynomial kernel estimator increases with k, and (iii) the bias of a higher order local polynomial kernel estimator is smaller than the bias of a lower order estimator. By these properties, the order of a local polynomial kernel estimator should be chosen as odd because a $(2k + 1)$-th order estimator is generally better than a $(2k)$-th order estimator. In most cases, the local linear kernel estimator should be good enough because the bias reduction is limited by using a higher order estimator and because the variability of a higher order estimator is often much larger.

The bandwidths used in the local polynomial kernel smoothing procedures control the size of the neighborhood around a given point used for data smoothing. If they are chosen larger, then the neighborhood is larger, implying that more observations are averaged for estimating the regression function at the given point, and vice versa. Therefore, the bandwidths also control the amount of observations used in data smoothing. If they are chosen too large, then the estimated regression function would become too smooth to capture local features of the true regression function, in which case we say that the estimator *oversmooths* the data. If the bandwidths are chosen too small, then the estimated regression function would capture too many details of the data, some of which are just noise, in which case the estimator *undersmooths* the data.

There are several existing procedures in the literature for choosing the bandwidths properly. We next introduce three of them: the cross-validation (CV) procedure, the Mallow's C_p criterion, and the plug-in algorithm. For simplicity, they are introduced only in the case of local linear kernel smoothing. Actually, these procedures are quite general and can be used for selecting bandwidths and other parameters used in most data smoothing procedures.

For bandwidth selection of the local linear kernel estimator $\widehat{f}_{LK}$, a natural criterion is the following residual mean squares (RMS):

$$\text{RMS}(h_n) = \frac{1}{n} \sum_{i=1}^{n} \left(Y_i - \widehat{f}_{LK}(x_i) \right)^2. \tag{2.56}$$

Since $\text{RMS}(h_n)$ measures the distance between the observed data and the estimator, the bandwidth h_n can be chosen by minimizing $\text{RMS}(h_n)$. However, it can be easily checked that the best bandwidth is $h_n = 0$ by this criterion, because the corresponding estimator connects all the data points in such a case and thus sets $\text{RMS}(h_n) = 0$. Obviously, this result is not what we want since no data smoothing is actually involved.

We can amend this criterion to allow for data smoothing as follows. Because $\widehat{f}_{LK}(x_i)$ is constructed by all observations in the neighborhood $[x_i - h_n/2, x_i + h_n/2]$, including Y_i, the residual $Y_i - \widehat{f}_{LK}(x_i)$ tends to be small for measuring the distance between the estimator and the data. This is a major reason why the criterion (2.56) is inappropriate for choosing h_n. As a remedy, let $\widehat{f}_{LK,-i}(x_i)$ be the local linear kernel estimator of $f(x_i)$ constructed from all observations in $[x_i - h_n/2, x_i + h_n/2]$ except Y_i, for $i = 1, 2, \cdots, n$. Then, a similar criterion to (2.56) is the following CV score:

$$CV(h_n) = \frac{1}{n} \sum_{i=1}^{n} \left(Y_i - \widehat{f}_{LK,-i}(x_i) \right)^2. \tag{2.57}$$

The optimal value of h_n can be estimated by the minimizer of the CV score in (2.57). This method for choosing the bandwidths or other procedure parameters is called the *CV procedure* in the literature (Allen (1974)).

We next discuss the Mallows' C_p criterion. Let $\mathbf{Y} = (Y_1, Y_2, \ldots, Y_n)'$ and $(\widehat{f}_{LK}(x_1), \widehat{f}_{LK}(x_2), \ldots, \widehat{f}_{LK}(x_n))' = \mathbf{HY}$. Then, $\mathbf{H}$ is called the *hat matrix* in regression analysis. The Mallows' C_p criterion (Mallows, 1973) is defined by

$$C_p(h_n) = \frac{1}{\sigma^2} \| (I_{n \times n} - \mathbf{H})\mathbf{Y} \|^2 - n + 2 \, \text{tr}(\mathbf{H}),$$

where $I_{n \times n}$ is the $n \times n$ identity matrix, $\| \cdot \|$ is the Euclidean norm, and tr(**H**) is the trace of the matrix **H** (i.e., the summation of all diagonal elements of **H**). By this criterion, the optimal value of h_n is estimated by the minimizer of $C_p(h_n)$.

The plug-in algorithm is based on the MSE criterion (cf., Subsection 2.7.1). It can be checked that the MSE of the local linear kernel estimator $\widehat{f}_{LK}(x)$ for estimating $f(x)$ is

$$\text{MSE}\left(\widehat{f}_{LK}(x), f(x)\right) \sim \left[h_n^2 f''(x) \frac{K_{21}^2 - K_{11}K_{31}}{2(K_{01}K_{21} - K_{11}^2)} \right]^2 + (nh_n)^{-1} C(K)\sigma^2, \quad (2.58)$$

where $K_{j1} = \int_{-1/2}^{1/2} u^j K(u) du$, for $j = 0, 1, 2, 3$, $C(K) = \frac{1}{(K_{01}K_{21} - K_{11})^2} \int_{-1/2}^{1/2} (K_{21} - K_{11}u)^2 K^2(u) du$, and "$\sim$" denotes equality when certain high-order terms have been ignored on the right-hand side of (2.58). By the criterion (2.58), the optimal value of h_n that minimizes $\text{MSE}\left(\widehat{f}_{LK}(x), f(x)\right)$ is asymptotically equal to

$$h_{n,opt} = \left(\frac{C_1(K)\sigma^2}{n[f''(x)]^2} \right)^{1/5}, \quad (2.59)$$

where $C_1(K) = C(K)(K_{01}K_{21} - K_{11}^2)^2 / (K_{21}^2 - K_{11}K_{21})^2$. Because the quantity $f''(x)$ is unknown, the criterion (2.59) can not be used directly for choosing h_n. One way out is the following iterative algorithm: (1) we assign an initial value for h_n and estimate $f''(x)$ by a local polynomial kernel estimator based on this initial bandwidth; (2) the value of h_n is updated by (2.59) after $f''(x)$ is replaced by its estimator; (3) we go back to step (1) using the updated bandwidth obtained from the previous step; and (4) steps (1)–(3) continue until some convergence criterion is satisfied. This *plug–in* bandwidth selection procedure has several different versions. For more discussions, read Härdle et al. (1992), Ruppert et al. (1995), Loader (1999), and the references cited therein.

2.9 Exercises

2.1 Assume that a random variable X has the following cdf

$$F(x) = \begin{cases} 0, & \text{if } x < a \\ \frac{x-a}{b-a}, & \text{if } a \leq x \leq b \\ 1, & \text{if } x > b, \end{cases}$$

where $a < b$ are two constants. A distribution with this cdf is called the *uniform distribution* on $[a, b]$, denoted as $X \sim Uniform(a, b)$.

(i) Find the pdf of X.

(ii) Using (2.6) and (2.7), find the mean and variance of X.

(iii) Describe the major features of this distribution.

(iv) If $X \sim Uniform(5, 10)$, find $P(7 \leq X \leq 9)$.

2.2 In Example 2.1, assume that each grade has a grade point associated with it. Let X be the grade and Y be the grade point of a randomly selected student of the introductory statistics class discussed in that example, and the relationship between grades and grade points is described by the table below.

Grade	A	B	C	D	F
Grade Points	4	3	2	1	0

(i) Does X have a cdf? If the answer is "yes", find the cdf of X.

(ii) Does Y have a cdf? If the answer is "yes", find the cdf of Y.

(iii) Using (2.4) and (2.5), find the mean and variance of Y.

2.3 Most statistical software packages can compute probabilities related to a standard normal distribution. For instance, in the software package R (see the appendix of the book for a brief description), the following commands compute $P(Z < 1)$ and $P(Z < 2)$, where Z denotes a random variable having the standard normal distribution:

```
> pnorm(1)                          > pnorm(2)
[1] 0.8413447                       [1] 0.9772499
```

From the above R printouts, we know that $P(Z < 1) = 0.8413447$ and $P(Z < 2) = 0.9772499$. Assume that $X \sim N(1.5, 0.5^2)$. Find the following probabilities:

(i) $P(X < 2)$

(ii) $P(X < 2.5)$

(iii) $P(2 < X < 2.5)$

(iv) $P(-1 < X < 2.5)$

2.4 For the binomial distribution discussed in Subsection 2.4.2, derive the formulas in (2.14).

2.5 If $X \sim Binomial(n, \pi)$, then $p = X/n$ can be interpreted as the proportion of successes out of n Bernoulli trials, and p is a good estimator of π (cf., the related discussion in Subsection 2.7.1). Derive the mean and variance of p, and give some reasonable explanations why p is a good estimator of π.

2.6 Suppose that 20% of all copies of a particular textbook fail a certain binding strength test. Let X denote the number of copies among 10 randomly selected copies that fail the test.

(i) Find the mean and variance of X.

(ii) Find the probability $P(3 \leq X \leq 5)$.

2.7 Consider a dataset with the following 11 observations:

$$4, 6, 7, 6, 3, 5, 8, 2, 9, 9, 100.$$

(i) Compute the sample mean of the data.

(ii) Compute the sample standard deviation of the data.

(iii) Find the sample median.

(iv) Explain why the sample mean and the sample median are very different in this case.

2.8 Consider a dataset with the following 20 observations:

$$10, 12, 15, 17, 20, 20, 23, 25, 28, 30,$$
$$40, 43, 45, 48, 50, 50, 59, 61, 64, 70.$$

(i) Find the first quartile, median, and the third quartile of the data.

(ii) Find the inter-quartile range.

(iii) Make a box plot of the data.

2.9 A small dataset about voters' political status contains the following 20 observations:

$$D, D, R, O, R, R, R, D, R, D, D, R, D, D, R, R, R, O, D, D,$$

where "D" denotes Democrats, "R" denotes Republicans, and "O" denotes others.

(i) Make a frequency table of the data.

(ii) Make a pie chart of the data.

(iii) Make a bar chart of the data. Is the shape of the bar chart meaningful? Why?

2.10 A *Consumer Reports* article on peanut butter reported the following scores for creamy peanut butter:

$$56, 44, 57, 62, 48, 53, 36, 50, 39, 53,$$
$$50, 50, 65, 45, 60, 40, 56, 68, 41, 22.$$

(i) Make a dot plot of the data.

(ii) Make a stem-and-leaf plot of the data.

(iii) Based on the plots made in parts (i) and (ii), describe the distribution of the data.

2.11 A consumer group has collected data on the gas mileage (denoted as X), in miles per gallon, of 325 cars manufactured in America. The data have been summarized in the partial frequency table below.

Class Interval	Frequency	Relative Frequency	Density
$[5, 10)$	20		
$[10, 15)$			0.0277
$[15, 20)$			
$[20, 25)$		0.2769	
$[25, 30)$	60		
$[30, 35)$		0.1077	

(i) Complete the table by filling all empty entries.

(ii) Use the completed table to make a density histogram and comment on its shape.

(iii) Estimate $P(5 \leq X < 20)$.

2.12 A random sample is to be selected from a population with mean $\mu = 50$ and standard deviation $\sigma = 10$.

(i) Determine the mean and the standard deviation of the sampling distribution of $\overline{X}$ when $n = 25$.

(ii) Determine the approximate probability that $\overline{X}$ will be larger than 51 when $n = 100$.

2.13 Assume that the mean speed of an automobile on highways of a given state is 60 miles per hour (mph) and the standard deviation is 4 mph. Let $X_1, X_2, \ldots, X_{100}$ denote the highway speeds of 100 randomly selected automobiles in that state.

(1) Determine the mean and the standard deviation of the sample mean $\overline{X}$.

(ii) What is the approximate probability that the sample mean $\overline{X}$ exceeds 61 mph?

(iii) What is the approximate probability that the sample mean $\overline{X}$ is between 59.2 mph and 60.4 mph?

2.14 A survey designed to obtain information on π, which denotes the proportion of registered voters who are in favor of a constitutional amendment, results in a sample of size $n = 400$. Of the 400 voters sampled, 272 are in favor of the constitutional amendment.

(i) Give a point estimate of π.

(ii) Calculate a 95% confidence interval for π.

2.15 The manager of the electronics department at a large department store is interested in knowing the mean size μ of the TV screens that customers purchase. Based on industry standards it is believed that the standard deviation $\sigma = 4$ inches.

(i) Determine how large a sample is needed in order to have the 90% confidence interval for μ shorter than or equal to 2 inches.

(ii) If a sample of size $n = 49$ yields a sample average TV screen size of 21.1 inches, calculate a 90% confidence interval for μ.

2.16 The following summary statistics resulted from a study of the relationship between the cost of a barrel of crude oil (x) and the price of a gallon of regular unleaded gasoline (Y):

$$n = 12, \quad \sum_{i=1}^{n} x_i = 241.1, \quad \sum_{i=1}^{n} x_i^2 = 4932.8, \quad \sum_{i=1}^{n} Y_i = 14.34,$$
$$\sum_{i=1}^{n} Y_i^2 = 17.288, \quad \sum_{i=1}^{n} x_i Y_i = 291.55.$$

(i) Compute the estimated linear regression model using (2.22).

(ii) In linear regression analysis, the *residual sum of squares (RSS)*, defined by

$$RSS = \sum_{i=1}^{n} \left[Y_i - \left(\hat{\beta}_0 + \hat{\beta}_1 x_i \right) \right]^2$$
$$= \sum_{i=1}^{n} (Y_i - \overline{Y})^2 - \hat{\beta}_1^2 \sum_{i=1}^{n} (x_i - \overline{x})^2,$$

is used for measuring the variability in the observed data that is not explained by the estimated regression model. The *sum of squares of total (SST)*, defined by

$$SST = \sum_{i=1}^{n} (Y_i - \overline{Y})^2,$$

is used for measuring the total variability in the observed data. And, the *coefficient of determination* R^2, defined by

$$R^2 = 1 - \frac{RSS}{SST},$$

is used for measuring the proportion of the variability in the observed data that is explained by the estimated regression model. Using the summary statistics given above, compute SST, RSS, and R^2.

2.17 A fishing hook manufacturer is interested in determining whether the hook packaging machine tends to overfill or underfill. Each package of hooks is supposed to contain 100 hooks. A random sample of 49 packages of hooks is selected, resulting in a sample average of 100.25 hooks and a sample standard deviation of 0.7 hooks. Using $\alpha = 0.05$, test

$$H_0 : \mu = 100 \qquad \text{versus} \qquad H_1 : \mu \neq 100,$$

where μ denotes the mean number of hooks in a randomly selected package.

2.18 Let μ denote the mean cholesterol level of heart attack patients under the age of 50. Some research reports have claimed that a cholesterol level of 240 or higher would dramatically increase the risk of heart attacks. A random sample of the cholesterol levels of 16 heart attack patients age 50 and under yields $\bar{x} = 247$ and $s = 14$. Using $\alpha = 0.05$, test

$$H_0 : \mu = 240 \qquad \text{versus} \qquad H_1 : \mu > 240.$$

In the test, what assumptions have actually been made so that the test is valid?

2.19 In a study to compare average snowfall in two different cities, measurements were taken in each of the cities for 9 randomly selected years. Snowfall, in inches, for the two cities is listed below. Assume the two population distributions are normal. Use $\alpha = 0.05$ to test whether there is a significant difference between the average snowfalls of the two cities.

Year	1942	1948	1954	1959	1967	1970	1975	1981	1983
City A	45	0	4	21	9	1	30	17	53
City B	40	10	2	20	7	9	33	17	50

2.20 A marching band needs to order hats for performances this summer. The hat sizes for the 100 band members are summarized in the chart below. Make a graph of the empirical cumulative distribution function of this data.

Size	6.7	6.8	6.9	7.0	7.1	7.2	7.3
Frequency	4	12	22	31	9	17	5

2.21 Assume that the following data are from a population with mean μ:

5.6, 2.5, 3.7, 7.9, 11.3, 6.4, 3.2, 2.7, 0.5, 5.7, 3.2, 2.4, 7.6, 3.8, 5.7, 5.8, 3.4, 5.1, 2.2, 4.0, 4.2, 4.2, 5.5, 1.5, 1.6, 6.3, 4.3, 5.0, 8.6, 6.3, 4.3, 6.4, 4.6, 2.0, 4.4.

(i) Using the delta method, construct a 95% confidence interval for $\theta = \mu(\mu+1)$.

(ii) Using the bootstrap method with the bootstrap sample size $B = 1000$, construct a 95% confidence interval for $\theta = \mu(\mu + 1)$. Compare the result with that in part (i).

2.22 For the data in the previous problem, test whether they are from a population with the distribution $N(5, 3^2)$, using both the X^2 and G^2 tests and the significance level $\alpha = 0.05$. In the tests, use five intervals to group the data.

2.23 For the data in Exercise 2.21, test $H_0 : \widetilde{\mu} = 5.5$ versus $H_1 : \widetilde{\mu} \neq 5.5$, using $\alpha = 0.05$ and

(i) the sign test,

(ii) the Wilcoxon signed-rank test.

2.24 Assume that two independent samples are obtained from two populations, respectively. Observations in the two samples are listed below.

Sample 1	1.2	8.0	0.9	2.1	2.7	7.3	7.6	
Sample 2	2.6	8.5	2.4	4.5	4.6	7.7	3.8	10.3

Assume that the two population distributions have the same shape and spread. Use the Wilcoxon rank-sum test and $\alpha = 0.05$ to test $H_0 : \mu_1 - \mu_2 = -1$ versus $H_1 : \mu_1 - \mu_2 < -1$, where μ_1 and μ_2 are the two population means.

2.25 For the data in Exercise 2.21, make a plot of the kernel density estimator of the population pdf, using the Epanechnikov kernel function and the bandwidth $h_n = 1.0$.

2.26 For the Nadaraya-Watson kernel estimator $\widehat{f}_{NW}(x)$ defined in (2.52), find its bias, variance, and MSE for estimating the true regression function $f(x)$.

2.27 Intuitively explain the major reason why the bias of the local linear kernel estimator $\widehat{f}_{LK}(x)$ is not substantially larger in the boundary regions $[0, h_n/2)$ and $(1 - h_n/2, 1]$, compared to its bias in the interior region $[h_n/2, 1 - h_n/2]$.

Chapter 3

Univariate Shewhart Charts and Process Capability

3.1 Introduction

As described briefly in Section 1.3, statistical process control (SPC) of a production process can be roughly divided into two phases. In the initial stage (i.e., phase I), we usually do not know much about the performance of the production process yet, and our major goal in this stage is to properly adjust the production process to make it run stably. To this end, we usually let the production process produce a given amount of products, and the values of the quality characteristic(s) of interest of the manufactured products are then recorded and analyzed. From the statistical analysis of the collected data, if we find that the production process does not seem to run stably, then the root causes responsible for that unfavorable performance should be figured out and the corresponding adjustment of the production process should be made as well. After the adjustment, another set of data needs to be collected and analyzed, and the production process should be adjusted again if necessary. This analysis-and-adjustment process is then repeated several times until it is confirmed by the data analysis that the performance of the production process is stable. Once all special causes have been accounted for and the production process is in-control (IC), we collect an IC dataset from the products manufactured under the stable operating conditions, and this IC dataset is then used for estimating the IC distribution of the quality characteristic(s) of interest. Based on the (estimated) IC distribution, a phase II SPC control chart is designed, and it is used for online monitoring of the production process. When it detects a significant shift in the distribution of the quality characteristic(s) from the IC distribution, a signal is delivered and the production process is stopped immediately for root cause identification and removal. This online monitoring stage of SPC is often called phase II SPC.

In both phase I and phase II SPC, many statistical tools, including histograms, stem-and-leaf plots, regression, design of experiment, and so forth (cf., Sections 2.6 and 2.7), are helpful. Among all these methods, control charts are especially useful because they are designed specifically for detecting any out-of-control (OC) performance of the production process. By using a control chart, a charting statistic determined by the observed data should be chosen, and it should contain as much of the information in the observed data about the distribution of the quality characteristic(s) as possible and be sensitive to any distributional shift as well. In the literature, several different types of control charts have been developed, including the Shewhart charts, the cumulative sum (CUSUM) control charts, the exponentially weighted moving average (EWMA) control charts, and the control charts based on change-point detection (CPD). These and some other control charts developed recently for handling various different situations will be described in the current and later chapters of the book.

As mentioned in Section 1.3, the first control chart in the literature was proposed by Walter A. Shewhart in 1931, and is called the *Shewhart chart*. Over the past about 80 years, many different versions of the Shewhart chart have been proposed for different purposes, and they are widely used in practice. In the first part of this chapter, some representative Shewhart charts are discussed in cases when the quality characteristic in question is univariate. Shewhart charts for monitoring multivariate quality characteristics in cases when their distributions are assumed normal will be discussed in Chapter 7. Shewhart charts in cases when the distribution of the quality characteristic(s) is nonparametric are discussed in Chapters 8 and 9. In the second part of this chapter, process capability analysis is briefly described.

3.2   Shewhart Charts for Numerical Variables

As discussed in Section 1.2, all quality characteristics can be classified into three categories: continuous numerical, discrete numerical, and categorical variables. Process control methodologies for different types of quality characteristics are different. In this section, we describe some basic Shewhart charts for monitoring numerical quality characteristics when their IC and OC distributions are assumed to be normal. These control charts are especially useful when the quality characteristics are continuous numerical, or when they are discrete numerical with a relatively large number of different values. In cases when the quality characteristics are discrete but the number of different values is small, we can consider using control charts for monitoring categorical quality characteristics, such as those described in Section 3.3.

3.2.1   The $\overline{X}$ and R charts

Let us first discuss phase I SPC. Assume that n random samples are collected from the products of a production process at n consecutive time points. Each sample contains m products. Measurements of the quality characteristic in question of the sampled products in the i-th sample are denoted as

$$X_{i1}, X_{i2}, \ldots, X_{im}, \qquad \text{for } i = 1, 2, \ldots, n.$$

Based on the observed data, we would like to know whether the process is IC. For the moment, let us assume that the IC mean μ_0 and the IC standard deviation σ of the quality characteristic X are both known. Let us further assume that the only possible shift is in the mean of X. That is, our focus is on monitoring the process mean. In such cases, to know whether the process is IC at the i-th time point, it is natural to consider testing the following hypotheses:

$$H_0 : \mu = \mu_0 \qquad \text{versus} \qquad H_1 : \mu \neq \mu_0,$$

where μ denotes the true process mean. By the discussion in Subsection 2.7.3, a good test statistic for testing the above hypotheses is

$$Z = \frac{\overline{X}_i - \mu_0}{\sigma/\sqrt{m}} \stackrel{H_0}{\sim} N(0,1). \tag{3.1}$$

The null hypothesis H_0 should be rejected at a pre-specified significance level α if the observed value of Z, denoted as Z^*, satisfies the condition that

$$|Z^*| > Z_{1-\alpha/2}, \tag{3.2}$$

where $Z_{1-\alpha/2}$ denotes the $(1-\alpha/2)$-th quantile of the distribution $N(0,1)$, which is also called the $\alpha/2$ critical value of the $N(0,1)$ distribution. Based on (3.1) and (3.2), the observed data provide us significant evidence to conclude that the process is OC at the i-th time point if

$$\overline{X}_i > \mu_0 + Z_{1-\alpha/2}\frac{\sigma}{\sqrt{m}} \quad \text{or} \quad \overline{X}_i < \mu_0 - Z_{1-\alpha/2}\frac{\sigma}{\sqrt{m}}. \tag{3.3}$$

In practice, the IC parameters μ_0 and σ are usually unknown. In such cases, they need to be estimated from the observed data beforehand in order to make decisions regarding the process performance using decision rules similar to (3.3). To this end, it is natural to estimate μ_0 by the grand sample mean

$$\widehat{\mu}_0 = \overline{\overline{X}} = \frac{1}{n}\sum_{i=1}^{n} \overline{X}_i, \tag{3.4}$$

where $\{\overline{X}_1, \overline{X}_2, \ldots, \overline{X}_n\}$ are the sample means of the n samples. To estimate σ, instead of the sample standard deviations, people traditionally use the *sample ranges*

$$R_i = X_{i(m)} - X_{i(1)}, \quad \text{for } i = 1, 2, \ldots, n,$$

where $X_{i(1)}$ and $X_{i(m)}$ denote the first and last order statistics (cf., Section 2.5) of the i-th sample $\{X_{i1}, X_{i2}, \ldots, X_{im}\}$, because sample ranges are easier to compute compared to sample standard deviations. This computational advantage of the sample ranges might be negligible nowadays; but, it was substantial several decades ago when Shewhart charts were first proposed and computers were unavailable. If the process is IC at the i-th time point (i.e., the i-th sample is from the population distribution $N(\mu_0, \sigma^2)$), then

$$d_1(m) = \mathrm{E}\left(\frac{R_i}{\sigma}\right)$$

is a constant that depends on m, which can be determined from the joint distribution of $(X_{i(1)}, X_{i(m)})$ (cf., the related discussion in Subsection 2.8.1). When $2 \leq m \leq 25$, the values of $d_1(m)$ are given in Table 3.1. Then, a natural estimator of σ is

$$\widehat{\sigma} = \frac{\overline{R}}{d_1(m)}, \tag{3.5}$$

Table 3.1 *Constants $d_1(m)$ and $d_2(m)$ used in constructing the $\overline{X}$ and R charts when the sample size $2 \le m \le 25$.*

m	$d_1(m)$	$d_2(m)$	m	$d_1(m)$	$d_2(m)$
2	1.128	0.853	14	3.407	0.763
3	1.693	0.888	15	3.472	0.756
4	2.059	0.880	16	3.532	0.750
5	2.326	0.864	17	3.588	0.744
6	2.534	0.848	18	3.640	0.739
7	2.704	0.833	19	3.689	0.734
8	2.847	0.820	20	3.735	0.729
9	2.970	0.808	21	3.778	0.724
10	3.078	0.797	22	3.819	0.720
11	3.173	0.787	23	3.858	0.716
12	3.258	0.778	24	3.895	0.712
13	3.336	0.770	25	3.931	0.708

where $\overline{R} = \frac{1}{n}\sum_{i=1}^{n} R_i$.

Based on (3.3)–(3.5), we can evaluate the process performance by comparing the sample means with the control limits given in the box below, and the resulting control chart is called the $\overline{X}$ *chart*.

Control Limits of the $\overline{X}$ Chart Using Sample Ranges

$$U = \overline{\overline{X}} + \frac{Z_{1-\alpha/2}}{d_1(m)\sqrt{m}}\overline{R}$$

$$C = \overline{\overline{X}} \qquad (3.6)$$

$$L = \overline{\overline{X}} - \frac{Z_{1-\alpha/2}}{d_1(m)\sqrt{m}}\overline{R}$$

More specifically, the $\overline{X}$ chart consists of a center line C, an upper control limit U, and a lower control limit L. At the i-th time point, if the i-th sample mean $\overline{X}_i$ is beyond the two control limits, i.e.,

$$\overline{X}_i < L \quad \text{or} \quad \overline{X}_i > U,$$

then we claim that the process is OC at that time point. Otherwise, the process is considered IC. In the formulas of the control limits U and L given in the above box, the critical value $Z_{1-\alpha/2}$ is usually chosen to be 3 in practice, which corresponds to $\alpha = 0.0027$. Namely, by using this critical value, the $\overline{X}$ chart would have a 0.27% chance to give a false signal of process distributional shift when the process is actually IC.

The $\overline{X}$ chart described above is mainly for monitoring the process mean. In practice, we are often concerned about the process variability as well. To monitor the process variability, it is natural to use the sample range as the charting statistic. To this end, for any $1 \leq i \leq n$, let

$$d_2(m) = \sqrt{\text{Var}\left(\frac{R_i}{\sigma}\right)}.$$

Then, $d_2(m)$ is a constant depending on the sample size m only, and its values are listed in Table 3.1 in cases when $2 \leq m \leq 25$. Based on (3.5), a natural estimator of σ_{R_i} is

$$\widehat{\sigma}_{R_i} = \frac{d_2(m)\overline{R}}{d_1(m)}.$$

So, we can monitor the process variability by using the so-called *R chart* with the control limits given in the box below. We claim that the process variability is OC at the *i*-th time point if

$$R_i < L \qquad \text{or} \qquad R_i > U,$$

where L and U are the lower and upper control limits, respectively.

Control Limits of the R Chart

$$U = \overline{R} + \frac{Z_{1-\alpha/2}d_2(m)}{d_1(m)}\overline{R} = \left(1 + \frac{Z_{1-\alpha/2}d_2(m)}{d_1(m)}\right)\overline{R}$$

$$C = \overline{R} \qquad\qquad\qquad\qquad\qquad\qquad\qquad\qquad (3.7)$$

$$L = \overline{R} - \frac{Z_{1-\alpha/2}d_2(m)}{d_1(m)}\overline{R} = \left(1 - \frac{Z_{1-\alpha/2}d_2(m)}{d_1(m)}\right)\overline{R}$$

In practice, the $\overline{X}$ and R charts are often used together, because both the process mean and process variability are important to the quality of products. If one or both charts give OC signals, then the special causes of the OC signals should be figured out and removed. If the special causes are transient in the sense that they only affect the products that are manufactured in a short period of time, then we can simply delete the samples observed in that time period and compute the control limits of the $\overline{X}$ and R charts again. Otherwise, new samples should be collected from products produced after the special causes are removed, and new $\overline{X}$ and R charts should be constructed from the new data. This analysis-and-adjustment process usually needs to be repeated several times until the production process becomes stable. Then, we can use the control charts constructed using a phase I dataset collected under the stable condition for phase II online process monitoring. In phase II SPC, samples are collected sequentially over time, and the sample means and sample ranges are plotted in the $\overline{X}$ and R charts, respectively, whose control limits are determined in phase I analysis. If the sample mean or the sample range of the sample collected at the current time point is outside their corresponding control limits, then we claim that the process is OC and it should be stopped immediately. The process can be re-started only after the special causes of the OC signal are figured out and removed and the process is properly adjusted.

Example 3.1 *To monitor an injection molding process, 20 samples of 5 manufactured parts each are collected. The major quality characteristic of the parts is their compressive strengths (in psi), and the summary statistics of the 5 observations of the compressive strength in each sample are listed in the top part of Table 3.2. The original data used in this example are from Tables 6E.11 and 6E.12 in Montgomery (2009). To monitor this process using the $\overline{X}$ and R charts, let us assume that we use $Z_{1-\alpha/2} = 3$. From Table 3.1, for $m = 5$, $d_1(m) = 2.326$, and $d_2(m) = 0.864$. Also, from Table 3.2, it can be computed that*

$$\overline{\overline{X}} = 79.533, \qquad \overline{R} = 8.745.$$

So, by (3.6), the control limits of the $\overline{X}$ chart are

$$
\begin{aligned}
U &= 79.533 + \frac{3}{2.326 * \sqrt{5}} * 8.745 = 84.577 \\
C &= 79.533 \\
L &= 79.533 - \frac{3}{2.326 * \sqrt{5}} * 8.745 = 74.489.
\end{aligned}
$$

By (3.7), the control limits of the R chart are

$$
\begin{aligned}
U &= 8.745 + \frac{3 * 0.864}{2.326} * 8.745 = 18.490 \\
C &= 8.745 \\
L &= 8.745 - \frac{3 * 0.864}{2.326} * 8.745 = -1.000.
\end{aligned}
$$

Because the sample range cannot be negative, we can modify the lower control limit of the R chart to be 0.

    The $\overline{X}$ and R charts are demonstrated in plots (a) and (b) of Figure 3.1, respectively, in which the solid dots, connected by solid lines, denote the sample means and sample ranges of the first 20 samples shown in Table 3.2. From the plots, it can be seen that both the process mean and process variability seem IC in this case, although the sample ranges shown in plot (b) seem to have a decreasing trend, which implies quality improvement. Then, we finish our phase I analysis, and use these two charts for online monitoring of the process (i.e., phase II SPC). In phase II SPC, new samples are collected sequentially. The summary statistics of the first 4 phase II samples are presented in the bottom part of Table 3.2. Their sample means and sample ranges are plotted in the $\overline{X}$ and R charts, which are denoted by little circles in the plots connected by dotted lines. At the 24th time point, we get a signal of mean shift. So, the process is stopped at that time point for investigation of possible special causes of the signal. It can be restored for manufacturing new products only in cases when any possible special causes are removed, the production process is adjusted properly, and we are confident that it is back to IC again. For the process adjustment at this stage, certain phase I SPC methodologies might still be useful.

Table 3.2 *Summary statistics of the 24 samples collected from an injection molding process. The first 20 samples are for setting up the control charts, and the last 4 samples are for online process monitoring. The quality characteristic in this example is the compressive strength of the manufactured parts.*

i	$\overline{X}_i$	R_i	s_i	i	$\overline{X}_i$	R_i	s_i
1	79.12	7.3	2.99	11	78.36	7.7	2.89
2	80.18	17.6	6.65	12	79.44	8	3.31
3	80.4	10.5	4.79	13	80.92	3.6	1.57
4	77.5	8.4	3.88	14	77.14	4.3	1.94
5	80.3	5.2	2.49	15	79.1	13.8	6.14
6	82.8	14.4	5.78	16	79.7	2	0.81
7	77.28	7.4	3.22	17	79.22	6.6	2.85
8	81.12	11.5	4.53	18	81.06	7.6	3.11
9	81.44	10	3.86	19	80.8	6.6	2.55
10	75.68	11	4.01	20	79.1	11.4	4.12
21	78.18	12.6	5.36	23	80.82	6.8	2.8
22	77.4	15.7	7.65	24	85.64	17.5	6.28

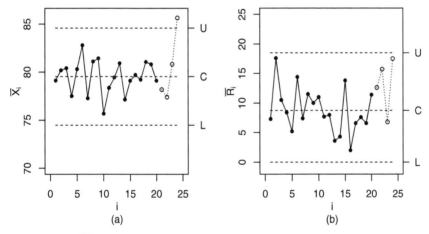

(a) (b)

Figure 3.1 *The $\overline{X}$ chart (plot (a)) and the R chart (plot (b)) constructed from the data of the first 20 samples summarized in Table 3.2 about the injection molding process. The sample means and sample ranges of the first 20 samples are shown by solid dots in the plots connected by solid lines. Since the process seems IC at the first 20 time points, the charts are then used for phase II process monitoring. The phase II sample means and sample ranges are shown by little circles connected by dotted lines. We get a signal of mean shift at the 24th time point, and the process is stopped at that time point for investigation of possible special causes of the signal.*

The performance of a control chart, such as the $\overline{X}$ and R charts, is traditionally measured by the so-called averaged run length (ARL), described below. Let us first discuss the case when the production process is IC. In such cases, because the observed data are random and our control charts use control limits as threshold values for making decisions regarding the process performance (i.e., IC versus OC), the control charts could give *false* signals of process distributional shifts. In the hypothesis testing context, this is the so-called type I error. The number of samples collected from the initial time point of consideration to the occurrence of the false OC signal is called the *run length*. In this definition, we have assumed that samples are collected at equally spaced time points. In such cases, the run length is equivalent to the length from the initial time point of consideration to the occurrence of a false OC signal. Obviously, the run length is a random variable, because it is determined by the collected samples, which are random. Further, its distribution does not depend on the selection of the initial time point of consideration, as long as the process is IC during all the time points considered and the samples at different time points are independent of each other or they are correlated but the correlation is homogeneous over time. The mean of the run length when the process is IC is called the IC ARL, which is often denoted as ARL_0. In the case when the process becomes OC at a given time point, the control chart in question would give us a signal of the shift after a certain time period. The number of samples collected from the time of shift occurrence to the time of signal is called OC run length, and its mean value is the OC ARL, which is often denoted as ARL_1.

For a given control chart, of course, the ideal situation is that its ARL_0 value is large and its ARL_1 value is small. But, similar to the Type I and Type II error probabilities in the hypothesis testing context (cf., Section 2.7.3), this is difficult to achieve. In most situations, when the ARL_0 value is large, the ARL_1 value would also be relatively large, and vice versa. To handle this issue, in the SPC literature, we usually fix the ARL_0 value at a given level, and try to make the ARL_1 value as small as possible, which is similar to the strategy used in the hypothesis testing context to fix the Type I error probability at a given level (e.g., 0.05) and try to make the Type II error probability as small as possible.

Next, we use the $\overline{X}$ chart as an example to discuss computation of its ARL_0 and ARL_1 values in cases of phase II SPC. Computation of the ARL_0 and ARL_1 values of other Shewhart charts (e.g., the R chart) can be discussed similarly. In phase II SPC, the IC mean μ_0 and the IC standard deviation σ can both be assumed known. In such cases, the upper and lower control limits of the $\overline{X}$ chart are those in expression (3.3), instead of those in (3.6). It is easy to know that the distribution of the IC run length is the geometric distribution $Geom(\alpha)$ (cf., Section 2.4.3). By the formulas of its mean and variance given in Section 2.4.3, we know that

$$ARL_0 = \frac{1}{\alpha}, \qquad \sigma_{RL}^{(0)} = \frac{\sqrt{1-\alpha}}{\alpha}, \qquad (3.8)$$

where $\sigma_{RL}^{(0)}$ denotes the standard deviation of the IC run length. In most applications, a small value of α is chosen. In such cases, $\sqrt{1-\alpha}$ is close to 1. Consequently,

$$ARL_0 \approx \sigma_{RL}^{(0)}.$$

For instance, by (3.8), when $\alpha = 0.0027$, $ARL_0 = 1/0.0027 = 370.37$ and $\sigma_{RL}^{(0)} = 369.87$. The two numbers are almost the same.

For computing ARL_1, let us assume that the mean of the quality characteristic has a shift of size $\delta = k\sigma$ (i.e., the mean shifts from μ_0 to $\mu_1 = \mu_0 + k\sigma$) at a given time point. Then, the probability that the sample mean $\overline{X}$ of a sample collected at a later time point is within the upper and lower control limits is

$$
\begin{aligned}
\beta &= P\left(\mu_0 - Z_{1-\alpha/2}\frac{\sigma}{\sqrt{m}} \leq \overline{X} \leq \mu_0 + Z_{1-\alpha/2}\frac{\sigma}{\sqrt{m}}\right) \\
&= P\left(-k\sqrt{m} - Z_{1-\alpha/2} \leq \frac{\overline{X} - \mu_1}{\sigma/\sqrt{m}} \leq -k\sqrt{m} + Z_{1-\alpha/2}\right) \\
&= \Phi\left(-k\sqrt{m} + Z_{1-\alpha/2}\right) - \Phi\left(-k\sqrt{m} - Z_{1-\alpha/2}\right),
\end{aligned}
$$

where Φ is the cumulative distribution function (cdf) of the standard normal distribution. From the above expression, it can be seen that the value of β depends on m and k. In cases when $m = 5, 10$, and 20, and $k \in [0,3]$, the values of β are shown in Figure 3.2. From the plot, it can be seen that β decreases when k or m increases. This is intuitively reasonable, because (i) the shift size of the process mean gets larger when k increases and thus the probability that $\overline{X}$ still stays within the control limits would become smaller, and (ii) $\overline{X}$ would be closer to the true process mean μ_1 when m gets larger and thus the probability value β, which measures the chance that $\overline{X}$ is within an interval centered at μ_0, would be smaller for a given shift size.

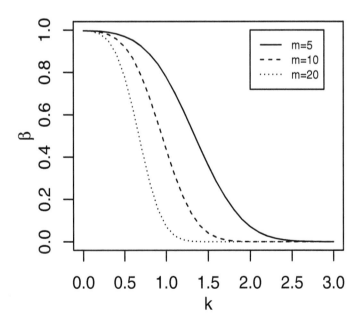

Figure 3.2: *The value of β in cases when $m = 5, 10$, and 20, and $k \in [0,3]$.*

It is obvious that the distribution of the OC run length is the geometric distribution $Geom(1-\beta)$. So,

$$ARL_1 = \frac{1}{1-\beta}, \qquad \sigma_{RL}^{(1)} = \frac{\sqrt{\beta}}{1-\beta}, \tag{3.9}$$

where $\sigma_{RL}^{(1)}$ denotes the standard deviation of the OC run length. In cases when $m = 5, 10,$ and 20, and $k \in [0,3]$, the values of ARL_1 and $\sigma_{RL}^{(1)}$ by (3.9) are shown in the two plots of Figure 3.3, respectively. From plots (a) and (b) of Figure 3.3, it can be seen that ARL_1 decreases when k or m increases, and the value of $\sigma_{RL}^{(1)}$ changes in a similar way.

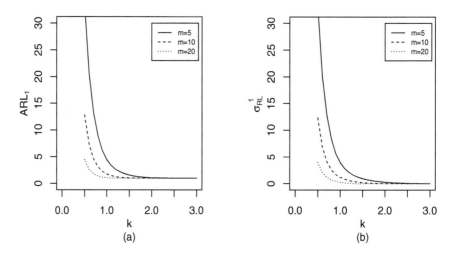

Figure 3.3 *The values of ARL_1 (plot (a)) and $\sigma_{RL}^{(1)}$ (plot (b)) in cases when $m = 5, 10,$ and 20, and $k \in [0,3]$.*

For the $\overline{X}$ chart with the control limits in (3.6), the probability for it to make a Type I error at a single time point (i.e., the process is declared OC when it is actually IC) is α, which is sometimes called the *false alarm rate (FAR)* in the literature. If we consider an IC dataset with n samples collected at n different time points, then the overall probability of Type I error, or the overall FAR, defined to be the probability that the process is IC at all n time points but it is declared OC at one or more such time points, would be

$$\tilde{\alpha} = 1 - (1-\alpha)^n. \tag{3.10}$$

Table 3.3 lists the $\tilde{\alpha}$ values when $n = 1, 2, 3, 4, 5, 6, 7, 8, 9, 10, 15, 20, 50,$ and 100, and $\alpha = 0.0001, 0.001, 0.0027,$ and 0.005. From the table, we can see that $\tilde{\alpha}$ increases quite fast when n increases. For instance, in cases when $\alpha = 0.0027$ (i.e., the FAR is 27 per 10,000), $\tilde{\alpha} = 0.0267$ when $n = 10$, and 0.2369 when $n = 100$, which are much larger than the value of α. If we need to control the overall probability of

Table 3.3 *The overall probability of Type I error, $\tilde{\alpha}$, of the $\overline{X}$ chart with FAR α when it is applied to an IC dataset with n samples, in cases when n and α take various values.*

n	$\alpha = 0.0001$	$\alpha = 0.001$	$\alpha = 0.0027$	$\alpha = 0.005$
1	0.0001	0.0010	0.0027	0.0050
2	0.0002	0.0020	0.0054	0.0100
3	0.0003	0.0030	0.0081	0.0149
4	0.0004	0.0040	0.0108	0.0199
5	0.0005	0.0050	0.0134	0.0248
6	0.0006	0.0060	0.0161	0.0296
7	0.0007	0.0070	0.0187	0.0345
8	0.0008	0.0080	0.0214	0.0393
9	0.0009	0.0090	0.0240	0.0441
10	0.0010	0.0100	0.0267	0.0489
15	0.0015	0.0149	0.0397	0.0724
20	0.0020	0.0198	0.0526	0.0954
50	0.0050	0.0488	0.1264	0.2217
100	0.0100	0.0952	0.2369	0.3942

Type I error in a given application, then we should use equation (3.10) to choose α properly. For instance, if we want $\tilde{\alpha} = 0.01$ when the $\overline{X}$ chart is applied to an IC dataset with $n = 50$ samples, by (3.10), α should be chosen as follows:

$$\alpha = 1 - (1 - \tilde{\alpha})^{1/n} = 1 - (1 - 0.01)^{1/50} = 0.0002.$$

Now, let us consider the $\overline{X}$ chart presented in Figure 3.4. From the plot, it seems that the process is IC because all the sample means are within the lower and upper control limits. But, there are a number of obvious patterns. For instance, the first six sample means have an increasing trend. In the literature, such a pattern is often called a *run*. So, the first six sample means form a run of length 6, the 4th up to the 8th sample means are all above the center line and they form a run of length 5, and so forth. Besides the runs, there is an obvious periodic pattern in Figure 3.4. If the production process only has common cause variation involved, then the $\overline{X}$ chart should demonstrate a random pattern, and the chance to have runs (especially those with relatively long lengths) and other patterns should be small. So, when we find a non-random pattern in a control chart, we should try to figure out its possible special causes, even in cases when all the charting statistic values are within the control limits.

The Western Electric Handbook (1956) suggests a set of decision rules for detecting non-random patterns on the control chart. By these rules, the process is declared OC if one of the following cases happens:

(i) One point is outside the three-sigma control limits,

(ii) Two out of three consecutive points are outside the two-sigma control limits,

(iii) Four out of five consecutive points are at least one sigma away from the center line, and

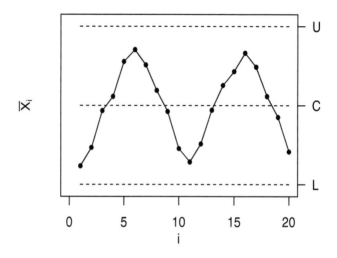

Figure 3.4: *The $\overline{X}$ chart contains some non-random patterns.*

(iv) Eight consecutive points are located on one side of the center line.

In the literature, some researchers have proposed alternative decision rules based on non-random patterns. For instance, Nelson (1984) and Champ and Woodall (1987) discussed various OC conditions when k out of ℓ consecutive points fall beyond one-, two-, or three-sigma control limits, where $2 \leq k \leq \ell$. Derman and Ross (1997) proposed two additional rules. By the first one, an OC signal is given if two consecutive points fall outside either one of the two three-sigma control limits. By the second one, an OC signal is given if two out of three consecutive points fall outside different three-sigma control limits. Klein (2000) modified these two rules by requiring the related points to exceed a same control limit.

3.2.2   The $\overline{X}$ and s charts

In the $\overline{X}$ and R charts discussed in the previous subsection (cf., (3.6) and (3.7)), the standard deviation σ of the quality characteristic is estimated by the simple average of the sample ranges, $\overline{R}$. As mentioned there, a major reason to use that estimator is its simple computation. However, this advantage of the sample ranges has become negligible nowadays with modern computing facilities. Therefore, it is natural to consider a more efficient estimator of σ based on the sample standard deviations.

As in the previous subsection, we assume that n samples are collected at n different time points, and each sample contains m observations of the quality characteristic. For the i-th sample, its sample standard deviation can be computed by

$$s_i = \sqrt{\frac{1}{m-1} \sum_{j=1}^{m} \left(X_{ij} - \overline{X}_i\right)^2}, \qquad \text{for } i = 1, 2, \ldots, n.$$

Then, a natural estimator of σ is

$$\bar{s} = \frac{1}{n} \sum_{i=1}^{n} s_i.$$

From the discussion in Section 2.7, for each i, the sample variance s_i^2 is an unbiased estimator of the variance σ^2. However, the sample standard deviation s_i is a biased estimator of σ, and the bias depends on the sample size m (cf. Kenney and Keeping, 1951, sec. 7.8). More specifically, let

$$E(s_i) = d_3(m)\sigma,$$

where $d_3(m)$ is a constant that depends on m. Then, $d_3(m)$ has the following expression:

$$d_3(m) = \frac{\Gamma\left(\frac{m}{2}\right)}{\Gamma\left(\frac{m-1}{2}\right)} \sqrt{\frac{2}{m-1}},$$

where $\Gamma(x) = \int_0^\infty u^{x-1} e^{-u}\, du$ is the Gamma function. By using the properties of the Gamma function, we have

$$d_3(m) = \begin{cases} \dfrac{2(k-1)\left(2^{k-2}(k-2)!\right)^2}{(2k-3)!} \sqrt{\dfrac{2}{\pi(2k-1)}}, & \text{if } m = 2k \\[3mm] \dfrac{(2k-1)!}{2\left(2^{k-1}(k-1)!\right)^2} \sqrt{\dfrac{\pi}{k}}, & \text{if } m = 2k+1. \end{cases} \tag{3.11}$$

Then, the bias of s_i for estimating σ is $(d_3(m)-1)\sigma$, and $s_i/d_3(m)$ is an unbiased estimator of σ. By combining all the unbiased estimators $\{s_i/d_3(m), i = 1,2,\ldots,n\}$ of σ, we come up with the following unbiased estimator of σ:

$$\hat{\sigma} = \frac{\bar{s}}{d_3(m)}. \tag{3.12}$$

The $\bar{X}$ chart based on (3.11) and (3.12) would have the control limits presented in the box below.

Control Limits of the $\bar{X}$ Chart Using Sample Standard Deviations

$$\begin{aligned} U &= \bar{\bar{X}} + \frac{Z_{1-\alpha/2}}{d_3(m)\sqrt{m}}\bar{s} \\[2mm] C &= \bar{\bar{X}} \\[2mm] L &= \bar{\bar{X}} - \frac{Z_{1-\alpha/2}}{d_3(m)\sqrt{m}}\bar{s} \end{aligned} \tag{3.13}$$

The process is declared to have a mean shift at the i-th time point if the i-th sample mean $\bar{X}_i$ is beyond the two control limits U and L.

To monitor the process variability, it is natural to use the sample standard deviations $\{s_i, \ i = 1, 2, \ldots, n\}$. When the process is IC, the mean of s_i can be estimated by $\bar{s}$. By Kenney and Keeping (1951), the standard deviation of s_i has the expression

$$\sigma_{s_i} = \sqrt{1 - d_3^2(m)}\,\sigma.$$

It can be estimated by

$$\hat{\sigma}_{s_i} = \sqrt{1 - d_3^2(m)}\,\frac{\bar{s}}{d_3(m)}.$$

Therefore, if s_i is beyond the control limits given in the box below, then the process variability can be declared OC. The resulting control chart is called the s *chart*.

Control Limits of the s Chart

$$U = \bar{s} + \frac{Z_{1-\alpha/2}\sqrt{1 - d_3^2(m)}}{d_3(m)}\bar{s} = \left(1 + \frac{Z_{1-\alpha/2}\sqrt{1 - d_3^2(m)}}{d_3(m)}\right)\bar{s}$$

$$C = \bar{s} \tag{3.14}$$

$$L = \bar{s} - \frac{Z_{1-\alpha/2}\sqrt{1 - d_3^2(m)}}{d_3(m)}\bar{s} = \left(1 - \frac{Z_{1-\alpha/2}\sqrt{1 - d_3^2(m)}}{d_3(m)}\right)\bar{s}$$

Example 3.2 *For the data discussed in Example 3.1, if we construct the $\overline{X}$ and s charts with control limits in (3.13) and (3.14), using the first 20 samples, then*

$$\overline{\overline{X}} = 79.533, \qquad \bar{s} = 3.575,$$

and

$$d_3(5) = \frac{3!}{2 * (2^1 * 1!)^2}\sqrt{\frac{\pi}{2}} = 0.940.$$

When $Z_{1-\alpha/2}$ is chosen to be 3, the control limits of the $\overline{X}$ chart are

$$U = 79.533 + \frac{3}{0.940 * \sqrt{5}} * 3.575 = 84.636$$

$$C = 79.533$$

$$L = 79.533 - \frac{3}{0.940 * \sqrt{5}} * 3.575 = 74.430,$$

and the control limits of the s chart are

$$U = \left(1 + \frac{3 * \sqrt{1 - 0.940^2}}{0.940}\right) * 3.575 = 7.468$$

$$C = 3.575$$

$$L = \left(1 - \frac{3 * \sqrt{1 - 0.940^2}}{0.940}\right) * 3.575 = -0.318.$$

We notice that the lower control limit of the s chart is −0.318, which is a negative number. However, the sample standard deviations s_i cannot be negative. Therefore, we can simply replace it by 0 as the lower control limit value. The two control charts are demonstrated in plots (a) and (b) of Figure 3.5, respectively, in which the solid dots denote the sample means and sample standard deviations of the first 20 samples shown in Table 3.2, and they are connected by solid lines. From the plots, it seems that both the process mean and the process variability are stable at the first 20 time points, which is consistent with the results found in Figure 3.1. Then, we use these two charts for online monitoring of the process (i.e., the phase II SPC). The sample means and sample standard deviations of the first four samples collected during the phase II SPC are denoted by little circles in the two plots connected by dotted lines. From the plots, we can see that a signal of process variability shift is given in plot (b) at the 22nd time point. So, the process should be stopped at that time point for investigation of possible special causes of the signal. This signal is two time units earlier than the one given by the $\overline{X}$ and R charts shown in Figure 3.1.

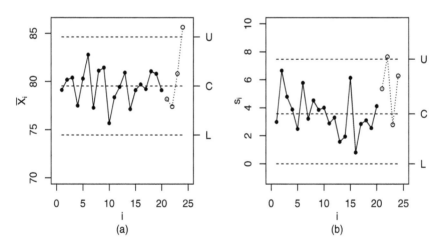

Figure 3.5 *The $\overline{X}$ chart (plot (a)) and the s chart (plot (b)) constructed from the data of the first 20 samples summarized in Table 3.2 about the injection molding process. The sample means and sample standard deviations of the first 20 samples are connected by solid lines. Since the process is IC at the first 20 time points, the charts are used for phase II process monitoring. The phase II sample means and sample standard deviations are shown by little circles connected by dotted lines. We get a signal of process variability shift at the 22nd time point, and the process should be stopped at that time point for investigation of possible special causes of the signal.*

To monitor the process variability, we can also use sample variances $\{s_i^2, i = 1, 2, \ldots, n\}$. Note that each individual sample variance s_i^2 is an unbiased estimator of σ^2. So, a natural unbiased estimator of σ^2 is

$$\overline{s^2} = \frac{1}{n} \sum_{i=1}^{n} s_i^2.$$

For the i-th sample variance s_i^2, from the discussion in Section 2.7, we know that

$$\frac{(m-1)s_i^2}{\sigma^2} \sim \chi_{m-1}^2$$

in cases when the process is IC at the i-th time point. Therefore, the process variability can be declared OC at the i-th time point if s_i^2 is beyond the control limits given in the box below, and the resulting control chart is called the s^2 *chart*.

Control Limits of the s^2 Chart

$$U = \frac{\overline{s^2}\chi_{1-\alpha/2,m-1}^2}{m-1}$$

$$C = \overline{s^2} \qquad\qquad (3.15)$$

$$L = \frac{\overline{s^2}\chi_{\alpha/2,m-1}^2}{m-1},$$

where $\chi_{\alpha/2,m-1}^2$ and $\chi_{1-\alpha/2,m-1}^2$ are the $(\alpha/2)$-th and $(1-\alpha/2)$-th quantiles of the χ_{m-1}^2 distribution.

3.2.3   The $\overline{X}$ and R charts for monitoring individual observations

In the previous two subsections, we assume that the sample collected at a given time point has m observations of the quality characteristic in question, and $m > 1$. In the literature, such data are often called *batch data*, or *grouped data*. In certain applications, however, it is more convenient to collect one observation at each time point. Namely, the sample size m is 1 in such cases. The resulting data are often called *individual observation data*. When the observed data are individual observation data, all the Shewhart charts discussed in the previous two subsections cannot be used because the sample ranges R_i or sample standard deviations s_i are all 0 in such cases and consequently the upper and lower control limits of each of these Shewhart charts would be the same. Obviously, such control charts do not have any power for either phase I or phase II SPC. For applications with individual observation data, one option is to use alternative control charts that will be described in later chapters, including the CUSUM charts, the EWMA charts, and the CPD charts. Another option is to modify the Shewhart charts for batch data so that the modified Shewhart charts can also be used for analyzing individual observation data, which is the focus of this subsection.

To use Shewhart charts for analyzing individual observation data, one natural idea is to "create" grouped data, by grouping observations collected at consecutive time points. Assume that the sample size of each group is $\tilde{m} > 1$. Then, the first $\tilde{m}$ observations can form the first group, the second $\tilde{m}$ observations can form the second group, and so forth. If we "create" grouped data in this way, then the Shewhart charts discussed before can be used as usual. One problem with this idea is that the gap between two consecutive groups is $\tilde{m}$ time points apart. Therefore, by analyzing such grouped data, it is difficult for us to know the process behavior at each time point. To overcome this difficulty, in the literature, most people adopt the idea of moving windows. With the window size $\tilde{m}$, this idea "creates" the grouped data as follows. Assume that the original observations are

$$X_1, X_2, \ldots, X_n.$$

Then, the grouped data are defined by

Group 1:	$X_1, X_2, \ldots, X_{\tilde{m}}$
Group 2:	$X_2, X_3, \ldots, X_{\tilde{m}+1}$
$\vdots$	$\vdots$
Group $n - \tilde{m} + 1$:	$X_{n-\tilde{m}+1}, X_{n-\tilde{m}+2}, \ldots, X_n.$

Let $\overline{X}$ be the sample mean of the original observations, $MR_1, MR_2, \ldots, MR_{n-\tilde{m}+1}$ be the sample ranges of the $n - \tilde{m} + 1$ groups of data, and $\overline{MR} = \frac{1}{n-\tilde{m}+1} \sum_{i=1}^{n-\tilde{m}+1} MR_i$. Then, by (3.5), we can estimate σ by

$$\frac{\overline{MR}}{d_1(\tilde{m})},$$

and the control limits of the $\overline{X}$ chart for monitoring individual observations become the ones in the box below.

Control Limits of the $\overline{X}$ Chart for Monitoring Individual Observations

$$
\begin{aligned}
U &= \overline{X} + \frac{Z_{1-\alpha/2}}{d_1(\tilde{m})}\overline{MR} \\
C &= \overline{X} \qquad\qquad\qquad\qquad (3.16) \\
L &= \overline{X} - \frac{Z_{1-\alpha/2}}{d_1(\tilde{m})}\overline{MR}
\end{aligned}
$$

By this chart, we get a signal of process mean shift at the i-th time point if

$$X_i < L \quad \text{or} \quad X_i > U.$$

Similarly, the control limits of the R chart for monitoring individual observations can be defined by the ones in the box below.

Control Limits of the R Chart for Monitoring Individual Observations

$$U \ = \ \left(1 + \frac{Z_{1-\alpha/2}d_2(\tilde{m})}{d_1(\tilde{m})}\right)\overline{MR}$$

$$C \ = \ \overline{MR} \qquad\qquad (3.17)$$

$$L \ = \ \left(1 - \frac{Z_{1-\alpha/2}d_2(\tilde{m})}{d_1(\tilde{m})}\right)\overline{MR}$$

By this chart, we get a signal of process variability shift at the i-th time point if

$$MR_i < L \qquad \text{or} \qquad MR_i > U.$$

In such cases, we should check the production process at all time points that belong to the i-th created group (i.e., from the i-th to the $(i+\tilde{m}-1)$-th time points) for special causes of the shift. Other Shewhart charts can be modified similarly for monitoring individual observations.

Example 3.3 *Twenty observations on concentration (in g/l) of the active ingredient in a liquid cleaner produced by a chemical process are given in the second column of Table 3.4. This is an individual observation dataset, and it is modified from the data in Table 6E.24 of Montgomery (2009). To construct the $\overline{X}$ and R charts using the idea of moving windows, let us consider the window size $\tilde{m} = 2$. In such cases, the sample ranges of the 19 "created" groups are presented in the third column of Table 3.4. To construct the control charts, let us assume that we use $Z_{1-\alpha/2} = 3$. From Table 3.1, for window size $\tilde{m} = 2$, $d_1(\tilde{m}) = 1.128$, and $d_2(\tilde{m}) = 0.853$. From Table 3.4, $\overline{X} = 72.38$ and $\overline{MR} = 8.72$.*
Then, the control limits of the $\overline{X}$ chart are

$$U \ = \ 72.38 + \frac{3}{1.128}*8.72 = 95.571$$

$$C \ = \ 72.38$$

$$L \ = \ 72.38 - \frac{3}{1.128}*8.72 = 49.189.$$

The control limits of the R chart are

$$U \ = \ \left(1 + \frac{3*0.853}{1.128}\right)*8.72 = 28.502$$

$$C \ = \ 8.72$$

$$L \ = \ \left(1 - \frac{3*0.853}{1.128}\right)*8.72 = -11.062.$$

Table 3.4 *Twenty observations (labeled by X_i) on concentration (in g/l) of the active ingredient in a liquid cleaner produced by a chemical process, and the sample ranges (labeled by MR_i) of the moving windows with window size $\tilde{m} = 2$.*

i	X_i	MR_i
1	78.7	5.9
2	72.8	5.6
3	78.4	1.2
4	79.6	19.2
5	60.4	9.1
6	69.5	4.6
7	64.9	10.6
8	75.5	5.1
9	70.4	2.3
10	68.1	10.3
11	78.4	0.2
12	78.2	18.2
13	60	14.7
14	74.7	1.1
15	75.8	0.8
16	76.6	8.2
17	68.4	14.7
18	83.1	22.0
19	61.1	11.9
20	73.0	
	$\overline{X} = 72.38$	$\overline{MR} = 8.72$

Since the sample ranges are always nonnegative, we change the lower control limit of the R chart from -11.062 to 0. The $\overline{X}$ and R charts are shown in Figure 3.6, which do not give any signals of process mean shift or process variability shift.

3.3 Shewhart Charts for Categorical Variables

The Shewhart charts discussed in the previous section are for monitoring production processes with numerical quality characteristics. In many applications, the quality characteristics of interest are categorical. In this section, we describe several Shewhart charts for monitoring production processes with categorical quality characteristics.

3.3.1 The p chart and mp chart

In certain applications, it is inconvenient to monitor the original quality characteristics of products directly. Instead, after certain products are randomly chosen for monitoring purposes, they can be classified into conforming and non-conforming products based on the designed requirements on the original quality characteristics, and then we can monitor the proportion of non-conforming products over time. Assume that the true proportion of non-conforming products of a production process is

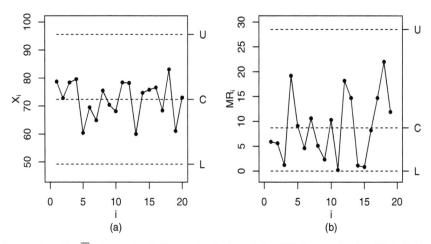

Figure 3.6 *The $\overline{X}$ chart (plot (a)) and the R chart (plot (b)) for monitoring 20 individual observations on concentration (in g/l) of the active ingredient in a liquid cleaner produced by a chemical process.*

π when the process is IC, and we obtain a random sample of m products at a given time point. Let X be the number of non-conforming products in the sample. Then, by the discussion in Subsection 2.4.1, it is obvious that

$$X \sim Binomial(m, \pi).$$

Namely, X has a binomial distribution with two parameters m and π. By (2.14), we have

$$\mu_X = m\pi, \qquad \sigma_X^2 = m\pi(1-\pi).$$

The distribution of the sample proportion of non-conforming products, denoted as $p = X/m$, can be determined accordingly, and it is obvious that

$$\mu_p = \pi, \qquad \sigma_p^2 = \frac{\pi(1-\pi)}{m}.$$

From the discussion in Subsection 2.7.1, in cases when the sample size is large, the distribution of p can be well approximated by the following normal distribution:

$$N\left(\pi, \frac{\pi(1-\pi)}{m}\right).$$

In such cases, the process can be declared OC if the observed value of p is beyond the control limits

$$L = \pi - Z_{1-\alpha/2}\sqrt{\frac{\pi(1-\pi)}{m}}, \qquad U = \pi + Z_{1-\alpha/2}\sqrt{\frac{\pi(1-\pi)}{m}}, \qquad (3.18)$$

where $Z_{1-\alpha/2}$ is the $(1-\alpha/2)$-th quantile of the $N(0,1)$ distribution.

In practice, the value of π is often unknown and it should be estimated from the observed data. To this end, assume that we have collected n random samples at n different time points, and each sample contains m products. The numbers of non-conforming products in the n samples are denoted as $X_1, X_2, \ldots, X_n$, respectively, and the corresponding sample proportions are $p_1 = X_1/m, p_2 = X_2/m, \ldots, p_n = X_n/m$. Then, π can be estimated by

$$\overline{p} = \frac{1}{n} \sum_{i=1}^{n} p_i.$$

Therefore, in large sample cases, we can evaluate the process performance at the n time points by comparing the sample proportions $\{p_i, i = 1, 2, \ldots, n\}$ with the control limits given in the box below, and the resulting control chart is called the p chart.

Control Limits of the p Chart in Large-Sample Cases

$$
\begin{aligned}
U &= \overline{p} + Z_{1-\alpha/2} \sqrt{\frac{\overline{p}(1-\overline{p})}{m}} \\
C &= \overline{p} \\
L &= \overline{p} - Z_{1-\alpha/2} \sqrt{\frac{\overline{p}(1-\overline{p})}{m}}
\end{aligned}
\tag{3.19}
$$

At the i-th time point, for $i = 1, 2, \ldots, n$, the production process is declared OC if

$$p_i < L \qquad \text{or} \qquad p_i > U.$$

The control limits in (3.19) are appropriate to use only in cases when the sample size m is large. In Subsection 2.7.1, it is mentioned that the sample size can be regarded as "large" if $m\pi \geq 10$ and $m(1 - \pi) \geq 10$. Namely, the expected numbers of "successes" and "failures" are both at least 10 in each sample. In the literature, some statisticians think that the normal distribution approximation to the binomial distribution should be reliable even in cases when $m\pi \geq 5$ and $m(1 - \pi) \geq 5$. See related discussion in Brown et al. (2001). In practice, π is often unknown, and it can be estimated by $\overline{p}$. Then, the resulting large-sample conditions become

$$m\overline{p} \geq 5, \qquad m(1 - \overline{p}) \geq 5. \tag{3.20}$$

In this chapter, we will use the less restrictive conditions in (3.20) to judge whether control charts based on the normal distribution approximation are appropriate to use or not in a specific application.

Example 3.4 *Assume that 20 random samples are obtained from the products of a production process, and each sample contains 50 randomly chosen products. The numbers of non-conforming products in the 20 samples are presented in Table 3.5, along with the sample proportions of non-conforming products. From the table, it*

Table 3.5 *Numbers of non-conforming products in 20 samples along with the sample propor-tions of non-conforming products.*

i	1	2	3	4	5	6	7	8	9	10
X_i	7	3	10	1	8	5	4	9	3	9
p_i	0.14	0.06	0.20	0.02	0.16	0.10	0.08	0.18	0.06	0.18
i	11	12	13	14	15	16	17	18	19	20
X_i	5	7	2	10	4	6	9	3	11	5
p_i	0.10	0.14	0.04	0.20	0.08	0.12	0.18	0.06	0.22	0.10

can be computed that $\bar{p} = 0.121$. So, $m\bar{p} = 6.05 > 5$ and $m(1 - \bar{p}) = 43.95 > 5$. The large-sample conditions in (3.20) are satisfied, and the p chart (3.19) is appropriate to use in this case. If we choose $Z_{1-\alpha/2} = 3$ in the chart, then its control limits are

$$U = 0.121 + 3 * \sqrt{\frac{0.121 * (1 - 0.121)}{50}} = 0.259$$

$$C = 0.121$$

$$L = 0.121 - 3 * \sqrt{\frac{0.121 * (1 - 0.121)}{50}} = -0.017.$$

Because the sample proportions are nonnegative, the lower control limit L can be changed to 0. The resulting p chart is shown in Figure 3.7. It can be seen from the plot that the process seems to be IC at all 20 time points.

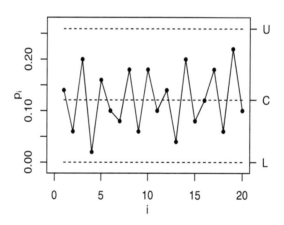

Figure 3.7: *The p chart for monitoring the data in Table 3.5.*

In cases when the sample size of a random sample collected at a given time point is small, but we still use a control chart based on the normal distribution approxi-mation, a direct consequence is that the actual Type I error probability, denoted as $\tilde{\alpha}$, could be quite different from the nominal Type I error probability α, which is demonstrated in the example below.

Example 3.5 *Assume that the sample size m equals 25 or 50, the true proportion of non-conforming products from a production process is π, which changes its value from 0.01 to 0.99 with a step size of 0.01, the nominal Type I error probability α is 0.01 or 0.001, and we use the large-sample control limits in (3.18) to monitor the process by assuming π is known beforehand. Then, the values of the actual Type I error probability $\widetilde{\alpha}$ in different cases are shown in Figure 3.8. From the four plots of the figure, it can be seen that $\widetilde{\alpha}$ could be substantially different from α, especially in cases when m is small and π is close to 0 or 1. As an example, assume that $\alpha = 0.01$ and $\widetilde{\alpha} = 0.02$. Then, a direct implication of the difference between α and $\widetilde{\alpha}$ in this case is that the actual false alarm rate of the control chart is two times the nominal false alarm rate. Consequently, the production process would be mistakenly stopped for investigating potential special causes of the false signals of shift twice as often as it should be, and much time and resources would be wasted.*

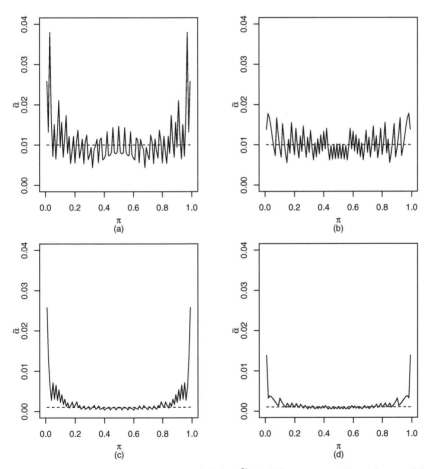

Figure 3.8 *The actual Type I error probability $\widetilde{\alpha}$ in different cases. Plot (a): $m = 25$ and $\alpha = 0.01$. Plot (b): $m = 50$ and $\alpha = 0.01$. Plot (c): $m = 25$ and $\alpha = 0.001$. Plot (d): $m = 50$ and $\alpha = 0.001$. The dashed horizontal line in each plot denotes the value of α.*

In cases when the large-sample conditions in (3.20) are not satisfied, one option to set up more appropriate control limits of the p chart is to use the discrete distribution of the sample proportion of non-conforming products directly. In cases when the true proportion of non-conforming products π is known and the nominal Type I error probability is α, let

$$L^* = \frac{\max\{a : P(X \le a) \le \alpha/2\}}{m}, \qquad U^* = \frac{\min\{a : P(X \ge a) \le \alpha/2\}}{m}, \quad (3.21)$$

where X is a random variable having the *Binomial*(n, π) distribution. Then, L^* and U^* in (3.21) can be used as the lower and upper control limits of the p chart, and the process can be declared OC at the i-th time point if

$$p_i < L^* \qquad \text{or} \qquad p_i > U^*.$$

Because of the discreteness of the binomial distribution, the actual Type I error probability $\tilde{\alpha}$ may not be exactly α. But, $\tilde{\alpha}$ can be computed beforehand to be

$$\tilde{\alpha} = P\left(X < mL^*\right) + P\left(X > mU^*\right).$$

Therefore, we should be able to know the actual Type I error probability before process monitoring, which is much better than the situation in which $\tilde{\alpha}$ is mistakenly assumed to be α. In cases when π is unknown, then it should be replaced by its estimator $\bar{p}$ in all the related computation.

In the literature, there are some alternative approximations to the binomial distribution besides the normal approximation. For related discussion, see Agresti and Coull (1998), Brown et al. (2001), Ross (2003), and the references cited therein. For instance, by using the approximation suggested by Agresti and Coull (1998), the control limits of the p chart should be

$$
\begin{aligned}
U &= \tilde{p} + Z_{1-\alpha/2}\sqrt{\frac{\tilde{p}(1-\tilde{p})}{\tilde{m}}} \\
C &= \tilde{p} \\
L &= \tilde{p} - Z_{1-\alpha/2}\sqrt{\frac{\tilde{p}(1-\tilde{p})}{\tilde{m}}}
\end{aligned}
\qquad (3.22)
$$

where

$$\tilde{m} = m + Z_{1-\alpha/2}^2, \qquad \tilde{p} = \frac{\bar{X} + Z_{1-\alpha/2}^2/2}{\tilde{m}}.$$

Instead of monitoring the sample proportions of non-conforming products $\{p_i, i = 1, 2, \ldots, n\}$, people sometimes prefer to monitor the frequencies of non-conforming products $\{X_i, i = 1, 2, \ldots, n\}$, although the two different versions of process monitoring should be theoretically equivalent. In large-sample cases, the corresponding control limits for monitoring frequencies are given in the box below, and the resulting control chart is often called the *mp chart*.

Control Limits of the mp Chart in Large-Sample Cases

$$
\begin{aligned}
U &= \overline{X} + Z_{1-\alpha/2}\sqrt{m\overline{p}(1-\overline{p})} \\
C &= \overline{X} \\
L &= \overline{X} - Z_{1-\alpha/2}\sqrt{m\overline{p}(1-\overline{p})},
\end{aligned}
\qquad (3.23)
$$

where $\overline{X} = \frac{1}{n}\sum_{i=1}^{n} X_i$.

At the i-th time point, for $i = 1, 2, \ldots, n$, the production process is declared OC if

$$
X_i < L \quad \text{or} \quad X_i > U.
$$

It is obvious that $\overline{X} = m\overline{p}$. Therefore, the large-sample conditions in (3.20) become

$$
\overline{X} \geq 5, \qquad m - \overline{X} \geq 5.
$$

Control limits in small-sample cases can be discussed in a similar way to that for monitoring the sample proportions, using the relationships that $X_i = mp_i$, for $i = 1, 2, \ldots, n$, and $\overline{X} = m\overline{p}$.

3.3.2 The c chart, u chart, and D chart

The p chart and mp chart discussed in the previous subsection are for monitoring proportions or numbers of non-conforming products in samples collected over different time points. In some applications, a product containing a certain number of nonessential defects would not be labeled as a non-conforming product as long as the quality of the product is still good enough to meet the customer's requirement. For instance, most new cars have small scratches on the surface and other minor defects. As long as a new car's major functions work well and its price is reasonable, customers would still be happy to buy it. However, from the manufacturers' viewpoint, it is important to monitor the occurrence of defects over time in order to keep and improve the quality of the products, which is the focus of this subsection.

In the new car example mentioned above, the number of scratches on the surface of a new car obviously depends on the area of the surface that is inspected: the number would be larger if the area of inspection is larger. For instance, the number of scratches found in the front part of a new car is not comparable with the number of scratches found on the entire body of another new car. Therefore, to properly discuss the occurrence of defects, the *inspection unit* should be relevant. An inspection unit may be comprised of one or more than one product. It may also be comprised of certain given parts of one or more than one product.

Let c be the number of defects found in an inspection unit, then it is appropriate to describe the distribution of c by a Poisson distribution (cf., the related discussion in Subsection 2.4.5). Assume that

$$
c \sim Poisson(\lambda),
$$

where $\lambda > 0$ is a parameter. Then, by (2.16), we have

$$\mu_c = \lambda, \qquad \sigma_c^2 = \lambda.$$

The above Poisson distribution is appropriate to use in cases when there is only one type of defect involved. In many applications, however, people are concerned about multiple types of defects. By the properties of the Poisson distribution (cf., Subsection 2.4.5), as long as different types of defects are well specified beforehand so that the number of different types of defects is given and unchanged over time and the occurrence of different types of defects is independent, the total number of all types of defects still follows a Poisson distribution. In such cases, the above Poisson distribution is still appropriate to use.

To make a decision whether the number of defects c is OC, the control limits usually used in the literature are

$$L = \lambda - Z_{1-\alpha/2}\sqrt{\lambda}, \qquad U = \lambda + Z_{1-\alpha/2}\sqrt{\lambda}.$$

In practice, the value of λ is often unknown, and it needs to be estimated from the observed data. In such cases, assume that n inspection units of the same size are randomly selected at n time points, and the observed numbers of defects are $c_1, c_2, \ldots, c_n$, respectively. Let $\bar{c} = \frac{1}{n}\sum_{i=1}^{n} c_i$. Then, after λ is replaced by $\bar{c}$, the corresponding control limits are given in the box below, and the resulting control chart is called the c *chart*.

Control Limits of the c Chart

$$
\begin{aligned}
U &= \bar{c} + Z_{1-\alpha/2}\sqrt{\bar{c}} \\
C &= \bar{c} \\
L &= \bar{c} - Z_{1-\alpha/2}\sqrt{\bar{c}}
\end{aligned}
\qquad (3.24)
$$

By (3.24), the production process is declared OC at the i-th time point if

$$c_i < L \qquad \text{or} \qquad c_i > U.$$

Example 3.6 *Surface defects have been counted on 15 rectangular steel plates, and the numbers of defects found on the 15 plates are*

$$2, 7, 4, 3, 9, 2, 5, 2, 6, 1, 8, 3, 5, 10, 2.$$

From the observed data, $\bar{c}$ is computed to be 4.6. So, by (3.24), if we choose $Z_{1-\alpha/2} = 3$, the control limits of the c chart are

$$
\begin{aligned}
U &= 4.6 + 3 * \sqrt{4.6} = 11.034 \\
C &= 4.6 \\
L &= 4.6 - 3 * \sqrt{4.6} = -1.834.
\end{aligned}
$$

Since the number of defects is nonnegative, the lower control limit can be changed from -1.834 to 0. The graph of the c chart is shown in Figure 3.9, from which it can be seen that the process producing the steel plates is IC at all 15 time points.

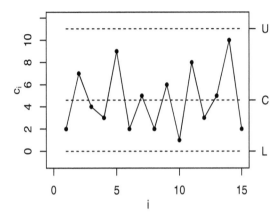

Figure 3.9 *The c chart for monitoring the numbers of defects found on 15 selected steel plates.*

Similar to the control limits of the p chart given in (3.19), the control limits of the c chart given in (3.24) are based on the normal approximation of the Poisson distribution. Theoretically speaking, when λ increases, the *Poisson*(λ) distribution gets closer and closer to the $N(\lambda, \lambda)$ distribution. For this reason, people often approximate the *Poisson*(λ) distribution by the $N(\lambda, \lambda)$ distribution. However, this approximation is reasonably good only in cases when λ is reasonably large (e.g., $\lambda \geq 10$). In cases when λ is small, the *Poisson*(λ) distribution is heavily skewed to the right. In such cases, the control limits given in (3.24) may not be appropriate to use, because the actual Type I error probability could be substantially different from the nominal Type I error probability α, similar to the phenomenon demonstrated in Figure 3.8. In such cases, we can use the ($\alpha/2$)-th quantile and the ($1 - \alpha/2$)-th quantile of the *Poisson*($\bar{c}$) distribution as the lower and upper control limits, respectively. We can also use some alternative approximations to the Poisson distribution. For more discussion on the latter topic, see Lesch and Jeske (2009).

The derivation of the control limits of the c chart given in (3.24) is based on the assumption that all n inspection units have the same size. In certain applications, however, different inspection units may have different sizes. For instance, assume that the inspection unit in an application is the shipment of products. Then, it is common in practice that different shipments may contain different numbers of products. In certain other applications, the size of the inspection unit might be related to its spatial area, and different inspection units might have different areas. In all such cases, the c chart with the control limits specified in (3.24) would not be appropriate to use because the numbers of defects found in different inspection units are not comparable. To overcome this difficulty, one natural idea is to consider

$$u_i = \frac{c_i}{m_i}, \qquad \text{for } i = 1, 2, \ldots, n,$$

where m_i is the number of products contained in the i-th inspection unit or another size metric of the i-th inspection unit. Obviously, u_i is the averaged number of defects per size unit within the i-th inspection unit. Assume that the number of defects per size unit follows the $Poisson(\widetilde{\lambda})$ distribution. Then, the number of defects within the i-th inspection unit, c_i, would follow the $Poisson(m_i\widetilde{\lambda})$ distribution. Therefore, $\mu_{c_i} = m_i\widetilde{\lambda}$ and $\sigma_{c_i}^2 = m_i\widetilde{\lambda}$. Consequently, we have

$$\mu_{u_i} = \widetilde{\lambda}, \qquad \sigma_{u_i}^2 = \frac{\widetilde{\lambda}}{m_i}.$$

After $\widetilde{\lambda}$ is replaced by its estimator

$$\bar{u} = \frac{\sum_{i=1}^{n} c_i}{\sum_{i=1}^{n} m_i},$$

it is natural to consider the control limits presented in the box below for monitoring $\{u_i, i = 1, 2, \ldots, n\}$, and the resulting control chart is called the u chart.

Control Limits of the u Chart

$$
\begin{aligned}
U &= \bar{u} + Z_{1-\alpha/2}\sqrt{\frac{\bar{u}}{m_i}} \\
C &= \bar{u} \\
L &= \bar{u} - Z_{1-\alpha/2}\sqrt{\frac{\bar{u}}{m_i}}
\end{aligned}
\qquad (3.25)
$$

Again, the control limits in (3.25) are appropriate to use only in cases when $\{\mu_{c_i} = m_i\widetilde{\lambda}, i = 1, 2, \ldots, n\}$ are all large, because they are based on the normal approximation to the related Poisson distributions. In cases when some or all of $\{\mu_{c_i} = m_i\widetilde{\lambda}, i = 1, 2, \ldots, n\}$ are small, the alternative approaches described above for constructing the control limits of the c chart can also be considered here.

In cases when different types of defects are present in the products, as mentioned above, the Poisson distribution is still appropriate for describing the total number of defects, c, found in an inspection unit, under some regularity conditions. However, when computing the value of c, all types of defects are treated equally. In practice, however, certain types of defects might be more essential to the quality of products, compared to some others. In the new car example mentioned at the beginning of this subsection, the defects in the engine system might be more serious than the minor scratches on the car surface. In such cases, it is natural to treat different types of defects differently. To this end, one option is to build separate control charts for different groups of defects. The defects within a group are believed to have similar impact on the quality of products, and thus they can be treated equally. Another option is to combine the defects in different groups by a weighted sum. Assume that all defects are classified into k groups. The numbers of defects in the k groups are denoted as $c_1^*, c_2^*, \ldots, c_k^*$. Then, we define the *weighted number of defects* to be

$$D = \sum_{j=1}^{k} w_j c_j^*, \tag{3.26}$$

where $\{w_j, j = 1, 2, \ldots, k\}$ are the nonnegative and pre-specified weights. Specification of the weights should reflect the relative importance of different types of defects. Obviously, it is reasonable to assume that c_j^*, for $j = 1, 2, \ldots, k$, follows the *Poisson*(λ_j^*) distribution, where $\lambda_j^* > 0$ is a parameter, and that $\{c_j^*, j = 1, 2, \ldots, k\}$ are independent of each other. Therefore, the mean and variance of D defined in (3.26) would be

$$\mu_D = \sum_{j=1}^{k} w_j \lambda_j^*, \qquad \sigma_D^2 = \sum_{j=1}^{k} w_j^2 \lambda_j^*. \tag{3.27}$$

In cases when we have n inspection units of the same size involved, and the observed number of defects belonging to the j-th group within the i-th inspection unit is c_{ij}^*, for $i = 1, 2, \ldots, n$ and $j = 1, 2, \ldots, k$. Then, the weighted number of defects in the i-th inspection unit is

$$D_i = \sum_{j=1}^{k} w_j c_{ij}^*, \qquad \text{for } i = 1, 2, \ldots, n.$$

From (3.27), we need to estimate $\{\lambda_j^*, j = 1, 2, \ldots, k\}$ properly in order to know the mean and variance of each D_i. An obvious estimator of λ_j^*, for $j = 1, 2, \ldots, k$, is

$$\bar{c}_j^* = \frac{1}{n} \sum_{i=1}^{n} c_{ij}^*.$$

Then, by (3.27), μ_D can be estimated by

$$\bar{D} = \frac{1}{n} \sum_{i=1}^{n} D_i = \frac{1}{n} \sum_{i=1}^{n} \sum_{j=1}^{k} w_j c_{ij}^* = \sum_{j=1}^{k} w_j \bar{c}_j^*,$$

and σ_D^2 can be estimated by

$$\sum_{j=1}^{k} w_j^2 \bar{c}_j^*.$$

To monitor the weighted numbers of defects $\{D_i, i = 1, 2, \ldots, n\}$, it is therefore reasonable to use the control limits in the box below, and the corresponding control chart is often called the *D chart*.

Control Limits of the D Chart

$$U = \overline{D} + Z_{1-\alpha/2} \sqrt{\sum_{j=1}^{k} w_j^2 \overline{c}_j^*}$$

$$C = \overline{D} \hspace{4cm} (3.28)$$

$$L = \overline{D} - Z_{1-\alpha/2} \sqrt{\sum_{j=1}^{k} w_j^2 \overline{c}_j^*}$$

It should be pointed out that the control limits in (3.28) are appropriate to use only in cases when $\{\lambda_j^*, j = 1, 2, \ldots, k\}$ are all reasonably large, and the n inspection units have the same size. Setup of the control limits of the D chart in other cases can be discussed similarly to the discussion about the control limits of the c chart, and is thus omitted here.

3.4   Process Capability Analysis

3.4.1   Process capability and its measurement

As described in Section 1.2, for a given product, we often have various requirements on its quality, based on customers' needs, engineering tolerances, and so forth. For simplicity, assume that X is the quality characteristic of interest, and it is univariate. Then, our requirement on the product quality is usually specified by the *lower specification limit (LSL)* and the *upper specification limit (USL)*. If the value of X is between the two limits, i.e., $LSL \leq X \leq USL$, then the product is acceptable and it is classified as a conforming product. Otherwise, it is classified as a non-conforming product. *Process capability analysis* of a production process is mainly for measuring its capability of manufacturing conforming products, by analyzing certain observed data that are representative of its production.

To perform a process capability analysis, the values of LSL and USL should be determined beforehand. Also, a dataset reflecting the IC production of the process should be available. In order to reflect the IC production of the process, the data should be collected after the process is adjusted properly during the phase I SPC so that the process is believed to be in statistical control. Distribution of the data should also estimate the population distribution of all products of the IC process well. To this end, observations in the data should be obtained by a random sampling scheme, and the sample size should be reasonably large as well, especially in cases when the data distribution is obviously non-normal (e.g., skewed). The following example demonstrates the process capability analysis of an injection molding process.

Example 3.7 *For an injection molding process, compressive strength (in psi) of its manufactured parts is our major concern about the quality of products. Based on*

*various considerations, it is determined that $LSL = 75$ and $USL = 85$ for this vari-
able. After a phase I SPC, the process is believed to be IC. Then, 50 manufactured
parts are randomly selected, and their compressive strengths are recorded. The density histogram of the 50 observations is shown in Figure 3.10, from which it can be
seen that the data are roughly normally distributed. The sample mean and sample
standard deviation are computed to be*

$$\overline{X} = 80.194, \qquad s = 2.775.$$

*Then, the process capability can be measured by the probability $P(LSL \leq X \leq USL)$,
where X denotes the compressive strength of a randomly selected manufactured part.
This probability can be computed as follows:*

$$
\begin{aligned}
P(LSL \leq X \leq USL) &= P\left(\frac{LSL - 80.194}{2.775} \leq \frac{X - 80.194}{2.775} \leq \frac{USL - 80.194}{2.775}\right) \\
&\approx P(-1.872 \leq Z \leq 1.732) \\
&= 0.928,
\end{aligned}
$$

*where Z denotes a random variable with the standard normal distribution. Based on
this measure, we know that about 92.8% of manufactured parts of the process are
conforming products, and about 7.2% of manufactured parts are non-conforming
products.*

The process capability analysis can also be made based on a control chart directly
in cases when the control chart confirms that the related production process is IC. For
instance, in the case of Example 3.2 in Subsection 3.2.2, the $\overline{X}$ chart constructed from
the first 20 samples of size 5 each confirms that the related injection molding process
is IC at the first 20 time points. In Example 3.2, it has been computed that

$$\overline{\overline{X}} = 79.533, \qquad \bar{s} = 3.575, \qquad d_3(5) = 0.940.$$

So, by (3.12), the mean and standard deviation of the quality characteristic X can be
estimated by

$$\widehat{\mu}_X = \overline{\overline{X}} = 79.533, \qquad \widehat{\sigma}_X = \frac{\bar{s}}{d_3(5)} = \frac{3.575}{0.940} = 3.803.$$

If we still use $LSL = 75$ and $USL = 85$, as in Example 3.7, and assume that X has a
normal distribution, then

$$
\begin{aligned}
P(LSL \leq X \leq USL) &= P\left(\frac{LSL - 79.533}{3.803} \leq \frac{X - 79.533}{3.803} \leq \frac{USL - 79.533}{3.803}\right) \\
&= P(-1.192 \leq Z \leq 1.438) \\
&= 0.808.
\end{aligned}
$$

3.4.2 Process capability ratios

To measure the process capability, many *process capability ratios (PCRs)* have been
proposed in the literature. Next, we describe a number of basic PCRs and their major

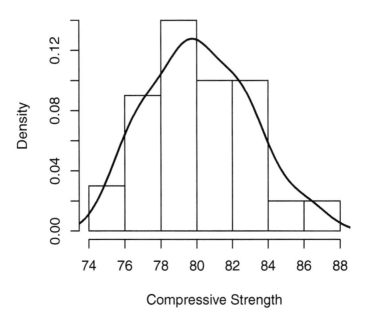

Figure 3.10 *Density histogram of the compressive strengths of 50 randomly selected manu-factured parts of an injection molding process. The solid curve is the estimated density curve of the data.*

properties in cases when the distribution of the quality characteristic X is assumed normal. For overviews and bibliographies on this topic, see Kotz and Johnson (2002), Spiring et al. (2003), and Yum and Kim (2011).

As in the previous subsection, assume that the mean and standard deviation of X are μ and σ, and the lower and upper specification limits are LSL and USL. Then, one of the earliest PCRs is defined by

$$C_p = \frac{USL - LSL}{6\sigma}. \tag{3.29}$$

In cases when $X \sim N(\mu, \sigma^2)$, we know that

$$P(\mu - 3\sigma \leq X \leq \mu + 3\sigma) = 0.9977.$$

In such cases, only about 0.23% of observations of X would be outside the interval $(\mu - 3\sigma, \mu + 3\sigma)$. Obviously, 6σ in (3.29) is the length of this interval, and C_p is the ratio of the length of the specification interval (LSL, USL) to 6σ. In cases when the center T of the specification interval (LSL, USL) is the same as the center μ of the distribution of X, if the value of C_p is larger, then the chance for the observations of X to be outside the specification interval (LSL, USL) would be smaller, which is demonstrated by Figure 3.11(a). In the plot, two distributions of X are considered and they are shown in the plot by the solid and dashed curves, respectively. It can be seen that the two distributions have the same center, but the one shown by the dashed curve has a smaller spread. In such cases, the C_p value for the distribution shown by the dashed curve would be larger than the C_p value for the distribution shown by the solid curve (see Example 3.8 below for details), which implies that the quality of the production process in the former scenario would be better than the quality of the production process in the latter scenario.

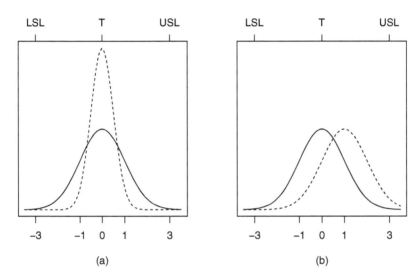

(a)   (b)

Figure 3.11 *(a) The distributions shown by the solid and dashed curves have the same center, but the one shown by the dashed curve has a smaller spread than the one shown by the solid curve. (b) The two distributions shown by the solid and dashed curves have different centers, but they have the same spread. In each plot, LSL and USL on the top denote the lower and upper specification limits, and T is their center.*

From its definition in (3.29), it can be seen that the PCR C_p considers the spread of the distribution of X. However, it does not take into account the center of the distribution. Let us consider the two cases shown in Figure 3.11(b) where the two distributions of X shown by the solid and dashed curves have the same spread but they have different centers. By the definition in (3.29), the C_p values in the two cases are the same. But, it is obvious that the probability $P(LSL \leq X \leq USL)$ would be smaller in the case with the distribution shown by the dashed curve, compared to the case with the distribution shown by the solid curve.

To overcome the limitation of C_p described above, some alternative PCRs have been proposed. One of them is defined as follows.

$$C_{pk} = \min\left(C_{pl}, C_{pu}\right),$$ (3.30)

where

$$C_{pl} = \frac{\mu - LSL}{3\sigma}, \qquad C_{pu} = \frac{USL - \mu}{3\sigma}.$$ (3.31)

The PCRs C_{pl} and C_{pu} defined in (3.31) are the two *one-sided* PCRs: C_{pl} is the lower-side PCR and C_{pu} is the upper-side PCR. The PCR C_{pk} defined in (3.30) takes the smaller value of them. Obviously, C_{pk} has taken into account both the center and spread of the distribution of X.

In mathematics, we have the simple fact that, for any two nonnegative numbers a and b, we have

$$\min(a,b) = \frac{(a+b) - |a-b|}{2}.$$

By using this fact, we have the following results:

$$
\begin{aligned}
C_{pk} &= \min\left(C_{pl}, C_{pu}\right) \\
&= \frac{1}{2}\left[(C_{pl} + C_{pu}) - |C_{pl} - C_{pu}|\right] \\
&= \frac{1}{2}\left[\frac{USL - LSL}{3\sigma} - \frac{|USL + LSL - 2\mu|}{3\sigma}\right] \\
&= C_p - \frac{|T - \mu|}{3\sigma} \\
&= \left[1 - \frac{|T - \mu|}{d}\right]C_p,
\end{aligned}
\tag{3.32}
$$

where $d = (USL - LSL)/2$ is the half length of the specification interval $[LSL, USL]$. The expression (3.32) implies that

(i) $C_{pk} \leq C_p$ and the equality holds if and only if $\mu = T$, and

(ii) C_{pk} decreases when $|\mu - T|$ increases.

By these results, the C_{pk} value in the case with the distribution of X shown by the dashed curve in Figure 3.11(b) would be smaller than its value in the case with the distribution shown by the solid curve in the same plot, which is intuitively reasonable.

Example 3.8 *In Figure 3.11(a), we consider two distributions of X, $N(0,1)$ and $N(0, 0.5^2)$, shown by the solid and dashed curves, respectively. In both cases, assume that $LSL = -3$ and $USL = 3$. Then, it is easy to compute the following values: In the case with $N(0,1)$,*

$$
\begin{aligned}
C_p &= [3 - (-3)]/(6 * 1) = 1, \\
C_{pl} &= C_{pu} = C_{pk} = 3/3 = 1.
\end{aligned}
$$

In the case with $N(0, 0.5^2)$,

$$
\begin{aligned}
C_p &= [3 - (-3)]/(6 * 0.5) = 2, \\
C_{pl} &= C_{pu} = C_{pk} = 3/(3 * 0.5) = 2.
\end{aligned}
$$

In both cases, we have

$$C_p = C_{pl} = C_{pu} = C_{pk}. \tag{3.33}$$

The results in (3.33) are not just a coincidence. As a matter of fact, as long as $\mu = T$, they must be true. Therefore, in such cases, all the PCRs C_p, C_{pl}, C_{pu}, and C_{pk}

are equivalent, and any one of them is reasonable to use for measuring the process capability. To compare the two cases, all four PCRs show that the process is more capable of producing conforming products in the case with $N(0, 0.5^2)$, compared to the case with $N(0, 1)$.

In Figure 3.11(b), the two distributions of X are $N(0, 1)$ and $N(1, 1)$. They have the same spread, but their centers are different. Let us still assume that $LSL = -3$ and $USL = 3$. Then, it is easy to compute the following values: In the case with $N(0, 1)$,

$$C_p = [3 - (-3)]/(6*1) = 1,$$
$$C_{pl} = C_{pu} = C_{pk} = 3/3 = 1.$$

In the case with $N(1, 1)$,

$$C_p = [3 - (-3)]/(6*1) = 1,$$
$$C_{pl} = [1 - (-3)]/(3*1) = 1.333,$$
$$C_{pu} = (3 - 1)/(3*1) = 0.667,$$
$$C_{pk} = \min(C_{pl}, C_{pu}) = 0.667.$$

It can be seen that the value of C_p equals 1 in both cases. So, this PCR is not appropriate to use in cases when μ and T are different. On the other hand, the value of C_{pk} is smaller in the case with $N(1, 1)$ than its value in the case with $N(0, 1)$. Therefore, it has indeed taken into account the center of the distribution of X, and is thus more appropriate to use than C_p in cases when μ and T are different. As a verification, by using the relationship between C_p and C_{pk} established in (3.32), in the case with $N(1, 1)$, we have $T = 0$, $\mu = 1$, and $d = 3$. Thus, $C_{pk} = (1 - |T - \mu|/d)C_p = 2C_p/3 = 0.667$, which is the same as its value computed above from the definition of C_{pk}.

Another alternative PCR that takes into account the center of the distribution of X is defined by

$$C_{pm} = \frac{USL - LSL}{6\tau}, \tag{3.34}$$

where

$$\tau = \sqrt{E[(X - T)^2]}.$$

Obviously,

$$
\begin{aligned}
\tau^2 &= E[(X - T)^2] \\
&= E[(X - \mu)^2] + (\mu - T)^2 \\
&= \sigma^2 + (\mu - T)^2. \tag{3.35}
\end{aligned}
$$

Namely, τ^2 is the mean squared distance between X and its target T. By combining (3.34) and (3.35), we have

$$
\begin{aligned}
C_{pm} &= \frac{USL - LSL}{6\sqrt{\sigma^2 + (\mu - T)^2}} \\
&= \frac{C_p}{\sqrt{1 + \left(\frac{\mu - T}{\sigma}\right)^2}}. \tag{3.36}
\end{aligned}
$$

From (3.36), it can be seen that

(i) $C_{pm} \leq C_p$ and the equality holds if and only if $\mu = T$, and

(ii) C_{pm} decreases when $|\mu - T|$ increases.

These two properties confirm that C_{pm} indeed takes into account the center μ of the distribution of X, and its value would be smaller when μ is farther away from the target value T.

To increase the sensitivity to the difference between μ and T, we can also combine the two PCRs C_{pk} and C_{pm}, by replacing C_p in (3.36) with C_{pk}. The resulting PCR is

$$C_{pkm} = \frac{C_{pk}}{\sqrt{1 + \left(\frac{\mu-T}{\sigma}\right)^2}}. \tag{3.37}$$

Example 3.8 (continued) *In the case when $X \sim N(1,1)$, $LSL = -3$, and $USL = 3$ considered in Example 3.8 (i.e., the case shown by the dashed curve in Figure 3.11(b)), it can be computed by (3.36) and (3.37) that*

$$C_{pm} = \frac{1}{\sqrt{1 + \left(\frac{1-0}{1}\right)^2}} = 0.707$$

and

$$C_{pkm} = \frac{0.667}{\sqrt{1 + \left(\frac{1-0}{1}\right)^2}} = 0.471.$$

Indeed, C_{pkm} is more sensitive to the departure of μ from T, compared to both C_{pk} and C_{pm}.

In the definitions of all the PCRs $C_p, C_{pl}, C_{pu}, C_{pk}, C_{pm}$, and C_{pkm} described above, the population parameters μ and σ are involved, which are often unknown in practice. In such cases, to use these PCRs for measuring process capability, a dataset reflecting the IC performance of the production process should be collected beforehand, as in cases considered in the previous subsection. Then, μ and σ can be estimated by the sample mean $\bar{x}$ and the sample standard deviation s of the collected IC dataset, respectively. After they are replaced by their estimators, the resulting estimators of $C_p, C_{pl}, C_{pu}, C_{pk}, C_{pm}$, and C_{pkm} can be obtained, which are denoted as $\widehat{C}_p, \widehat{C}_{pl}, \widehat{C}_{pu}, \widehat{C}_{pk}, \widehat{C}_{pm}$, and $\widehat{C}_{pkm}$, respectively. In the case of Example 3.7, it can be computed that

$$\widehat{C}_p = \frac{USL - LSL}{6s} = \frac{85 - 75}{6 * 2.775} = 0.601,$$

$$\widehat{C}_{pl} = \frac{\bar{x} - LSL}{3s} = \frac{80.194 - 75}{3 * 2.775} = 0.624,$$

$$\widehat{C}_{pu} = \frac{USL - \bar{x}}{3s} = \frac{85 - 80.194}{3 * 2.775} = 0.577,$$

$$\widehat{C}_{pk} = \min\left(\widehat{C}_{pl}, \widehat{C}_{pu}\right) = 0.577,$$

$$\widehat{C}_{pm} = \frac{\widehat{C}_p}{\sqrt{1 + \left(\frac{\bar{x}-T}{s}\right)^2}} = \frac{0.601}{\sqrt{1 + \left(\frac{80.194-80}{2.775}\right)^2}} = 0.600,$$

$$\widehat{C}_{pkm} = \frac{\widehat{C}_{pk}}{\sqrt{1 + \left(\frac{\bar{x}-T}{s}\right)^2}} = \frac{0.577}{\sqrt{1 + \left(\frac{80.194-80}{2.775}\right)^2}} = 0.576.$$

Besides point estimators, it is also possible to derive confidence intervals (CIs) for the PCRs. For instance, from Subsection 2.7.1, we know that

$$\frac{(m-1)s^2}{\sigma^2} \sim \chi^2_{m-1},$$

where s^2 is the sample variance of an IC dataset of size m. Then, by the relationship $s/\sigma = C_p/\widehat{C}_p$, we have

$$P\left(\chi^2_{\alpha/2, m-1} \le \frac{(m-1)s^2}{\sigma^2} \le \chi^2_{1-\alpha/2, m-1}\right) = 1 - \alpha$$

$$\Longleftrightarrow \quad P\left(\widehat{C}_p \sqrt{\frac{\chi^2_{\alpha/2, m-1}}{m-1}} \le C_p \le \widehat{C}_p \sqrt{\frac{\chi^2_{1-\alpha/2, m-1}}{m-1}}\right) = 1 - \alpha$$

where $\chi^2_{\alpha/2, m-1}$ and $\chi^2_{1-\alpha/2, m-1}$ are the $(\alpha/2)$-th and $(1 - \alpha/2)$-th quantiles of the χ^2_{m-1} distribution. So, the $100(1 - \alpha)\%$ CI for C_p is

$$\left(\widehat{C}_p \sqrt{\frac{\chi^2_{\alpha/2, m-1}}{m-1}}, \ \widehat{C}_p \sqrt{\frac{\chi^2_{1-\alpha/2, m-1}}{m-1}}\right). \tag{3.38}$$

The CIs for C_{pk} and C_{pm} can be obtained from (3.38), by using the relationships between C_p and both C_{pk} and C_{pm} established in (3.32) and (3.36), after μ and σ in the related formulas are replaced by $\overline{X}$ and s. Such CIs should be reliable to use when the sample size m of the IC data is reasonably large and the distribution of X is normal. In the literature, alternative CIs have been proposed for C_{pk} and C_{pm}, along with CIs for some other PCRs. Interested readers can see Kotz and Lovelace (1998), Pearn et al. (1992), Zhang et al. (1990), and the references cited therein.

Readers are reminded that all the PCRs described above are based on the assumption that the distribution of X is normal. In cases when the distribution of X is not normal, it might be misleading to use these PCRs to measure the process capability, because they do not take into account the shape of the distribution of X in their definitions. In such cases, one possible approach is to transform X first such that the distribution of the transformed X is close to normal, and then we can compute the PCRs in the transformed scale. Such a transformation can be searched based on an IC data of the process collected beforehand. Another option is to develop PCRs that are appropriate for non-normal distributions. In the literature, there has been much existing research on this latter approach. See, for instance, Levinson (2010), Luceño (1996), Rodriguez (1992), Yeh and Bhattacharya (1998), and Yeh and Chen (2001).

3.5   Some Discussions

We have described some basic Shewhart charts for monitoring production processes when the quality characteristic of interest X is univariate. These Shewhart charts make decisions about the process performance at a given time point using the observed data at that time point only, under a similar framework to hypothesis testing. In cases when X is numerical, the related Shewhart charts (e.g., the $\overline{X}, R$, and s charts) are based on the assumption that the distribution of X is normal. In cases when X is categorical, the related Shewhart charts (e.g., the p, c, and u charts) are constructed based on the binomial or Poisson distribution model. They are simple to construct, and convenient to use. Therefore, they are popular in practice.

Because the Shewhart charts evaluate the process performance based on the observed data at each individual time point, they are good at detecting relatively large and transient shifts in the distribution of X. They are less efficient in detecting relatively small and persistent shifts, compared to the CUSUM, EWMA, and other control charts that will be discussed in later chapters. For this reason, the Shewhart charts are especially popular in phase I SPC, because relatively large and transient shifts are common in phase I SPC and less common in phase II SPC.

The $\overline{X}, R$, and s charts discussed in Section 3.2 are appropriate to use only in cases when the distribution of X is normal. In cases when the distribution of X is non-normal, their actual Type I error probabilities could be substantially different from the nominal level α. If the actual Type I error probability of a Shewhart chart is larger than α, then the control chart would give false signals of shift more often than expected. Consequently, much time and many resources would be wasted in investigating possible root causes of the false signals and in adjusting the related production process. If the actual Type I error probability of a Shewhart chart is smaller than α, then real shifts would be missed by the control chart more often than expected and consequently many non-conforming products could be manufactured. In cases when the distribution of X is non-normal but the sample size m of the sample collected at each time point is large, the problem described above would not be serious, because the central limit theorem (cf., Subsection 2.7.1) guarantees that the distributions of the related statistics (e.g., the sample mean) would be approximately normal. In cases when the distribution of X is non-normal and m is small, one option to use the related Shewhart charts properly is to transform the observed data to normal and then apply the Shewhart charts to the transformed data. To this end, there is some existing research. See, for instance, Chou et al. (1998) and Yourstone and Zimmer (1992). Another option is to use the Shewhart charts that are designed for monitoring processes with non-normal data, which will be discussed in Chapter 8 in detail.

If the performance of the Shewhart charts is evaluated by the IC ARL value ARL_0 and OC ARL value ARL_1, then these performance measures are also affected by the possible correlation among the data collected at different time points. Note that the formulas (3.8) and (3.9) for computing ARL_0 and ARL_1 are derived based on the assumption that samples obtained at different time points are independent of each other. In cases when this assumption is violated, these formulas become invalid, and the actual ARL_0 and ARL_1 values could be affected by the correlation among different samples, as demonstrated by many authors, including Alwan (1992), Black et al. (2011), and English et al. (2000). In such cases, the correlation should be handled properly when computing the actual ARL_0 and ARL_1 values.

The p and mp charts discussed in Subsection 3.3.1 are for monitoring the proportion of non-conforming products over time. In this case, the status of each product is binary: conforming or non-conforming. In certain applications, however, the status of a product might have more than two possible categories. For instance, the quality of a product can be classified into three or more categories, and the products in different categories of quality can be sold at different prices. In such cases, the observed numbers of products in a sample that belong to different categories have a multinomial distribution. Much research has been done in recent years on SPC of multinomial data. See Topalidou and Psarakis (2009) for an overview. In our description of the p, mp, c, u, and D charts in Section 3.3, the control charts are constructed based solely on the related binomial or Poisson distribution model. In practice, however, the variability of the observed data is often larger than that computed from these distribution models. This is the so-called *overdispersion* phenomenon. An intuitive explanation of this phenomenon is that the quality characteristic X is often affected by many variables or factors that are not considered in our study. Such variables or factors would contribute to the variability of X, making the variance of X larger. In the literature, there is a limited discussion on this issue. See, for instance, Albers (2011) and Grigg et al. (2009).

In Section 3.4, we briefly discuss the topic of process capability analysis. From the discussion, it can be seen that the process capability could be poor even when the related process is IC. Therefore, although the major goal of most control charts described in this book is to detect distributional shifts in the quality characteristic(s) of interest and to make sure that the related process runs stably, the ultimate goal of quality control and management is to improve the quality of manufactured products, by keeping the mean of the quality characteristic(s) close to its target and by constantly reducing the variability of the quality characteristic(s). To achieve the goal of quality control and improvement, SPC charts play an important role. However, many other statistical tools, such as design of experiment, regression, acceptance sampling, and so forth, are also important, although they are not discussed in detail in this book.

3.6   Exercises

3.1 For the hypothesis testing problem discussed at the beginning of Subsection 3.2.1 with the hypotheses

$$H_0 : \mu = \mu_0 \qquad \text{versus} \qquad H_1 : \mu \neq \mu_0,$$

verify that, for the testing procedure with the test statistic in (3.1), H_0 can be rejected at the significance level α if $\overline{X}_i$ satisfies the conditions in (3.3). See related discussion in Subsection 2.7.3 about this testing procedure.

3.2 For the point estimator $\widehat{\mu}_0$ of the IC mean μ_0 defined in (3.4),

(i) show that $\widehat{\mu}_0$ is an unbiased estimator of μ_0 in cases when all samples are collected from an IC process, and

(ii) specify the sampling distribution of $\widehat{\mu}_0$ in such cases.

3.3 Assume that the process mean shifts from μ_0 to $\mu_1 = \mu_0 + \delta$ at the τ-th time point, with $\delta \neq 0$ and $1 < \tau \leq n$. Namely, the process mean is μ_0 at the i-th time point when $i < \tau$, and is μ_1 when $i \geq \tau$. All other properties of the distribution of the quality characteristic X are unchanged before and after the shift. For the point estimator $\widehat{\mu}_0$ of the IC mean μ_0 defined in (3.4), do the following problems.

(i) Is $\widehat{\mu}_0$ still an unbiased estimator of μ_0? Why?

(ii) Specify the sampling distribution of $\widehat{\mu}_0$ in such cases.

3.4 The values of $d_1(m)$ and $d_2(m)$ presented in Table 3.1 can be obtained by a simulation study described below. First, generate n random samples from the distribution $N(0, 1)$ with each sample having m observations. Second, compute the sample ranges of the n samples. Third, the value of $d_1(m)$ can be estimated by the sample mean of the n sample ranges, and the value of $d_2(m)$ can be estimated by their sample standard deviation.

(i) Using the simulation study described above, verify the values of $d_1(m)$ and $d_2(m)$ that are presented in Table 3.1 in cases when $m = 2, 5, 10$, and 20. In your simulation study, use $n = 1,000$ and $n = 10,000$, respectively, to obtain two sets of results. Compare these two sets of results, and summarize your findings.

(ii) In cases when the IC distribution of the quality characteristic X is $N(\mu_0, \sigma^2)$, show that $d_1(m)$ and $d_2(m)$ do not depend on μ_0 and σ.

3.5 Assume that we have collected 10 samples of size 5 each from a process producing bearings. The quality characteristic of interest X is the inside diameter measurements of the bearings. The original observed data of X along with the sample means and sample ranges are given in the table below.

Sample	x_1	x_2	x_3	x_4	x_5	$\overline{X}$	R
1	34.09	36.30	35.76	35.01	36.50	35.53	2.41
2	36.11	34.39	35.15	36.76	37.63	36.01	3.24
3	33.43	35.41	34.00	35.20	35.67	34.74	2.24
4	36.79	35.96	35.62	34.48	33.63	35.30	3.16
5	36.46	35.89	35.83	35.43	35.40	35.80	1.06
6	33.59	34.76	33.98	34.35	35.39	34.41	1.80
7	36.17	36.20	34.60	34.97	34.83	35.35	1.60
8	34.66	35.05	36.08	34.99	35.15	35.19	1.42
9	35.95	34.18	35.02	35.32	34.77	35.05	1.77
10	35.62	35.18	34.93	36.35	36.24	35.66	1.42

(i) Construct the $\overline{X}$ and R charts using $\alpha = 0.0027$. Does the process seem to be in statistical control?

(ii) Assume that the manufacturing process is IC, and the observed dataset shown in the above table is an IC dataset. Provide estimates for the IC mean μ_0 and the IC standard deviation σ of X.

(iii) Using the results in parts (i) and (ii), compute the probability that the sample mean of a new sample of size 5 would give a signal of mean shift in the $\overline{X}$ chart constructed in part (i) in cases when the true process mean is not shifted.

(iv) Compute the ARL_0 value of the $\overline{X}$ chart constructed in part (i), and its ARL_1 value when detecting a mean shift of size 1 (i.e., the shifted mean $\mu_1 = \mu_0 + 1$).

3.6 The sample standard deviations of the observed data presented in Exercise 3.5 are computed to be

$$0.99, 1.28, 0.97, 1.25, 0.43, 0.70, 0.77, 0.53, 0.66, 0.63$$

(i) Construct the $\overline{X}$ and s charts using $\alpha = 0.0027$ (cf., (3.13) and (3.14)). Does the process seem to be in statistical control?

(ii) Compare the control charts in part (i) and the control charts obtained in part (i) of Exercise 3.5, and summarize your findings.

(iii) The standard deviation σ of the quality characteristic X can be estimated by (3.5) based on the sample ranges. It can also be estimated by (3.12) based on the sample standard deviations. Further, it can be estimated by the sample standard deviation of the combined sample (i.e., all samples are combined into a single sample). Discuss the strengths and limitations of the three estimators of σ for the purpose of SPC, and compute their values using the observed data given in Exercise 3.5.

3.7 To monitor a production process, assume that n random samples of size m each are collected at n different time points. Based on the observed data, an $\overline{X}$ chart is constructed for monitoring the process mean. In order to guarantee the overall probability of Type I error at all n time points to be $\widetilde{\alpha}$, the significance level α used at a single time point when constructing the $\overline{X}$ chart should be chosen by the formula (3.10).

(i) When $\tilde{\alpha} = 0.001$, determine the values of α in cases when (a) $n = 10$ and (b) $n = 100$. Discuss the difference between the values of α in the two different cases, and its impact on the $\overline{X}$ chart.

(ii) When $\tilde{\alpha} = 0.0001$, redo the problem in part (i). Compare the results here with those obtained in part (i), and summarize your major findings.

3.8 Assume that we have collected 14 samples of size 5 each from an injection molding process. The sample means and sample standard deviations of the compressive strength measurements of the sampled parts are listed below.

Sample	1	2	3	4	5	6	7
$\overline{X}_i$	80.22	78.31	81.40	78.53	81.32	80.54	77.33
s_i	3.99	5.35	4.79	4.68	3.79	5.78	3.52
Sample	8	9	10	11	12	13	14
$\overline{X}_i$	79.24	81.44	77.76	79.48	76.74	81.12	86.79
s_i	4.51	5.16	4.81	4.56	7.35	3.91	6.31

(i) Use the first 10 samples to construct the $\overline{X}$ and s charts (cf., (3.13) and (3.14)). Does the process seem to be in statistical control? If the answer is positive, then use the constructed control charts to monitor the last 4 samples, which are treated as phase II data. Describe your findings of the phase II SPC.

(ii) Use the first 10 samples to construct the s^2 chart (cf., (3.15)). If the process variability seems to be in statistical control at the first 10 time points, then use the constructed control chart to monitor the process variability at the last 4 time points by treating the last 4 samples as phase II data. Describe your findings of the phase II SPC.

(iii) Note that by taking the squared root of the lower and upper control limits of the s^2 chart, we obtain the lower and upper control limits of a new version of the s chart. Compare this version of the s chart with the s chart constructed in part (i) for both phase I and phase II analyses.

3.9 The data given below are measurements of the tensile strength of the sampled papers manufactured by a production process. This is an individual observation dataset. At each time point, only one observation is obtained, and different observations are obtained at equally spaced time points.

25, 24, 39, 26, 25, 22, 24, 21, 28, 24, 24, 22, 16, 26, 25, 26, 21, 25, 23, 24

Construct the $\overline{X}$ and R charts (cf., (3.16) and (3.17)) using moving windows of size $\tilde{m}$ each with (i) $\tilde{m} = 2$, and (ii) $\tilde{m} = 5$. Compare the two sets of control charts, and summarize your major findings.

3.10 Assume that $X \sim Binomial(m, \pi)$. In practice, people often approximate the distribution $Binomial(m, \pi)$ using the normal distribution $N(m\pi, m\pi(1 - \pi))$. Use the original binomial distribution and its normal approximation to compute the following probabilities, respectively. Compare the two sets of results, and summarize your major findings.

(i) $P(X \leq 3)$ when $m = 20$ and $\pi = 0.1$.

(ii) $P(X \leq 3)$ when $m = 20$ and $\pi = 0.4$.

(iii) $P(X \leq 3)$ when $m = 100$ and $\pi = 0.1$.

(iv) $P(X \leq 3)$ when $m = 100$ and $\pi = 0.4$.

3.11 To control the proportion of non-conforming products manufactured by a production process, 20 samples of size $m = 100$ each are obtained. The numbers of nonconforming products in the 20 samples are listed below.

10, 15, 31, 18, 24, 12, 23, 15, 19, 21, 16, 24, 28, 15, 23, 19, 14, 27, 20, 18

(i) Construct the p chart using the above data and $\alpha = 0.0027$. Does the process seem to be in statistical control?

(ii) If the true proportion of non-conforming products is $\pi = 0.19$ when the process is IC, what would be the minimum value of the sample size m in order to detect a shift in p from 0.19 to 0.29 with a 90% chance, using a p chart with the control limits specified in (3.19) and $\alpha = 0.0027$?

3.12 Reproduce the results shown in Figure 3.8.

3.13 The numbers of non-conforming products in 20 samples of size $m = 50$ each are listed below.

4, 3, 4, 2, 4, 4, 4, 2, 3, 5, 5, 5, 8, 1, 4, 3, 6, 7, 4, 4

(i) Construct the p chart using the control limits specified in (3.19) and $\alpha = 0.0027$. Does the process seem to be in statistical control?

(ii) For the above data, it can be computed that $m\bar{p} = 4.1$. So, the large-sample conditions in (3.20) are not satisfied. Consequently, the p chart constructed in part (i) may not be appropriate to use. In such cases, one alternative method is to use (3.21) to specify the control limits of the p chart, after π is estimated by $\bar{p} = 0.082$. Construct the p chart using this method, and compare this chart with the chart constructed in part (i).

(iii) Because of the discreteness of the binomial distribution, the actual Type I error probability α' of the chart constructed in part (ii) may not be exactly α. Compute the value of α' of that control chart.

(iv) Construct the p chart using the control limits specified in (3.22) and $\alpha = 0.0027$. Compare this chart with the charts constructed in parts (i) and (ii), and summarize your major findings.

3.14 For the data in Exercise 3.11, construct the mp chart using the control limits specified in (3.23) and $\alpha = 0.0027$. Is this control chart equivalent to the control chart constructed in part (i) of Exercise 3.11?

3.15 Assume that the number of defects, c, found in an inspection unit follows the $Poisson(\lambda)$ distribution. When deriving the control limits of the c chart listed in (3.24), the $Poisson(\lambda)$ distribution is approximated by the $N(\lambda, \lambda)$ distribution. Use the original Poisson distribution and its normal approximation to compute the following probabilities. Compare the two sets of results, and summarize your major findings.

(i) $P(c \le 3)$ when $\lambda = 5$.

(ii) $P(c \le 3)$ when $\lambda = 10$.

(iii) $P(c \le 3)$ when $\lambda = 100$.

(iv) $P(c \le 20)$ when $\lambda = 100$.

3.16 Surface defects have been counted on 10 rectangular steel plates, and the numbers of defects on the 10 plates are as follows.

$$10, 10, 5, 10, 14, 9, 16, 12, 8, 10, 6, 9, 13, 10, 8, 9, 10, 13, 10, 7$$

Set up a c chart for monitoring the number of defects using these data and $\alpha = 0.0027$. Does the process producing the plates appear to be in statistical control?

3.17 To monitor a production process, 10 inspection units are chosen. The number of products in the i-th inspection unit, m_i, and the number of defects found on those products, c_i, for $i = 1, 2, \ldots, 10$, are listed below.

i	1	2	3	4	5	6	7	8	9	10
m_i	79	57	83	67	59	73	65	59	66	75
c_i	44	32	59	35	24	34	51	32	38	43

Set up a u chart using these data and $\alpha = 0.0027$. Does the production process appear to be in statistical control?

3.18 Assume that we are mainly concerned about three types of defects on the electric components manufactured by a production process. Twenty inspection units of the same size are randomly chosen, and the numbers of defects of the three different types found in these units are listed below.

i	1	2	3	4	5	6	7	8	9	10
c_{i1}^*	5	6	3	4	2	4	8	7	1	5
c_{i2}^*	11	11	7	17	10	8	8	15	9	7
c_{i3}^*	21	19	17	23	29	30	14	36	26	24
i	11	12	13	14	15	16	17	18	19	20
c_{i1}^*	4	8	8	6	2	3	1	4	3	6
c_{i2}^*	6	7	12	11	6	9	9	8	10	13
c_{i3}^*	15	18	23	28	21	22	18	17	27	28

Considering different impacts of the three types of defects on the product quality, the weights $w_1 = 0.6, w_2 = 0.3$, and $w_3 = 0.1$ are used when defining the weighted numbers of detects in the 20 inspection units using (3.26). Construct the D chart using the control limits specified in (3.28) and $\alpha = 0.0027$. Does the production process appear to be in statistical control?

3.19 Assume that the observed data in Exercise 3.5 are obtained from an IC process, each of the 10 samples is a simple random sample, and the 10 samples are independent of each other. So, for the process capability analysis, the 10 samples can be combined into a single sample.

(i) Check for the normality of the combined sample.

(ii) Assume that the lower and upper specification limits are $LSL = 33$ and $USL =$

38 for the quality characteristic X. Estimate the probability $P(LSL \leq X \leq USL)$ using the combined sample.

(iii) With the same lower and upper specification limits as those in part (ii), estimate the probability $P(LSL \leq X \leq USL)$ based on the control charts constructed in part (i) of Exercise 3.5.

3.20 Assume that a quality characteristic X has a normal distribution when a related production process is IC, and the specification limits on X are $LSL = 2100$ and $USL = 2300$. A random sample of size 50 obtained from the IC process results in $\overline{X} = 2250$ and $s = 50$.

(i) Compute point estimates of C_p, C_{pk}, and C_{pm}. Between C_p and C_{pk}, which one is more appropriate to use in this case? Why?

(ii) Construct a 95% confidence interval for C_p.

(iii) Based on the provided information, compute the probability that a randomly selected product from the IC process would be a defective product.

3.21 Using the confidence interval formula (3.38) and the relationships between C_p and both C_{pk} and C_{pm} established in (3.32) and (3.36), derive the confidence interval formulas for C_{pk} and C_{pm}. Then, construct 95% confidence intervals for C_{pk} and C_{pm} in the case considered in Exercise 3.20.

Chapter 4

Univariate CUSUM Charts

4.1 Introduction

Assume that we want to monitor a univariate quality characteristic X of a production process. The Shewhart charts described in the previous chapter make decisions about the process performance at a given time point based solely on the observed data at that single time point. Their decision rules are constructed in a similar way to those of the hypothesis testing procedures, by which the process is declared OC if the observed value of the charting statistic of a Shewhart chart is beyond its lower and upper control limits at the given time point, and the process is declared IC otherwise. For simplicity of perception, let us further assume that we are interested in monitoring the process mean in a given application, and other distributional properties of X would not change after a mean shift. In such cases, the distribution of X is F_0 before the mean shift, it changes to another distribution F_1 after the shift, and the only difference between F_0 and F_1 is their means. At a given time point n, if there is no shift before n, then all observations obtained before n should provide useful information about the IC process performance. In cases when the mean shift occurs before n, all observations obtained between the occurrence of the shift and the time n should all provide useful information about the OC process performance. In both cases, the history data obtained before n contain useful information for process monitoring; but, the Shewhart charts do not make use of such information at all. For this reason, they would not be effective for detecting persistent shifts, especially in cases when the shifts are relatively small, which is demonstrated in the following example.

Example 4.1 *Assume that 10 samples of size 5 each are generated from the $N(0,1)$ distribution, and another 10 samples of size 5 each are generated from the $N(0.2,1)$ distribution. These samples can be regarded as observed data collected at 20 consecutive time points from a production process, the IC distribution of the process is $N(0,1)$, and the process has a mean shift of size 0.2 at the 11th time point. Figure 4.1(a) shows the $\overline{X}$ chart constructed from the 20 samples with the control limits computed by the formulas in (3.6). In the plot, the first 10 sample means are denoted by solid dots and they are connected by solid lines, and the second 10 sample means are denoted by little circles and they are connected by dotted lines. This chart can be used for a phase I SPC. From the plot, it can be seen that the true mean shift at the 11th time point is not detected by the chart. Figure 4.1(b) shows an alternative $\overline{X}$ chart for a phase II SPC, in which the IC process distribution is assumed to be known $N(0,1)$, the lower and upper control limits of the chart are set to $-3/\sqrt{5}$ and*

$3/\sqrt{5}$, respectively, and all the symbols and notations in the plot have the same interpretation as those in Figure 4.1(a). From the plot, it can be seen that the alternative
$\overline{X}$ chart cannot detect the true mean shift at the 11th time point either.

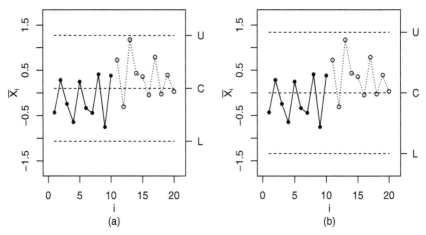

Figure 4.1 (a) The $\overline{X}$ chart constructed from all 20 samples with the control limits computed
by the formulas in (3.6). (b) An alternative $\overline{X}$ chart for a phase II SPC, in which the IC process
distribution is assumed to be known $N(0, 1)$ and the lower and upper control limits of the
chart are set to be $-3/\sqrt{5}$ and $3/\sqrt{5}$, respectively. In each plot, the first 10 sample means are
denoted by solid dots and they are connected by solid lines, and the second 10 sample means
are denoted by little circles and they are connected by dotted lines. The dashed horizontal lines
labeled U, C, and L denote the upper control limit, the center line, and the lower control limit,
respectively.

From the above example, it can be seen that the $\overline{X}$ chart is indeed ineffective in
detecting small and persistent mean shifts. As mentioned above, the major reason
for this limitation of the Shewhart charts is that they use observed data at a single
time point alone to evaluate the process performance at that time point. To overcome
this limitation, many alternative control charts have been proposed in the literature.
The major ones will be described in this and later chapters of the book. This chapter focuses on cumulative sum (CUSUM) control charts that were first proposed by
Page (1954) and then discussed by many authors, including Bissell (1969, 1984b),
Ewan (1963), Gan (1991a, 1993a), Goel and Wu (1971), Hawkins (1981, 1993a),
Hawkins and Olwell (1998), Johnson and Leone (1962), Lucas (1973), Montgomery
(2009), Page (1961), and Woodall and Adams (1993). Our discussion here focuses
on cases when the quality characteristic of interest is univariate, its IC distribution
has a parametric form, and the potential shift is in one or more distribution parameters. CUSUM charts for monitoring multiple quality characteristics will be discussed
in Chapter 7. CUSUM charts for handling cases when the IC process distribution is
nonparametric will be discussed in Chapters 8 and 9.

At the end of this section, we would like to mention that the Shewhart charts described in Chapter 3 are popular in practice, mainly because of their simplicity. Due

to the facts that they are good at detecting relatively large and transient shifts and such shifts are common in phase I SPC, they provide a reasonably good statistical tool for phase I SPC. As a comparison, the CUSUM charts discussed in this chapter and some other alternative control charts discussed in later chapters are mainly for phase II SPC, because relatively small and persistent shifts are often our major concern in phase II SPC. For this reason, our description about the CUSUM charts and the other alternative charts is mainly in the context of phase II SPC, although on several occasions we will also briefly discuss how to use some of them for phase I SPC. Also, when constructing Shewhart charts, we often use batch data (see the related discussion in Subsection 3.2.3). When constructing the CUSUM and other alternative control charts, however, individual observation data are often used. Although these control charts can generally be applied to batch data as well, it is usually more efficient to use them with the individual observation data, in the sense that fewer observations are required in the latter case to detect a given shift with given levels of IC and OC ARLs.

4.2 Monitoring the Mean of a Normal Process

In this section, we discuss cases when the quality characteristic of interest X is univariate, its IC distribution is $N(\mu_0, \sigma^2)$, and the potential shift in the production process is in the mean of X only. This might be the most popular SPC problem in the literature because the mean of X is often the most relevant quantity to the quality of products and people in the SPC community routinely assume that X has normal distributions before and after a mean shift. Our discussion in this section is divided into four parts. In Subsection 4.2.1, different forms of the CUSUM chart are described. Then, its design and implementation are discussed in Subsection 4.2.2. In these two subsections, observations at different time points are assumed to be independent. The case when observations at different time points are correlated is discussed in Subsection 4.2.3. Finally, the optimality properties of the CUSUM chart and a general method for constructing a CUSUM chart in different cases are discussed in Subsection 4.2.4.

4.2.1 The V-mask and decision interval forms of the CUSUM chart

Assume that $\{X_1, X_2, \ldots\}$ are individual observation data obtained online at consecutive time points from a production process for phase II SPC, the observations are independent and identically distributed (i.i.d.) with a common IC distribution $N(\mu_0, \sigma^2)$ before a process mean shift, and they are i.i.d. with a common OC distribution $N(\mu_1, \sigma^2)$ after the mean shift, where $\mu_1 \neq \mu_0$ and $\delta = \mu_1 - \mu_0$ is the shift size. It has been demonstrated in Example 4.1 that it is inefficient to detect the mean shift at a given time point using the observed data at that time point alone. A natural idea to overcome this limitation is to use the observation at the current time point and all history data as well for shift detection, as described in the first paragraph of

Section 4.1. To this end, it is natural to consider the charting statistic

$$C_n = \sum_{i=1}^{n}(X_i - \mu_0),\tag{4.1}$$

where $n \geq 1$ denotes the current time point. It is obvious that

$$C_n = C_{n-1} + (X_n - \mu_0),$$

where $C_0 = 0$. Therefore, C_n is a cumulative sum of the deviations $\{X_i - \mu_0, i = 1, 2, \ldots, n\}$. When the process is IC up to the time point n, it is easy to check that

$$C_n \sim N\left(0, n\sigma^2\right).\tag{4.2}$$

On the other hand, when the process has a mean shift from μ_0 to μ_1 at the time point $1 \leq \tau \leq n$ (i.e., the process mean becomes μ_1 starting from the time point τ), we have, for $n \geq \tau$,

$$C_n \sim N\left((n-\tau+1)\delta, n\sigma^2\right).\tag{4.3}$$

From (4.1)–(4.3), it can be seen that C_n is a summation of i.i.d. zero-mean random variables in cases when there is no mean shift up to the time point n (i.e., C_n is the so-called *random walk* in such cases), and its mean starts to change linearly with the slope δ after a process mean shift of size δ at τ. Therefore, C_n contains useful information for shift detection, and it is an indication of a positive (or negative) shift if its mean increases (or decreases) after a specific time point.

Example 4.2 *For the data considered in Example 4.1, let $\overline{X}_n$ denote the sample mean of the n-th sample. Then, the IC distribution of $\overline{X}_n$ is $N(0, 1/5)$, its OC distribution is $N(0.2, 1/5)$, and the mean shift occurs at $n = 11$. Figure 4.2 shows the sequence*

$$C_n = C_{n-1} + \left(\overline{X}_n - 0\right), \qquad for\ n = 1, 2, \ldots, 20.$$

From the plot, it can be seen that C_n does increase linearly with a slope around 0.2 after the shift time $\tau = 11$.

   In practice, we usually do not know whether a mean shift has occurred or not by the current time point. A major goal of SPC is to detect the first potential mean shift from all observed data. From (4.3) and Figure 4.2, it can be seen that, if the statistic C_n changes linearly with a slope δ after a time point, then it is an indication of a mean shift of size δ. However, the expression (4.3) shows that the variance of C_n increases with n as well, making the identification of a linear trend in the mean of C_n challenging. To overcome this difficulty, the *V-mask form* of the cumulative sum (CUSUM) control chart is constructed as follows. Intuitively, if a positive mean shift of size δ occurs at τ, then C_n, when $n \geq \tau$, should be all above a half-line starting at $(\tau, C_\tau - h)$ with a slope $k = \delta/2$, where $h > 0$ is a constant, as demonstrated in Figure 4.3(a). The inclusion of the constant h and the slope k (instead of the slope $\delta = 2k$) is for accommodating the increasing variability of C_n over n. In practice, τ is unknown; so, the above half-line cannot actually be used for shift detection.

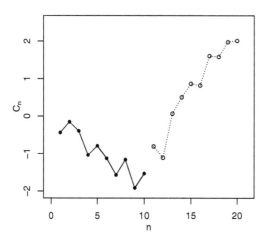

Figure 4.2 *Values of the statistic C_n computed from the data considered in Example 4.1. In the plot, the first 10 values of C_n are denoted by solid dots and they are connected by solid lines, and the second 10 values of C_n are denoted by little circles and they are connected by dotted lines. A process mean shift of size 0.2 occurs at the time $\tau = 11$.*

But, we can consider a half-line starting at $(n, C_n - h)$ and going backwards with a slope $k = \delta/2$, as demonstrated in Figure 4.3(b) by the bottom solid line of the shaded region around $n = 20$. If a mean shift of size δ occurs before n, then there should be some values of the charting statistic falling below that half-line. Similarly, for detecting a negative mean shift of size $-\delta$, we can consider another half-line starting at $(n, C_n + h)$ and going backwards with a slope $-k$, as demonstrated by the upper dashed line of the shaded region in Figure 4.3(b). Therefore, to detect a mean shift at the current time point n, we can use a truncated V-mask, as demonstrated by the shaded region in Figure 4.3(b). If there are some values of the charting statistic located outside that mask, then we can conclude that a mean shift has occurred before n. In the case of Figure 4.3(b), some values of the charting statistic are located below the V-mask, indicating an upward mean shift.

From the V-mask form of the CUSUM chart, after obtaining a signal of mean shift, we can also obtain point estimates of the shift time τ and the shift size δ as follows. If there is only one point of the CUSUM charting statistic located below the V-mask, then the corresponding time point of that point can be used for estimating τ. If there is more than one point below the V-mask, then we can choose the time point of the point that is farthest away from the mask as the estimate of τ. After the estimate of τ, denoted as $\widehat{\tau}$, is obtained, δ can be estimated by

$$\widehat{\delta} = \frac{C_n - C_{\widehat{\tau}}}{n - \widehat{\tau}}. \tag{4.4}$$

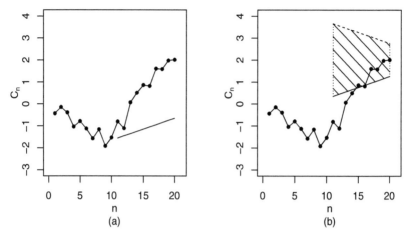

Figure 4.3 *(a) Values of the charting statistic C_n computed from the data considered in Example 4.1 are denoted by solid dots that are connected by solid lines, where an upward process mean shift of size $\delta = 0.2$ occurs at $\tau = 11$. When $n \geq \tau$, C_n are all above the solid half-line that starts at $(\tau, C_\tau - h)$ with a slope of $k = \delta/2$. (b) Values of the charting statistic C_n are the same as those in plot (a). The shaded region denotes a truncated V-mask that can be used for mean shift detection at $n = 20$. A mean shift occurred before n is detected if some values of $\{C_i, i < n\}$ are located outside the V-mask.*

Of course, a better estimate of δ would be the slope of the least squares line obtained from all observed values of the charting statistic in the time range $[\hat{\tau}, n]$. But, the one given by (4.4) is easier to compute; thus, it is commonly used in practice.

Page (1954) might be the first paper that formally discussed the CUSUM chart and its ability to detect small and persistent shifts. To detect an upward (positive) mean shift, Page's CUSUM charting statistic is defined by

$$P_n^+ = P_{n-1}^+ + (X_n - \mu_0) - k, \qquad \text{for } n \geq 1, \tag{4.5}$$

where $P_0^+ = 0$, and $k > 0$ is a parameter called the *reference value* or *allowance*. This chart indicates an upward mean shift if

$$P_n^+ - \min_{0 \leq m < n} P_m^+ > h, \tag{4.6}$$

where $h > 0$ is a parameter called the *control limit* or *decision interval*. It is not difficult to check that the CUSUM chart defined by (4.5) and (4.6) is equivalent to the CUSUM chart with the charting statistic

$$C_n^+ = \max\left(0, C_{n-1}^+ + (X_n - \mu_0) - k\right), \tag{4.7}$$

where $C_0^+ = 0$. The decision rule of this chart is that it gives a signal of an upward mean shift when

$$C_n^+ > h. \tag{4.8}$$

The form of the CUSUM chart defined by (4.7) and (4.8) is called the *decision interval (DI) form*. This form reveals the re-starting mechanism of the CUSUM chart that its charting statistic C_n^+ is reset to 0 every time when $C_{n-1}^+ + (X_n - \mu_0) < k$. In such cases, there is little evidence in the observed data that is in favor of an upward mean shift. Van Dobben de Bruyn (1968) showed that the DI form of the CUSUM chart is equivalent to the V-mask form of the CUSUM chart described earlier (cf., Figure 4.3(b)) in cases when the two forms use the same values of h and k.

It is obvious that detection of a downward shift of size $-\delta$ in the original quality characteristic X is equivalent to detection of an upward shift of size δ in $-X$. By this relationship, to detect a downward (or negative) mean shift in X, the DI form of the CUSUM chart would have the charting statistic

$$C_n^- = \min\left(0, C_{n-1}^- + (X_n - \mu_0) + k\right), \qquad (4.9)$$

where $C_0^- = 0$. This chart gives a signal of a downward mean shift if

$$C_n^- < -h. \qquad (4.10)$$

Both the CUSUM chart (4.7)–(4.8) and the CUSUM chart (4.9)–(4.10) are one-sided: the chart (4.7)–(4.8) is for detecting an upward mean shift and the chart (4.9)–(4.10) is for detecting a downward mean shift. To detect an arbitrary mean shift, we can consider a two-sided version of the CUSUM chart, which is a combination of the two one-sided CUSUM charts. By the two-sided version, a signal of mean shift is given once (4.8) or (4.10) or both hold.

The CUSUM charts (4.7)–(4.8) and (4.9)–(4.10) are designed for analyzing individual-observation data. If batch data are involved and we have m observations at each time point, then the CUSUM charts can be constructed in the same way, except that the sample mean $\overline{X}(n)$ should be used to replace the individual observation X_n at each observation time.

In the literature, the DI form of the CUSUM chart is often preferred because it is easier to use, compared to the V-mask form. The control limits h and $-h$ for C_n^+ and C_n^- are constants over time, similar to the control limits of the Shewhart charts. However, the V-mask form has its own advantages. For instance, if mean shifts of different sizes need to be detected, then we can simply use V-masks of different slopes in the same CUSUM chart for that purpose. With the DI form, to detect different mean shifts, the value of k should be chosen differently (see related discussion in the next subsection). However, if the value of k changes, the entire CUSUM chart needs to be re-constructed, making the DI form inconvenient to use in such cases. As in most other books and research papers, we mainly use the DI form in our discussion about different CUSUM charts in the remaining part of this book.

Example 4.2 (continued) *For the data considered in Example 4.1, values of the charting statistic C_n^+ computed by (4.7) are shown in Figure 4.4 by the black dots. In the chart, k is set at 0.25 and h is set at 5.597, so that the ARL_0 value of the chart is 200. See Table 4.1 and the related discussion in the next subsection for the relationship among k, h, and ARL_0. By the way, the original data are batch data with batch size of 5. Therefore, when computing C_n^+, the standardized sample means $\overline{X}(n)/(1/\sqrt{5})$ have been used in (4.7).*

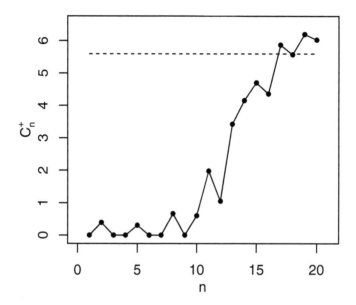

Figure 4.4 *Values of the charting statistic C_n^+, computed by (4.7) from the data considered in Example 4.1, are shown by the black dots. In the chart, k and h are chosen to be 0.25 and 5.597, respectively, so that the ARL_0 value of the chart is 200.*

4.2.2 Design and implementation of the CUSUM chart

In the CUSUM chart (4.7)–(4.8), there are two parameters h and k to choose before the chart can actually be used for process monitoring. In this subsection, we discuss the selection of h and k for the chart (4.7)–(4.8) in cases when $\mu_0 = 0$ and $\sigma = 1$. This is sufficient because in cases with arbitrary values of μ_0 and σ we can consider the corresponding CUSUM chart for the standardized observations

$$\frac{X_n - \mu_0}{\sigma}, \qquad \text{for } n \geq 1.$$

Parameter selection of the chart (4.9)–(4.10) and of the two-sided version of the CUSUM chart can be discussed in a similar way.

Obviously, the values of h and k would affect the performance of the chart (4.7)–(4.8). As discussed in Subsection 3.2.1, performance of a control chart can be measured by the IC ARL value ARL_0 and the OC ARL value ARL_1. Usually, the value of ARL_0 is specified beforehand, to be the minimum tolerable ARL_0 value. Then, for detecting a given shift δ, the chart performs better if its ARL_1 value is smaller.

For given values of h and k, there are some theoretical methods for studying the properties of the CUSUM chart (see the related discussion in Section 4.6), based on which some formulas have been derived in the literature for approximating the ARL_0

value. For instance, Siegmund (1985) developed the following formula:

$$ARL_0 \approx \frac{\exp\left(2k(h+1.166)\right) - 2k(h+1.166) - 1}{2k^2}. \tag{4.11}$$

It has been shown that this formula provides quite an accurate approximation to the true value of ARL_0 in cases when k is small to moderate (e.g., $k \leq 1$). Its approximation could be poor in cases when k is large. Related discussion can be found in papers such as Brook and Evans (1972), Fellner (1990), Goel and Wu (1971), Hawkins (1992a), and Woodall and Adams (1993).

To compute ARL_0 values of the CUSUM chart (4.7)–(4.8), we can also use Monte Carlo simulations. Because various powerful computing facilities are available nowadays, this might be a more appropriate approach to use, compared to the approach using approximation formulas (e.g., (4.11)). Results from the simulation approach are usually more accurate, and this approach is also flexible enough to handle a variety of different scenarios. For given values of h and k, the ARL_0 value of the chart (4.7)–(4.8) can be computed by an algorithm described by the pseudo code in the box below.

Pseudo Code to Compute an Estimate of the ARL_0 Value

Let M be the number of replicated simulations. It can be chosen as a large positive integer (e.g., $M = 10^5$).

Step 1 In the j-th replicated simulation, for $1 \leq j \leq M$, compute the run length $RL(j)$ by the following loop: for $n \geq 1$

- generate the observation X_n from $N(0,1)$

- compute the value of C_n^+ by (4.7)

- if (4.8) holds, then let $RL(j) = n$ and stop the loop; otherwise, let $n = n+1$ and continue the loop.

Step 2 The ARL_0 value can be estimated by the simple average of the M run length values $\{RL(j), j = 1, 2, \ldots, M\}$.

In the above pseudo code, observations at different time points within a simulation should be independent of each other. They are independent of each other among different simulations as well. Although it is rare, theoretically it can happen that a small number of simulations might require extremely long sequences of observations, which could cause some practical inconvenience. To prevent the occurrence of this situation, we can set the maximum length of the sequences to be a given positive integer N. If the j-th simulation does not stop by the time point $n = N$, then we can simply stop the corresponding loop and ignore that simulation when computing the final estimate of the ARL_0 value. To minimize the impact of this modification, a large N should be chosen (e.g., $N = 10^4$). In the literature, a number of numerical algorithms based on Monte Carlo simulations for estimating the value of ARL_0 have been developed (e.g., Hawkins and Olwell (1998)). Some existing R packages for that purpose are described in Appendix A.2.

In cases when $k = 0, 0.25, 0.5, 0.75$ and $h = 1, 1.5, 2, 2.5, 3, 3.5, 4$, the computed ARL_0 values of the CUSUM chart (4.7)–(4.8) by the R package spc (see its introduction in the appendix) are shown in Figure 4.5. From the plot, it can be seen that (i) ARL_0 increases with the value of k, and (ii) ARL_0 increases with the value of h as well. These results can be easily explained using the V-mask form of the CUSUM chart. From Figure 4.3(b), when k is larger or h is larger, the chance that the charting statistic C_n^+ is located below the bottom solid line of the truncated V-mask would be smaller. Consequently, the ARL_0 value would be larger in such cases.

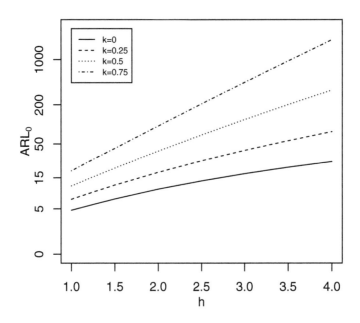

Figure 4.5 *ARL_0 values of the CUSUM chart (4.7)–(4.8) in cases when $k = 0, 0.25, 0.5, 0.75$, and $h = 1, 1.5, 2, 2.5, 3, 3.5, 4$. To better demonstrate the difference among different cases, the y-axis is in natural log scale.*

From Figure 4.5, it can be seen that the value of ARL_0 indeed depends on the values of h and k. As mentioned earlier, in practice, the value of ARL_0 is often pre-specified. Therefore, to use the control chart (4.7)–(4.8), its parameters h and k should be chosen to reach the pre-specified ARL_0 value. To this end, Page (1954) pointed out that to detect a mean shift of size δ, the chart (4.7)–(4.8) would have an optimal performance if k was chosen as $\delta/2$. So, if we have a target shift size δ in mind in a specific application, then we should choose $k = \delta/2$. In certain applications, it is possible to figure out a meaningful target shift for this purpose. As an example, assume that a machine consists of 10 components. Its productivity would lose 10% if one component goes out of order. When all components work stably, assume that the productivity of the machine can be figured out to be a, based on our past experience or engineering knowledge. Then, a meaningful target shift in this case would be $0.1 * a$ when one component of the machine goes out of order. However, in most applications, the shift size can be any value in an interval, and a specific target shift cannot be obtained using our engineering or scientific knowledge about the production process. In such cases, we can specify a target shift that is desired to be detected as soon as possible, which is usually the one that is small enough that it cannot be noticed easily by our naked eyes and large enough to have a meaningful impact on the quality of products.

After the values of ARL_0 and k are given, the value of h can be determined by a searching algorithm such that the given ARL_0 value is reached to a certain accuracy. A pseudo code of such a searching algorithm based on the bisection searching is described in the box below.

Pseudo Code to Search for the Value of h

Let A_0 be the pre-specified ARL_0 value, $[h_L, h_U]$ be the interval from which h is searched, $\rho > 0$ be a small number denoting the required estimation accuracy, and M be the maximum number of iterations involved in the search. Then, the search is performed iteratively as follows. In the j-th iteration, for $1 \leq j \leq M$, do the following two steps:

Step 1 Let $h = (h_L + h_U)/2$. Compute the ARL_0 value of the chart (4.7)–(4.8) with this h value and the given k value.
Step 2 If the ARL_0 value computed in Step 1 is included in $[A_0 - \rho, A_0 + \rho]$, then the algorithm is stopped, and the searched value of h is set to be the one used in Step 1. Otherwise, define

$$\begin{cases} h_L = h_L, h_U = (h_L + h_U)/2 & \text{if } ARL_0 > A_0 \\ h_L = (h_L + h_U)/2, h_U = h_U & \text{if } ARL_0 < A_0, \end{cases}$$

and return to Step 1.
If the above algorithm does not stop before or at the M-th iteration and the ARL_0 value computed in the M-th iteration is still outside the interval $[A_0 - \rho, A_0 + \rho]$, print a statement that the estimation accuracy specified by ρ cannot be reached, and set the h value defined in the M-th iteration as the searched value of h.

In the above pseudo code, the initial value of h_L should be chosen small enough such that the ARL_0 value when $h = h_L$ is smaller than A_0, and the initial value of h_U should be chosen large enough such that the ARL_0 value when $h = h_U$ is larger than A_0. For most ARL_0 and k values, it should be good enough to choose $h_L = 0$ and $h_U = 15$. In the pseudo code, we have used the bisection searching method. Some modifications of this method for a faster computation are possible. For instance, instead of defining $h = (h_L + h_U)/2$ in Step 1, we can use the linear interpolation method to define

$$h = h_L + \frac{A_0 - ARL_0(h_L)}{ARL_0(h_U) - ARL_0(h_L)}(h_U - h_L),$$

Table 4.1 *Computed h values of the CUSUM chart (4.7)–(4.8) for some commonly used ARL_0 and k values.*

	k						
ARL_0	0.1	0.25	0.5	0.75	1.0	1.25	1.5
50	4.567	3.340	2.225	1.601	1.181	0.854	0.570
100	6.361	4.418	2.849	2.037	1.532	1.164	0.860
200	8.520	5.597	3.502	2.481	1.874	1.458	1.131
300	9.943	6.324	3.892	2.745	2.073	1.624	1.282
370	10.722	6.708	4.095	2.882	2.175	1.709	1.359
400	11.019	6.852	4.171	2.933	2.214	1.741	1.387
500	11.890	7.267	4.389	3.080	2.323	1.830	1.466
1000	14.764	8.585	5.071	3.538	2.665	2.105	1.708

where $ARL_0(h_L)$ and $ARL_0(h_U)$ denote the ARL_0 values of the chart (4.7)–(4.8) with $h = h_L$ and $h = h_U$, respectively. In the pseudo code, ρ controls the estimation error of the algorithm. A small positive number, such as 0.5, 0.1, and 0.01, should be good enough to use in practice. Because the bisection search is fast to converge, the maximum number of iterations, M, can be chosen as a relatively small integer number like 20 or 50.

By the R-package spc, the computed h values for some commonly used ARL_0 and k values are presented in Table 4.1. From the table, we can see that, as a function of ARL_0 and k, h decreases with k and increases with ARL_0.

To design the control chart (4.7)–(4.8), only the ARL_0, h, and k values are involved. To evaluate the performance of the chart when detecting a specific shift, we need to compute its ARL_1 value, as described earlier. Assume that the process distribution changes from $N(\mu_0, \sigma^2)$ to

$$N\left(\mu_0 + \delta, \lambda^2\sigma^2\right),$$

starting at the initial observation time point, where $\delta, \lambda > 0$ are two given numbers. Then, the mean of the process distribution has a shift of size δ, and its standard deviation has a shift of size $(\lambda - 1)\sigma$. In such cases, equation (4.7) can be written as

$$\frac{C_n^+}{\lambda\sigma} = \max\left[0, \frac{C_{n-1}^+}{\lambda\sigma} + \left(\frac{X_n}{\lambda\sigma} - \frac{\mu_0 + \delta}{\lambda\sigma}\right) - \frac{k - \delta}{\lambda\sigma}\right].$$

If we denote $C_n^* = C_n^+/(\lambda\sigma)$, then it is easy to see that C_n^* is just the conventional charting statistic of the upward CUSUM chart when its allowance is $k^* = (k - \delta)/(\lambda\sigma)$, its control limit is $h^* = h/(\lambda\sigma)$, and the process IC distribution is $N(0, 1)$. Based on this relationship, we have the conclusion stated in the box below.

Relationship Between ARL_0 and ARL_1 Values

Assume that the process distribution shifts at the initial time point from $N(\mu_0, \sigma^2)$ to $N(\mu_0 + \delta, \lambda^2 \sigma^2)$. Then, the ARL_1 value of the chart (4.7)–(4.8) with parameters h and k equals the ARL_0 value of the chart when the allowance is chosen to be $k^* = (k - \delta)/(\lambda \sigma)$, the control limit is chosen to be $h^* = h/(\lambda \sigma)$, and the process IC distribution is $N(0, 1)$.

Example 4.3 *Assume that the IC process distribution is $N(0, 1)$. We would like to compute the ARL_1 values of the chart (4.7)–(4.8) with $k = 0.5$ and $h = 3.502$, whose ARL_0 value is 200 by Table 4.1, in cases when the process distribution shifts at the initial observation time point to one of the following four distributions:*

(i) $N(0.25, 1)$,

(ii) $N(-0.25, 1)$,

(iii) $N(0, 2^2)$, *and*

(iv) $N(0.25, 2^2)$.

In case (i), by the results given in the above box, the ARL_1 value of the chart (4.7)–(4.8) equals the ARL_0 value of the same chart when the allowance is $k^* = k - \delta = 0.5 - 0.25 = 0.25$, the control limit is $h^* = h = 3.502$, and the IC distribution is $N(0, 1)$. By the approximation formula (4.11), this ARL_1 value can be approximated by

$$\frac{\exp(2k^*(h^* + 1.166)) - 2k^*(h^* + 1.166) - 1}{2(k^*)^2} = 55.881.$$

*In case (ii), the ARL_1 value of the chart (4.7)–(4.8) equals the ARL_0 value of the same chart when the allowance is $k^* = 0.5 + 0.25 = 0.75$, the control limit is $h^* = h = 3.502$, and the IC distribution is $N(0, 1)$. Again, by (4.11), this ARL_1 value can be approximated by 969.624. Obviously, this ARL_1 value is much larger than the assumed ARL_0 value (i.e., 200). This situation often happens when a mean shift is downward but the control chart used is designed for detecting upward shifts (e.g., the chart (4.7)–(4.8)), or the mean shift is upward but the control chart is designed for detecting downward shifts (e.g., the chart (4.9)–(4.10)). This is the so-called **biasness** phenomenon of the one-sided control charts.*

In case (iii), the ARL_1 value of the chart (4.7)–(4.8) equals the ARL_0 value of the same chart when the allowance is $k^* = k/(\lambda \sigma) = 0.5/2 = 0.25$, the control limit is $h^* = h/(\lambda \sigma) = 3.502/2 = 1.751$, and the IC distribution is $N(0, 1)$. By (4.11), this value can be approximated by 14.728. Similarly, in case (iv), the ARL_1 value of the chart (4.7)–(4.8) equals the ARL_0 value of the same chart when the allowance is $k^* = (k - \delta)/(\lambda \sigma) = (0.5 - 0.25)/2 = 0.125$, the control limit is $h^* = h/(\lambda \sigma) = 3.502/2 = 1.751$, and the IC distribution is $N(0, 1)$. This ARL_1 value can be approximated by 11.017.

In the above discussion of the ARL_1 value, we assume that the potential shift occurs at the initial observation time point. In cases when the shift occurs at a later time point, then the computed ARL_1 value could be different from the one when the shift occurs at the initial observation time point, because the distribution of the charting statistic C_n^+ depends on n. This is also true for the ARL_0 value. In the literature, this phenomenon has been well discussed (e.g., Hawkins and Olwell, 1998, Subsection 3.3.4). From the definition of C_n^+, $C_0^+ = 0$, C_1^+ has the probability of $P(X_1 \leq k)$ to be 0 and the probability of $P(X_1 > k)$ to be in the interval $(0, \infty)$, C_2^+ has the probability of

$$P(X_1 \leq k)P(X_2 \leq k) + P(X_1 > k)P(X_1 + X_2 \leq 2k | X_1 > k)$$

to be 0 and the remaining probability to be in the interval $(0, \infty)$, and so forth. In cases when the IC process distribution is $N(0, 1)$ and $k = 0$, C_1^+ has the probability of 0.5 to be 0, C_2^+ has the probability of $3/8$ to be 0, and so forth, when the process is IC. So, the distributions of C_n^+ when n is small are all different. However, when n increases, the distribution of C_n^+ tends to a *steady-state* distribution, although we cannot derive its closed-form expression yet. For this reason, the ARL_0 value computed from the run lengths recorded from the initial observation time point and the one computed from the run lengths recorded from a given time point n^*, with n^* being a large positive integer, could be different. To distinguish them, the former is often called the *zero-state ARL_0* value, and the latter the *steady-state ARL_0* value. Similarly, the ARL_1 value is called the zero-state ARL_1 if the shift occurs at the initial observation time point, and it is called the steady-state ARL_1 if the shift occurs at the time n^* with n^* being a relatively large positive integer. In practice, n^* can be chosen as 100, or as small as 50, to approximate the steady-state distribution by the distribution of $C_{n^*}^+$. Without further specification, the ARL_0 and ARL_1 values discussed in this and other chapters, including the ARL_0 values presented in Figure 4.5 and Table 4.1, are computed under the zero-state condition.

In Table 4.2, the left part presents the zero-state (ZS) ARL_0 values, the steady-state (SS) ARL_0 values, and their relative differences (i.e., $(ZS - SS)/ZS$) of the chart (4.7)–(4.8) in cases when $k = 0.5$ and $h = 2, 2.5, 3, 3.5, 4, 4.5, 5$, which are computed using the functions xcusum.ad() and xcusum.arl() in the R-package spc. The right part of the table presents the ZS and SS ARL_1 values and their relative differences of the same chart for detecting a mean shift of size $\delta = 1$. From the table, it can be seen that (i) the ZS and the corresponding SS ARL_0 (or ARL_1) values are indeed different, and (ii) the relative difference seems bigger for the ARL_1 values. The intuitive explanation of the second result is that ARL_0 values are usually large. Consequently, they are not affected much by different ways of computation (i.e., ZS versus SS). As a comparison, ARL_1 values are usually small. So, they are affected more seriously by different ways of computation. The fact that the relative differences for both the ARL_0 and ARL_1 values decrease when h increases (except the case for the ARL_1 values when $h = 2.0$) can be explained in a similar way.

Example 4.4 *For the control chart (4.7)–(4.8), let us consider cases when $k = 0.25, 0.5$, or 0.75, and the assumed ARL_0 value is 200. By Table 4.1, the h values in these cases are 5.597, 3.502, and 2.481, respectively. When the chart is used for detecting a mean shift of size δ occurring at the initial observation time, where δ changes its value from 0 to 2 with a step of 0.1, its ARL_1 values computed by the function xcusum.arl() in the R-package spc are shown in Figure 4.6. From the figure, it can be seen that the chart with $k = 0.25$ performs the best when δ is small (e.g., $\delta \leq 0.6$), it performs the best in a region around $\delta = 1$ when k is 0.5, and its*

Table 4.2 *The left part presents the zero-state (ZS) ARL_0 values, the steady-state (SS) ARL_0 values, and their relative differences $(ZS - SS)/ZS$ of the chart (4.7)–(4.8) when $k = 0.5$ and $h = 2, 2.5, 3, 3.5, 4, 4.5, 5$. The right part presents the ZS ARL_1 values, the SS ARL_1 values, and their relative differences of the same chart for detecting a mean shift of size $\delta = 1$.*

| | ARL_0 | | | ARL_1 | | |
h	ZS	SS	$(ZS-SS)/ZS$	ZS	SS	$(ZS-SS)/ZS$
2.0	38.548	37.262	0.033	4.449	4.079	0.083
2.5	68.186	66.253	0.028	5.423	4.951	0.087
3.0	117.596	114.953	0.022	6.404	5.853	0.086
3.5	199.574	196.167	0.017	7.391	6.778	0.083
4.0	335.368	331.143	0.013	8.383	7.722	0.079
4.5	559.947	554.863	0.009	9.379	8.680	0.074
5.0	930.887	924.908	0.006	10.376	9.650	0.070

performance for detecting large shifts (e.g., $\delta \geq 1.4$) is the best when k is 0.75. This example confirms the general conclusions stated in the box below.

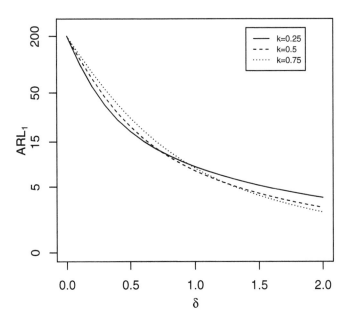

Figure 4.6 *ARL_1 values of the CUSUM chart (4.7)–(4.8) in cases when $k = 0.25, 0.5, 0.75$, the assumed ARL_0 value is 200, and the shift size δ changes its value from 0 to 2 with a step of 0.1. To better demonstrate the difference among different cases, the y-axis is in natural log scale.*

General Guidelines for Selecting k

For the CUSUM chart (4.7)–(4.8), small k values are ideal for detecting relatively small upward mean shifts, and large k values are ideal for detecting relatively large upward mean shifts. The chart is optimal for detecting a target shift of size $\delta = 2k$, in the sense that its ARL_1 value is the shortest among the ARL_1 values of all control charts with the same ARL_0 value.

So far, our discussion in this subsection focuses on the CUSUM chart (4.7)–(4.8), which is designed for detecting upward mean shifts. By the relationship between the upward CUSUM chart (4.7)–(4.8) and the downward CUSUM chart (4.9)–(4.10) (as described in the paragraph containing the equation (4.9) in Subsection 4.2.1), design, implementation, and computation of the ARL_0 and ARL_1 values of the downward CUSUM chart (4.9)–(4.10) can be done in a similar way to that with the upward CUSUM chart (4.7)–(4.8). Regarding the two-sided version of the CUSUM chart described in Subsection 4.2.1, it is a combination of the upward CUSUM chart and the downward CUSUM chart. For given h and k values, assume that the ARL values of the two one-sided charts are ARL^+ and ARL^-, respectively. Then, Van Dobben de Bruyn (1968) showed that the ARL value of the two-sided version, denoted as ARL^*, would follow the equation

$$\frac{1}{ARL^*} = \frac{1}{ARL^+} + \frac{1}{ARL^-}. \tag{4.12}$$

The equation (4.12) is true for both IC and OC ARL values. So, it can be used for the design of the two-sided CUSUM chart and for computing its ARL_0 and ARL_1 values. For instance, if we would like to use $k = 0.5$ in the two-sided CUSUM chart and its ARL_0 is specified to be 200, then by (4.12), each of the two one-sided CUSUM charts should have ARL_0 value of 400. By Table 4.1, the upward CUSUM chart should use the control limit $h = 4.171$ and the downward CUSUM chart should use the control limit $-h = -4.171$. As another example, in Example 4.3, the upward CUSUM chart is used for detecting both the upward shift of size 0.25 and the downward shift of size -0.25. The ARL_1 values in these two cases are computed to be 55.881 and 969.624, respectively. By (4.12) and the relationship between the two one-sided CUSUM charts, the ARL_1 value of the two-sided CUSUM chart for detecting a shift of size 0.25 or -0.25 would be

$$\left(\frac{1}{55.881} + \frac{1}{969.624} \right)^{-1} = 52.836.$$

This result shows that the ARL_1 value of the two-sided CUSUM chart is smaller than the ARL_1 values of the two one-sided CUSUM charts. As a matter of fact, this is always true because the equation (4.12) implies the following results:

$$\frac{1}{ARL^*} > \frac{1}{ARL^+},$$

$$\frac{1}{ARL^*} > \frac{1}{ARL^-},$$

$$ARL^* < \min\left(ARL^+, ARL^-\right),$$

and

$$\frac{\min\left(ARL^+, ARL^-\right)}{2} \leq ARL^* \leq \frac{\max\left(ARL^+, ARL^-\right)}{2}.$$

At the end of this subsection, we would like to mention that the Shewhart charts discussed in the previous chapter can be considered as special CUSUM charts. To demonstrate the connection between the two types of control charts, let us consider the case when the $\overline{X}$ chart is used for detecting process mean shifts, we have m observations at each time point, and the IC process mean μ_0 and the IC process standard deviation σ are both known. In such cases, by the $\overline{X}$ chart, a signal of an upward mean shift is delivered at the n-th time point, for any $n \geq 1$, if

$$\frac{\overline{X}_n - \mu_0}{\sigma/\sqrt{m}} > Z_{1-\alpha/2}, \tag{4.13}$$

where the critical value $Z_{1-\alpha/2}$ is often chosen to be 3, corresponding to the significance level of $\alpha = 0.0027$. Obviously, the decision rule (4.13) is equivalent to the rule that a signal is given at the n-th time point if

$$S_n^+ > 0, \tag{4.14}$$

where

$$S_n^+ = \max\left(0, S_{n-1}^+ + \frac{\overline{X}_n - \mu_0}{\sigma/\sqrt{m}} - Z_{1-\alpha/2}\right)$$

and $S_0^+ = 0$. The decision rule (4.14) is exactly an upward CUSUM chart applied to the standardized data $\{(\overline{X}_n - \mu_0)/(\sigma/\sqrt{m}), n \geq 1\}$ with $k = Z_{1-\alpha/2}$ and $h = 0$. From the discussion in Subsection 3.2.1, the $\overline{X}$ chart (4.13) has an ARL_0 value of $2/\alpha$. By the equivalence of (4.13) and (4.14) and by the optimality properties of the CUSUM chart (4.14), we know that the Shewhart chart (4.13) also has the optimality properties for detecting a mean shift of size $2Z_{1-\alpha/2}$ among all control charts with $ARL_0 = 2/\alpha$. In cases when $Z_{1-\alpha/2} = 3$, the corresponding $\overline{X}$ chart would be optimal for detecting a mean shift of size 6 among all control charts with $ARL_0 = 2/0.0027 = 740.741$. So, it is true that Shewhart charts are good for detecting large shifts.

4.2.3 Cases with correlated observations

Design and implementation of the CUSUM chart (4.7)–(4.8) discussed in the previous subsection is based on the following assumptions:

(i) Process observations are independent and identically distributed (i.i.d.) before and after a potential mean shift,

(ii) Both the IC and OC process distributions are normal, and

(iii) Parameters of the IC distribution are known.

In practice, all these assumptions can be invalid. In this subsection, we discuss the cases when process observations collected at different time points are correlated (i.e., the assumption (i) is invalid). Cases when the process distributions are non-normal (i.e., the assumption (ii) is invalid) will be discussed in Section 4.4 and Chapter 8. Cases when the parameters of the IC distribution are unknown (i.e., the assumption (iii) is invalid) will be discussed in Section 4.5.

As discussed in Section 3.5, possible correlation among observations collected at different time points would affect the actual IC and OC ARL values of the Shewhart charts. Similarly, such possible correlation could affect the IC and OC ARL values of the CUSUM chart (4.7)–(4.8), as demonstrated by the example below.

Example 4.5 *For the upward CUSUM chart (4.7)–(4.8), let us consider cases when $(k,h) = (0.25, 5.597), (0.5, 3.502),$ and $(0.75, 2.481)$. From Table 4.1, it can be seen that the ARL_0 values of the chart in these three different cases are all 200 when process observations are independent of each other. Now, let us consider the scenario when process observations are actually generated from the following auto-regressive model of order 1, denoted as AR(1): for $n \geq 1$,*

$$X_n = \mu_0 + \phi (X_{n-1} - \mu_0) + e_n, \tag{4.15}$$

where $X_0 = \mu_0$, $\mu_0 = 0$ is the IC mean, ϕ is the model coefficient, and $\{e_n, n \geq 1\}$ are the white noise with $\sigma_e = 1$ (i.e., they are i.i.d. with a common distribution $N(0, \sigma_e^2)$). In such cases, we know that the correlation coefficient between X_n and X_{n-j} is ϕ^j, for $j = 1, 2, \ldots, n-1$. When $\phi = -0.5, -0.25, -0.1, 0, 0.1, 0.25,$ or 0.5, the actual ARL_0 values of the chart (4.7)–(4.8) are presented in the left part of Table 4.3. These values are computed by an algorithm written by the author in R, which is the same as the one described by the pseudo code in the previous subsection, except that process observations are generated from the model (4.15) in this example. It can be seen that they are quite different from each other in cases with a same set of (k, h) but different values of ϕ. For instance, when $(k, h) = (0.25, 5.597)$ and $\phi = -0.5$, the actual ARL_0 value is about 4 times the nominal ARL_0 value of 200, while the actual ARL_0 value is only about 20% of the nominal ARL_0 in the case when (k, h) remain the same but $\phi = 0.5$. Now, let us consider the scenario when a mean shift of size $\delta = 0.5$ occurs at the initial observation time. In this scenario, the actual ARL_1 values of the chart (4.7)–(4.8) in various cases are presented in the right half of Table 4.3, which are calculated by a numerical algorithm that is similar to the one for calculating the ARL_0 values. From the table, we can see that the actual ARL_1 values are also quite different in cases with a same set of (k, h) but different values of ϕ.

Example 4.5 demonstrates that correlation among observed data could have a substantial impact on the IC and OC behavior of the CUSUM charts. Therefore, it should be taken into account when designing the CUSUM charts. Otherwise, results from a CUSUM chart designed for independent observations could be misleading when the chart is applied to a production process with correlated data. As an example,

Table 4.3 *The left part presents the actual ARL_0 values of the chart (4.7)–(4.8) when $(k,h) = (0.25, 5.597), (0.5, 3.502)$, or $(0.75, 2.481)$ (so that its nominal ARL_0 values are 200 in all three cases) and when the process observations are correlated and follow the AR(1) model (4.15) with the coefficient ϕ. The right part presents the actual ARL_1 values of the chart when a mean shift of size $\delta = 0.5$ occurs at the initial observation time in various cases considered.*

		$\delta = 0$			$\delta = 0.5$	
ϕ	$k = 0.25$	$k = 0.5$	$k = 0.75$	$k = 0.25$	$k = 0.5$	$k = 0.75$
−0.5	818.620	743.026	326.828	19.575	28.169	39.633
−0.25	509.875	537.801	410.570	19.686	25.481	35.447
−0.1	295.254	308.936	284.428	19.417	23.210	30.094
0	199.623	199.258	202.063	19.188	21.681	26.908
0.1	139.672	134.461	139.602	18.815	20.276	24.067
0.25	85.464	78.179	83.084	18.134	18.108	20.411
0.5	42.827	38.390	40.020	16.847	15.652	16.386

in the case when (k,h) are chosen to be $(0.5, 3.502)$ in the CUSUM chart (4.7)–(4.8), its nominal ARL_0 value is 200 when process observations are independent of each other, which is confirmed by the calculated actual ARL_0 value of 199.258 presented in Table 4.3 in the entry with $\delta = 0, k = 0.5$, and $\phi = 0$. However, by the same CUSUM chart, if the observed data have a negative auto-correlation with $\phi = -0.5$, for example, its actual ARL_0 value becomes 743.026, which is more than 3 times the nominal ARL_0 value. A direct consequence of this difference between the actual and nominal ARL_0 values is that the chart would not be sensitive enough to process mean shifts. As shown in the right part of Table 4.3, when detecting a mean shift of size $\delta = 0.5$, the chart is supposed to have an ARL_1 value of 21.681 when the observations are independent. But, its actual ARL_1 value is 28.169 when the data are negatively correlated with $\phi = -0.5$, which is a delay of about one week if the time unit is one day. Remember that products produced after the mean shift are mostly defective products and usually cannot be sold. Therefore, many resources would be wasted in this scenario. Now, let us consider the case when the same chart is applied to a process having positively correlated observations with $\phi = 0.5$. In such a case, the actual ARL_0 value is 38.390, which is substantially smaller than the nominal ARL_0 value of 200. A direct consequence of this difference is that the chart would deliver too many false signals and the production process would be unnecessarily stopped too many times for engineers to figure out the root causes of the signals. Thus, many resources would be wasted in this scenario as well.

To accommodate the possible correlation among observed data, one commonly used approach in practice is to group neighboring observations into batches and then apply the conventional CUSUM charts for independent data to the batch means. More specifically, if the batch size is m, then the observed data are grouped in the following way:

Batch 1: $X_1, X_2, \ldots, X_m$;
Batch 2: $X_{m+1}, X_{m+2}, \ldots, X_{2m}$;

$\vdots$ $\vdots$

Let $Y_j = \frac{1}{m}\sum_{i=1}^{m} X_{(j-1)m+i}$ be the sample mean of the j-th batch, for $j = 1, 2, \ldots$. Then, the conventional CUSUM charts can be applied to the data $\{Y_j, j = 1, 2, \ldots\}$. A major reasoning behind this method is that possible correlation in the original data could be mostly eliminated in the new data $\{Y_j, j = 1, 2, \ldots\}$ when m is large (cf., Runger and Willemain, 1996). Although this approach is intuitively appealing, it should be used with care, as explained below. Let us reconsider the scenarios discussed in Example 4.5, where the original observations are generated from the AR(1) model (4.15) with the coefficient ϕ. Instead of working on the original observations, this time let us group them into batches of size $m = 5$ each, and apply the conventional CUSUM chart (4.7)–(4.8) to the standardized batch means

$$\frac{Y_j}{1/\sqrt{m}}, \qquad \text{for } j \geq 1.$$

In the CUSUM chart (4.7)–(4.8), (k, h) are still chosen to be (0.25,5.597), (0.5,3.502), or (0.75,2.481), so that its nominal ARL_0 value is 200 when $\phi = 0$ (i.e., no correlation in the observed data) in each of the three cases. Other setups remain the same as those in Example 4.5. Then, the actual ARL_0 values of the chart (4.7)–(4.8) when $\phi = -0.5, -0.25, -0.1, 0, 0.1, 0.25$, or 0.5 are presented in Table 4.4. From the table, it can be seen that the actual ARL_0 values are close to 200 when $\phi = 0$ in all three cases, as expected. But, in cases when $\phi \neq 0$, we can see that the actual ARL_0 values are even farther away from the nominal ARL_0 value of 200, compared to the corresponding actual ARL_0 values in Table 4.3, which is contradictory to our intuition that the difference between the actual and nominal ARL_0 values should be smaller in this case, compared to the case considered in Table 4.3. To further investigate this phenomenon, let us focus on the case when $(k, h) = (0.25, 5.597)$ and $\phi = 0.5$. In such a case, the sample correlation coefficient between two consecutive observations in the original data (i.e., the sample correlation computed from the pairs $\{(X_{n-1}, X_n), n \geq 2\}$) is about 0.5, and the corresponding sample correlation coefficient in the batch data (i.e., the sample correlation computed from the pairs $\{(Y_{j-1}, Y_j), j \geq 2\}$) is about 0.148. Therefore, it is true that the pairwise correlation between two consecutive data points is greatly reduced by grouping. On the other hand, the sample standard deviation of the original data is about 1.155, which is different from 1 due to the positive autocorrelation in the data and which partially explains the reason why the actual ARL_0 value in the entry of $k = 0.25$ and $\phi = 0.5$ in Table 4.3 is different from the nominal ARL_0 value. For the batch data, the sample standard deviation of the standardized group means $\{Y_j/(1/\sqrt{m}), j = 1, 2, \ldots\}$ is about 1.723, which is much larger than the sample standard deviation of the original data. This explains the reason the actual ARL_0 value computed from the batch data is even farther away from the nominal ARL_0 value. To overcome this limitation of the grouping approach, we need to scale the group means properly, which is difficult to achieve in practice unless we know how observations in the original data are correlated. See Kim et al. (2007) and Runger and Willemain (1995, 1996) for a related discussion.

Besides the limitation described above, the grouping approach has another disadvantage that the resulting control chart may not be able to react to a shift promptly,

Table 4.4 *Actual ARL_0 values of the chart (4.7)–(4.8) when $(k,h) = (0.25, 5.597)$, $(0.5, 3.502)$, or $(0.75, 2.481)$ (so that its nominal ARL_0 values are 200 in all three cases) and when the process observations are grouped into batches of size $m = 5$ each. The original process observations are correlated and follow the AR(1) model (4.15) with the coefficient ϕ.*

ϕ	$k = 0.25$	$k = 0.5$	$k = 0.75$
-0.5	926.588	956.284	953.297
-0.25	607.130	722.284	743.046
-0.1	330.990	372.706	389.072
0	200.505	199.555	200.630
0.1	127.086	111.815	106.165
0.25	67.261	52.725	46.801
0.5	26.307	18.839	15.981

due to the data averaging and due to the fact that a signal of shift can only be delivered after all observations within a group are obtained. Another related approach to handle correlated data is to use a subset of the original data, and two consecutive observations in the subset are w time points apart in the original data, where $w \geq 1$ is an integer. By this approach, the correlation among observations in the subset could be smaller than the correlation among observations in the original data, especially in cases when w is chosen to be relatively large. However, much information in the original data is ignored, and the process performance between two consecutive observations in the subset cannot be evaluated.

An alternative strategy to handle the possible correlation among process observations is to describe the correlation by a statistical model. Because most data involved in SPC are time series, statistical models for time series analysis are especially useful here (e.g., Box et al., 2008; Brockwell and Davis, 2009). A commonly used and also quite flexible time series model is the following auto-regressive moving average (ARMA) model:

$$X_n = \xi + \phi_1 X_{n-1} + \phi_2 X_{n-2} + \ldots + \phi_p X_{n-p} +$$
$$e_n + \psi_1 e_{n-1} + \psi_2 e_{n-2} + \ldots + \psi_q e_{n-q}, \tag{4.16}$$

where $\xi, \phi_1, \phi_2, \ldots, \phi_p, \psi_1, \psi_2, \ldots, \psi_q$ are coefficients, and $\{e_n, n \geq 1\}$ are the white noise with variance σ_e^2. This model, which is often denoted as $ARMA(p,q)$, assumes that X_n depends on all previous data through the previous p observations and through the random noise at the previous q time points as well. When $q = 0$, the model (4.16) becomes the AR model of order p, denoted as $AR(p)$. When $p = 0$, it becomes the so-called moving average (MA) model of order q, denoted as $MA(q)$. In time series analysis, model estimation, model selection, and model goodness-of-fit testing of the model (4.16) have been well discussed, and they can be easily accomplished in most statistical software packages. For instance, in R, functions such as ar(), arima(), acf(), pacf(), Box.test(), and so forth are all for these purposes.

A special case of the model (4.16) is the AR(1) model (4.15), in which we assume that X_n depends on all previous data through X_{n-1} only. For the model (4.15), it is easy to check that X_n has the mean μ_0 and the variance

$$\sigma_{X_n}^2 = \frac{\sigma_e^2}{1 - \phi^2}. \tag{4.17}$$

By (4.17), the variance of X_n would be augmented by any autocorrelation in the data. It is also easy to check that the correlation coefficient between X_n and X_{n-j} is ϕ^j, for $j = 1, 2, \ldots, n-1$. Namely, the autocorrelation between X_n and X_{n-j} would exponentially decay when j increases, which could describe the autocorrelation in the data of many real applications well. Because of its simplicity, the AR(1) model is popular in the SPC literature (e.g., Lu and Reynolds, 1999; Runger and Willemain, 1995).

After a time series model (e.g., the AR(1) model (4.15)) is chosen and estimated by certain routine model selection, model estimation, and model diagnostics procedures in time series analysis, we can define the residuals of the fitted model by

$$\widehat{e}_n = X_n - \widehat{X}_n, \qquad \text{for } n \geq 1,$$

where $\widehat{X}_n$ is the fitted value of X_n. If the chosen time series model describes the observed IC data adequately and the production process is IC up to the time point n, then $\{\widehat{e}_n, n \geq 1\}$ should be approximately i.i.d. with a common normal distribution $N(0, \widehat{\sigma}_e^2)$, where $\widehat{\sigma}_e^2$ is an appropriate estimator of σ_e^2. Then, we can apply the conventional CUSUM chart (4.7)–(4.8) (or other alternative control charts) to these residuals. A signal from the control chart would detect a shift from the IC model (i.e., the time series model chosen for describing the IC observations of a production process). However, it should be pointed out that the signal may not be caused solely by a process mean shift. To explain this, let us use the AR(1) model as an example. In this model, the mean of X_n, the mean of the noise term e_n, and the coefficient ϕ can all shift either simultaneously or separately, besides the possible shifts in the variances of X_n and e_n. If the mean of X_n shifts from μ_0 to μ_1 with the size $\delta = \mu_1 - \mu_0$, then the mean of e_n would shift from 0 to δ, and the means of all subsequent noise terms would shift from 0 to $\delta^* = (1 - \phi)\delta$. If e_n has a mean shift from 0 to δ^* at the time point n, then the mean of X_{n+j}, for $j \geq 0$, would shift from μ_0 to

$$\mu_0 + \delta^* \sum_{k=0}^{j} \phi^k = \mu_0 + \delta^* \frac{1 - \phi^{j+1}}{1 - \phi},$$

which is approximately $\mu_0 + \delta^*/(1 - \phi)$. Therefore, the CUSUM chart (4.7)–(4.8) with $k = \delta^*/2$ would be asymptotically optimal for detecting a mean shift in X_n of size $\delta^*/(1 - \phi)$, instead of just δ^*. Related discussion on model-based control charts for monitoring autocorrelated processes can be found in Jiang et al. (2000), Johnson and Bagshaw (1974), Lu and Reynolds (1999, 2001), Montgomery and Mastrangelo (1991), Timmer et al. (1998), Yashchin (1993a), and so forth.

Example 4.6 *To monitor the production of an aluminum smelter, 189 observations of the content of MgO in its products are collected over a period of time. The original* *observations are shown in Figure 4.7(a). A more detailed description of this data can be found in Qiu and Hawkins (2001). By using the* ar() *function in* R, *it can be found that the data are auto-correlated, and the following AR(1) model is appropriate to describe the autocorrelation:*

$$X_n = 12.972 + 0.546 (X_{n-1} - 12.972) + e_n, \qquad \text{for } 1 \leq n \leq 189,$$

where $X_0 = 12.972$. The residuals $\{\hat{e}_n, 2 \leq n \leq 189\}$ from this fitted model are presented in Figure 4.7(b). The Shapiro-Wilk test (cf., Shapiro and Wilk, 1965) for testing the normality of the original data gives a p-value of 0.044, and the same test on the residuals gives a p-value of 0.276, which confirms that the original data may not be normally distributed but the distribution of the residuals is not significantly different from a normal distribution. After the data standardization (i.e., first subtracting the sample mean and then dividing by the sample standard deviation), the CUSUM chart (4.7)–(4.8) with $(k, h) = (0.25, 5.597)$ (note: its nominal ARL_0 is 200) is applied to both the original data and the residuals. The two control charts are shown in Figure 4.7(c)–(d), respectively. From the plots, it can be seen that the chart for the original data has signals very early (at the 15th and 16th time points), and the chart for the residuals does not give signals until the 64th time point. Based on our discussion above, the early signal of the former chart might be due to the autocorrelation in the original data that is not accommodated in that chart.

4.2.4 Optimality of the CUSUM chart

Assume that the probability density function (pdf) or the probability mass function (pmf) (see Section 2.2 for their definitions) of a population is $f(x; \theta)$, where θ is a parameter. And, we are interested in testing the hypotheses

$$H_0 : \theta = \theta_0 \qquad \text{versus} \qquad H_1 : \theta = \theta_1, \qquad (4.18)$$

where $\theta_0 \neq \theta_1$ are two given numbers. Then, the well-known sequential probability ratio test (SPRT) is constructed as follows. Let $\{X_1, X_2, \ldots, X_n\}$ be n i.i.d. observations collected from the population for testing the hypotheses in (4.18). Then, the likelihood function (see its definition in Subsection 2.7.2) based on the simple random sample $\{X_1, X_2, \ldots, X_n\}$ is

$$L(\theta; X_1, X_2, \ldots, X_n) = \Pi_{i=1}^n f(X_i; \theta),$$

which provides a measure of the likelihood that the parameter θ takes different values on the number line. To compare the likelihoods of $\theta = \theta_0$ and $\theta = \theta_1$, let us consider the likelihood ratio

$$\Lambda_n = \frac{L(\theta_1; X_1, X_2, \ldots, X_n)}{L(\theta_0; X_1, X_2, \ldots, X_n)} = \Pi_{i=1}^n \frac{f(X_i; \theta_1)}{f(X_i; \theta_0)}.$$

In practice, it is often more convenient to work on the logarithm of Λ_n

$$\log(\Lambda_n) = \sum_{i=1}^n \Psi_i,$$

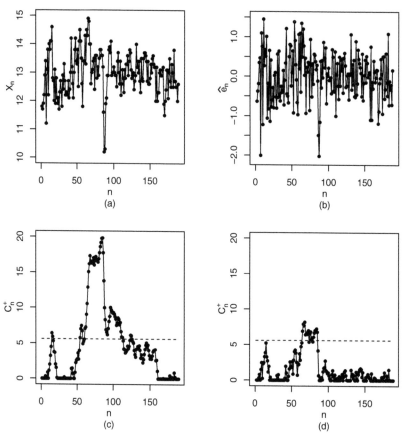

Figure 4.7 *(a) Original data. (b) Residuals of the fitted AR(1) model. (c) Control chart (4.7)–(4.8) with $(k,h) = (0.25, 5.597)$ when it is applied to the standardized original data. (d) Control chart (4.7)–(4.8) with $(k,h) = (0.25, 5.597)$ when it is applied to the standardized residuals.*

where

$$\Psi_i = \log\left(\frac{f(X_i; \theta_1)}{f(X_i; \theta_0)}\right).$$

Without loss of generality, let us assume that $\theta_1 > \theta_0$. Then, in cases when $L(\theta; X_1, X_2, \ldots, X_n)$ is a non-decreasing function of θ, the SPRT has the following decision rule:

(i) fail to reject H_0 if $\log(\Lambda_n) \leq A$,

(ii) reject H_0 if $\log(\Lambda_n) \geq B$, and

(iii) need one more observation to make a decision if $A < \log(\Lambda_n) < B$,

where $A < B$ are two threshold values. The SPRT was first suggested by Wald (1945). It was proved in Wald and Wolfowitz (1948) that this test has the following optimality property: among all testing procedures whose Type I error probabilities are below a given level, under some regularity conditions, the SPRT requires a minimum expected number of observations to reach a decision (i.e., to reject H_0 or fail to reject H_0). Related discussion can be found in Ferguson (1967) and Simons (1976).

Now, let us focus on the case when the population distribution is $N(\mu, \sigma^2)$, and we are interested in testing

$$H_0 : \mu = \mu_0 \qquad \text{versus} \qquad H_1 : \mu = \mu_1,$$

where $\mu_0 < \mu_1$ are two given values. In such cases,

$$\Psi_i = \frac{\mu_1 - \mu_0}{\sigma^2} \left[(X_i - \mu_0) - k \right],$$

and

$$\log(\Lambda_n) = \frac{\mu_1 - \mu_0}{\sigma^2} \sum_{i=1}^{n} \left[(X_i - \mu_0) - k \right],$$

where $k = (\mu_1 - \mu_0)/2$. Let $\widetilde{C}_n^+ = \frac{\sigma^2}{\mu_1 - \mu_0} \log(\Lambda_n)$. Then, we have, for $n \geq 1$,

$$
\begin{aligned}
\widetilde{C}_n^+ &= \sum_{i=1}^{n} \left[(X_i - \mu_0) - k \right] \\
&= \widetilde{C}_{n-1}^+ + \left[(X_n - \mu_0) - k \right],
\end{aligned}
\tag{4.19}
$$

where $\widetilde{C}_0^+ = 0$. By comparing $\widetilde{C}_n^+$ in (4.19) with C_n^+ in (4.7), we can see that they are very similar. The only difference is that when $C_{n-1}^+ + [(X_n - \mu_0) - k]$ is negative, which is in favor of H_0 so that H_0 should not be rejected at the n-th time point, C_n^+ is re-started from 0, while $\widetilde{C}_n^+$ does not have this re-starting mechanism involved. The re-starting mechanism is reasonable to use in the SPC problem, because the sequential process in SPC should be stopped only when H_0 is rejected (i.e., H_1 is accepted). In the case when H_0 is accepted, the process should be continued. By the re-starting mechanism, the charting statistic C_n^+ of the CUSUM chart (4.7)–(4.8) is reset to 0 each time when it is negative, making the chart more effective in detecting upward process mean shifts.

Based on the connection described above between the SPRT test and the CUSUM chart, several authors, including Lorden (1971), Moustakides (1986), Ritov (1990), and Yashchin (1993b), derived various optimality properties of the CUSUM chart. For instance, Moustakides (1986) proved that among all control charts with a fixed ARL_0 value, the CUSUM chart with an allowance k would have the shortest ARL_1 value for detecting a persistent shift of size $\delta = 2k$.

The SPRT connection also provides a general way to construct CUSUM charts, summarized in the box below.

General Formula for the DI Form of the CUSUM Chart

Assume that the IC and OC pdf's (or pmf's) of the process observations are f_0 and f_1, respectively, and $\{X_1, X_2, \ldots\}$ are independent observations obtained from the process. Then, the charting statistic of the DI form of the CUSUM chart for detecting the distributional shift from f_0 to f_1 is defined as follows: for $n \geq 1$,

$$G_n = \max\left(0, G_{n-1} + \widetilde{\Psi}_n\right), \qquad (4.20)$$

where $G_0 = 0$ and $\widetilde{\Psi}_n = \log(f_1(X_n)/f_0(X_n))$.

Note that the description in the above box about the CUSUM chart is more general than the description in the remaining part of this subsection, in that both f_0 and f_1 are not restricted to parametric functions here, while they are assumed parametric and their only difference is in the parameter values in the remaining part of this subsection. The major purpose of this generalization is that the resulting CUSUM chart (4.20) is also available to certain applications in which the process observation distributions are nonparametric, or f_0 and f_1 are parametric but their difference is not just in the parameter values.

4.3   Monitoring the Variance of a Normal Process

Variability of process observations is another key characteristic that affects the quality of products. In this section, we discuss monitoring of the process variability when the process is univariate and the process distribution is normal. Our discussion is divided into three parts. In Subsection 4.3.1, we explain why the process variability is important to the quality of products and the possible impact of a shift in process variability on the performance of a CUSUM chart for monitoring the process mean. Then, some CUSUM charts for monitoring the process variability are described in Subsection 4.3.2. The problem of monitoring both the process mean and the process variability is discussed in Subsection 4.3.3.

4.3.1   Process variability and quality of products

In Section 3.4, we mentioned that our requirement on product quality is usually specified by a lower specification limit (LSL) and an upper specification limit (USL) in cases when a univariate numerical quality characteristic X is our only concern. A product is classified as a conforming product when its value of X is between LSL and USL. Otherwise, it is classified as a non-conforming product. The overall quality of the related production process can be measured by the probability $P(LSL \leq X \leq USL)$, which is also a measure of the process capability. (Note that the term "quality" here actually means the quality of conformance, instead of the quality of design. See related description in Section 1.2 for their difference). In this chapter, we assume that the distribution of X is normal. In such cases, the quality of the production process is uniquely determined by the mean and standard deviation of

X. Therefore, the product quality would be affected if the mean or standard deviation of X shifts at a certain time point during the manufacturing of the products, which is demonstrated in the example below.

Example 4.7 *For an injection molding process, our major concern about the quality of its products is the compressive strength (in psi), denoted as X. Assume that when the production process works stably, X would follow the distribution of $N(80, 2.5^2)$. To define conforming and non-conforming products, LSL=75 and USL=85 have been specified, so that the production process has the probability of $P(75 \leq X \leq 85) = 0.954$ to produce conforming products when it works stably. Now, assume that at a certain time point of production, the mean or the standard deviation of X shifts, and the shift belongs to the following four cases:*

(i) *only the mean shifts to the value of 82.5,*

(ii) *only the standard deviation shifts to the value of 4,*

(iii) *the mean shifts to the value of 82.5, and the standard deviation shifts to the value of 4, and*

(iv) *only the standard deviation shifts to the value of 2.*

In case (i), after the mean shift, the production process has the following probability to manufacture conforming products:

$$
\begin{aligned}
&P(75 \leq X \leq 85) \\
=\ &P\left(\frac{75 - 82.5}{2.5} \leq Z \leq \frac{85 - 82.5}{2.5}\right) \\
=\ &P(-3 \leq Z \leq 1) = 0.840,
\end{aligned}
$$

where Z denotes a random variable with the standard normal distribution. In cases (ii)–(iv), the corresponding probability values can be computed in a similar way to be 0.789, 0.704, and 0.988, respectively.

From case (i) in the above example, it can be seen that the process capability would decrease if the process mean shifts and the process standard deviation remains unchanged. When the process mean remains unchanged, but the process standard deviation increases (i.e., case (ii) in the example), the process capability would decrease as well. Therefore, the quality of products is indeed affected by the shift in the process variability. From case (iii), it can be seen that the situation would become worse if the process standard deviation increases and the process mean changes at the same time. However, case (iv) demonstrates that the process capability does not always decrease when the process variance shifts. As a matter of fact, it is even improved in cases when the process mean remains unchanged but the process standard deviation decreases, which is intuitively reasonable because more products would meet the designed requirements in such cases. Therefore, in practice, we are usually concerned about upward shifts in process variability only. Downward shifts in process variability would become an issue only when the cost of production needs to be reduced, because good quality of products often implies high cost of production and downward shifts in process variability mean that the cost of production might have been elevated, or the cost has room to be reduced. These points are summarized in the box below.

Process Variance Shifts and the Quality of a Production Process

Upward process variance shifts would generally downgrade the quality of conformance of the related production process, and downward shifts would generally improve the quality of conformance. Therefore, we are mainly concerned about upward variance shifts in practice. Downward shifts are concerned only in cases when we want to cut the cost of production and so forth.

4.3.2   CUSUM charts for monitoring process variance

In this subsection, we describe several control charts for detecting shifts in process variance. Assume that the distribution of the quality characteristic of concern X is $N(\mu_0, \sigma_0^2)$ when the production process is in IC, and it changes to $N(\mu_0, \sigma_1^2)$ after the process becomes OC at an unknown time point, where μ_0 is the IC mean, $\sigma_0^2 \neq \sigma_1^2$ are the IC and OC variances. First, we investigate the performance of the CUSUM chart (4.7)–(4.8), which is designed for detecting process mean shifts, in cases when the process variance shifts. To this end, let us discuss the example below.

Example 4.8 *Figure 4.8(a) shows 100 observations, the first 50 observations are i.i.d. from the distribution $N(0,1)$, the second 50 observations are i.i.d. from the distribution $N(0,2^2)$, and the two halves of the data are independent of each other. So, these data simulate a production process with an IC distribution of $N(0,1)$ which has an upward variance shift at the 51st time point. We then apply the conventional CUSUM chart (4.7)–(4.8) to these data, in which (k,h) are chosen to be $(0.25, 5.597)$ so that its ARL_0 is 200 (cf., Table 4.1). The CUSUM chart is shown in Figure 4.8(b), from which it can be seen that the charting statistic crosses the control limit h several times after the variance shift with the first signal at the 53rd time point, and the variability of the charting statistic is much larger for the second half of the data. Figure 4.8(c) shows another 100 observations, generated in the same way as that for the observations in Figure 4.8(a), except that the second half of the data in Figure 4.8(c) are i.i.d. from the distribution $N(0,0.5^2)$. Therefore, these data simulate a production process with an IC distribution of $N(0,1)$, which has a downward variance shift at the 51st time point. The CUSUM chart (4.7)–(4.8) with $(k,h) = (0.25, 5.597)$ for these data is shown in Figure 4.8(d). Apparently, the variability of the charting statistic is much smaller for the second half of the data in this case.*

Example 4.8 shows that the process variance shift would change the distribution of the charting statistic of the CUSUM chart (4.7)–(4.8). An upward variance shift would increase the variability of the charting statistic and consequently the chart (4.7)–(4.8) has a certain ability to detect the shift, while a downward shift would decrease the variability of the charting statistic and the chart (4.7)–(4.8) does not have any ability to detect the shift. By the relationship among the upward CUSUM chart (4.7)–(4.8), the downward CUSUM chart (4.9)–(4.10), and their two-sided version, these results should also be true for the other two versions of the CUSUM chart. A direct implication of these results is that when a process mean shift is our major concern and the process variability changes at an early time point (e.g., at the initial observation time point), the actual ARL_0 value of a control chart designed for detect-

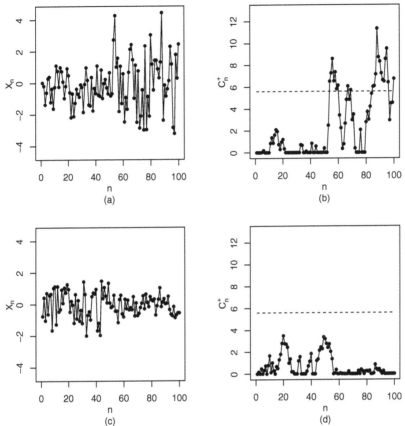

Figure 4.8 *(a) A dataset with 100 observations among which the first 50 observations are from the distribution $N(0,1)$, the second 50 observations are from the distribution $N(0,2^2)$, and all observations are independent. (b) Control chart (4.7)–(4.8) with $(k,h)=(0.25,5.597)$ when it is applied to the data in plot (a). (c) A dataset with 100 observations among which the first 50 observations are from the distribution $N(0,1)$, the second 50 observations are from the distribution $N(0,0.5^2)$, and all observations are independent. (d) Control chart (4.7)–(4.8) with $(k,h)=(0.25,5.597)$ when it is applied to the data in plot (c).*

ing mean shifts, such as the chart (4.7)–(4.8), would be affected by the variance shift. Usually, when the variance shift is upward, the actual ARL_0 value would be smaller than the nominal ARL_0 value; the actual ARL_0 value would be larger than the nominal ARL_0 value if the variance shift is downward.

It should be pointed out that although the CUSUM charts designed for detecting process mean shifts have a certain ability in detecting upward process variability shifts, they are not effective enough in detecting variance shifts. Therefore, CUSUM charts that are designed specifically for detecting variability shifts are still needed, and are described below. Let us first discuss the case when the process variance shifts upward from σ_0^2 to σ_1^2 (i.e., $\sigma_1^2 > \sigma_0^2$). In such cases, according to the general formula

(4.20), the DI form of the optimal CUSUM chart is

$$G_n = \max\left(0, G_{n-1} + \widetilde{\Psi}_n\right),$$

where $G_0 = 0$, $\widetilde{\Psi}_n = \log(f_1(X_n)/f_0(X_n))$, and f_0 and f_1 are the pdf's of the IC and OC distributions, respectively. By the pdf formula (2.12) of a normal distribution in Subsection 2.3.1, it can be checked that

$$\widetilde{\Psi}_n = \log\left(\frac{\sigma_0}{\sigma_1}\right) + \frac{1}{2}\left[1 - \left(\frac{\sigma_0}{\sigma_1}\right)^2\right]\left(\frac{X_n - \mu_0}{\sigma_0}\right)^2.$$

For $n \geq 0$, let

$$C_n^+ = G_n \left/ \left\{\frac{1}{2}\left[1 - \left(\frac{\sigma_0}{\sigma_1}\right)^2\right]\right\}\right..$$

Then, the CUSUM chart is

$$C_n^+ = \max\left(0, C_{n-1}^+ + \left(\frac{X_n - \mu_0}{\sigma_0}\right)^2 - k^+\right), \tag{4.21}$$

where $C_0^+ = 0$, and

$$k^+ = \frac{2\log(\sigma_0/\sigma_1)}{(\sigma_0/\sigma_1)^2 - 1}.$$

This chart gives a signal of upward variance shift if

$$C_n^+ > h_U, \tag{4.22}$$

where $h_U > 0$ is a control limit that is chosen to reach a pre-specified ARL_0 value.

Similarly, we can derive an optimal CUSUM chart for detecting a downward variance shift from σ_0^2 to σ_1^2 (i.e., the case when $\sigma_1^2 < \sigma_0^2$) as follows. By the general formula (4.20), the DI form of the optimal CUSUM chart can be written as

$$\begin{aligned}
-G_n &= \min\left(0, -G_{n-1} - \widetilde{\Psi}_n\right) \\
&= \min\left(0, -G_{n-1} - \log\left(\frac{\sigma_0}{\sigma_1}\right) + \frac{1}{2}\left[\left(\frac{\sigma_0}{\sigma_1}\right)^2 - 1\right]\left(\frac{X_n - \mu_0}{\sigma_0}\right)^2\right).
\end{aligned}$$

In the second equation of the above expression, the expression for $\widetilde{\Psi}_n$ given in the previous paragraph has been used. For $n \geq 0$, let

$$C_n^- = -G_n \left/ \left\{\frac{1}{2}\left[\left(\frac{\sigma_0}{\sigma_1}\right)^2 - 1\right]\right\}\right..$$

Then, the charting statistic of the optimal CUSUM chart for detecting the downward variance shift is defined by

$$C_n^- = \min\left(0, C_{n-1}^- + \left(\frac{X_n - \mu_0}{\sigma_0}\right)^2 - k^-\right), \qquad \text{for } n \geq 1, \tag{4.23}$$

where $C_0^- = 0$, and k^- has the same expression as that of k^+. This chart gives a signal of downward variance shift when

$$C_n^- < h_L, \tag{4.24}$$

where $h_L < 0$ is a control limit that is chosen to reach a pre-specified ARL_0 value.

Example 4.8 (continued) *For the data shown in Figure 4.8(a), an upward process variance shift occurs at the 51st time point, and the IC and OC process variances are $\sigma_0^2 = 1$ and $\sigma_1^2 = 2^2$, respectively. To apply the optimal CUSUM chart (4.21)–(4.22), k^+ should be chosen as $2\log(1/2)/[(1/2)^2 - 1] = 1.848$. If we choose $ARL_0 = 200$, then by the software package ANYGETH.EXE provided by Hawkins and Olwell (1998) (see its description in the appendix), the control limit is computed to be $h_U = 7.416$. The CUSUM chart is shown in Figure 4.9(a). It can be seen that the first signal is given by this chart at the 54th time point. For the data shown in Figure 4.8(c), a downward process variance shift occurs at the 51st time point, and the IC and OC process variances are $\sigma_0^2 = 1$ and $\sigma_1^2 = 0.5^2$, respectively. To apply the optimal CUSUM chart (4.23)–(4.24), k^- should be chosen as $2\log(1/0.5)/[(1/0.5)^2 - 1] = 0.462$. When $ARL_0 = 200$, by the software package ANYGETH.EXE, the control limit is computed to be $h_L = -2.446$. The corresponding CUSUM chart is shown in Figure 4.9(b), and the first signal is given by this chart at the 63rd time point.*

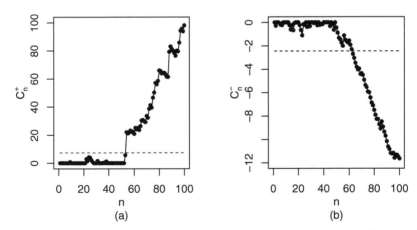

Figure 4.9 *(a) CUSUM chart (4.21)–(4.22) when it is applied to the data shown in Figure 4.8(a) and when $k^+ = 1.848$ and $h_U = 7.416$ (shown by the dashed horizontal line). (b) CUSUM chart (4.23)–(4.24) when it is applied to the data shown in Figure 4.8(c) and when $k^- = 0.462$ and $h_L = -2.446$ (shown by the dashed horizontal line).*

To detect process variance shifts, batch data are commonly used. Assume that the batch size is m at each time point. Then, for $n \geq 1$, we have m observations at the n-th time point

$$X_{n1}, X_{n2}, \ldots, X_{nm}.$$

Let $\overline{X}_n$ and s_n^2 be the sample mean and sample variance of these observations. Then, we know that, for estimating (μ, σ^2) of the population distribution

$N(\mu, \sigma^2)$, we would not lose any useful information to replace the original sample $\{X_{n1}, X_{n2}, \ldots, X_{nm}\}$ by $(\overline{X}_n, s_n^2)$ (i.e., $(\overline{X}_n, s_n^2)$ are sufficient statistics for estimating (μ, σ^2), see Lehmann and Casella (1998)). To monitor the process variability, we can simply use the sample variance s_n^2, which has the following distribution:

$$\frac{(m-1)s_n^2}{\sigma^2} \sim \chi_{m-1}^2.$$

By the pdf formula of the chi-square distribution given in Subsection 2.3.2, it can be checked that the pdf of s_n^2 is

$$f(s_n^2) = \frac{1}{2^{(m-1)/2}\Gamma((m-1)/2)} \left[\frac{(m-1)}{\sigma^2} s_n^2\right]^{(m-1)/2-1} \frac{(m-1)}{\sigma^2} e^{-\frac{(m-1)}{2\sigma^2} s_n^2}.$$

By the general formula (4.20), the DI form of the optimal CUSUM chart is

$$G_n = \max\left(0, G_{n-1} + \tilde{\Psi}_n\right),$$

where $G_0 = 0$,

$$\begin{aligned}
\tilde{\Psi}_n &= \log\left(\frac{f_1(s_n^2)}{f_0(s_n^2)}\right) \\
&= \frac{m-1}{2}\log\left(\sigma_0^2/\sigma_1^2\right) + \frac{(\sigma_1^2 - \sigma_0^2)(m-1)}{2\sigma_0^2\sigma_1^2} s_n^2,
\end{aligned}$$

and f_0 and f_1 are the pdf's of the IC and OC distributions of s_n^2, respectively. Assume that the shift in process variability is upward (i.e., $\sigma_1^2 > \sigma_0^2$). In such cases, $\sigma_1^2 - \sigma_0^2 > 0$. For $n \geq 0$, let

$$C_n^+ = G_n \left/ \left\{\frac{(\sigma_1^2 - \sigma_0^2)(m-1)}{2\sigma_0^2\sigma_1^2}\right\}\right..$$

Then, the CUSUM charting statistic for detecting the upward process variance shift is

$$C_n^+ = \max\left(0, C_{n-1}^+ + s_n^2 - k^+\right), \tag{4.25}$$

where $C_0^+ = 0$, and

$$k^+ = \frac{2\log\left(\sigma_1/\sigma_0\right)\sigma_0^2\sigma_1^2}{\sigma_1^2 - \sigma_0^2}.$$

This chart gives a signal of upward variance shift if

$$C_n^+ > h_U, \tag{4.26}$$

where $h_U > 0$ is a control limit that is chosen to reach a pre-specified ARL_0 value. We can define a CUSUM chart for detecting a downward process variance shift based on batch data in a similar way.

Comparing the chart (4.21)–(4.22) with the chart (4.25)–(4.26), the former would be sensitive to process mean shifts besides its ability in detecting process variance shifts, while the latter is for detecting process variance shifts only and it is immune to possible process mean shifts. That is because the charting statistic defined in (4.21) depends on the true process mean, but the one defined in (4.25) does not.

4.3.3   Joint monitoring of process mean and variance

As mentioned earlier, shifts in both the process mean and process variance would have an impact on the quality of products. Therefore, in practice, we often need to monitor process mean and process variance simultaneously. To this end, with the Shewhart charts discussed in Chapter 3, a Shewhart chart designed for monitoring the process mean (e.g., the $\overline{X}$ chart) and a Shewhart chart designed for monitoring the process variance (e.g., the R chart) are usually used as a pair. With the CUSUM charts, we can jointly monitor the process mean and variance in a similar way, as briefly described below.

Let us first focus on the case discussed at the end of the previous subsection about batch data. At each time point, we have a sample of size m, $\{X_{n1}, X_{n2}, \ldots, X_{nm}\}$, for $n \geq 1$. Then, a CUSUM chart can be constructed based on the sample means $\{\overline{X}_n, n \geq 1\}$ for monitoring the process mean (cf., Example 4.2(continued) at the end of Subsection 4.2.1), and another CUSUM chart can be constructed based on the sample variances $\{s_n^2, n \geq 1\}$ for monitoring the process variance (cf., formulas (4.25) and (4.26)). For simplicity of presentation, in this subsection, the CUSUM chart for monitoring the process mean is called the CUSUM-M chart, and the one for monitoring the process variance is called the CUSUM-V chart. To jointly monitor the process mean and variance, the two charts CUSUM-M and CUSUM-V can be used as a pair. A signal for joint monitoring is given when either one of the two charts gives a signal, and the status of a potential shift (i.e., a mean shift versus a variance shift) can be judged from the specific chart that gives the signal. If the signal of joint monitoring is due to a signal from the CUSUM-M chart, then we can conclude that the process mean has shifted. Otherwise, we can conclude that process variance has shifted. The two control charts in the pair can be designed separately, although it is often a good idea to make their ARL_0 values the same, unless we have different preference regarding detection of the mean shift versus detection of the variance shift. In the current setup with batch data, the charting statistics of the CUSUM-M and CUSUM-V charts are independent, because the sample mean and sample variance are independent at each time point. Let $S_M(n)$ and $S_V(n)$ be the events that the CUSUM-M and CUSUM-V charts have signals at the n-th time point, respectively. Then, the probability that the joint monitoring scheme gives a signal is

$$
\begin{aligned}
P(S_M(n) \text{ or } S_V(n)) &= P(S_M(n)) + P(S_V(n)) - P(S_M(n))P(S_V(n)) \\
&\approx P(S_M(n)) + P(S_V(n)),
\end{aligned}
\tag{4.27}
$$

where "$\approx$" means "approximately equal," and the term $P(S_M(n))P(S_V(n))$ is omitted in the second line because it is often much smaller than both $P(S_M(n))$ and $P(S_V(n))$. In Subsection 4.2.2, it is mentioned that in cases when a process is IC, the distribution of the charting statistic of a CUSUM chart would approach a steady-state distribution when n gets large. Therefore, the three probabilities $P(S_M(n) \text{ or } S_V(n))$, $P(S_M(n))$, and $P(S_V(n))$ in (4.27) will approach three constants. By these results and the result that the IC ARL of a CUSUM chart can be reasonably approximated by the mean of a geometric distribution (cf., Gan, 1993a; Goel and Wu, 1971; Hawkins, 1992b; Vance, 1986; Waldmann, 1986; Woodall, 1983), the ARL_0 value of the joint monitoring scheme can be approximated well using the following formula:

$$\frac{1}{ARL_{0,J}} \approx \frac{1}{ARL_{0,M}} + \frac{1}{ARL_{0,V}}, \tag{4.28}$$

where $ARL_{0,J}, ARL_{0,M}$, and $ARL_{0,V}$ denote the ARL_0 values of the joint monitoring scheme, the CUSUM-M, and the CUSUM-V charts, respectively. A considerable amount of approximation error could get involved if a similar formula to (4.28) is used for approximating the ARL_1 values of the joint monitoring scheme.

Example 4.9 *Assume that 10 samples of size 5 each are generated from the $N(0,1)$ distribution, and another 10 samples of size 5 each are generated from the $N(1,1)$ distribution. These samples can be regarded as observed batch data from a produc-tion process that has a mean shift of size 1 at the 11th time point. The sample means $\{\overline{X}_n, n = 1, 2, \ldots, 20\}$ are shown in Figure 4.10(a). To monitor the production pro-cess, a joint monitoring scheme for monitoring both the process mean and the process variance is used, in which the CUSUM-M chart is the CUSUM chart (4.7)–(4.8) for detecting upward mean shifts with X_n in (4.7) replaced by $\overline{X}_n$, and the CUSUM-V chart is the CUSUM chart (4.25)–(4.26) for detecting upward variance shifts. In the CUSUM-M chart, (k, h) are chosen to be $(0.5, 0.881)$ such that its ARL_0 is 400. This chart is optimal for detecting a mean shift of size 1. In the CUSUM-V chart, (k^+, h_U) are chosen to be $(1.848, 2.533)$ so that its ARL_0 is 400 as well. This chart is optimal for detecting a variance shift from $\sigma_0^2 = 1$ to $\sigma_1^2 = 2^2$. By (4.28), the ARL_0 value of the joint monitoring scheme is about 200. The two control charts are shown in Fig-ure 4.10(b)–(c), respectively. It can be seen that the CUSUM-M chart gives a signal at the 11th time point, and consequently the joint monitoring scheme signals at this time point as well. We then consider another scenario in which the first 10 samples of size 5 each are generated from the $N(0,1)$ distribution, and the second 10 samples of size 5 each are generated from the $N(0,2^2)$ distribution. These samples simulate the observed batch data from a production process that has a variance shift from $\sigma_0^2 = 1$ to $\sigma_1^2 = 2^2$ at the 11th time point. The sample means are shown in Figure 4.10(d), and the CUSUM-M and CUSUM-V charts are shown in Figure 4.10(e)–(f). It can be seen that the CUSUM-M chart does not give any signal, although its charting statistic has a larger variability after the variance shift, the CUSUM-V chart gives a signal at the 12th time point, and consequently the joint monitoring scheme signals at the 12th time point as well.*

For individual observation data, a CUSUM chart designed mainly for monitoring the process mean (e.g., the one (4.7)–(4.8)) and a CUSUM chart designed mainly for monitoring the process variance can also be used as a pair to jointly monitor the process mean and variance. In such cases, the formula (4.28) for approximating the ARL_0 value of the joint monitoring scheme may not be valid, because the charting statistics of the two individual CUSUM charts are usually correlated. However, the ARL_0 value of the joint monitoring scheme can be computed easily by a numeri-cal algorithm similar to the one described by the pseudo code in the box below the expression (4.11).

Some authors have proposed some specific control charts for joint monitoring of the process mean and variance. For instance, Hawkins (1981) proposed a joint mon-itoring scheme described as follows. Assume that the IC distribution of a production

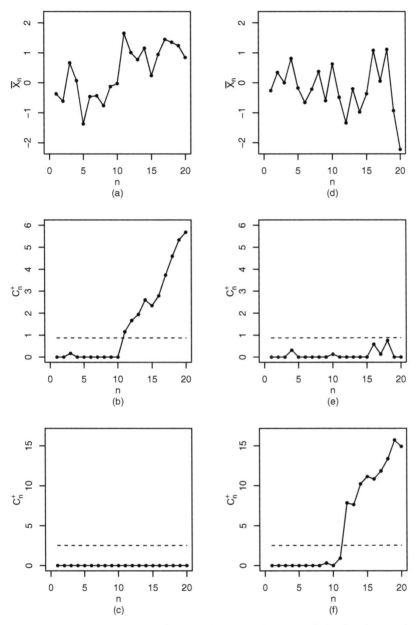

Figure 4.10 *(a) Sample means of 20 samples of size 5 each with the first 10 samples from the $N(0,1)$ distribution, the second 10 samples from the $N(1,1)$ distribution, and all samples independent of each other. (b) CUSUM-M chart with $(k,h) = (0.5, 0.881)$ for the data shown in plot (a). (c) CUSUM-V chart with $(k^+, h_U) = (1.848, 2.533)$ for the data shown in plot (a). (d) Sample means of 20 samples of size 5 each with the first 10 samples from the $N(0,1)$ distribution, the second 10 samples from the $N(0, 2^2)$ distribution, and all samples independent of each other. (e) CUSUM-M chart with $(k,h) = (0.5, 0.881)$ for the data shown in plot (d). (f) CUSUM-V chart with $(k^+, h_U) = (1.848, 2.533)$ for the data shown in plot (d). The dashed horizontal lines in plots (b), (c), (e), and (f) are control limits of the related control charts.*

process is $N(\mu_0, \sigma_0^2)$, and an individual observation X_n is observed at the n-th time point, for $n \geq 1$. Then, we can construct a CUSUM chart for detecting process mean shifts based on the standardized observations

$$Z_n = \frac{X_n - \mu_0}{\sigma_0}, \qquad \text{for } n \geq 1,$$

using formulas (4.7) and (4.8) with $X_n - \mu_0$ in (4.7) replaced by $(X_n - \mu_0)/\sigma_0$. This is the CUSUM-M chart used in the joint monitoring scheme. To monitor the process variability, Hawkins proposed using the transformed observations

$$W_n = \frac{\sqrt{|Z_n|} - 0.822}{0.349}.$$

When the process is IC at the n-th time point, the distribution of W_n is close to $N(0, 1)$. If the process has an upward (downward) variance shift at the n-th time point, then the mean of W_n would increase (decrease). Therefore, detection of variance shift in X_n can be achieved by detecting the mean shift in W_n. The latter task can be achieved by using a CUSUM chart constructed using the formulas (4.7) and (4.8) with $X_n - \mu_0$ in (4.7) replaced by W_n. This chart is used as the CUSUM-V chart in the joint monitoring scheme. Note that Z_n and W_n have approximately the same IC distribution. Therefore, the two charting statistics would have similar scales as well when the process is IC, which makes it possible to include the two charts in a single plot. Yeh et al. (2004b) proposed another joint monitoring scheme based on batch data. The CUSUM-M and CUSUM-V charts in that scheme are based on two statistics that both have the $Uniform(0, 1)$ distribution when the process is IC, so that they can be shown in a single plot as well.

As discussed in Sections 3.5 and 4.1, the Shewhart charts are good for detecting relatively large and transient shifts in the distribution of the quality characteristic X, and the CUSUM charts are good for detecting relatively small and persistent shifts in the distribution of X. In practice, it is often our interest to detect both transient and persistent shifts. To this end, it has become a common practice to use a joint monitoring scheme that consists of a Shewhart chart and a CUSUM chart (e.g., Lucas, 1982; Wu et al., 2008b). For instance, if batch data are available, then the $\overline{X}$-chart defined in (3.6) can be used for detecting transient process mean shifts and the CUSUM chart (4.7)–(4.8) constructed from sample means can be used for detecting persistent process mean shifts. Detailed discussion about the design of such joint monitoring schemes can be found in Wu et al. (2008b) and the references cited there.

4.4 CUSUM Charts for Distributions in Exponential Family

In the previous section, CUSUM charts for monitoring process mean and/or variance are described in cases when the distribution of the quality characteristic of interest X is normal. In practice, the distribution of X is often not normal. In Subsection 4.2.4, we have provided a formula for constructing optimal CUSUM charts in general cases when the IC and OC process distributions are specified. In cases when the process distributions are parametric and they belong to the so-called exponential family (see its definition in Subsection 4.4.1 below), the optimal CUSUM charts can be computed using simpler formulas. Because the exponential family contains many commonly used parametric distributions, this section discusses construction of optimal CUSUM charts for some important parametric distributions in this family. Our discussion is divided into two parts. Some cases when X is continuous numerical and its distribution belongs to the exponential family (e.g., the gamma and Weibull distributions) are discussed in Subsection 4.4.1. Cases when X follows several commonly used discrete distributions that belong to the exponential family (e.g., the binomial and Poisson distributions) are discussed in Subsection 4.4.2.

4.4.1   Cases with some continuous distributions in the exponential family

Assume that the distribution of X belongs to the one-parameter exponential family. Then, its probability density function (pdf) or probability mass function (pmf) has the expression

$$f(x; \theta) = h(x) \exp\left(\eta(\theta) T(x) + A(\theta)\right), \tag{4.29}$$

where $h(x), T(x), \eta(\theta)$, and $A(\theta)$ are known functions. If we are interested in detecting a shift in θ from θ_0 to θ_1 based on the independent observations $\{X_1, X_2, \ldots\}$ obtained from a related production process, then by the general formula (4.20) given in Subsection 4.2.4, the optimal CUSUM chart has the charting statistic

$$G_n = \max\left(0, G_{n-1} + \widetilde{\Psi}_n\right), \tag{4.30}$$

where $G_0 = 0$, $\widetilde{\Psi}_n = \log(f_1(X_n)/f_0(X_n))$, and f_0 and f_1 are the pdf's (or pmf's) of the IC and OC distributions of X, respectively. From (4.29), when the distribution of X belongs to the one-parameter exponential family, we have

$$\widetilde{\Psi}_n = [\eta(\theta_1) - \eta(\theta_0)] T(X_n) + [A(\theta_1) - A(\theta_0)]. \tag{4.31}$$

In cases when $\eta(\theta_1) - \eta(\theta_0) > 0$, let us define

$$C_n^+ = G_n / [\eta(\theta_1) - \eta(\theta_0)].$$

Then, by (4.30) and (4.31), the charting statistic of the optimal CUSUM chart can be defined by

$$C_n^+ = \max\left(0, C_{n-1}^+ + T(X_n) - k^+\right), \tag{4.32}$$

where $C_0^+ = 0$, and

$$k^+ = -\frac{A(\theta_1) - A(\theta_0)}{\eta(\theta_1) - \eta(\theta_0)}. \tag{4.33}$$

The chart gives a signal when $C_n^+ > h^+$, where h^+ is a control limit chosen to achieve a pre-specified ARL_0 value. In cases when $\eta(\theta_1) - \eta(\theta_0) < 0$, let us define

$$C_n^- = -G_n / [\eta(\theta_0) - \eta(\theta_1)].$$

Then, the charting statistic of the optimal CUSUM chart can be defined by

$$C_n^- = \min\left(0, C_{n-1}^- + T(X_n) - k^-\right),\tag{4.34}$$

where $C_0^- = 0$, and k^- has the same expression as that of k^+ in (4.33). This chart gives a signal when $C_n^- < h^-$, where h^- is a control limit chosen to achieve a pre-specified ARL_0 value as well.

Next, we discuss several special continuous distributions that belong to the one-parameter exponential family (4.29), and give the formulas of the corresponding optimal CUSUM charts. It should be pointed out that monitoring the mean of a normal process and monitoring its variability are also special cases of the monitoring problem discussed here, because the normal distribution is an important distribution in the exponential family. They have been discussed in detail in Subsections 4.2.4 and 4.3.2, respectively.

Let us first discuss the case when X has a gamma distribution with the following pdf: for $x > 0$,

$$\begin{aligned} f(x; \alpha, \beta) &= \frac{x^{\alpha-1}}{\beta^\alpha \Gamma(\alpha)} e^{-x/\beta} \\ &= \exp\left[(\alpha-1)\log(x) - \frac{x}{\beta} - \alpha \log(\beta) - \log(\Gamma(\alpha))\right], \end{aligned}\tag{4.35}$$

where $\alpha > 0$ is a shape parameter, and $\beta > 0$ is a scale parameter. This distribution is often denoted as $\Gamma(\alpha, \beta)$. It can be checked (cf., (2.6) and (2.7)) that, if $X \sim \Gamma(\alpha, \beta)$, then

$$\mu_X = \alpha\beta, \qquad \sigma_X^2 = \alpha\beta^2.$$

If we are interested in detecting shifts in the shape parameter α, then β can be regarded as known and thus $f(x; \alpha, \beta)$ in (4.35) can be regarded as a special case of the one-parameter exponential family (4.29) with

$$\begin{aligned} \theta = \alpha, \qquad \eta(\theta) = \alpha - 1, \qquad T(x) = \log(x), \\ A(\theta) = -\alpha\log(\beta) - \log(\Gamma(\alpha)). \end{aligned}$$

The optimal CUSUM chart can be constructed by (4.32) or (4.34). For instance, if we are interested in detecting an upward shift in α from α_0 to α_1 with $\alpha_0 < \alpha_1$, then the charting statistic of the optimal CUSUM chart is defined by

$$C_n^+ = \max\left(0, C_{n-1}^+ + \log(X_n) - k^+\right),\tag{4.36}$$

where $C_0^+ = 0$, and

$$k^+ = \log\left(\frac{1}{\beta}\right) + \frac{\log\left(\Gamma(\alpha_0)/\Gamma(\alpha_1)\right)}{\alpha_1 - \alpha_0}.\tag{4.37}$$

The formulas (4.36) and (4.37) are derived from (4.32) and (4.33), respectively. The optimal CUSUM chart for detecting a downward shift in α can be constructed by (4.34) in a similar way.

It is easy to check that the distribution of $\Gamma(k/2,2)$ is the same as the chi-square distribution χ_k^2 (cf., its definition in Subsection 2.3.2). By this connection, the CUSUM charts for detecting shifts in α of the $\Gamma(\alpha,\beta)$ distribution can also be used for detecting shifts in k of the related chi-square distribution, as long as we set $\alpha = k/2$ and $\beta = 2$.

To detect shifts in the scale parameter β of the $\Gamma(\alpha,\beta)$ distribution, the shape parameter α can be regarded as known and thus $f(x;\alpha,\beta)$ in (4.35) can also be regarded as a special case of the one-parameter exponential family (4.29) with

$$\theta = \beta, \qquad \eta(\theta) = -1/\beta, \qquad T(x) = x,$$
$$A(\theta) = -\alpha \log(\beta).$$

Because $\eta(\theta)$ is an increasing function of θ in this case, the optimal CUSUM chart for detecting upward shifts in β can be constructed by (4.32) and the one for detecting downward shifts can be constructed by (4.34). For instance, if we are interested in detecting an upward shift in β from β_0 to β_1 with $\beta_0 < \beta_1$, then the charting statistic of the optimal CUSUM chart is defined by

$$C_n^+ = \max\left(0, C_{n-1}^+ + X_n - k^+\right), \tag{4.38}$$

where $C_0^+ = 0$, and

$$k^+ = \frac{\alpha\beta_0\beta_1 \log\left(\beta_0/\beta_1\right)}{\beta_1 - \beta_0}. \tag{4.39}$$

When $\alpha = 1$, the distribution of $\Gamma(1,\beta)$ is the same as the exponential distribution with a rate parameter $\lambda = 1/\beta$, denoted as *Exponential*(λ), which is commonly used in practice to describe the survival time of a biological or mechanical system. Therefore, the CUSUM chart for detecting shifts in β of the $\Gamma(\alpha,\beta)$ distribution can also be used for detecting shifts in λ of the related exponential distribution, as long as we set $\alpha = 1$ and $\beta = 1/\lambda$.

Detection of normal variance shifts can also be regarded as a special case of detection of shifts in the scale parameter β of the $\Gamma(\alpha,\beta)$ distribution. Let us first look at the CUSUM chart (4.21)–(4.22) for detecting upward normal variance shifts. From (4.21), it can be seen that the chart actually uses the transformed data

$$Y_n = \left(\frac{X_n - \mu_0}{\sigma_0}\right)^2.$$

When the process is IC, the distribution of Y_n is χ_1^2, which is equivalent to $\Gamma(1/2,2)$. After an upward shift of the process variance from σ_0^2 to σ_1^2 with $\sigma_0^2 < \sigma_1^2$, the distribution of Y_n becomes $\Gamma(1/2, 2\sigma_1^2/\sigma_0^2)$. Therefore, detection of the upward normal variance shift from σ_0^2 to σ_1^2 can also be achieved by detecting the upward shift in β from 2 to $2\sigma_1^2/\sigma_0^2$ in the $\Gamma(1/2,\beta)$ distribution. It can be checked that the control chart (4.38)–(4.39) in such a case is exactly the same as the one defined by (4.21)–(4.22). In a similar way, the CUSUM chart (4.25)–(4.26) that is constructed from batch data can also be derived as a CUSUM chart for detecting shifts in β of the $\Gamma((m-1)/2,\beta)$ distribution.

As described in Subsection 2.3.5, in life science and reliability analysis, the Weibull distribution plays an important role for describing the distribution of the lifetimes of products. If $X \sim Weibull(a,b)$, then its pdf has the expression

$$
\begin{aligned}
f(x;a,b) &= \frac{b}{a}\left(\frac{x}{a}\right)^{b-1} e^{-(x/a)^b}, \qquad \text{for } x \geq 0, \\
&= \exp\left[(b-1)\log(x) - \frac{x^b}{a^b} + \log(b) - b\log(a)\right], \qquad (4.40)
\end{aligned}
$$

where $a > 0$ is a scale parameter and $b > 0$ is a shape parameter. If we are interested in detecting shifts in the scale parameter a, then b can be regarded as known and $f(x;a,b)$ in (4.40) is obviously a member of the one-parameter exponential family (4.29) with

$$
\theta = a, \qquad \eta(\theta) = -a^{-b}, \qquad T(x) = x^b,
$$
$$
A(\theta) = \log(b) - b\log(a).
$$

The function $\eta(\theta)$ is an increasing function of θ. So, to detect an upward shift in a from a_0 to a_1 with $a_0 < a_1$, the optimal CUSUM chart can be constructed by (4.32) with the charting statistic

$$
C_n^+ = \max\left(0, C_{n-1}^+ + X_n^b - k^+\right),
$$

where $C_0^+ = 0$, and

$$
k^+ = \frac{b\log(a_1/a_0)}{a_1^{-b} - a_0^{-b}}.
$$

The optimal CUSUM chart for detecting a downward shift can be constructed by (4.34) in a similar way.

It is obvious that the distribution of $Weibull(a,1)$ is equivalent to the distribution of $Exponential(\lambda)$ with $\lambda = 1/a$. Therefore, the CUSUM charts discussed above for detecting shifts in the scale parameter a of the distribution $Weibull(a,b)$ can also be used for detecting shifts in λ of the $Exponential(\lambda)$ distribution.

CUSUM charts for cases with other continuous distributions in the exponential family, such as the beta, the log-normal, and the inverse normal distributions (cf., Exercise 4.17 in Section 4.8 below), can be discussed in a similar way. Olteanu (2010) discussed construction of CUSUM charts for monitoring censored reliability data, where Weibull and log-normal distributions are widely used. Related discussion can also be found in Steiner and MacKay (2000) and Zhang and Chen (2004).

4.4.2   Cases with discrete distributions in the exponential family

If we randomly select m products at each of a sequence of equally spaced time points of a production process and classify each product as either a conforming or a non-conforming product by certain design requirements, then the number of non-conforming products X of the sample collected at a single time point would have the *Binomial*(m, π) distribution, where π is the true proportion of non-conforming products. See Subsection 2.4.2 for a detailed description about the binomial distribution and its properties. For many applications, it is important to monitor the related production processes to make sure that π is stable over time. For these applications, we are interested in detecting shifts in π.

By (2.13), when $X \sim Binomial(m, \pi)$, its probability mass function (pmf) is, for $x = 0, 1, 2, \ldots, m$,

$$
\begin{aligned}
f(x; m, \pi) &= P(X = x) \\
&= \binom{m}{x} \pi^x (1 - \pi)^{m-x} \\
&= \exp\left[\log\binom{m}{x} + x \log\left(\frac{\pi}{1 - \pi} \right) + m \log(1 - \pi) \right].
\end{aligned}
$$

Obviously, this is a member of the one-parameter exponential family (4.29) with

$$
\theta = \pi, \qquad \eta(\theta) = \log\left(\frac{\pi}{1 - \pi} \right), \qquad T(x) = x,
$$
$$
A(\theta) = m \log(1 - \pi).
$$

Now, assume that we would like to monitor a production process based on its observed data $\{X_1, X_2, \ldots\}$, and X_n is the number of non-conforming products out of m products collected at the n-th time point, for $n \geq 1$. If we are interested in detecting an upward shift in π from π_0 to π_1 with $0 \leq \pi_0 < \pi_1 \leq 1$, then, because $\eta(\theta)$ is an increasing function of θ, the optimal CUSUM chart can be constructed by (4.32) with the charting statistic

$$
C_n^+ = \max\left(0, C_{n-1}^+ + X_n - k^+\right), \tag{4.41}
$$

where $C_0^+ = 0$, and

$$
k^+ = -\frac{m \log\left((1 - \pi_1)/(1 - \pi_0)\right)}{\log\left[\left(\frac{\pi_1}{1 - \pi_1} \right) \Big/ \left(\frac{\pi_0}{1 - \pi_0} \right) \right]}.
$$

This chart signals a shift when

$$
C_n^+ > h^+, \tag{4.42}
$$

where $h^+ > 0$ is a control limit chosen to achieve a pre-specified ARL_0 value. Similarly, to detect a downward shift in π from π_0 to π_1 with $0 \leq \pi_1 < \pi_0 \leq 1$, the optimal CUSUM chart can be constructed by (4.34) with the charting statistic

$$
C_n^- = \min\left(0, C_{n-1}^- + X_n - k^-\right), \tag{4.43}
$$

where $C_0^- = 0$, and k^- has the same expression as that of k^+.

Example 4.10 *Assume that the proportion of non-conforming products is known to be $\pi_0 = 0.1$ when a production process is IC, and we are interested in detecting an upward shift from $\pi_0 = 0.1$ to $\pi_1 = 0.2$ based on samples of size $m = 100$ each*

Table 4.5 *The actual ARL_0 values of the chart (4.41)–(4.42) for detecting an upward shift from $\pi_0 = 0.1$ to $\pi_1 = 0.2$ based on samples of size $m = 100$ each collected at consecutive time points. The value of k^+ is computed to be 14.524 by the formula given immediately below (4.41).*

h^+	ARL_0	h^+	ARL_0	h^+	ARL_0
3.0	93.94	3.7	171.21	4.4	193.42
3.1	93.94	3.8	171.21	4.5	193.42
3.2	93.94	3.9	171.21	4.6	350.32
3.3	93.94	4.0	193.42	4.7	350.32
3.4	93.94	4.1	193.42	4.8	350.32
3.5	93.94	4.2	193.42	4.9	350.32
3.6	171.21	4.3	193.42	5.0	412.93

collected at consecutive time points. In this case, we should use the upward CUSUM chart (4.41)–(4.42) with

$$k^+ = -\frac{100 \log\left((1-0.2)/(1-0.1)\right)}{\log\left[\left(\frac{0.2}{1-0.2}\right) \Big/ \left(\frac{0.1}{1-0.1}\right)\right]} = 14.524.$$

When h^+ changes its value from 3.0 to 5.0 with a step of 0.1, the actual ARL_0 value of this CUSUM chart is presented in Table 4.5. From the table, it can be seen that the CUSUM chart cannot have all possible ARL_0 values. As a matter of fact, when $h^+ \in [3.0, 5.0]$, it can only have 5 discrete ARL_0 values 93.94, 171.21, 193.42, 350.32, and 412.93.

Based on Table 4.5, assume that we decide to use $h^+ = 4.0$ with $ARL_0 = 193.42$. If the observed numbers of non-conforming products in the first 15 samples are

$$8, 17, 10, 8, 11, 14, 8, 16, 8, 13, 21, 15, 20, 18, 17,$$

then the CUSUM chart (4.41)–(4.42) is presented in Figure 4.11. From the plot, it can be seen that a signal is given by the chart at the 11th time point.

Example 4.10 demonstrates that the CUSUM chart (4.41)–(4.42) can only take certain discrete ARL_0 values when the value of k^+ is given. That is because its charting statistic C_n^+ defined in (4.41) can only take discrete values $\{C_{n-1}^+ - k^+, C_{n-1}^+ + 1 - k^+, \ldots, C_{n-1}^+ + n - k^+\}$ when both C_{n-1} and k^+ are given. This is true for the downward CUSUM chart with the charting statistic C_n^- defined in (4.43) as well. As a matter of fact, this phenomenon is generally true for cases with other discrete distributions and other types of control charts (e.g., Shewhart charts). Therefore, when we monitor processes with discrete distributions, we should be aware of this phenomenon and only specify the ARL_0 values that are achievable. These issues are summarized in the box below.

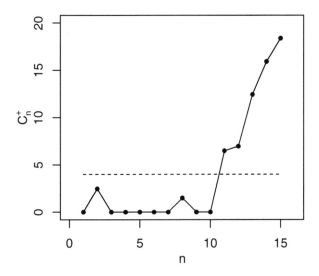

Figure 4.11 *The CUSUM chart (4.41)–(4.42) with $k^+ = 14.524$ and $h^+ = 4.0$ when it is applied to the data in Example 4.10*

Discreteness of the ARL_0 Values

When monitoring a production process with a discrete distribution, the ARL_0 value of a related control chart is usually discrete as well. In such cases, when designing the control chart, we should specify an ARL_0 value that is achievable.

Another widely used discrete distribution in practice is the Poisson distribution. See Subsection 2.4.5 for its definition and major properties. As described in Subsection 3.3.2, if X denotes the number of defects found in a randomly selected inspection unit, then the distribution of X can be described reasonably well by the *Poisson*(λ) distribution, where $\lambda > 0$ is a rate parameter, denoting the average number of defects in a randomly selected inspection unit. More generally, the *Poisson*(λ) distribution is often a reasonable probability model for describing the number of events (e.g., defects in a product, traffic accidents on a road) that occurred in a unit time interval or within a unit space.

By (2.15), if $X \sim Poisson(\lambda)$, then its pmf is, for $x = 0, 1, 2, \ldots,$

$$
\begin{aligned}
f(x; \lambda) &= P(X = x) \\
&= \frac{\lambda^x e^{-x}}{x!} \\
&= \exp\left[x \log(\lambda) - \lambda - \log(x!)\right].
\end{aligned}
$$

This pmf is also a member of the one-parameter exponential family (4.29) with

$$\theta = \lambda, \qquad \eta(\theta) = \log(\lambda), \qquad T(x) = x,$$
$$A(\theta) = -\lambda.$$

Assume that we are interested in detecting a shift in λ of a production process with the $Poisson(\lambda)$ distribution. The observed data from the process are denoted as $\{X_1, X_2, \ldots\}$, as usual. Then, because $\eta(\theta)$ is an increasing function of θ in this case, the optimal CUSUM chart for detecting an upward shift in λ from λ_0 to λ_1 with $0 < \lambda_0 < \lambda_1$ can be constructed by (4.32) with the charting statistic

$$C_n^+ = \max\left(0, C_{n-1}^+ + X_n - k^+\right), \tag{4.44}$$

where $C_0^+ = 0$, and

$$k^+ = \frac{\lambda_1 - \lambda_0}{\log(\lambda_1) - \log(\lambda_0)}.$$

This chart gives a signal when

$$C_n^+ > h^+, \tag{4.45}$$

where $h^+ > 0$ is a control limit chosen to achieve a pre-specified ARL_0 value. Because X_n takes nonnegative integer values only, for given values of C_{n-1}^+ and k^+, the possible values of C_n^+ are discrete. Therefore, the possible ARL_0 values of the CUSUM chart (4.44)–(4.45) are discrete as well. Optimal CUSUM charts for detecting downward shifts in λ can be constructed similarly by (4.34).

In the literature, there has been much discussion on monitoring of processes with the above two or some other discrete distributions that belong to the exponential family. Interested readers are referred to papers such as Bourke (2001), Gan (1993b), Megahed et al. (2011a), Mousavi and Reynolds (2009), Reynolds and Stoumbos (2000), Wu et al. (2008a), and Xie et al. (2002, 1998).

4.5 Self-Starting and Adaptive CUSUM Charts

This section discusses two topics. In Subsection 4.5.1, we describe the first topic about self-starting CUSUM charts for handling cases when the IC parameters are unknown. Then, in Subsection 4.5.2, we describe the second topic, about adaptive CUSUM charts that choose the allowance parameter k adaptively based on recursive estimation of the size of a potential shift.

4.5.1 Self-Starting CUSUM charts

In the previous sections when we describe different CUSUM charts, the IC mean μ_0 and the IC variance σ_0^2 are assumed to be known. In practice, μ_0 and σ_0^2 are rarely known. Instead, they need to be estimated from a sample obtained at the end of the phase I analysis after it is assured that the related process is IC. This sample is called *an IC dataset* hereafter. Therefore, instead of the true values of μ_0 and σ_0^2, we only have their estimators $\widehat{\mu}_0$ and $\widehat{\sigma}_0^2$ in most real cases. As discussed in

Subsection 2.7.1, estimators constructed from a random sample are random variables. Does the randomness in $\widehat{\mu}_0$ and $\widehat{\sigma}_0^2$ affect the performance of the CUSUM charts in a substantial way? If the answer is positive, how can we eliminate such an impact? Hawkins (1987) explored these research questions in detail. His major findings are described in this section. The first question is addressed in the example below.

Example 4.11 *Assume that the true IC distribution of a production process is $N(0,1)$, which is unknown to us. Instead, it needs to be estimated from an IC dataset of size 100. From Subsection 2.7.1, we know that μ_0 and σ_0^2 can be estimated by the sample mean and sample variance of the IC dataset, denoted as $\widehat{\mu}_0$ and $\widehat{\sigma}_0^2$, and their sampling distributions are*

$$\widehat{\mu}_0 \sim N(0, 0.01), \qquad 99\widehat{\sigma}_0^2 \sim \chi_{99}^2.$$

On one occasion, the values of $\widehat{\mu}_0$ and $\widehat{\sigma}_0^2$ computed from an IC dataset of size 100 are 0.1 and 1.21, respectively. We then use these two values as the assumed IC mean and IC variance in designing a CUSUM chart for detecting upward mean shifts. To use the results in Table 4.1, we consider the normalized version of the CUSUM chart (4.7)–(4.8), with the charting statistic

$$C_n^+ = \max\left(0, C_{n-1}^+ + \frac{X_n - \widehat{\mu}_0}{\widehat{\sigma}_0} - k^+\right), \qquad \text{for } n \geq 1, \qquad (4.46)$$

where $C_0^+ = 0$ and $k^+ > 0$ is the allowance. The chart gives a signal of an upward mean shift when

$$C_n^+ > h^+, \qquad (4.47)$$

where $h^+ > 0$ is a control limit. When we choose $ARL_0 = 200$ and $k^+ = 0.5$, by Table 4.1, h^+ should be 3.502. Now, we use the chart (4.46)–(4.47) to monitor the process in question. Remember that the actual IC distribution of the process is $N(0,1)$. When the process is IC, the actual distribution of $(X_n - \widehat{\mu}_0)/\widehat{\sigma}_0$ is $N(-\widehat{\mu}_0/\widehat{\sigma}_0, 1/\widehat{\sigma}_0^2) = N(-0.091, 0.826)$. So, the chart (4.46)–(4.47) in such a case is equivalent to the chart (4.7)–(4.8) for detecting a distributional shift from $N(0,1)$ to $N(-0.091, 0.826)$. By this connection, the actual ARL_0 value of the chart (4.46)–(4.47) is computed to be 785.49. Now, if there is a true upward mean shift of size $\delta = 0.2$, the actual ARL_1 value of the chart (4.46)–(4.47) is the same as the ARL_1 value of the chart (4.7)–(4.8) for detecting a distributional shift from $N(0,1)$ to $N(0.091, 0.826)$, which is computed to be 210.99. These results and the results in cases with several other values of $\widehat{\mu}_0$ and $\widehat{\sigma}_0$ are presented in Table 4.6. Note that all the values of $\widehat{\mu}_0$ and $\widehat{\sigma}_0$ listed in Table 4.6 are typical ones in the current setup.

From Example 4.11, it can be seen that the randomness of the estimators $\widehat{\mu}_0$ and $\widehat{\sigma}_0^2$ could have a substantial impact on the performance of the CUSUM chart (4.46)–(4.47). For instance, in cases when μ_0 and σ_0 are known beforehand, which corresponds to the case when $\widehat{\mu}_0 = 0$ and $\widehat{\sigma}_0 = 1.0$ in Table 4.6, the ARL_0 value is 199.99, and the ARL_1 value for detecting a mean shift of size 0.2 is 70.12. However, in the case when the values of $\widehat{\mu}_0$ and $\widehat{\sigma}_0$ are 0.1 and 1.1, respectively, the actual ARL_0 value of the chart becomes 785.49, which is much larger than the nominal ARL_0 value of 200. Consequently, the chart is ineffective for detecting the real upward mean shift

Table 4.6 *The actual ARL_0 values of the chart (4.46)–(4.47) in cases with several estimated values of the IC mean μ_0 and IC standard deviation σ_0, along with its actual ARL_1 values for detecting a mean shift of size 0.2. In the chart (4.46)–(4.47), the parameters (k^+, h^+) are chosen to be $(0.5, 3.502)$, so that its nominal ARL_0 is 200.*

$\widehat{\mu}_0$	$\widehat{\sigma}_0$	ARL_0	ARL_1
−0.1	0.9	67.22	30.15
−0.1	1.0	115.46	44.96
−0.1	1.1	210.99	70.06
0	0.9	107.58	43.97
0	1.0	199.99	70.12
0	1.1	398.01	117.98
0.1	0.9	179.58	67.22
0.1	1.0	362.26	115.46
0.1	1.1	785.49	210.99

of 0.2 in such a case, because its actual ARL_1 value of 210.99 is much larger than the ARL_1 value of 70.12 that it is supposed to have. In the case when values of $\widehat{\mu}_0$ and $\widehat{\sigma}_0$ are −0.1 and 0.9, respectively, the actual ARL_0 value of the chart becomes 67.22, indicating that the actual false alarm rate of the chart is much higher than its nominal false alarm rate and therefore results of the chart may not be reliable. Several authors, including Jensen et al. (2006), Jones et al. (2004), and Castagliola and Maravelakis (2011), have studied the impact of the randomness of the estimated IC parameters on the performance of the CUSUM charts more systematically. These authors confirmed that the estimated IC parameters could have a substantial impact on the performance of the CUSUM charts, unless we have a large IC dataset.

In many applications, however, we cannot have a large IC dataset. To overcome this difficulty, Hawkins (1987) suggested the so-called *self-starting CUSUM charts*, which try to estimate the IC parameters from the data observed in phase II SPC. They are described below in cases when we are interested in detecting upward shifts in the process mean and the process IC and OC distributions are normal. In such cases, the IC process distribution is $N(\mu_0, \sigma^2)$, the OC process distribution is $N(\mu_1, \sigma^2)$, $\mu_0 < \mu_1$, and the IC parameters μ_0 and σ^2 are unknown. In cases when μ_0 and σ^2 can be estimated accurately from an IC dataset, the CUSUM chart (4.46)–(4.47) is natural to use. When they cannot be estimated accurately from an IC dataset due to the unavailability of a large IC dataset, we try to estimate them from the available observations collected in the phase II SPC as follows. Assume that at the $(n-1)$-th time point, no signal of mean shift is given by the related control chart. A new observation X_n is thus collected at the n-th time point and we want to make a decision whether a signal of mean shift should be given at this time point. Because no signal is given at the $(n-1)$-th time point, all observations collected at that time point and before, i.e., the observations $X_{n-1}, X_{n-2}, \ldots, X_1$, can be regarded as IC observations. Therefore, μ_0 and σ^2 can be estimated by their sample mean and sample variance, denoted as $\overline{X}_{n-1}$ and s_{n-1}^2, respectively, as long as $n \geq 3$. Then, it is natural to use

$$T_n = \frac{X_n - \overline{X}_{n-1}}{s_{n-1}}$$

to replace $(X_n - \widehat{\mu}_0)/\widehat{\sigma}_0$ in (4.46) for constructing a CUSUM chart. In cases when $\{X_1, X_2, \ldots, X_n\}$ are i.i.d., which is true when the process is IC up to the n-th time point, it is easy to check that

$$\sqrt{\frac{n-1}{n}} T_n \sim t_{n-2},$$

because $X_n \sim N(\mu_0, \sigma^2)$, $\overline{X}_{n-1} \sim N(\mu_0, \sigma^2/(n-1))$, $(n-2)s_{n-1}^2/\sigma^2 \sim \chi_{n-2}^2$, and $(X_n, \overline{X}_{n-1}, s_{n-1}^2)$ are independent of each other (cf., the related discussion in Subsection 2.7.1). Also, Hawkins (1969) showed that $\{T_n, n \geq 1\}$ are independent of each other. Therefore, in such cases,

$$Z_n = \Phi^{-1}\left[\Upsilon_{n-2}\left(\sqrt{\frac{n-1}{n}} T_n \right) \right] \qquad (4.48)$$

would be i.i.d. with the $N(0,1)$ distribution, where Φ is the cdf of the $N(0,1)$ distribution, and Υ_{n-2} is the cdf of the t_{n-2} distribution. Because both Φ^{-1} and Υ_{n-2} are increasing functions, a mean shift in the original observations X_n occurs if and only if a mean shift in the transformed observations Z_n occurs. Therefore, detection of a normal mean shift in the original observations X_n can be accomplished by detecting a mean shift in the transformed observations Z_n, and the latter are i.i.d. with the $N(0,1)$ distribution when the process is IC. The self-starting CUSUM chart for detecting normal mean shifts is then summarized in the box below.

Self-Starting CUSUM Chart for Detecting Normal Mean Shifts

For detecting normal mean shifts when the IC parameters are unknown, the self-starting CUSUM chart is just the conventional CUSUM chart constructed from the transformed observations $\{Z_n, n \geq 3\}$ that are i.i.d. with the $N(0,1)$ distribution when the process is IC. The self-starting CUSUM chart requires at least 2 IC observations for phase II process monitoring.

Example 4.12 *Assume that $\{X_n, n = 1, 2, \ldots, 10\}$ listed in the 2nd column of Table 4.7 are the first 10 observations obtained from a production process for phase II SPC. In this application, we are interested in detecting upward mean shifts. Based on the historical data, it is reasonable to assume the process distribution to be normal, but the IC parameters μ_0 and σ^2 are unknown. So, in this example, we want to construct a self-starting CUSUM chart for detecting upward mean shifts. The computed sample means $\overline{X}_n$, for $n \geq 1$, the sample standard deviations s_n, for $n \geq 2$, and the transformed observations Z_n, for $n \geq 3$, are listed in the 3rd, 4th, and 5th columns of*

Table 4.7: *The first 10 values of X_n, $\overline{X}_n$, s_n, Z_n, and $C_{n,SS}^+$, respectively.*

n	X_n	$\overline{X}_n$	s_n	Z_n	$C_{n,SS}^+$
1	−0.502	−0.502	0.000	0.000	0.000
2	0.132	−0.185	0.448	0.000	0.000
3	−0.079	−0.150	0.323	0.153	0.000
4	0.887	0.109	0.581	1.605	1.355
5	0.117	0.111	0.504	0.011	1.115
6	0.319	0.145	0.458	0.351	1.216
7	−0.582	0.042	0.501	−1.277	0.000
8	0.715	0.126	0.521	1.138	0.888
9	−0.825	0.020	0.581	−1.518	0.000
10	−0.360	−0.018	0.561	−0.594	0.000

Table 4.7. Then, the charting statistic of the self-starting CUSUM chart for detecting upward mean shifts is defined by

$$C_{n,SS}^+ = \max\left(0, C_{n-1,SS}^+ + Z_n - k\right), \qquad for\ n \geq 3, \qquad (4.49)$$

where $C_{1,SS}^+ = C_{2,SS}^+ = 0$, and $k > 0$ is the allowance parameter. When $k = 0.25$, $\{C_{n,SS}^+, n = 1, 2, \ldots, 10\}$ are presented in the last column of Table 4.7.

We then collect another 30 observations from the same production process. The self-starting CUSUM chart (4.49) with $(k, h) = (0.25, 5.597)$ (note: its ARL_0 value is 200 according to Table 4.1) constructed from all 40 observations is presented in Figure 4.12 by the solid curve. It gives a signal at the 20th time point. The first 10 observations used in this example are actually generated from the $N(0, 1)$ distribution, and the remaining 30 observations are generated from the $N(0.5, 1)$ distribution. Therefore, a true mean shift occurs at the 11th time point. For comparison purposes, the conventional CUSUM chart (4.7)–(4.8) using the same values of (k, h) is also plotted in Figure 4.12 by the dotted curve. In this chart, both μ_0 and σ are assumed to be known with values 0 and 1, respectively. This chart gives a signal at the 22nd time point.

Example 4.12 demonstrates that the self-starting CUSUM chart can indeed detect mean shifts in an effective way without knowing the IC parameter values. However, Figure 4.12 also shows that, unlike the conventional CUSUM chart whose signal of mean shift is persistent and gets stronger and stronger over time, the signal of mean shifts from the self-starting CUSUM chart cannot last quite as long, which can be explained intuitively as follows. Assume that a mean shift occurs at the time point τ. Then, when $n \geq \tau$, the mean of $X_n - \overline{X}_{n-1}$ is

$$E\left(X_n - \overline{X}_{n-1}\right) = \frac{\tau - 1}{n - 1}(\mu_1 - \mu_0). \qquad (4.50)$$

From (4.48) and (4.49), it can be seen that the mean shift affects the performance of the self-starting CUSUM chart through $X_n - \overline{X}_{n-1}$. Therefore, from the expression

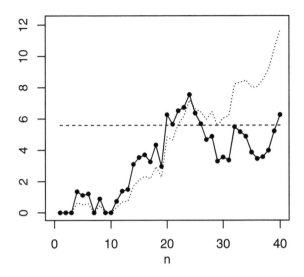

Figure 4.12 *The self-starting CUSUM chart (4.49) with* $(k, h) = (0.25, 5.597)$ *is shown by the solid curve. The conventional CUSUM chart (4.7)–(4.8) with the same values of* (k, h) *is shown by the dotted curve. The dashed horizontal line denotes the control limit h for both charts.*

(4.50), it can be seen that such an effect is strong when $n - \tau$ is relatively small. As n increases, the mean of $X_n - \overline{X}_{n-1}$ tends to 0, making the signal of mean shifts from the self-starting CUSUM chart disappear gradually. Thus, in practice, it is important to react to the signal from the self-starting CUSUM chart quickly. Otherwise, even a persistent mean shift can be missed by it. The expression (4.50) also implies that the impact of the mean shift on the self-starting CUSUM chart would be weak when τ is small. Therefore, it is a good idea to collect a dozen or more IC observations before the self-starting CUSUM chart is used.

If we are interested in detecting shifts in both process mean and process variance, then we can use a pair of self-starting CUSUM charts. One is constructed from Z_n, as in Example 4.12, for detecting mean shifts, and the other one is constructed from Z_n^2 for detecting variance shifts. These two CUSUM charts are correlated. A mean shift would affect not only the first chart, but also the second chart. Similarly, a variance shift would affect both charts as well. For detailed discussion on construction of self-starting CUSUM charts for detecting normal variance shifts, see Section 7.3 of Hawkins and Olwell (1998).

Nowadays, self-starting control charts are popular in the literature. See, for instance, Chatterjee and Qiu (2009), Hawkins and Maboudou-Tchao (2007), and Sullivan and Jones (2002). Some of the control charts discussed in these papers will be described in later chapters.

4.5.2 Adaptive CUSUM charts

The conventional CUSUM chart (4.7)–(4.8) has an allowance parameter k involved. For a potential upward mean shift of size δ, it has been mentioned in Subsection 4.2.2 that the chart is optimal for detecting that shift when k is set to equal $\delta/2$. In practice, however, δ is often unknown at the time when we design the CUSUM chart. In such cases, how can k be chosen properly? Sparks (2000) addressed this issue carefully, and proposed two approaches to solve the problem, which are described below. For simplicity of presentation, our discussion here is on the specific case for detecting upward normal mean shifts by the CUSUM chart (4.7)–(4.8), although Sparks' adaptive CUSUM idea is quite general and it can be applied to other cases as well. Also, without loss of generality, we assume that the IC mean $\mu_0 = 0$ and the IC variance $\sigma^2 = 1$. Otherwise, standardized observations $\{(X_n - \mu_0)/\sigma, n \geq 1\}$ can be used for constructing the related CUSUM charts.

Sparks' first proposal is to use several CUSUM charts simultaneously, and these charts use different k values so that mean shifts of different sizes are targeted simultaneously by the joint control scheme. Assume that p CUSUM charts are used jointly, and the j-th CUSUM chart uses (k_j, h_j) as its allowance and control limit, for $j = 1, 2, \ldots, p$. Then, the joint control scheme has a signal if and only if at least one of the p individual CUSUM charts has a signal. In his paper, Sparks suggested using $p = 3$ if we are interested in detecting moderate shifts in the range $\delta \in [0.5, 2]$. In such cases, he suggested using $k_1 = 0.375, k_2 = 0.5$, and $k_3 = 0.75$. Then, $\{h_j, j = 1, 2, 3\}$ can be chosen accordingly, such that the ARL_0 values of the three charts are the same and the overall ARL_0 of the joint control scheme equals a pre-specified value. Based on his numerical results, such a joint control scheme is especially effective for detecting shifts in the range $[0.6, 1.75]$. Of course, if we have some prior information about the potential mean shifts, then such prior information should be accommodated in choosing k_j and ARL_0 values of the individual CUSUM charts.

Sparks' second proposal is to estimate the shift size δ recursively at each time point, and then choose k accordingly. To this end, let us first re-write the charting statistic of the conventional CUSUM chart (4.7)–(4.8) as

$$\widetilde{C}_n^+ = \max\left[0, \widetilde{C}_{n-1}^+ + (X_n - k)/h\right], \qquad \text{for } n \geq 1,$$

where $\widetilde{C}_n^+ = C_n^+/h$, and $\widetilde{C}_0^+ = 0$. Then, the chart gives a signal when $\widetilde{C}_n^+ > 1$. At the n-th time point, if δ can be estimated by $\widehat{\delta}_n$, then it is natural to define the charting statistic of the upward CUSUM chart by

$$C_{n,A}^+ = \max\left[0, C_{n-1,A}^+ + \left(X_n - \widehat{\delta}_n/2\right)\Big/h(\widehat{\delta}_n)\right], \qquad \text{for } n \geq 1, \qquad (4.51)$$

where $C_{0,A}^+ = 0$ and $h(\widehat{\delta}_n)$ is the control limit. Namely, in the control chart (4.51), k is chosen as $\widehat{\delta}_n/2$, which can change its value at different time points. So can the control limit $h(\widehat{\delta}_n)$. Sparks proposed the following recursive formula for estimating δ:

$$\widehat{\delta}_n = \max\left(\delta_{min}, \lambda X_n + (1-\lambda)\widehat{\delta}_{n-1}\right), \qquad \text{for } n \geq 1, \qquad (4.52)$$

where $\widehat{\delta}_0 = \delta_{min}$, $\delta_{min} \geq 0$ is the minimum shift size that we are interested in detecting, and $\lambda \in (0,1]$ is a weighting parameter. In (4.52), we need to specify the minimum shift size δ_{min}, so that the selected k could make the CUSUM chart (4.51) effective in detecting upward mean shifts with sizes equal to or larger than δ_{min}. In practice, if shifts of all sizes need to be detected, then we can simply set $\delta_{min} = 0$. From (4.52), it can be seen that process observations contribute to the estimator $\widehat{\delta}_n$ through the weighted average $\lambda X_n + (1 - \lambda)\widehat{\delta}_{n-1}$. This weighted average, which is the so-called exponentially weighted moving average and which will be described in detail in the next chapter, guarantees that more recent observations receive more weight in the average. Based on the numerical study in Sparks (2000), the value of λ can be chosen between 0.1 and 0.2.

Because the chosen k depends on $\widehat{\delta}_n$, the control limit $h(\widehat{\delta}_n)$ of the chart (4.51) should also depend on $\widehat{\delta}_n$, in order to reach a pre-specified ARL_0 value. Then, the chart (4.51) gives a signal of an upward mean shift at the n-th time point if

$$C_{n,A}^+ > 1. \tag{4.53}$$

Based on a large simulation study, Sparks provided formulas for approximating $h(\widehat{\delta}_n)$ in cases with different ARL_0 values. For instance, when $ARL_0 = 200$, he suggested approximating $h(\widehat{\delta}_n)$ by

$$-4.3883 + 25.4353 \bigg/ \sqrt{1 + 8.895\widehat{\delta}_n - 0.8525\widehat{\delta}_n^2 + 0.7295\widehat{\delta}_n}$$

$$-1.5652\widehat{\delta}_n^2 + 1.5065\widehat{\delta}_n^3 - 0.7262\widehat{\delta}_n^4 + 0.1730\widehat{\delta}_n^5 - 0.0163\widehat{\delta}_n^6.$$

Based on a Markov chain model, Shu and Jiang (2006) provided the following alternative formula for approximating $h(\widehat{\delta}_n)$:

$$h(\widehat{\delta}_n) \approx \frac{\log\left(1 + 2\widehat{\delta}_n^2 ARL_0 + 2.332\widehat{\delta}_n\right)}{2\widehat{\delta}_n} - 1.166. \tag{4.54}$$

Note that the formula (4.54) can be used for all values of ARL_0. In their paper, Shu and Jiang pointed out that this formula provided an accurate approximation to $h(\widehat{\delta}_n)$ in cases when $\widehat{\delta}_n ARL_0 \gg h(\widehat{\delta}_n)$, which is true in most applications.

4.6 Some Theory for Computing ARL Values

In the literature, there are some theoretical results for computing ARL values of CUSUM charts, based on which some numerical algorithms have been developed (e.g., some functions in the R package spc). In this section, two major theoretical methodologies for computing the ARL values are described. The method by the Markov chain approach is discussed in Subsection 4.6.1, and the method by the integral equations is discussed in Subsection 4.6.2. Our discussion focuses on the upward CUSUM chart (4.7)–(4.8), although similar theoretical results are also available for some other CUSUM charts.

4.6.1 The Markov chain approach

A sequence of discrete random variables $\xi_1, \xi_2, \ldots$ is called a *Markov chain* if

(i) all possible values of $\{\xi_n, n \geq 1\}$ form a countable set Ω, called the *state space* of the chain, and

(ii) the sequence has the *Markov property* that for any $n \geq 1$,

$$P(\xi_{n+1} = u_{n+1} | \xi_1 = u_1, \xi_2 = u_2, \ldots, \xi_n = u_n) = P(\xi_{n+1} = u_{n+1} | \xi_n = u_n),$$

for any $u_1, u_2, \ldots, u_n, u_{n+1} \in \Omega$.

The Markov property says that ξ_{n+1} depends on all previous observations through ξ_n. In other words, conditional on ξ_n, ξ_{n+1} and all other history data $\{\xi_1, \xi_2, \ldots, \xi_{n-1}\}$ are independent. The Markov chain is called a *stationary Markov chain* if it has the time-homogeneous property that, for any $n \geq 2$ and any $u, v \in \Omega$,

$$P(\xi_{n+1} = u | \xi_n = v) = P(\xi_n = u | \xi_{n-1} = v).$$

If $\{\xi_1, \xi_2, \ldots\}$ is a stationary Markov chain and its state space Ω has N states $\{\omega_1, \omega_2, \ldots, \omega_N\}$, then the probability

$$p_{ij} = P(\xi_{n+1} = \omega_j | \xi_n = \omega_i), \qquad \text{for any } n \geq 1$$

is called a (single-step) transition probability from the i-th state to the j-th state, for all $i, j = 1, 2, \ldots, N$. The $N \times N$ matrix $\mathbf{P} = (p_{ij})$ is called the (single-step) transition probability matrix. For detailed discussion about Markov chains and their properties, see the classic references Doob (1953) and Hoel et al. (1972).

To compute the ARL values of the CUSUM chart (4.7)–(4.8), Brook and Evans (1972) made a connection between this chart and a Markov chain. Let us first assume that process observations $\{X_1, X_2, \ldots\}$ take integer values, and so do the parameters k and h. In such cases, the charting statistic C_n^+, for any $n \geq 1$, can only take $h + 1$ integer values in the state space $\Omega = \{0, 1, 2, \ldots, h\}$. When $C_n^+ = h$, a signal is delivered and the process is stopped. So, the last state h is also called the absorbing state. When the value of C_n^+ is given, C_{n+1}^+ depends only on X_{n+1}, and thus it is independent of $\{C_1^+, C_2^+, \ldots, C_{n-1}^+\}$, for any $n \geq 1$. Also, when the process is IC, it is obvious that the conditional distribution of $C_{n+1}^+ | C_n^+$ does not depend on n. Therefore, the sequence $\{C_n^+, n \geq 1\}$ is a stationary Markov chain with the state space Ω. Its (single-step) transition probabilities are, for $i = 0, 1, 2, \ldots, h-1$, and $j = 1, 2, \ldots, h-1$,

$$
\begin{aligned}
p_{i0} &= P(C_n^+ = 0 | C_{n-1}^+ = i) = P(X_n \leq k - i), \\
p_{ij} &= P(C_n^+ = j | C_{n-1}^+ = i) = P(X_n = k + j - i), \\
p_{ih} &= P(C_n^+ = h | C_{n-1}^+ = i) = P(X_n \geq k + h - i), \\
p_{h0} &= P(C_n^+ = 0 | C_{n-1}^+ = h) = 0, \\
p_{hj} &= P(C_n^+ = j | C_{n-1}^+ = h) = 0, \\
p_{hh} &= P(C_n^+ = h | C_{n-1}^+ = h) = 1.
\end{aligned}
\qquad (4.55)
$$

Let $\mathbf{P} = (p_{ij})$ be the $(h+1) \times (h+1)$ transition matrix, $\mathbf{R}$ be the sub-matrix of $\mathbf{P}$ consisting of the first h rows and the first h columns of $\mathbf{P}$, and T_i be the number of steps taken from the state i to the absorbing state h, for $i = 0, 1, \ldots, h-1$. Then, for $i = 0, 1, \ldots, h-1$,

$$
\begin{aligned}
\mu_{T_i} &= \mathrm{E}(T_i) \\
&= \sum_{r=1}^{\infty} r P(T_i = r) \\
&= \sum_{r=1}^{\infty} r \sum_{j=0}^{h} p_{ij} P(T_j = r - 1) \\
&= \sum_{j=0}^{h-1} p_{ij} \left(\mu_{T_j} + 1 \right) + p_{ih}.
\end{aligned}
\tag{4.56}
$$

The last equation of (4.56) can be explained intuitively as follows. The event from the state i to the absorbing state h can be accomplished in two stages. First, the state i at the current time point is transited to the state j at the next time point; and then the state j at the next time point is transited to the absorbing state h. So, if μ_{T_j} denotes the average number of steps taken from the state j at the next time point to the absorbing state h, then $\mu_{T_j} + 1$ is the average number of steps taken from the state i at the current time point to the absorbing state h, given the condition that the chain is in the state j at the next time point. When $j = h$, it is obvious that $\mu_{T_j} + 1 = 1$. Thus, that equation is true. The last equation of (4.56) can be written in matrix notation as

$$
(\mathbf{I} - \mathbf{R}) \mu_{\mathbf{T}} = \mathbf{1},
\tag{4.57}
$$

where $\mathbf{I}$ is the $h \times h$ identity matrix, $\mu_{\mathbf{T}} = (\mu_{T_0}, \mu_{T_1}, \ldots, \mu_{T_{h-1}})'$, and $\mathbf{1} = (1, 1, \ldots, 1)'$.

It can be seen that μ_{T_0} is just the *ARL* value of the chart (4.7)–(4.8) in cases when process observations and both k and h are discrete. If the probabilities defined in (4.55) are calculated from the IC process distribution, then μ_{T_0} is the ARL_0 value of that chart. Therefore, the *ARL* value of the chart (4.7)–(4.8) can be computed from (4.57). As a by-product, we can also compute μ_{T_i}, for $i = 1, 2, \ldots, h-1$, from (4.57). These are the *ARL* values of the chart (4.7)–(4.8) when the initial value of its charting statistic is defined to be $C_0^+ = i$. The resulting CUSUM chart is the so-called *fast initial response (FIR) CUSUM chart* that was systematically studied by Lucas and Crosier (1982a).

When the process distribution is continuous with a cdf F, the charting statistic C_n^+ can take values in $[0, \infty)$. In such cases, we can categorize C_n^+ by dividing $[0, \infty)$ into $M + 2$ groups:

$$
0, (0, h/M], (h/M, 2h/M], \ldots, ((M-1)h/M, h], (h, \infty).
$$

These $M + 2$ groups constitute the discrete states of C_n^+, among which the last state (h, ∞) is the absorbing state. Then, the transition probabilities can be computed similarly to those in (4.55), by various different approximations. For instance, to handle the condition that $C_{n-1}^+ \in ((i-1)h/M, ih/M]$, for $i = 1, 2, \ldots, M$, when computing the transition probabilities, Brook and Evans (1972) suggested an approximation by replacing C_{n-1}^+ by the mid-point of the interval $((i-1)h/M, ih/M]$. Hawkins (1992a) suggested a more accurate approximation, by which the conditional probability $P(a < C_n^+ \le b | c < C_{n-1}^+ \le d)$, for any intervals $(a,b]$ and $(c,d]$, that needs to be handled when computing the transition probabilities, is approximated by

$$\frac{1}{6}\{[F(b-c+k)+4F(b-(c+d)/2+k)+F(b-d+k)] \\ -[F(a-c+k)+4F(a-(c+d)/2+k)+F(a-d+k)]\}.$$

4.6.2   The integral equations approach

The integral equations approach for computing the ARL of a CUSUM chart was first suggested by Page (1954). This approach is also discussed quite extensively by Van Dobben de Bruyn (1968). Let $L(z)$ be the ARL of the CUSUM chart (4.7)–(4.8) with the allowance parameter k and the control limit h when the chart starts with $C_0^+ = z$, where $z \in (0,h]$ is the so-called *head start* value of the chart. As mentioned in the previous subsection, this is the FIR version of the CUSUM chart. When $z = 0$, the FIR version is just the conventional CUSUM chart defined in (4.7)–(4.8). To make the distinction between the two different versions, sometimes the conventional CUSUM chart is called the CUSUM chart with *zero-start*. Let us further assume that the process distribution has the cdf $F(x)$ and the pdf $f(x)$. Then, after the first observation X_1 is obtained, one of the following three events will happen:

$$E_1 = \left(C_1^+ = 0\right), \qquad E_2 = \left(C_1^+ \in (0,h]\right), \qquad E_3 = \left(C_1^+ > h\right).$$

Then, it is easy to see that

$$L(z) = 1 + L(0)P(E_1) + \int_0^h L(x)dF_{C_1^+}(x),$$

where $F_{C_1^+}(x)$ denotes the cdf of C_1^+ given E_2. On the right-hand side of the above expression, the first term "1" denotes the one-unit time lag from C_0^+ to C_1^+, $L(0)$ in the second term denotes the ARL value counting from X_1 in cases when E_1 occurs, $L(x)$ in the third term denotes the ARL value counting from X_1 in cases when E_2 occurs and when $C_1^+ = x$, and $dF_{C_1^+}(x)$ in the third term provides a measure of likelihood of the event $C_1^+ = x$ given E_2. It is obvious that

$$P(E_1) = P(C_1^+ = 0) = P(z + X_1 - k \le 0) = P(X_1 \le k - z) = F(k - z),$$

and for $x \in (0,h]$

$$F_{C_1^+}(x) = P(C_1^+ \le x | C_1^+ \in (0,h]) = P(z + X_1 - k \le x) = F(x + k - z).$$

Therefore, we have the integral equation

$$L(z) = 1 + L(0)F(k - z) + \int_0^h L(x)f(x + k - z)dx. \qquad (4.58)$$

To compute $L(z)$, we need to solve the integral equation (4.58). To this end, some numerical algorithms are available. See, for instance, Gan (1992a), Goel and Wu (1971), Luceño and Puig-Pey (2000), and Woodall (1983) for related discussion.

4.7 Some Discussions

In this chapter, we have described some CUSUM charts for monitoring the mean and/or variance of a univariate normal process. These CUSUM charts enjoy certain optimality properties, as described in Subsection 4.2.4, when detecting persistent mean and/or variance shifts. However, the optimality properties are achievable only when the IC parameters and the mean/variance shift sizes are known. When the IC parameters are unknown, we can use the self-starting CUSUM charts described in Subsection 4.5.1, which recursively estimate the IC parameters during online monitoring of the process. They perform reasonably well as long as there are one dozen or more IC observations collected before the phase II process monitoring. Also, we need to react to their signals quickly. Otherwise, even a persistent shift could be missed. The adaptive CUSUM charts described in Subsection 4.5.2 are designed for handling the cases when the target shift size is unknown. They try to estimate the size of a potential shift at each time point, and constantly update the procedure parameters based on the estimated shift size.

When the IC and OC pdf's (or pmf's) of process observations can be specified properly, a general formula for constructing a CUSUM chart was given in Subsection 4.2.4. By this formula, some optimal CUSUM charts were derived in Sections 4.3 and 4.4 in certain cases when the process distributions belong to the one-parameter exponential family. However, most of these CUSUM charts are designed for detecting shifts in one specific parameter of the process distribution. In many cases, a shift in that parameter would change both the mean and variance of the process distribution. As an example, assume that the process distribution is χ_k^2. Then, the mean and variance of this distribution are k and $2k$, respectively. Therefore, the optimal CUSUM chart discussed in Subsection 4.4.1 for detecting a shift in k is actually not designed for detecting a shift in either the mean alone or the variance alone of the process distribution. In applications, however, we are often concerned about shifts in the process mean and/or variance. To this end, we need to introduce a location or scale parameter when it is necessary, and derive the CUSUM chart using the general formula (4.20) based on that parameter. In such cases, the process distribution, in terms of the introduced parameter, may not belong to the one-parameter exponential family any more. Construction of appropriate CUSUM charts for handling such cases requires further research.

Some interesting univariate CUSUM charting techniques have not been discussed in this chapter yet. For instance, when designing the upward CUSUM chart (4.7)–(4.8), we need to specify a target shift size δ such that its allowance constant k can be chosen as $\delta/2$. In many applications, however, the single target shift size δ is difficult to specify. Instead, based on our past experience and engineering or scientific knowledge about the related production process, it is more convenient to specify a range $[a, b]$ for the potential shift size δ. In the literature, there is much discussion on designing appropriate CUSUM monitoring schemes for jointly detecting a range of mean/variance shifts. Most of such joint monitoring schemes are combinations of two or more individual CUSUM charts that are properly designed to achieve good overall performance. Interested readers are referred to papers including Han et al. (2007), Lorden (1971), Sparks (2000), and Zhao et al. (2005). A related methodology constructs a CUSUM chart by accommodating a prior distribution on the potential shift size δ. See, for instance, Liu (2010) and Ryu et al. (1995).

In this chapter, we only discuss how to detect step shifts in the process mean and/or variance. In practice, the process mean and/or variance often changes gradually with or without a parametric pattern, after the process becomes OC. For instance, the surface finish or physical size of the parts produced by a machine could change gradually due to tool wear. When the mean shift follows a linear model (i.e., the case with the so-called *linear drift*), design and implementation of the CUSUM charts have been discussed by several authors, including Bissell (1984a,b), Davis and Woodall (1988), and Gan (1992b). Shu et al. (2008) proposed a CUSUM chart for detecting gradual mean shifts that follow a parametric model.

In recent years, CUSUM charts with variable sampling rate (VSR) have received much attention in the literature (e.g., Costa, 1998; Luo et al., 2009; Reynolds and Arnold, 2001; Reynolds et al., 1990; Wu et al., 2007). By using a VSR CUSUM chart, the sampling rate changes over time based on all observed data. There are several possible ways to change the sampling rate, including the variable sampling intervals (VSI), the variable sample sizes (VSS), and the variable sample sizes and sampling intervals (VSSI) approaches. One major advantage of the VSR CUSUM charts, compared to the fixed sampling rate (FSR) CUSUM charts, is that VSR charts often provide faster detection of small to moderate process mean/variance shifts, for a given ARL_0 value and a given IC average sampling rate. The sampling scheme of a conventional VSI CUSUM chart involves sampling intervals of two different lengths. A longer sampling interval should be used to collect the next sample in cases when the charting statistic value at the current time point is far away from the control limits and falls into the so-called central region, and a shorter sampling interval should be used to collect the next sample in cases when the charting statistic value is close to but does not exceed the control limits and is within the so-called warning region. In cases when the charting statistic value exceeds the control limits (i.e., it falls into the so-called action region), the process is considered OC and stopped immediately. Recently, Li et al. (2013b) and Li and Qiu (2013) suggested implementing a CUSUM chart using statistical p-values (cf., Subsection 2.7.3), based on which the concept of dynamic sampling was proposed. By a dynamic sampling scheme, the sampling interval to collect the next sample can have an arbitrary length (instead of just two different lengths), determined by the p-value of the charting statistic at the current time point.

4.8   Exercises

4.1 Generate 10 samples of size 5 each from the $N(0,1)$ distribution and another 10 samples of size 5 each from the $N(0.2,1)$ distribution. Repeat the analysis described in Example 4.1. Are your results similar to those presented in Example 4.1? Note that your results would not be exactly the same as those in Example 4.1, due to the randomness of the samples.

4.2 Assume that we obtain the following 20 observations from a production process whose IC distribution is $N(0,1)$:

n	1	2	3	4	5	6	7	8	9	10
X_n	1.68	−0.15	−0.76	−1.31	−1.23	0.83	0.51	−0.26	−0.50	−0.62

n	11	12	13	14	15	16	17	18	19	20
X_n	1.20	2.19	1.77	1.40	2.61	−0.19	1.88	2.45	1.55	1.29

(i) Use the V-mask form of the CUSUM chart with the charting statistic C_n defined in (4.1), $k = 0.5$, and $h = 3.502$ to detect a potential upward mean shift. Does your CUSUM chart give any signal? If yes, provide estimates of the shift location and shift size?

(ii) Use the CUSUM chart (4.7)–(4.8) to detect a potential upward mean shift. Compare the results with those in part (i).

4.3 Show that the CUSUM chart (4.5)–(4.6) and the CUSUM chart (4.7)–(4.8) are equivalent to each other.

4.4 For the CUSUM chart (4.7)–(4.8), assume that k and h take several different pairs of values listed below. Compute the ARL_0 values of the chart in these cases when the IC process distribution is assumed to be $N(0,1)$ using an algorithm described by the pseudo code given in Subsection 4.2.2. In the algorithm, use $M = 10^5$.

(i) $k = 0.25$, $h = 3.5$,

(ii) $k = 0.25$, $h = 5.5$,

(iii) $k = 0.5$, $h = 3.5$.

Based on the computed results, describe your major findings about the relationship among ARL_0, k, and h.

4.5 For the CUSUM chart (4.7)–(4.8), assume that the IC process distribution is $N(0,1)$, its ARL_0 and k take the values listed below. Search for the values of its control limit h in these cases, using an algorithm described by the pseudo code given in Subsection 4.2.2. In the algorithm, use $\rho = 0.5$.

(i) $ARL_0 = 350$, $k = 0.25$,

(ii) $ARL_0 = 350$, $k = 0.5$,

(iii) $ARL_0 = 550$, $k = 0.5$.

Based on the computed results, describe your major findings about the relationship among ARL_0, k, and h.

4.6 Redo the calculation in Exercise 4.4 using the R package spc or other software packages that are available to you.

4.7 Redo the calculation in Exercise 4.5 using the R package spc or other software packages that are available to you.

4.8 Redo the calculation in Exercise 4.4 using the approximation formula (4.11). Compare the results here with those obtained in Exercise 4.4.

4.9 Assume that the IC process distribution is $N(0,1)$. Using the relationship between the ARL_0 and ARL_1 values that is summarized in Subsection 4.2.2 and the approximation formula (4.11), compute the ARL_1 values of the chart (4.7)–(4.8) with $k = 0.5$ and $h = 4.095$ (note: its ARL_0 value is 200 by Table 4.1), in cases when the process distribution shifts at the initial observation time point to one of the following six distributions:

(i) $N(0.5,1)$,

(ii) $N(1,1)$,

(iii) $N(-1,1)$,

(iv) $N(0,2^2)$,

(v) $N(0,0.5^2)$,

(vi) $N(1,2^2)$.

Summarize your results in terms of the relationship between ARL_1 and the shift size in the process mean and/or the shift size in the process variance.

4.10 Using the relationship between ARL_0 and ARL_1 values that is summarized in Subsection 4.2.2 and the approximation formula (4.11), reproduce the results in Example 4.4.

4.11 In the two-sided version of the CUSUM chart for detecting arbitrary mean shifts, assume that we use $(k,h) = (0.5, 3.08)$ in both the upward CUSUM chart (4.7)–(4.8) and the downward CUSUM chart (4.9)–(4.10).

(i) Compute the ARL_0 value of the two-sided CUSUM chart.

(ii) To detect a mean shift of size $\delta = 1.2$, what will be the zero-state ARL_1 values of the two-sided CUSUM chart and the two one-sided CUSUM charts?

(iii) To detect a mean shift of size $\delta = -1.2$, what will be the zero-state ARL_1 values of the two-sided CUSUM chart and the two one-sided CUSUM charts?

(iv) Summarize your major findings from the results in parts (ii) and (iii) regarding the ARL_1 values of the three different versions of the CUSUM charting techniques.

4.12 Assume that process observations follow the AR(1) model (4.15). Use a numerical method to compute the actual ARL_0 values of the CUSUM chart (4.7)–(4.8) with $(k,h) = (0.5, 4.095)$ (note: by Table 4.1, the nominal ARL_0 value of the chart is 370) in cases when $\phi = -0.5, 0$, and 0.5, respectively.

4.13 Assume that the IC process distribution is $N(0, \sigma_0^2)$, the process variance changes to σ_1^2 after the process becomes OC, and the process mean remains unchanged. Then, the CUSUM charts (4.21)–(4.22) and (4.23)–(4.24) are optimal for detecting upward and downward variance shifts, respectively. In each of the following three cases, design an optimal CUSUM chart (i.e., determine the optimal values of (k^+, h_U) or (k^-, h_L)):

(i) $\sigma_0 = 1$, $\sigma_1 = 2$, and $ARL_0 = 370$,

(ii) $\sigma_0 = 1$, $\sigma_1 = 0.5$, and $ARL_0 = 370$,

(iii) $\sigma_0 = 0.5$, $\sigma_1 = 1$, and $ARL_0 = 370$.

Note that the h_U or h_L value in each of the three cases can be approximated by a numerical algorithm.

4.14 The data given in Exercise 3.5 contain 10 samples of size 5 each from a process producing bearings. Based on our experience, it is assumed that the IC process distribution is $N(35.5, 1)$. For this process, we are mainly concerned about downward variance shifts. Design a CUSUM chart for this purpose with $ARL_0 = 200$ and a target OC variance of $\sigma_1^2 = 0.25$. Then, apply the designed CUSUM chart to the observed data, and summarize your results.

4.15 For a production process, its IC distribution is assumed to be $N(1, 0.5^2)$, and we are concerned about both upward mean shifts and upward variance shifts. The sample means and sample standard deviations of 24 independent random samples of size $m = 5$ each collected at 24 consecutive time points from the process are listed in the table below. Design a CUSUM joint monitoring scheme with the target OC mean of $\mu_1 = 2$, the target OC variance of $\sigma_1^2 = 1$, and the ARL_0 value of the joint monitoring scheme to be 200. Apply the joint monitoring scheme to the observed data, and summarize your results.

n	1	2	3	4	5	6	7	8
$\overline{X}_n$	1.06	0.93	1.08	1.15	1.05	0.82	1.07	1.24
s_n	0.25	0.32	0.17	0.62	0.36	0.31	0.47	0.27
n	9	10	11	12	13	14	15	16
$\overline{X}_n$	0.96	1.04	0.76	1.19	2.40	1.99	1.56	1.63
s_n	0.60	0.67	0.82	0.71	1.23	0.49	1.47	1.21
n	17	18	19	20	21	22	23	24
$\overline{X}_n$	1.66	2.50	1.66	1.93	1.99	1.92	1.61	2.44
s_n	1.25	0.79	1.02	1.55	1.12	0.78	0.68	0.84

4.16 The exponential distribution with a rate parameter $\lambda > 0$, denoted as $Exponential(\lambda)$, has a pdf with the following expression:

$$f(x; \lambda) = \lambda e^{-\lambda x}, \qquad \text{for } x \geq 0.$$

In Subsection 4.4.1, it is pointed out that this distribution is equivalent to the $\Gamma(1, 1/\lambda)$ distribution. It is also equivalent to the $Weibull(1/\lambda, 1)$ distribution. In Subsection 4.4.1, we have derived optimal CUSUM charts for detecting shifts in β of the $\Gamma(\alpha, \beta)$ distribution and shifts in a of the $Weibull(a, b)$ distribution in cases when observations of a production process follow one of these two distributions. Therefore, if observations of a production process follow the $Exponential(\lambda)$ distribution and we are interested in detecting an upward shift in λ, then the optimal CUSUM chart can be derived using either the connection between the exponential and Gamma distributions or the connection between the exponential and Weibull distributions. Show that the resulting CUSUM charts by the two approaches are actually the same.

4.17 A distribution is called a *log-normal distribution* if it has a pdf with the following expression:

$$f(x;\mu,\sigma) = \frac{1}{x\sigma\sqrt{2\pi}} e^{-\frac{[\log(x)-\mu]^2}{2\sigma^2}}, \qquad \text{for } x > 0,$$

where μ and σ are two parameters. If a random variable X has the above log-normal distribution, denoted as $X \sim LN(\mu,\sigma^2)$, then $\log(X)$ has the $N(\mu,\sigma^2)$ distribution. The log-normal distribution is often appropriate for describing the distribution of a variable that is a multiplicative product of many independent and positive random variables. For a production process, assume that its observations follow the $LN(\mu_0,\sigma_0^2)$ distribution when it is IC.

(i) Derive an optimal CUSUM chart for detecting an upward shift in μ, from μ_0 to μ_1 with $\mu_1 > \mu_0$, while the other parameter stays unchanged.

(ii) Derive an optimal CUSUM chart for detecting a downward shift in σ^2, from σ_0^2 to σ_1^2 with $\sigma_1^2 < \sigma_0^2$, while the other parameter stays unchanged.

4.18 To control the proportion of non-conforming products manufactured by a production process, 20 samples of size $m = 100$ each are obtained. The numbers of nonconforming products in the 20 samples are listed below.

10, 15, 31, 18, 24, 12, 23, 15, 19, 21, 16, 24, 28, 15, 23, 19, 14, 27, 20, 18

When the process is IC, it is known that the true proportion of non-conforming products is $\pi = 0.19$.

(i) Design an optimal CUSUM chart for detecting an upward shift in π from 0.19 to 0.29, with the ARL_0 value to be the smallest achievable one that is bigger than or equal to 370.

(i) The data considered here are also considered in Exercise 3.11, where a p chart is used for a similar purpose. Compare the two charting techniques, and discuss their strengths and limitations.

4.19 A faculty member in a university usually receives an average of 10 junk emails per day. He complains that this rate has doubled recently. To support his complaint, the numbers of junk emails that he received in the past 30 days were recorded as follows.

14, 10, 6, 10, 9, 7, 17, 9, 13, 10, 22, 21, 15, 25, 27,
26, 20, 26, 26, 17, 23, 16, 18, 22, 15, 23, 19, 15, 18, 13

Design a CUSUM chart to detect an upward shift in the average number of junk emails received by this faculty member, using an ARL_0 value around 200. Summarize your findings after applying the CUSUM chart to the observed data.

4.20 The IC distribution of a production process is assumed to be $N(\mu_0,\sigma_0^2)$. Because both μ_0 and σ_0 are unknown, an IC dataset has been collected in the phase I SPC. The estimates of μ_0 and σ_0 from the IC data are 11.5 and 1.5, respectively. Then, these values are regarded as the true values of μ_0 and σ_0, and they are used for standardizing the phase II observations for online process monitoring. The CUSUM chart (4.46)–(4.47) with $(k^+, h^+) = (0.5, 3.502)$ is then used for detecting an upward mean shift, with a nominal ARL_0 value of 200. We know that the estimates of μ_0 and σ_0 from the IC data would not equal the true values of μ_0 and σ_0 due to the randomness involved in the observed IC data. Consequently, the actual ARL_0 value of the chart (4.46)–(4.47) would be different from the nominal value of 200. Compute the actual ARL_0 values of the chart in cases when μ_0 and σ_0 take the following true values:

(i) $\mu_0 = 10$ and $\sigma_0 = 1.0$,

(ii) $\mu_0 = 10$ and $\sigma_0 = 1.5$,

(iii) $\mu_0 = 11.5$ and $\sigma_0 = 1.0$,

(iv) $\mu_0 = 13$ and $\sigma_0 = 2.0$.

Summarize your major findings from the computed actual ARL_0 values.

4.21 Observations from a production process are believed to follow a normal distribution when it is IC. But, the IC mean and variance are unknown. For online monitoring, the process has been adjusted properly in the phase I analysis, and 10 IC observations have been collected at the end of the phase I analysis. These 10 observations along with 20 phase II observations are listed below. Use the self-starting CUSUM chart (4.49) with $(k, h) = (0.5, 3.502)$ to monitor the process. Summarize your results.

$$8.9, \ 9.4, \ 10.8, \ 8.5, \ 9.3, \ 8.2, \ 9.2, \ 10.9, \ 8.5, \ 7.9$$

$$10.0, \ 12.0, \ 9.1, \ 8.3, \ 11.6, \ 9.8, \ 9.9, \ 12.6, \ 11.0, \ 8.8$$

$$13.8, \ 14.4, \ 12.4, \ 17.0, \ 14.2, \ 15.1, \ 12.3, \ 16.6, \ 14.1, \ 16.4$$

4.22 For the production process discussed in Exercise 4.21, assume that the IC distribution is $N(10, 1.5^2)$. Use the adaptive CUSUM chart (4.51)–(4.54) with $ARL_0 = 200$ to monitor the 30 observations listed in that exercise. In (4.51) and (4.52), you need to replace X_n by the standardized observations $(X_n - 10)/1.5$. Also, you need to specify δ_{min} and λ in (4.52). Try the following three sets of values for δ_{min} and λ, and summarize the major differences among the three sets of results:

(i) $\delta_{min} = 0$ and $\lambda = 0.1$,

(ii) $\delta_{min} = 0$ and $\lambda = 0.5$,

(iii) $\delta_{min} = 2.0$ and $\lambda = 0.1$.

Chapter 5

Univariate EWMA Charts

5.1 Introduction

In Section 4.1, we have explained that Shewhart charts are not effective for detecting small and persistent shifts because they evaluate the process performance based on observed data at a single time point alone. To overcome this limitation, the CUSUM charts described in the previous chapter try to make use of all observed data available at the current time point, including those observed at the current time point and all historical data that are observed before the current time point, to evaluate the process performance. One major feature of the CUSUM charts is their re-starting mechanism, which is that their charting statistics reset to the initial state every time when evidence of a shift in the observed data is smaller than a threshold. As it has been described in Subsection 4.2.4, they have certain optimality properties under some regularity conditions. However, they are relatively complicated to construct and use. Is there a more convenient way to construct control charts that have similar performance? Roberts (1959) provided an answer to this question by proposing the so-called *exponentially weighted moving average (EWMA)* control chart. This chart is constructed based on a weighted average of all observed data available at the current time point. Thus, it is easy to perceive. In the literature, there has been an extensive discussion about its design, implementation, and properties. See, for instance, Capizzi and Masarotto (2003), Crowder (1987a,b, 1989), Gan (1995), Han and Tsung (2004), Knoth (2007), Lucas and Saccucci (1990), and Reynolds and Stoumbos (2005, 2006), among many others. From all these discussions, it can be seen that the EWMA charts have similar performance to the CUSUM charts.

In this chapter, we describe some major EWMA charts for detecting mean and/or variance shifts of univariate processes. Our discussion will mainly focus on the case when the IC and OC process distributions are normal, although cases with several major non-normal parametric distributions will also be briefly discussed in Section 5.5. Process monitoring by EWMA charts in cases when the processes are multivariate will be discussed in Chapter 7. Cases when the process distribution is nonparametric will be discussed in Chapters 8 and 9. Also, similar to the CUSUM charts, the EWMA charts are mainly for phase II SPC, because their major strength is in effectively detecting small and persistent shifts. It will be more economic to use them with individual observation data as well, although they can easily be applied to batch data.

5.2 Monitoring the Mean of a Normal Process

In this section, we discuss the basic idea of the EWMA charting technique, and the design and implementation of an EWMA control chart, in cases when the quality characteristic of interest X is univariate, its IC distribution is $N(\mu_0, \sigma^2)$, and the potential shift in its distribution is in the mean of X only. Our discussion is divided into three parts. In Subsection 5.2.1, formulation, design, and implementation of an EWMA control chart for detecting mean shifts of normal processes is discussed in the conventional case when observations at different time points are independent. The case when the observations at different time points are correlated is discussed in Subsection 5.2.2. Finally, comparison between the CUSUM and EWMA control charts is discussed in Subsection 5.2.3.

5.2.1 Design and implementation of the EWMA chart

Assume that $\{X_1, X_2, \ldots\}$ are independent observations obtained from a univariate process at consecutive time points for phase II SPC. They have a common IC distribution $N(\mu_0, \sigma^2)$ when the process is IC, and their distribution changes to $N(\mu_1, \sigma^2)$ after the process becomes OC, where $\mu_0 \neq \mu_1$ are the IC and OC process means. To detect the mean shift from μ_0 to μ_1, the CUSUM chart uses the cumulative sum $C_n = \sum_{i=1}^{n}(X_i - \mu_0)$. From (4.2) and (4.3), it can be seen that, when the mean shift occurs at $1 \leq \tau \leq n$, the mean of C_n would increase linearly with n. So would the variance of C_n. Therefore, C_n carries useful information about the mean shift; but, the increasing variance makes it difficult to detect the mean shift based on C_n directly. To overcome this difficulty, the CUSUM chart uses the V-masks to detect the mean shift in its V-mask form (cf., Figure 4.3(b)), or the re-starting mechanism in its DI form (cf., (4.7)). The re-starting mechanism makes the variance of the resulting charting statistic C_n^+ defined in (4.7) stable when the process is IC, because the IC distribution of C_n^+ converges to its steady-state distribution when n increases. See Subsection 4.2.2 for a related discussion. Is there an alternative charting statistic that makes use of all observed data, contains useful information about the mean shift, and has a stable variance when n increases? Roberts (1959) gave a positive answer to this question with the following charting statistic: for $n \geq 1$,

$$E_n = \lambda X_n + (1 - \lambda)E_{n-1}, \qquad (5.1)$$

where $E_0 = \mu_0$, and $\lambda \in (0, 1]$ is a weighting parameter. Obviously, the charting statistic E_n in (5.1) is a weighted average of the observation X_n at the current time point and the charting statistic at the previous time point, and the weight is controlled by the parameter λ.

From (5.1), it is easy to see that

$$
\begin{aligned}
E_n &= \lambda X_n + \lambda(1-\lambda)X_{n-1} + \cdots + \lambda(1-\lambda)^{n-1}X_1 + (1-\lambda)^n \mu_0 \\
&= \lambda \sum_{i=1}^{n}(1-\lambda)^{n-i}X_i + (1-\lambda)^n \mu_0.
\end{aligned}
$$

Further, we can check that

$$\lambda \sum_{i=1}^{n} (1-\lambda)^{n-i} + (1-\lambda)^n = 1.$$

Therefore, E_n is a weighted average of μ_0 and all available observations $\{X_n, X_{n-1}, \ldots, X_1\}$ up to the time point n, and the weight $\lambda(1-\lambda)^{n-i}$ received by the i-th observation decays exponentially when i moves away from n. That is the reason the control chart based on E_n is called the EWMA chart. From (5.1), we can see that if λ is chosen to be larger, then more weight is assigned to the current observation X_n and less weight is assigned to its previous observations. On the other hand, if λ is chosen to be smaller, then less weight is assigned to the current observation X_n and more weights are assigned to its previous observations. In the special case when $\lambda = 1$, $E_n = X_n$. In such cases, the control chart based on E_n is just a Shewhart chart.

In cases when the process is IC up to the time point n, we can check that

$$\mu_{E_n} = \mu_0,$$

and

$$\sigma_{E_n}^2 = \frac{\lambda}{2-\lambda} \left[1 - (1-\lambda)^{2n} \right] \sigma^2. \tag{5.2}$$

Therefore, in such cases,

$$E_n \sim N \left(\mu_0, \frac{\lambda}{2-\lambda} \left[1 - (1-\lambda)^{2n} \right] \sigma^2 \right). \tag{5.3}$$

In cases when the process has a mean shift from μ_0 to μ_1 at the time point $1 \leq \tau \leq n$, the variance of E_n is still the one in (5.2), but its mean becomes

$$\begin{aligned} \mu_{E_n, \tau} &= (1-\lambda)^{n-\tau+1} \mu_0 + \left[1 - (1-\lambda)^{n-\tau+1} \right] \mu_1 \\ &= \mu_0 + \left[1 - (1-\lambda)^{n-\tau+1} \right] (\mu_1 - \mu_0). \end{aligned} \tag{5.4}$$

From (5.4), we can see that the OC mean of E_n is a weighted average of μ_0 and μ_1, and the weight of μ_1 is larger when n is larger. Therefore, the statistic E_n indeed contains useful information about the mean shift. On the other hand, from (5.2), it can be seen that the variance of E_n converges to

$$\widetilde{\sigma}_{0,\lambda}^2 = \frac{\lambda}{2-\lambda} \sigma^2. \tag{5.5}$$

In cases when $\sigma^2 = 1$, $\lambda = 0.05, 0.1$, or 0.2, and n changes from 1 to 30, the values of $\sigma_{E_n}^2$ are shown in Figure 5.1. From the plot, we can see that $\sigma_{E_n}^2$ is small when n is small, and it converges to $\widetilde{\sigma}_{0,\lambda}^2$ quickly, especially when λ is relatively large. If $\lambda = 0.05$, which is regarded as small, then $\sigma_{E_n}^2$ is close to $\widetilde{\sigma}_{0,\lambda}^2$ when $n \geq 20$. If $\lambda = 0.2$, which is regarded as relatively large, $\sigma_{E_n}^2$ is already very close to $\widetilde{\sigma}_{0,\lambda}^2$ when $n \geq 10$. Therefore, the variability of the statistic E_n is stable when n is medium to large.

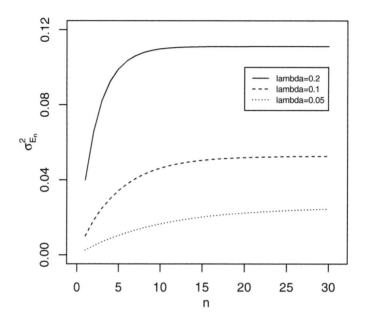

Figure 5.1 *Values of $\sigma_{E_n}^2$ in cases when $\sigma^2 = 1$, $\lambda = 0.05, 0.1$, or 0.2, and n changes from 1 to 30.*

From the above discussion, we can see that the statistic E_n makes use of all available observations up to the current time point n, it contains useful information about the mean shift, and its variance is stable. Therefore, it should be a reasonable charting statistic for detecting the mean shift. Based on its IC distribution given in (5.3), the resulting EWMA control chart for detecting an arbitrary mean shift is summarized in the box below.

Control Limits of the EWMA Chart for Detecting a Mean Shift

$$U = \mu_0 + \rho\sqrt{\frac{\lambda}{2-\lambda}[1-(1-\lambda)^{2n}]}\,\sigma$$
$$C = \mu_0$$
$$\tag{5.6}$$
$$L = \mu_0 - \rho\sqrt{\frac{\lambda}{2-\lambda}[1-(1-\lambda)^{2n}]}\,\sigma,$$

where $\rho > 0$ is a parameter.

More specifically, the EWMA chart defined by (5.1) and (5.6) has a center line C, an upper control limit U, and a lower control limit L. To use this chart, at the time point n, we compare the charting statistic E_n with the two control limits. If E_n is within the interval $[L, U]$, then we conclude that the process is IC up to the time point n. Otherwise, a signal of mean shift is given.

Example 5.1 *Assume that the IC distribution of a univariate production process is* $N(10, 2^2)$, *and the first 30 observations obtained at the first 30 consecutive time points are listed below.*

$$6.15, \ 11.36, \ 10.66, \ 9.16, \ 11.26, \ 7.45, \ 10.20, \ 13.20, \ 6.74, \ 11.19,$$
$$9.63, \ 5.43, \ 10.20, \ 8.60, \ 10.22, \ 14.36, \ 10.85, \ 13.70, \ 10.67, \ 14.40,$$
$$17.80, \ 12.22, \ 12.29, \ 12.72, \ 11.85, \ 12.91, \ 11.65, \ 14.17, \ 11.99, \ 12.48$$

In the EWMA chart defined by (5.1) and (5.6), assume that we choose $\lambda = 0.1$ *and* $\rho = 2.703$. *Then, its charting statistic* E_n *and the control limits are shown in Figure 5.2. From the plot, the chart gives a signal of mean shift at the 21st time point.*

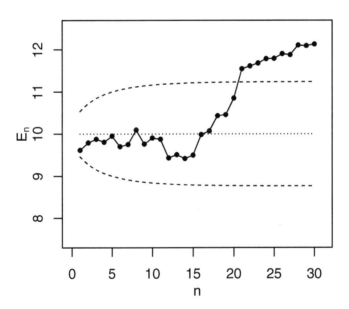

Figure 5.2 *The EWMA chart with* $\lambda = 0.1$ *and* $\rho = 2.703$ *(black dots connected by a solid line) when it is applied to the data in Example 5.1. The two dashed lines denote the upper control limit U and the lower control limit L, and the dotted line denotes the center line C.*

In practice, people often use the asymptotic variance of E_n, i.e., $\widetilde{\sigma}_{0,\lambda}^2$ given in (5.5), to construct the EWMA chart. In such cases, its control limits become

$$
\begin{aligned}
U &= \mu_0 + \rho\sqrt{\frac{\lambda}{2-\lambda}}\sigma \\
C &= \mu_0 \\
L &= \mu_0 - \rho\sqrt{\frac{\lambda}{2-\lambda}}\sigma.
\end{aligned}
\tag{5.7}
$$

It can be seen that both U and L in (5.7) are constants. Therefore, the resulting control chart would be easier to use, and its properties would also be easier to study. Because $\sigma_{E_n}^2$ converges to $\tilde{\sigma}_{0,\lambda}^2$ quickly for a given λ value, as demonstrated by Figure 5.1, the two versions of the EWMA chart would have similar performance, as long as the potential mean shift does not occur very early in the process monitoring. For simplicity, most software packages, including those used in this chapter, use the simpler control limits defined in (5.7). For applications, we recommend using the ones defined in (5.6) because they could give slightly more reliable results, although the chart would be a little more complicated to use.

The EWMA chart defined by (5.1) and (5.6) is a two-sided control chart for detecting an arbitrary mean shift. If we are interested in detecting upward mean shifts only in a given application, then we can just use the upper control limit U and the resulting upward EWMA chart gives a signal at the n-th time point if $E_n > U$. Similarly, if we are interested in detecting downward mean shifts only, then we can consider using a downward EWMA chart which gives a signal at the n-th time point if $E_n < L$. For constructing the upward or downward EWMA chart, we can also adopt the re-starting mechanism of the CUSUM charts. To detect upward mean shifts, the charting statistic of the upward EWMA chart with the re-starting mechanism can be defined by

$$E_n^+ = \max\left(0, \lambda(X_n - \mu_0) + (1-\lambda)E_{n-1}^+\right), \tag{5.8}$$

where $E_0^+ = 0$, and the chart gives a signal of upward mean shift when

$$E_n^+ > \rho_U \sqrt{\frac{\lambda}{2-\lambda}}\sigma, \tag{5.9}$$

where $\rho_U > 0$ is a parameter chosen to achieve a pre-specified ARL_0 value. Similarly, the charting statistic of the downward EWMA chart with the re-starting mechanism is defined by

$$E_n^- = \min\left(0, \lambda(X_n - \mu_0) + (1-\lambda)E_{n-1}^-\right), \tag{5.10}$$

where $E_0^- = 0$, and a signal of downward mean shift is given when

$$E_n^- < -\rho_L \sqrt{\frac{\lambda}{2-\lambda}}\sigma, \tag{5.11}$$

where $\rho_L > 0$ is a parameter chosen to achieve a given ARL_0 value.

In the above discussion, we assume that the observed data are single observation data. In some applications, if a random sample of size m is collected at each time point, then the EWMA chart defined in (5.1) and (5.6) can still be used, after X_n in (5.1) is replaced by the mean $\overline{X}(n)$ of the sample collected at the n-th time point and σ in (5.6) is replaced by $\sigma/\sqrt{m}$. The one-sided EWMA charts defined by (5.8)–(5.11) can be modified in a similar way, by replacing $X_n - \mu_0$ in (5.8) and (5.10) by $\overline{X}(n) - \mu_0$ and replacing σ in (5.9) and (5.11) by $\sigma/\sqrt{m}$.

Next, we discuss the design of the two-sided EWMA chart defined by (5.1) and (5.7). The design of the one-sided EWMA charts can be discussed similarly. From the construction of the two-sided EWMA chart, its performance would depend on the values of the two parameters λ and ρ. In cases when the IC process distribution is $N(0,1)$, $\lambda = 0.05, 0.1, 0.2$, or 0.3, and ρ changes from 1.0 to 3.0 with a step of 0.5, its zero-state ARL_0 values computed using the function xewma.arl() in the R-package spc that is described in the appendix are presented in Figure 5.3. From the plot, it can be seen that the ARL_0 value of the EWMA chart increases when ρ increases, and decreases when λ increases.

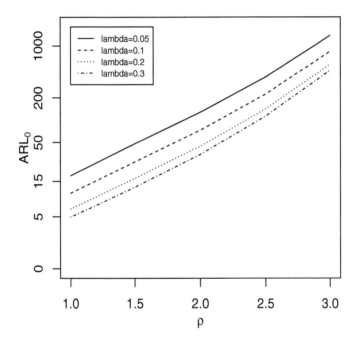

Figure 5.3 *ARL_0 values of the EWMA chart defined by (5.1) and (5.7) in cases when the IC process distribution is $N(0,1)$, $\lambda = 0.05, 0.1, 0.2$, or 0.3, and ρ changes from 1.0 to 3.0 with a step of 0.5.*

For the two-sided EWMA chart, the λ value is usually specified beforehand, and the value of ρ is then determined to reach a pre-specified ARL_0 value. For some commonly used ARL_0 and λ values, the corresponding ρ values computed by the function xewma.crit() in the R-package spc are presented in Table 5.1. From the table, it can be seen that a larger ρ should be chosen if ARL_0 or λ is larger.

It should be pointed out that Monte Carlo simulations by computer algorithms similar to those described by the pseudo codes given in Subsection 4.2.2 can also be used to compute the ARL_0 value of the EWMA chart when λ and ρ are given, or to search for the value of ρ when ARL_0 and λ are given. As a matter of fact, with various powerful computing facilities available nowadays, this might be a better

Table 5.1 *Computed ρ values of the EWMA chart defined by (5.1) and (5.7) for some commonly used ARL_0 and λ values.*

				λ				
ARL_0	0.01	0.05	0.1	0.2	0.3	0.4	0.5	0.75
50	0.845	1.520	1.811	2.054	2.166	2.229	2.268	2.315
100	1.152	1.879	2.148	2.360	2.453	2.504	2.534	2.568
200	1.500	2.216	2.454	2.635	2.713	2.754	2.777	2.802
300	1.710	2.399	2.619	2.785	2.854	2.890	2.911	2.931
370	1.819	2.490	2.701	2.859	2.925	2.959	2.978	2.996
400	1.859	2.523	2.731	2.886	2.950	2.984	3.002	3.020
500	1.973	2.615	2.814	2.962	3.023	3.054	3.071	3.087
1000	2.308	2.884	3.059	3.187	3.238	3.263	3.277	3.289

approach to use, as long as it properly simulates the scenario in question, because it often gives more accurate results and is more flexible for handling different scenarios. Functions in spc and other R-packages are convenient to use, but they usually make the computation using numerical approximations offered by the Markov chain, integral equations, and other theoretical approaches that have various model assumptions (cf., Section 4.6 and Section 5.6 for a related discussion). Besides the approximation error, their results may not be reliable when their model assumptions are violated. Therefore, we should use them with care.

To study the impact of the λ value on the performance of the EWMA chart, let us consider cases when the IC process distribution is $N(0, 1)$, $ARL_0 = 200$, $\lambda = 0.05, 0.1$, or 0.2, and the shift size $\delta = \mu_1 - \mu_0$ changes from 0 to 2.0 with a step of 0.1. By Table 5.1, the ρ values should be 2.216, 2.454, and 2.635, respectively. The computed ARL_1 values of the EWMA chart are presented in Figure 5.4. From the plot, it can be seen that the chart with $\lambda = 0.05$ performs the best for detecting small mean shifts (e.g., shifts with δ smaller than 0.5), the chart with $\lambda = 0.1$ performs the best for detecting medium mean shifts (e.g., shifts with δ between 0.6 and 0.8), and the chart with $\lambda = 0.2$ performs the best for detecting large mean shifts (e.g., shifts with δ larger than 1.0). Therefore, a relatively large λ should be chosen for detecting large shifts, and a relatively small λ should be chosen for detecting small shifts.

To use the EWMA chart in practice, we need to choose the ARL_0 and λ values properly beforehand and then search for the value of ρ such that the pre-specified ARL_0 value is reached. As mentioned in Subsection 4.4.2, the value of ARL_0 is often chosen to be the minimum tolerable ARL_0. Some economic considerations, including the cost related to a false alarm and the resulting process downtime, are often associated with this decision. To choose the value of λ properly, we need to specify a target shift size first. To this end, some practical guidelines have been described in the paragraph immediately below Figure 4.5 in Subsection 4.2.2. After a target shift size is chosen, we can search for a λ value and the corresponding ρ value such that the pre-specified ARL_0 value is reached while the ARL_1 value for detecting the target shift is minimized. This and the other practical guidelines on choosing λ are sum-

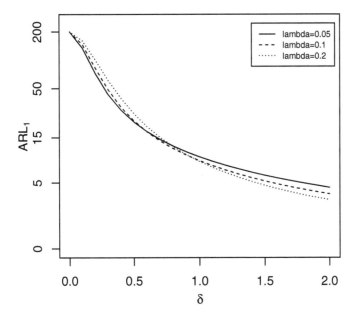

Figure 5.4 *ARL$_1$ values of the EWMA chart defined by (5.1) and (5.7) in cases when the IC process distribution is $N(0,1)$, $ARL_0 = 200$, $\lambda = 0.05, 0.1$, or 0.2, and the shift size δ changes from 0 to 2.0 with a step of 0.1.*

marized in the box below. For a more detailed discussion on this issue, see Crowder (1989).

General Guidelines for Selecting λ

For the EWMA chart defined in (5.1) and (5.7) (or (5.6)), small λ values are good for detecting relatively small mean shifts, and large λ values are good for detecting relatively large mean shifts. For a given target shift, λ can be chosen such that the ARL_1 value for detecting the target shift is minimized.

In the above discussion, we assume that the process distribution is normal. In the literature, some researchers have demonstrated that the EWMA chart is actually quite robust to the normality assumption (e.g., Borror et al., 1999). To see this, let us consider the following example.

Example 5.2 *The IC distribution of a production process is assumed to be $N(0,1)$, but it is actually the standardized version with mean 0 and variance 1 of the χ^2_{df} distribution, where df is its degrees of freedom. Therefore, the actual IC process distribution is skewed to the right, and it is closer to the $N(0,1)$ distribution when df gets larger. The EWMA chart defined by (5.1) and (5.7) is then applied to the process.*

The parameters (λ, ρ) are chosen to be $(0.05, 2.216), (0.1, 2.454)$, or $(0.2, 2.635)$. By Table 5.1, the nominal ARL_0 values of the chart are 200 in all three cases if the IC process distribution is $N(0, 1)$. The actual ARL_0 values of the chart are computed by a computer algorithm written by the author in cases when df changes from 1 to 10, and they are shown in Figure 5.5.

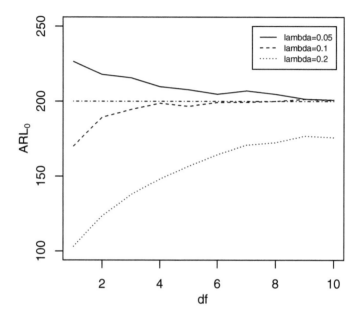

Figure 5.5 *Actual ARL_0 values of the EWMA chart defined by (5.1) and (5.7) in cases when the actual IC process distribution is the standardized version with mean 0 and variance 1 of the χ^2_{df} distribution, where df changes from 1 to 10. In the EWMA chart, the parameters (λ, ρ) are chosen to be $(0.05, 2.216), (0.1, 2.454)$, or $(0.2, 2.635)$ (so that the nominal ARL_0 values of the chart are 200 in all three cases when the IC process distribution is $N(0, 1)$), and the corresponding results are shown by the solid, dashed, and dotted curves, respectively. The dot-dashed horizontal line denotes the nominal ARL_0 value.*

From Figure 5.5, it can be seen that the actual ARL_0 values of the EWMA chart are quite different from the nominal ARL_0 value of 200 in cases when $\lambda = 0.2$ and when df is quite small (i.e., the actual IC process distribution is quite skewed). Such a difference gets smaller when λ is chosen smaller, which can be explained intuitively as follows. As pointed out earlier, when λ is chosen smaller, the charting statistic E_n would get more previous observations of X_n involved in a substantial way. Therefore, by the central limit theorem, the distribution of E_n would be closer to normal. Based on this result, many people believe that the EWMA chart is robust to the normality assumption, as long as a small λ is chosen.

We would like to remind readers that the robustness property of the EWMA chart should be used with care, for the following two major reasons. First, from the numerical results presented in Example 5.2 and in Borror et al. (1999) and other research papers, it seems that it depends on the magnitude of the difference between the actual IC process distribution and a normal distribution to decide how small the λ value should be chosen in order to have the robustness property. In many applications, it is difficult to measure such a difference. Second, if a small λ is chosen in order to keep the actual ARL_0 value of the EWMA chart close to the nominal ARL_0 value in many different cases, then the resulting chart would be ineffective for detecting relatively large shifts. For instance, Borror et al. (1999) suggested choosing $\lambda = 0.05$ to have the robustness property. In such cases, Figure 5.4 shows that the corresponding EWMA chart would be ineffective in detecting mean shifts that are larger than 1.0σ. In cases when the actual IC process distribution is non-normal, we suggest using the nonparametric (or distribution-free) control charts that will be discussed in Chapter 8, unless it is known beforehand that the IC process distribution is quite close to normal.

5.2.2 Cases with correlated observations

As discussed in Sections 3.5 and 4.2, possible correlation among observations collected at different time points could affect the actual IC and OC performance of the Shewhart and CUSUM charts. In this subsection, we discuss the impact of the possible correlation among observations on the EWMA charts. To this end, let us first discuss the example below.

Example 5.3 *As in Example 4.5, assume that process observations follow the AR(1) model*

$$X_n = \mu_0 + \phi \left(X_{n-1} - \mu_0 \right) + e_n, \qquad \text{for } n \geq 1,$$

where $X_0 = \mu_0$, $\mu_0 = 0$ is the IC process mean, ϕ is the model coefficient, and $\{e_n, n \geq 1\}$ are white noise with $\sigma_e = 1$ (i.e., they are i.i.d. with the common distribution $N(0,1)$). In such cases, the correlation coefficient between X_n and X_{n-j} is ϕ^j, for $j = 1, 2, \ldots, n-1$. To monitor the process, the EWMA chart defined by (5.1) and (5.7) is used, in which (λ, ρ) are chosen to be $(0.05, 2.216), (0.1, 2.454)$, or $(0.2, 2.635)$. By Table 5.1, the ARL_0 value of the chart in all three cases is 200 when $\phi = 0$ (i.e., when process observations are i.i.d. with the common distribution $N(0,1)$). In cases when $\phi = -0.5, -0.25, -0.1, 0, 0.1, 0.25$, and 0.5, the actual ARL_0 values of the chart are presented in the left part of Table 5.2. From the table, it can be seen that they are quite different from each other within each column of the table. Therefore, different correlation among observed data has a substantially different impact on the IC performance of the EWMA chart. By comparing the actual ARL_0 values presented in columns 2–4 of the table, it seems that the impact of the correlation among observed data is less serious when λ is chosen smaller. In Example 5.2, we have shown that the EWMA chart is more robust to the normality assumption in such cases. This example shows that the EWMA chart is more robust to the independence assumption as well when λ is chosen smaller.

To investigate the possible impact of the correlation among process observations on the OC performance of the EWMA chart defined by (5.1) and (5.7), we consider the case when a mean shift of size $\delta = 0.5$ occurs at the initial observation time and all other setups remain unchanged from those described in the previous paragraph.

Table 5.2 *The left part presents the actual ARL_0 values of the EWMA chart defined by (5.1) and (5.7) when (λ, ρ) are chosen to be $(0.05, 2.216), (0.1, 2.454)$, or $(0.2, 2.635)$ (so that its nominal ARL_0 value is 200 in all three cases) and when the process observations are correlated and follow the AR(1) model with the coefficient ϕ. The right part presents the actual ARL_1 values of the chart when a mean shift of size $\delta = 0.5$ occurs at the initial observation time in various cases considered.*

ϕ	$\delta = 0$			$\delta = 0.5$		
	$\lambda = 0.05$	$\lambda = 0.1$	$\lambda = 0.2$	$\lambda = 0.05$	$\lambda = 0.1$	$\lambda = 0.2$
−0.5	833.919	876.351	856.938	21.862	24.398	35.860
−0.25	532.279	582.474	591.485	22.085	24.013	32.091
−0.1	306.152	320.821	325.913	22.150	23.336	29.099
0	200.594	201.873	199.618	22.181	22.825	27.074
0.1	135.913	127.683	123.492	22.137	22.264	24.980
0.25	77.784	68.757	63.276	21.938	21.305	22.335
0.5	33.729	28.069	25.143	20.218	17.997	17.052

The actual ARL_1 values of the chart in various cases are presented in the right half of Table 5.2. From the table, it can be seen that the correlation among process observations does have an impact on the OC performance of the chart, and the impact seems less serious in cases when λ is chosen smaller.

Example 5.3 shows that the possible correlation among the observed data at different time points would have a substantial impact on the IC and OC performance of the EWMA chart defined by (5.1) and (5.7). To accommodate the correlation among the observed data, in Subsection 4.2.3 we discussed several possible approaches, including the one to group consecutive observations into batches and then apply the conventional control charts to the batch means. All these approaches can also be considered here. In Subsection 4.2.3, we already pointed out two major limitations of the grouping approach. One is that the variability of the batch means depends on how the original observations are correlated, which is hard to know in practice and which makes the grouping approach impractical. Another limitation is that the resulting control chart of this approach cannot react to a shift promptly because a signal of shift can only be delivered after all observations in a group are obtained. The second limitation is also shared by the alternative approach to use a subset of the original data in the way that two consecutive observations in the subset are certain time points apart in the original data. Further, much information in the original data is ignored by this alternative approach. For these reasons, the grouping approach and the subset approach are not recommended for practical use.

Another possible approach to accommodate the possible correlation among the observed data is to describe the association by a time series model, obtain residuals from the fitted time series model, and then apply the conventional control charts to the residuals, as discussed in Subsection 4.2.3. A major reasoning behind this approach is that the residuals are roughly i.i.d. when the original process is IC and any mean shift in the distribution of the original observations will result in a shift in the distribution of the residuals (e.g., Alwan and Roberts, 1988; Wardell et al., 1994; Zhang, 1997). However, some researchers have demonstrated that residuals are actually autocorrelated due to model estimation errors, which can have a substantial impact on the actual IC ARL value of the residual-based control charts (e.g., Adams and Tseng, 1998; Apley and Shi, 1999). To overcome this difficulty, Apley and Lee (2003) proposed a procedure to adjust the control limits of a residual-based EWMA chart properly in cases when an ARMA model (cf., the expression (4.16)) is used for describing the correlation among the observed data.

Instead of using residual-based EWMA charts in cases when process observations are correlated, some researchers suggest applying the EWMA charts directly to the original data, but adjusting their control limits properly to reflect the impact of the correlated observations (e.g., Wardell et al., 1994; Zhang, 1998). Next, we describe this approach by introducing the method proposed by Zhang (1998). Assume that when a process is IC, its observations are correlated with a constant mean μ_0 and the following stationary covariance function:

$$\xi(s) = \text{Cov}(X_n, X_{n-s}) = \text{E}[(X_n - \mu_0)(X_{n-s} - \mu_0)].$$

Note that $\xi(s)$ does not depend on n. Therefore, the correlation between X_n and X_{n-s} depends only on how many time points are between them. It is easy to check that the AR(1) model (4.15) when $|\phi| < 1$ has this property. Let

$$\eta(s) = \frac{\xi(s)}{\xi(0)} = \frac{\text{Cov}(X_n, X_{n-s})}{\sqrt{\text{Var}(X_n)\text{Var}(X_{n-s})}}$$

be the correlation function. Then, it is not difficult to check that the variance of the EWMA charting statistic E_n has the expression

$$\sigma_{E_n}^2 = \frac{\lambda}{2-\lambda}\left\{1 - (1-\lambda)^{2n} + 2\sum_{i=1}^{n-1}\eta(i)(1-\lambda)^i\left[1 - (1-\lambda)^{2(n-i)}\right]\right\}\xi(0),$$

which converges to a positive constant when n increases. Therefore, in practice, it can be approximated by

$$\sigma_{E_n,M}^2 = \frac{\lambda}{2-\lambda}\left\{1 + 2\sum_{i=1}^{M}\eta(i)(1-\lambda)^i\left[1 - (1-\lambda)^{2(M-i)}\right]\right\}\xi(0),$$

where $M > 0$ is an integer number. Then, the lower and upper control limits of the EWMA chart can be adjusted respectively to

$$\mu_0 - \rho\sigma_{E_n,M}, \qquad \mu_0 + \rho\sigma_{E_n,M}.$$

In practice, the correlation function $\eta(s)$ needs to be estimated from an IC dataset of size N. Zhang pointed out that N should be at least 50 to make the estimator of $\eta(s)$ reliable and M should be chosen near the upper end in the range $[1, N/4]$.

5.2.3 *Comparison with CUSUM charts*

As discussed earlier, both the CUSUM and EWMA charts are proposed as alternatives to the Shewhart charts for detecting small and persistent distributional shifts of a production process. In this subsection, we briefly summarize the major properties of the two types of control charts and make a comparison between them as well. For simplicity, our discussion will focus on the case when the process distribution is normal and we are interested in detecting process mean shifts, although the discussion in some other cases can be made in a similar way. Also, the CUSUM chart (4.7)–(4.8) and the EWMA chart defined by (5.1) and (5.7) are used as representatives of the two types of control charts in some of our discussion here for convenience. Most statements in this subsection should also be appropriate for other CUSUM and EWMA charts.

First, both the CUSUM and EWMA charts are mainly for phase II process monitoring, although they can also be used for phase I analysis. That is mainly because both types of control charts are good for detecting small and persistent shifts and detection of such shifts is our major concern in phase II process monitoring. As a comparison, in phase I analysis, shifts can be relatively large and transient. In such cases, the Shewhart charts are often preferred.

Second, both the CUSUM and EWMA charts are mainly for monitoring processes with individual observation data, although they can both be easily applied to cases with batch data. The main reason why individual observation data are preferred when using the CUSUM or EWMA charts, compared to batch data, is that less observations will be required to detect a given shift if individual observation data are used. To demonstrate this, let us consider the next example.

Example 5.4 *Assume that 50 observations are generated from the $N(0,1)$ distribution, and another 50 observations are generated from the $N(0.5,1)$ distribution. These observations can be regarded as observed data collected at 100 consecutive time points from a production process, the IC distribution is $N(0,1)$, and the process becomes OC at the 51st time point with the OC distribution $N(0.5,1)$. Let us first use the upward CUSUM chart (4.7)–(4.8) to monitor the process. In the chart, (k,h) are chosen to be $(0.25,5.597)$ so that its ARL_0 value is 200. The chart is presented in Figure 5.6(a), from which we can see that a signal is first given at the 64th time point. Now, let us treat the first 50 observations as 10 samples of size 5 each, and the last 50 observations as another 10 samples of size 5 each. Then, these 20 samples can be regarded as observed batch data at 20 consecutive time points from a production process whose IC distribution is $N(0,1)$. This process becomes OC at the 11th time point with the OC distribution $N(0.5,1)$. We then compute the 20 standardized sample means (i.e., each sample mean is standardized by dividing it by its standard deviation $1/\sqrt{5}$). The same CUSUM chart mentioned above is applied to these standardized sample means, and the chart is presented in Figure 5.6(b). From the plot, a signal is first given at the 18th time point. By comparing the two different sampling schemes, the CUSUM chart with the first sampling scheme requires 14 extra observations to give a signal that an upward mean shift of size 0.5 occurred at the 51st time point. With the second sampling scheme, the CUSUM chart requires $8 * 5 = 40$ extra observations to give a signal of the shift. Therefore, fewer extra observations are needed for the chart to give a signal in the case with the individual observation data. With the two sampling schemes described above, we also apply the EWMA chart defined by (5.1) and (5.7) with $(\lambda,\rho) = (0.1,2.454)$ to the related data. By Table 5.1, the EWMA chart has an ARL_0 value of 200 as well. It is shown in Figure 5.6(c)–(d) in the two cases considered. We can have similar conclusions* *in this scenario regarding the numbers of extra observations needed by the chart to give a signal of mean shift in cases with the two different sampling schemes.*

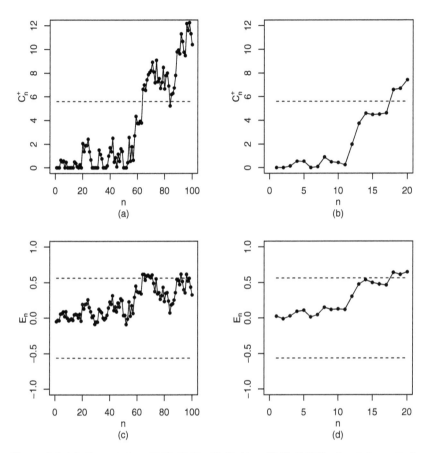

Figure 5.6 *(a) Control chart (4.7)–(4.8) with $(k, h) = (0.25, 5.597)$ when it is applied to 100 individual observations collected from a production process that has an upward mean shift of size 0.5 at the 51st time point. (b) The same chart when it is applied to the standardized sample means of 20 samples of size 5 each collected from a production process that has an upward mean shift of size 0.5 at the 11th time point. Plots (c) and (d) are the same as plots (a) and (b), except that the CUSUM chart is replaced by the EWMA chart defined by (5.1) and (5.7) with $(\lambda, \rho) = (0.1, 2.454)$.*

Third, the CUSUM chart (4.7)–(4.8) has two parameters k and h involved. For a given target shift of size δ, k is chosen to be $\delta/2$, and h is chosen to reach a pre-specified ARL_0 value. Such a CUSUM chart would have some optimality properties. See Subsection 4.2.4 for a related discussion. As a comparison, the EWMA chart defined by (5.1) and (5.7) also has two parameters λ and ρ involved. Usually, λ is chosen to be one of several commonly used values, such as 0.05, 0.1, and 0.2, and ρ is chosen to reach a pre-specified ARL_0 value. With the EWMA chart, we only know that the chart is good for detecting relatively small shifts when λ is chosen small, and it is good for detecting relatively large shifts when λ is chosen large. However, we do not have an explicit formula about the relationship between λ and a target shift size δ under which the chart would perform optimally, although such a relationship can be tabulated for certain special cases through Monte Carlo simulations (cf., Crowder, 1989).

Fourth, both the CUSUM chart (4.7)–(4.8) and the EWMA chart defined by (5.1) and (5.7) try to make a trade-off between the use of all historical data and the elimination of the negative impact of the historical data on the sensitivity of the control charts to process mean shifts. The CUSUM chart assigns equal weights to all previous observations when it accumulates all useful information in the historical data, and eliminates the dampening effect of the historical data by using a re-starting mechanism. As a comparison, the EWMA chart simply uses a weighted average of all available observations as its charting statistic, and the dampening effect of the historical data is mostly eliminated by the exponentially decaying weights assigned to the historical data. Its control limits have a similar form to those of the Shewhart charts. Therefore, to most users, the EWMA chart is easier to understand and implement. However, the CUSUM chart has been proven to have some optimality properties, as discussed in Subsection 4.2.4, and a similar theoretical background of the EWMA chart is still lacking. In the SPC literature, it has been demonstrated by several authors that the two types of control charts have similar performance (e.g., Lucas and Saccucci, 1990). To further investigate their relative performance, let us consider the cases when the IC process distribution is $N(0,1)$, and the process has an upward mean shift with size δ at the initial time point, where δ changes its value from 0.1 to 2.0 with a step of 0.1. The upward CUSUM chart (4.7)–(4.8) and the upward EWMA chart defined by (5.1) with the upper control limit in (5.7) are considered for monitoring the process. In the EWMA chart, the optimal λ value is searched with a precision of 0.001 for each shift size such that the ARL_1 value of the chart reaches the minimum. Its ρ value is chosen such that its ARL_0 value is 200 in each case. For the CUSUM chart, the allowance constant k is chosen to be $\delta/2$, and its h value is chosen such that $ARL_0 = 200$. All the computation is accomplished by using the related functions in the R-package spc. The ARL_1 values of the two charts and the corresponding parameter values are presented in Table 5.3. From the table, we can see that the optimal performance of the two charts is indeed quite similar in the cases considered.

Finally, we will demonstrate in Chapter 8 that the CUSUM charts are not robust to the normality assumption. In the literature, people tend to think that the EWMA charts are robust to the normality assumption because their charting statistics are weighted averages of the current and all previous observations. As a matter of fact, the charting statistic E_n defined in (5.1) gets many observations involved only when λ is very small. To demonstrate this, let us consider the case when $n = 100$. In such a case, from the description in Subsection 5.2.1, we know that the i-th observation X_i receives a weight of

$$w_i = \lambda(1-\lambda)^{n-i}$$

in E_n, for $i = 1, 2, \ldots, n$. In cases when $\lambda = 0.05, 0.1, 0.2$, and 0.5, these weights

Table 5.3 *Optimal ARL_1 values and the related parameter values of the upward CUSUM chart (4.7)–(4.8) and the upward EWMA chart defined by (5.1) with the upper control limit in (5.7) in cases when the IC process distribution is $N(0, 1)$, and the process has an upward mean shift of size δ at the initial time point, where δ changes its value from 0.1 to 2.0 with a step of 0.1.*

	EWMA			CUSUM		
δ	λ	ρ	ARL_1	k	h	ARL_1
0.1	0.009	1.397	92.916	0.05	10.297	92.815
0.2	0.020	1.764	54.303	0.1	8.520	54.063
0.3	0.034	1.991	36.062	0.15	7.262	35.746
0.4	0.051	2.147	25.937	0.2	6.325	25.592
0.5	0.069	2.252	19.694	0.25	5.597	19.343
0.6	0.089	2.332	15.550	0.3	5.015	15.206
0.7	0.111	2.394	12.647	0.35	4.537	12.314
0.8	0.135	2.444	10.526	0.4	4.136	10.207
0.9	0.159	2.482	8.925	0.45	3.796	8.621
1.0	0.185	2.514	7.684	0.5	3.502	7.395
1.1	0.212	2.540	6.701	0.55	3.246	6.426
1.2	0.239	2.560	5.908	0.6	3.020	5.646
1.3	0.267	2.577	5.258	0.65	2.820	5.009
1.4	0.297	2.592	4.718	0.7	2.641	4.481
1.5	0.327	2.603	4.264	0.75	2.481	4.038
1.6	0.358	2.613	3.877	0.8	2.336	3.662
1.7	0.391	2.621	3.545	0.85	2.205	3.340
1.8	0.425	2.627	3.257	0.9	2.085	3.063
1.9	0.460	2.632	3.004	0.95	1.975	2.821
2.0	0.496	2.636	2.781	1.0	1.874	2.610

are shown in Figure 5.7(a)–(d). From plot (d), it can be seen that E_n actually only gets 3 or 4 previous observations substantially involved in the case when $\lambda = 0.5$. When λ is chosen smaller, it is true that more previous observations are involved in E_n. From plots (a)–(c), it seems that about 40, 20, and 10 previous observations receive meaningful weights in E_n in cases when $\lambda = 0.05, 0.1$, and 0.2, respectively. However, the weights received by different previous observations are dramatically different in their magnitudes, which greatly reduces the effectiveness of the Central Limit Theorem (cf., Subsection 2.7.1 for a related discussion). Even when all the weights in a weighted average are the same, a conventional opinion in the statistical literature is that at least 30 independent observations are needed to make their simple average to be approximately normally distributed (cf., Peck and Devore, 2012, Section 8.2). Further, when λ is chosen small, as pointed out in Subsection 5.2.1, the resulting EWMA chart is good for detecting small mean shifts only. For instance, when $\lambda = 0.05$, from Table 5.3, the resulting EWMA chart is good for detecting mean shifts around 0.4, and it will not be good for detecting mean shifts as large as 1. Therefore, we think that the robustness property of the EWMA chart should be used with care.

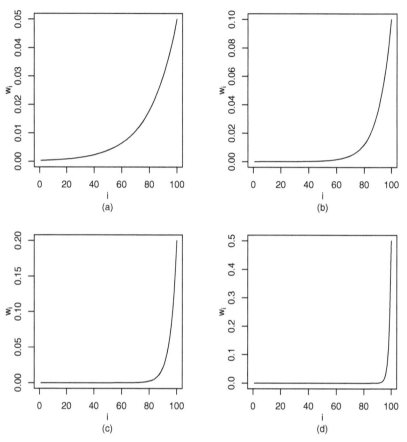

Figure 5.7 *The weights $w_i = \lambda(1-\lambda)^{n-i}$, for $i = 1,2,\ldots,n$, received by the i-th observation X_i in E_n when $n = 100$, $\lambda = 0.05$ (plot (a)), $\lambda = 0.1$ (plot (b)), $\lambda = 0.2$ (plot (c)), or $\lambda = 0.5$ (plot (d)).*

5.3   Monitoring the Variance of a Normal Process

As discussed in Subsection 4.3.1, process variability is related to the quality of conformance of the related production process. Generally speaking, an upward process variability shift would downgrade the quality of conformance, while a downward process variability shift would improve the quality of conformance. Therefore, in practice, we are also interested in detecting shifts in process variability. More specifically, we are mainly concerned about upward process variability shifts in practice. Downward shifts are of concern only in cases when we want to cut production cost or we have some other considerations. In this section, we discuss how to detect process variability shifts using the EWMA charts. Our discussion is divided into two parts: Monitoring of the process variability alone is discussed in Subsection 5.3.1, and joint monitoring of both the process mean and process variability is discussed in Subsection 5.3.2.

5.3.1 Monitoring the process variance

A shift in the process variability would change the distribution of the charting statistic E_n of the EWMA chart defined by (5.1) and (5.7) that is originally designed for detecting process mean shifts. Therefore, that EWMA chart also has a certain ability to detect process variance shifts, which can be seen from the example below.

Example 5.5 *Let us reconsider the two scenarios discussed in Example 4.8. In the first scenario, 100 observations are generated as follows. The first 50 observations are generated from the $N(0,1)$ distribution, and the second 50 observations are generated from the $N(0,2^2)$ distribution. This scenario simulates a production process with the IC distribution $N(0,1)$ and it becomes OC with an upward variance shift from 1 to 2^2 at the 51st time point. The observed data are shown in Figure 5.8(a). Then, we apply the EWMA chart defined by (5.1) and (5.7) with $\lambda = 0.2$ and $\rho = 2.635$ to the dataset. From Table 5.1, it can be seen that the ARL_0 value of the chart is 200. The EWMA chart is shown in Figure 5.8(b), from which we can see that its charting statistic E_n has a larger variability after the process variance shift, it briefly touches the lower control limit at the 25th time point, and gives several quite convincing signals at a number of time points after the variance shift. In the second scenario, the first 50 observations are still generated from the $N(0,1)$ distribution, but the remaining 50 observations are generated from the $N(0,0.5^2)$ distribution. This scenario simulates a production process with the IC distribution $N(0,1)$ that becomes OC with a downward variance shift from 1 to 0.5^2 at the 51st time point. The data are shown in Figure 5.8(c), and the EWMA chart with the same setup as that in Figure 5.8(b) when it is applied to the data here is shown in Figure 5.8(d). It can be seen from the plot that a downward variance shift would decrease the variability of the charting statistic E_n and thus it cannot be detected by the EWMA chart designed for detecting mean shifts.*

Although the EWMA chart defined by (5.1) and (5.7) that is designed for detecting process mean shifts has certain power for detecting process variance shifts, it may not be efficient for that purpose, as demonstrated in Subsection 4.3.2 with the CUSUM chart (4.7)–(4.8). In cases when the IC process distribution is $N(\mu_0, \sigma_0^2)$ and one observation is collected at each time point (i.e., the case with individual observation data), the optimal CUSUM chart (4.21)–(4.24) for detecting process variance shifts is based on squares of the standardized observations $[(X_n - \mu_0)/\sigma_0]^2$. It is therefore natural to consider the following EWMA charting statistic in such cases:

$$E_n = \lambda \left(\frac{X_n - \mu_0}{\sigma_0} \right)^2 + (1 - \lambda)E_{n-1}, \qquad (5.12)$$

where $E_0 = 1$, and $\lambda \in (0, 1]$ is a weighting parameter. In cases when the process is IC up to the n-th time point, we have

$$Y_i = \left(\frac{X_i - \mu_0}{\sigma_0} \right)^2 \sim \chi_1^2, \qquad \text{for } i \leq n.$$

Therefore, in such cases,

$$\mu_{Y_i} = 1, \qquad \sigma_{Y_i}^2 = 2.$$

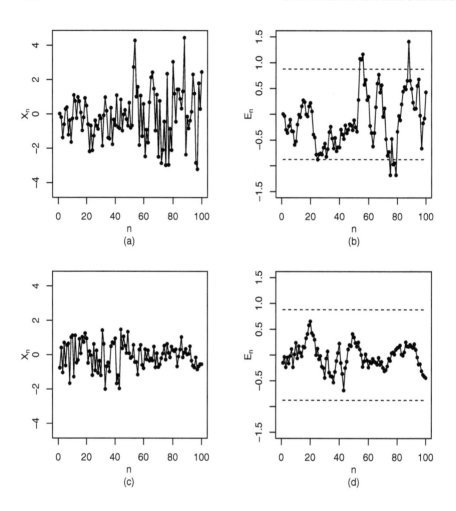

Figure 5.8 *(a) A dataset of 100 observations with the first 50 observations from the distribution $N(0,1)$, the second 50 observations from the distribution $N(0,2^2)$, and all observations being independent. (b) The EWMA chart defined by (5.1) and (5.7) with $(\lambda, \rho) = (0.2, 2.635)$ for the data in plot (a). (c) A dataset of 100 observations with the first 50 observations from the distribution $N(0,1)$, the second 50 observations from the distribution $N(0,0.5^2)$, and all observations being independent. (d) The EWMA chart defined by (5.1) and (5.7) with $(\lambda, \rho) = (0.2, 2.635)$ when it is applied to the data in plot (c).*

Similar to (5.2)–(5.5), it can be checked that, in cases when the process is IC up to the n-th time point, we have

$$\mu_{E_n} = 1, \qquad \sigma_{E_n}^2 = \frac{2\lambda}{2-\lambda}\left[1-(1-\lambda)^{2n}\right]. \tag{5.13}$$

In cases when the process has a variance shift from σ_0^2 to σ_1^2 at the time point $1 \leq \tau \leq n$ and its mean remains unchanged, we have

$$
\begin{aligned}
\mu_{E_n,\tau} &= \lambda \sum_{i=1}^{\tau-1}(1-\lambda)^{n-i} + \lambda \sum_{i=\tau}^{n}(1-\lambda)^{n-i}\left(\frac{\sigma_1}{\sigma_0}\right)^2 \\
&= 1 + \lambda \sum_{i=\tau}^{n}(1-\lambda)^{n-i}\left[\left(\frac{\sigma_1}{\sigma_0}\right)^2 - 1\right],
\end{aligned}
\tag{5.14}
$$

and

$$
\begin{aligned}
\sigma_{E_n,\tau}^2 & \\
&= 2\lambda^2 \sum_{i=1}^{\tau-1}(1-\lambda)^{2(n-i)} + 2\lambda^2 \sum_{i=\tau}^{n}(1-\lambda)^{2(n-i)}\left(\frac{\sigma_1}{\sigma_0}\right)^4 \\
&= \frac{2\lambda}{2-\lambda}\left[1-(1-\lambda)^{2n}\right] + 2\lambda^2 \sum_{i=\tau}^{n}(1-\lambda)^{2(n-i)}\left[\left(\frac{\sigma_1}{\sigma_0}\right)^4 - 1\right].
\end{aligned}
\tag{5.15}
$$

From (5.14), we can see that, if the process variance shifts upward (i.e., $\sigma_1^2 > \sigma_0^2$) at τ, then the mean of E_n also shifts upward at τ, the initial shift size is $\lambda[(\sigma_1/\sigma_0)^2 - 1]$, and the shift size increases with n and converges to $[(\sigma_1/\sigma_0)^2 - 1]$. From (5.15), we can see that the upward process variance shift will also cause an upward variance shift in E_n. Both results suggest the following decision rule for detecting upward process variance shifts. A signal of upward process variance shift is given at the n-th time point if

$$
E_n > U = 1 + \rho_U \sqrt{\frac{2\lambda}{2-\lambda}[1-(1-\lambda)^{2n}]},
\tag{5.16}
$$

where $\rho_U > 0$ is a parameter chosen to achieve a pre-specified ARL_0 value. Similarly, if the process variance shifts downward (i.e., $\sigma_1^2 < \sigma_0^2$) at τ, then both the mean and variance of E_n would shift downward. In such cases, a reasonable decision rule is that a signal of downward process variance shift is given at the n-th time point if

$$
E_n < L = 1 - \rho_L \sqrt{\frac{2\lambda}{2-\lambda}[1-(1-\lambda)^{2n}]},
\tag{5.17}
$$

where $\rho_L > 0$ is a parameter chosen to achieve a pre-specified ARL_0 value.

Regarding the upward EWMA chart defined by (5.12) and (5.16) and the downward EWMA chart defined by (5.12) and (5.17), we would like to make two remarks. First, the IC variance of E_n defined in (5.13) converges to

$$
\tilde{\sigma}_{0,\lambda}^2 = \frac{2\lambda}{2-\lambda}.
$$

If this asymptotic variance of E_n is used when defining the upper and lower control limits, then they become

$$
U = 1 + \rho_U \sqrt{\frac{2\lambda}{2-\lambda}}, \qquad L = 1 - \rho_L \sqrt{\frac{2\lambda}{2-\lambda}},
\tag{5.18}
$$

Table 5.4 *Computed ρ_U and ρ_L values of the upward and downward EWMA charts defined by (5.12) and (5.18) for some commonly used ARL_0 and λ values. In the entry with a "*", the actual ARL_0 value is moderately different from the assumed ARL_0. For the two entries with a "-" symbol, the assumed ARL_0 value cannot be reached.*

	ρ_U				ρ_L			
ARL_0	$\lambda = 0.05$	0.1	0.2	0.3	$\lambda = 0.05$	0.1	0.2	0.3
50	0.901	1.380	1.916	2.259	0.865	1.100	1.195	1.166
100	1.455	1.988	2.606	2.996	1.201	1.366	1.360	1.273
200	2.017	2.595	3.258	3.702	1.510	1.580	1.480	1.349
300	2.342	2.944	3.655	4.132	1.670	1.682	1.538	1.384
370	2.518	3.133	3.854	4.354	1.746	1.731	1.563	1.401
400	2.588	3.199	3.935	4.440	1.774	1.750	1.574	1.407
500	2.796	3.419	4.184	4.722	1.862	1.808	1.605	1.426
1000	5.122	5.822	7.788	8.467	2.569	2.452*	-	-

which do not depend on n. Second, for a given value of ARL_0, the two parameters ρ_U and ρ_L in the two charts are usually different. That is mainly because the IC distribution of E_n is skewed to the right. For some commonly used ARL_0 and λ values, the corresponding ρ_U and ρ_L values computed by an R code written by the author are presented in Table 5.4. In the code, each ρ_U or ρ_L value is computed based on 10,000 replicated simulations. Except the last three entries in the last row of the table, the computed actual ARL_0 values of the EWMA charts with the presented parameter values are all within 0.5 of the assumed ARL_0 values. In cases when $ARL_0 = 1000$ and $\lambda = 0.1$, although the computed ρ_L value is presented in the last row of the table to be 2.452, the computed actual ARL_0 value is 981.349, which is moderately different from the assumed ARL_0 value 1000. In cases with the last two entries in the last row of the table, the assumed ARL_0 value 1000 cannot be reached. From the table, we can make several conclusions. First, ρ_U increases when either ARL_0 or λ increases. Second, ρ_L is indeed quite different from ρ_U. It seems that ρ_L increases when ARL_0 increases. But, when λ increases, ρ_L first increases and then decreases in cases when $ARL_0 \leq 300$. In cases when $ARL_0 > 300$, ρ_U decreases when λ increases.

To use each of the two one-sided charts, we can first pre-specify the value of λ, and then search for the value of ρ_U or ρ_L to reach a pre-specified ARL_0 value. Table 5.4 should be helpful for this purpose. The two one-sided charts can also be used simultaneously as a two-sided EWMA chart for detecting arbitrary variance shifts. In the two-sided chart, if the ARL_0 values of the two one-sided charts are chosen to be the same, then the values of ρ_U and ρ_L are usually different, as demonstrated in Table 5.4.

Example 5.5 (continued) *The data shown in Figure 5.8(a) are collected from a production process that has an upward variance shift from 1 to 2^2 at the 51st time point. To detect the variance shift, we consider the upward EWMA chart defined by (5.12) and (5.18), in which λ is chosen to be 0.1 and ARL_0 is chosen to be 200. By Table 5.4, ρ_U is chosen to be 2.595. The chart is shown in Figure 5.9(a), and it gives a first signal of upward variance shift at the 54th time point. Similarly, for the data*

shown in Figure 5.8(c), we consider the downward EWMA chart defined by (5.12)
and (5.18), in which λ and ARL_0 are chosen respectively to be 0.1 and 200 as well.
By Table 5.4, ρ_L is chosen to be 1.580. The chart is shown in Figure 5.9(b), and it
gives a first signal of downward variance shift at the 63rd time point.

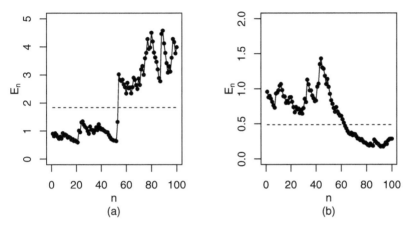

Figure 5.9 *(a) Upward EWMA chart defined by (5.12) and (5.18) when it is applied to the*
data shown in Figure 5.8(a) and when $\lambda = 0.1$ and $\rho_U = 2.595$ (the dashed horizontal line
denotes the upper control limit $U = 1.842$). (b) Downward EWMA chart defined by (5.12) and
(5.18) when it is applied to the data shown in Figure 5.8(c) and when $\lambda = 0.1$ and $\rho_L = 1.580$
(the dashed horizontal line denotes the lower control limit $L = 0.487$).

In the literature, there are some different versions of the EWMA charts based on
$Y_n = [(X_n - \mu_0)/\sigma_0]^2$. For instance, Domangue and Patch (1991) proposed a family
of the so-called *omnibus EWMA schemes* with the charting statistics

$$E_{n,a} = \lambda Y_n^{a/2} + (1 - \lambda)E_{n-1}, \tag{5.19}$$

where a is a given constant, and $\lambda \in (0, 1]$ is a weighting parameter. The initial value
$E_{0,a}$ is often chosen to be $\mu_{E,a}^* = \lim_{n \to \infty} \mu_{E_{n,a}}^{(0)}$, where $\mu_{E_{n,a}}^{(0)}$ is the IC mean of $E_{n,a}$.
Obviously, the family of the charting statistics defined by (5.19) includes the one
defined by (5.12) as a special case when $a = 2$. However, the numerical study in
Domangue and Patch (1991) shows that the chart when $a = 2$ often performs well,
compared to the other charts in the family. For related discussions, see papers including Huwang et al. (2009), MacGregor and Harris (1993), Reynolds and Stoumbos
(2006), Yeh et al. (2010), and the references cited therein.

As discussed in Subsection 4.3.2, batch data are commonly used for monitoring
the process variability. Assume that a sample of size m is collected at each time point.
At the n-th time point, for $n \geq 1$, let the sample mean and sample variance be $\overline{X}_n$ and
s_n^2, respectively. Then, when the process is IC at the n-th time point, we have

$$\frac{(m-1)s_n^2}{\sigma_0^2} \sim \chi_{m-1}^2.$$

In such cases, it is natural to consider the following EWMA charting statistic:

$$E_n = \lambda \frac{s_n^2}{\sigma_0^2} + (1 - \lambda)E_{n-1}, \tag{5.20}$$

where $E_0 = 1$ and $\lambda \in (0, 1]$ is a weighting parameter. If the process is IC up to the n-th time point, then we have

$$\mu_{E_n} = 1, \qquad \sigma_{E_n}^2 = \frac{2\lambda}{(2 - \lambda)(m - 1)} \left[1 - (1 - \lambda)^{2n}\right]. \tag{5.21}$$

By comparing the expressions in (5.21) and those in (5.13), it can be seen that everything is the same except that there is a factor $1/(m-1)$ in $\sigma_{E_n}^2$ when batch data with the batch size of m are used. Therefore, all the formulas (5.16)–(5.18) are still valid in this case, after the factor $1/(m-1)$ is included. For instance, to detect an upward variance shift, we can define the upward EWMA chart as follows. The chart gives a signal of an upward variance shift at the n-th time point if

$$E_n > U = 1 + \rho_U \sqrt{\frac{2\lambda}{(2 - \lambda)(m - 1)} \left[1 - (1 - \lambda)^{2n}\right]}, \tag{5.22}$$

where $\rho_U > 0$ is a parameter chosen to reach a pre-specified ARL_0 value. If the asymptotic variance of E_n is used, then the upper control limit U can be defined by

$$U = 1 + \rho_U \sqrt{\frac{2\lambda}{(2 - \lambda)(m - 1)}}. \tag{5.23}$$

The downward EWMA chart can be defined in a similar way.

Similar to the case with the CUSUM charts discussed in Subsection 4.3.2, the EWMA chart defined by (5.20) and (5.22) (or (5.23)) would not be affected by a possible mean shift in the process considered, while the EWMA chart defined by (5.12) and (5.16) (or (5.18)) is sensitive to such a mean shift. Therefore, if process variance shifts are our only concern, then the EWMA charts based on batch data are desirable. If both process mean shifts and process variance shifts are concerned, then the chart defined by (5.12) and (5.16) (or (5.18)) can be considered, although more effective charting schemes are available for that purpose. See Subsection 5.3.2 below for details.

From the above description, it can be seen that the upward and downward EWMA charts behave quite differently, because of the skewness of the distribution of $[(X_n - \mu_0)/\sigma_0]^2$ or s_n^2/σ_0^2. For instance, Table 5.4 shows that the values of ρ_U and ρ_L should be chosen quite differently for the two types of charts in cases when λ and ARL_0 are given. In the literature, there is much existing research in addressing this issue (e.g., Crowder and Hamilton, 1992; Huwang et al., 2010; Maravelakis and Castagliola, 2009; Shu and Jiang, 2008). Next, we briefly introduce the method proposed by Crowder and Hamilton (1992) in cases when batch data are available.

To handle the asymmetry of the distribution of s_n^2/σ_0^2, a natural idea is to consider $W_n = \log(s_n^2/\sigma_0^2)$, because the distribution of W_n would be closer to normal, compared to the distribution of s_n^2/σ_0^2. To this end, Crowder and Hamilton (1992) proposed the EWMA charting statistic

$$E_n^+ = \max\left(0, \lambda W_n + (1-\lambda)E_{n-1}\right)$$

for detecting upward process variance shifts, where $E_0^+ = 0$ and $\lambda \in (0, 1]$ is a weighting parameter. An intuitive explanation of this statistic is that, if the process variance shifts from σ_0^2 to σ_1^2 with $\sigma_1^2 > \sigma_0^2$ at the n-th time point, then $s_n^2/\sigma_0^2 \approx \sigma_1^2/\sigma_0^2 > 1$. Consequently, W_n would be larger than 0. On the other hand, in cases when the process is IC at the n-th time point, W_n would be close to 0. The restarting mechanism adopted here is similar to that used in the DI form of a CUSUM chart (cf., the expression (4.7) in Subsection 4.2.1). It can be checked that

$$\sigma_{W_n}^2 \approx \sigma_W^2 = \frac{2}{m-1} + \frac{2}{(m-1)^2} + \frac{4}{3(m-1)^3} - \frac{16}{15(m-1)^5}.$$

Therefore, the EWMA chart by Crowder and Hamilton (1992) gives a signal of an upward variance shift at the n-th time point if

$$E_n^+ > \rho_U \sqrt{\frac{\lambda}{2-\lambda}}\, \sigma_W,$$

where $\rho_U > 0$ is a parameter chosen to reach a pre-specified ARL_0 value. To detect downward variance shifts, Crowder and Hamilton (1992) suggested using

$$E_n^- = \min\left(0, \lambda W_n + (1-\lambda)E_{n-1}\right),$$

where $E_0^- = 0$. The chart gives a signal if

$$E_n^- < -\rho_L \sqrt{\frac{\lambda}{2-\lambda}}\, \sigma_W,$$

where $\rho_L > 0$ is a parameter chosen to reach a pre-specified ARL_0 value.

5.3.2 Joint monitoring of the process mean and variance

In applications, we are often interested in monitoring both the process mean and the process variance, because both of them are key to the quality of products. To this end, Example 5.5 shows that the conventional EWMA chart defined by (5.1) and (5.7), which is designed for detecting process mean shifts, has a certain ability to detect process variance shifts as well. However, that example also demonstrates that only the upward variance shifts can be signaled by that control chart. Further, with an OC signal from this chart alone, it is difficult to know whether the signal is due to a process mean shift or a process variance shift. For these reasons, that chart alone may not be appropriate for the joint monitoring of the process mean and variance.

To monitor both the process mean and variance, another possible approach is to use the omnibus EWMA schemes proposed by Domangue and Patch (1991), which have their charting statistics defined by (5.19). Among all omnibus EWMA schemes, most people focus on the two when $a = 0.5$ and 2. Between the two charts, Domangue and Patch (1991) and several other papers (e.g., Gan, 1995) show that the one with $a = 2$ often performs better in various cases. Therefore, here we focus on the one with $a = 2$, which is equivalent to the chart with the charting statistic E_n defined by (5.12). The main reason the chart (5.12) can detect variance shifts is that the mean of E_n would increase (decrease) if there is an upward (downward) variance shift at or before the n-th time point, as shown by (5.14). Now, if the process in question has an arbitrary mean shift from μ_0 to μ_1 at the n-th time point and the variance remains unchanged, we have

$$
\begin{aligned}
Y_n &= \left(\frac{X_n - \mu_0}{\sigma_0} \right)^2 \\
&= \left(\frac{X_n - \mu_1}{\sigma_0} \right)^2 + 2 \left(\frac{X_n - \mu_1}{\sigma_0} \right) \left(\frac{\mu_1 - \mu_0}{\sigma_0} \right) + \left(\frac{\mu_1 - \mu_0}{\sigma_0} \right)^2 .
\end{aligned}
$$

Therefore,

$$
\mu_{Y_n} = 1 + \left(\frac{\mu_1 - \mu_0}{\sigma_0} \right)^2 . \tag{5.24}
$$

The equation (5.24) implies that an arbitrary mean shift in the original observations would result in an upward mean shift in Y_n. For these reasons, if we are only concerned about arbitrary mean shifts and upward variance shifts, then we can consider using the upward EWMA chart with the charting statistic E_n defined by (5.12). Such a control chart would have no power for detecting downward variance shifts though, as justified by the expressions (5.14) and (5.15). Therefore, if downward variance shifts are also our concern, then we should consider using a two-sided version of this chart. All these results are demonstrated in the example below.

Example 5.6 *Let us consider four different situations in which the first 50 observations are generated from the $N(0,1)$ distribution and the remaining 50 observations are generated from one of the following four distributions:*

(i) $N(1,1)$,

(ii) $N(0,2^2)$,

(iii) $N(1,2^2)$, *and*

(iv) $N(0,0.5^2)$.

These four situations simulate four different production processes whose IC distributions are all $N(0,1)$ and who have mean and/or variance shifts at the 51st time point. More specifically, the first process has an upward mean shift of size 1, the second process has an upward variance shift of size $2^2 - 1 = 3$, the third process has both an upward mean shift of size 1 and an upward variance shift of size 3, and the fourth process has a downward variance shift of size $0.5^2 - 1 = -0.75$. One set of data in each situation is shown in Figure 5.10(a),(c),(e),(g), respectively. We then consider applying the upward EWMA chart defined by (5.12) and the upper control limit U in (5.18) with $\lambda = 0.1$ and $\rho_U = 2.595$ to the four datasets. By Table 5.4, this chart has an ARL_0 value of 200. The chart in the four situations is shown in Figure 5.10(b),(d),(f),(h), respectively. It can be seen that the chart can indeed detect the *mean shift in situation (i), the upward variance shift in situation (ii), and the mean and the upward variance shifts in situation (iii). In situation (iii) when both the mean shift and the upward variance shift are present, the signal from the chart is obviously much more convincing, compared to the signals in situations (i) and (ii). From plot (h), it can be seen that the chart has no power to detect the downward variance shift.*

Next, we consider applying the two-sided version of the EWMA chart defined by (5.12) and (5.18). In the chart, we choose $\lambda = 0.1$, $\rho_U = 3.199$, and $\rho_L = 1.750$. By Table 5.4, the two one-sided EWMA charts with these parameter values both have ARL_0 value of 400. Therefore, the two-sided EWMA chart would have an ARL_0 value of 200, which has been confirmed by a numerical study. The two-sided EWMA chart in the four situations considered is shown in the four plots of Figure 5.11. It can be seen that this chart can indeed detect arbitrary mean shifts and arbitrary variance shifts.

Although the upward or the two-sided version of the EWMA chart with the charting statistic defined by (5.12) can detect shifts in both the process mean and process variance, as demonstrated in Example 5.6, they have one fundamental limitation that they cannot distinguish mean shifts from variance shifts after signaling. To distinguish the two types of shifts, one possible approach is to use two control charts simultaneously, with at least one of them sensitive to the process mean or variance shifts only. To this end, for $n \geq 1$, assume that we have m observations at the n-th time point

$$X_{n1}, X_{n2}, \ldots, X_{nm},$$

and the sample mean and sample variance of these observations are $\overline{X}_n$ and s_n^2, respectively. Then, we can use an EWMA chart with the charting statistic

$$E_{n,M} = \lambda \overline{X}_n + (1-\lambda)E_{n-1,M},$$

to detect process mean shifts, where $E_{0,M} = \mu_0$. Similarly, we can use the EWMA chart with the charting statistic defined by (5.20) based on the sample variance s_n^2 for detecting process variance shifts. To distinguish the two charts, the charting statistic defined by (5.20) is denoted as $E_{n,V}$. Because $\overline{X}_n$ and s_n^2 are independent of each other, the two charts would also behave independently. Each chart can be designed properly for detecting upward, downward, or arbitrary mean or variance shifts. The joint monitoring scheme can be designed for detecting any combination of a specific type of mean shifts and a specific type of variance shifts. The next example provides a demonstration.

Example 5.7 *Let us reconsider the four cases that are discussed in Example 5.6. This time, instead of a single observation collected at each time point, we assume that a sample of size $m = 5$ is collected at each time point. The sample means of four datasets corresponding to the four cases are shown in the first column of Figure 5.12. For the EWMA chart based on $\overline{X}_n$, let us consider using a two-sided version with $\lambda = 0.1$ and $ARL_0 = 400$. In such cases, if the asymptotic variance of $E_{n,M}$ is used, then its upper and lower control limits are*

$$U = \mu_0 + \rho \sqrt{\frac{\lambda}{2-\lambda}} \frac{\sigma_0}{\sqrt{m}}, \qquad L = \mu_0 + \rho \sqrt{\frac{\lambda}{2-\lambda}} \frac{\sigma_0}{\sqrt{m}},$$

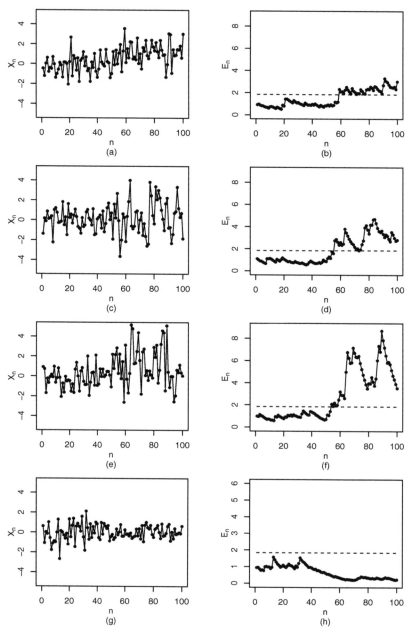

Figure 5.10 *Four different situations are considered, in each of which the first 50 observations are generated from the $N(0,1)$ distribution and the remaining 50 observations are generated from one of the following four distributions: $N(1,1)$, $N(0,2^2)$, $N(1,2^2)$, and $N(0,0.5^2)$. Plots (a), (c), (e), and (g) show four sets of data in the four situations, respectively. Plots (b), (d), (f), and (h) show the corresponding upward EWMA chart defined by (5.12) and the upper control limit U in (5.18) with $\lambda = 0.1$ and $\rho_U = 2.595$.*

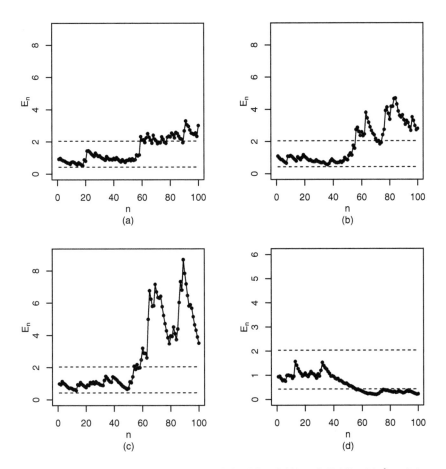

Figure 5.11 *The two-sided EWMA chart defined by (5.12) and (5.18) with* $\lambda = 0.1$, $\rho_U = 3.199$, *and* $\rho_L = 1.750$ *when it is applied to the four datasets shown in Figure 5.10(a), (c), (e), (g).*

where $\rho = 2.731$. *For the EWMA chart based on* s_n^2, *let us consider using the upward version with* $\lambda = 0.1$ *and* $ARL_0 = 400$. *In such cases, its upper control limit U is defined in (5.23) with* $\rho_U = 2.836$, *which is computed by the author using a numerical algorithm similar to the one used in the computation of Table 5.4. By using these two charts simultaneously, the joint monitoring scheme would have an* ARL_0 *value of 200, and it should be able to detect arbitrary mean shifts and upward variance shifts. The two charts when they are applied to the four datasets are shown in the second and third columns of Figure 5.12. From the plots in the figure, it can be seen that the mean shifts are all detected by the chart based on* $\overline{X}_n$, *and the upward variance shifts are all detected by the chart based on* s_n^2. *The downward variance shift in the fourth case is not detected by either chart. If downward variance shifts are also our concern, then we should use a two-sided version of the EWMA chart based on* s_n^2.

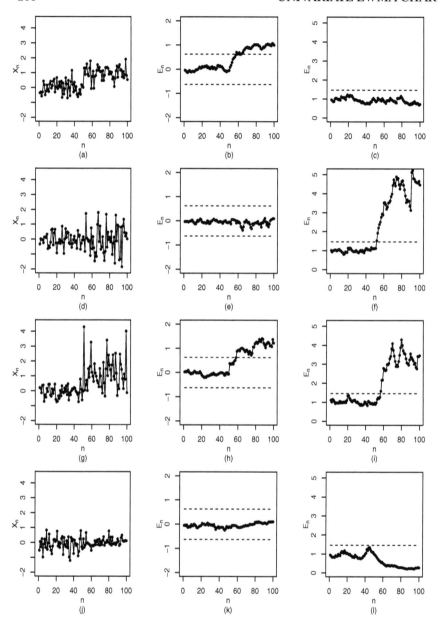

Figure 5.12 *Four different situations are considered, in each of which the first 50 samples of size 5 each are generated from the $N(0,1)$ distribution and the remaining 50 samples of size 5 each are generated from one of the following four distributions: $N(1,1)$ (plot (a)), $N(0,2^2)$ (plot (d)), $N(1,2^2)$ (plot (g)), and $N(0,0.5^2)$ (plot (j)). Plots in the first column show the sample means, plots in the second column show the two-sided EWMA charts with the charting statistic $E_{n,M}$ that is based on the sample means, and plots in the third column show the upward EWMA charts with the charting statistic $E_{n,V}$ that is based on the sample variances. Each EWMA chart is designed using $\lambda = 0.1$ and $ARL_0 = 400$, so that the joint monitoring scheme consisting of the mean chart and the variance chart has an ARL_0 value of 200.*

Gan (1995) provided many numerical results about the performance of the above joint monitoring EWMA scheme using $E_{n,M}$ and $E_{n,V}$, a corresponding joint monitoring CUSUM scheme (cf., Example 4.9 in Subsection 4.3.3), and several omnibus EWMA schemes including the ones with the charting statistic defined by (5.12) after $(X_n - \mu_0)/\sigma_0$ being replaced by $(\overline{X}_n - \mu_0)/(\sigma_0/\sqrt{m})$. Based on these results, we can conclude that (i) the joint monitoring EWMA scheme and the joint monitoring CUSUM scheme have similar performance in most cases, and (ii) they are often more effective than the omnibus EWMA schemes for the joint monitoring of both the process mean and process variance.

5.4 Self-Starting and Adaptive EWMA Charts

The EWMA charts discussed in the previous sections are based on the assumption that the IC mean μ_0 and the IC variance σ_0^2 are known. In practice, they are often unknown and need to be estimated from the observed data. Also, when designing an EWMA chart, the weighting parameter λ should be chosen beforehand. From the discussion in Section 5.2, we know that λ should be chosen large if the target shift is large, and it should be chosen small if the target shift is small. However, for a real production process, the size of a future shift is usually unknown. Therefore, it is still unclear how to choose λ properly in real applications. Possible solutions to these issues are discussed in this section.

5.4.1 Self-starting EWMA charts

When the IC mean μ_0 and the IC variance σ_0^2 are unknown, a natural idea is to estimate them from an IC dataset. In Subsection 4.5.1, we have demonstrated that the randomness in their estimators, denoted as $\widehat{\mu}_0$ and $\widehat{\sigma}_0^2$, respectively, would affect the performance of a CUSUM chart in a substantial way in cases when the sample size of the IC dataset is not large enough. Next, let us reconsider the scenarios discussed in Example 4.11 to see whether it is also true for an EWMA chart.

Example 5.8 *As in Example 4.11, let us assume that the true IC distribution of a production process is $N(0,1)$, which is unknown to us. Instead, it needs to be estimated from an IC dataset of size 100. The IC mean μ_0 and the IC variance σ_0^2 are estimated by the sample mean and sample variance of the IC dataset, denoted as $\widehat{\mu}_0$ and $\widehat{\sigma}_0^2$, and their sampling distributions are $\widehat{\mu}_0 \sim N(0,0.01)$ and $99\widehat{\sigma}_0^2 \sim \chi_{99}^2$ (cf., Section 2.7 and Example 4.11 for related discussions). Now, let us consider nine cases with the values of $\widehat{\mu}_0$ and $\widehat{\sigma}_0$ listed in the first two columns of Table 5.5. In each case, we treat the values of $\widehat{\mu}_0$ and $\widehat{\sigma}_0$ as the true values of μ_0 and σ_0, and then use the conventional upward EWMA chart defined by (5.8) and (5.9), except that the original observation X_n is replaced by its standardized value $(X_n - \widehat{\mu}_0)/\widehat{\sigma}_0$. In such cases, the actual distribution of $(X_n - \widehat{\mu}_0)/\widehat{\sigma}_0$ is $N(-\widehat{\mu}_0/\widehat{\sigma}_0, \widehat{\sigma}_0^{-2})$. In the control chart, λ and ARL_0 are chosen to be 0.1 and 200. Then, by the function xewma.crit() in the R-package spc, the value of ρ_U is computed to be 2.365. In the nine cases considered, the actual ARL_0 values of this EWMA chart are listed in the third column of Table 5.5. We then consider an upward mean shift of size 0.2 occurring at the initial observation time point. The corresponding ARL_1 values of the chart are listed in the fourth*

Table 5.5 *The actual ARL_0 values of the upward EWMA chart defined by (5.8) and (5.9) and the two-sided EWMA chart defined by (5.1) and (5.7) in cases with several estimated values of the IC mean μ_0 and IC standard deviation σ_0, along with their actual ARL_1 values for detecting a mean shift of size 0.2. In the upward chart, (λ, ρ_U) are chosen to be $(0.1, 2.365)$ so that its nominal ARL_0 value is 200. In the two-sided EWMA chart, (λ, ρ) are chosen to be $(0.1, 2.454)$, and its ARL_0 value is also 200.*

$\widehat{\mu}_0$	$\widehat{\sigma}_0$	Upward EWMA ARL_0	ARL_1	Two-Sided EWMA ARL_0	ARL_1
−0.1	0.9	70.276	29.478	90.828	36.375
−0.1	1.0	106.295	39.018	147.327	48.895
−0.1	1.1	168.837	53.562	247.974	67.531
0	0.9	120.751	43.487	114.030	57.254
0	1.0	200.449	61.918	201.873	83.542
0	1.1	348.040	90.295	361.680	123.971
0.1	0.9	224.109	70.276	90.828	90.828
0.1	1.0	403.217	106.295	147.327	147.327
0.1	1.1	632.519	168.837	247.974	247.974

column of Table 5.5. From the table, it can be seen that the actual ARL_0 values of the chart could be very different from the assumed ARL_0 value of 200 in various cases when the IC parameters are estimated, and its OC performance is often affected quite dramatically by the estimated values of the IC parameters as well.

In this example, we also consider the two-sided EWMA chart defined by (5.1) and (5.7), in which X_n is replaced by $(X_n - \widehat{\mu}_0)/\widehat{\sigma}_0$, the λ and ARL_0 values are chosen to be 0.1 and 200, respectively. By Table 5.1, the ρ value should be chosen 2.454. In the nine cases considered when the IC parameters are estimated by $\widehat{\mu}_0$ and $\widehat{\sigma}_0$, its actual ARL_0 and ARL_1 values are presented in the fifth and sixth columns of Table 5.5. It can be seen that these results are similar to those of the upward EWMA chart.

Example 5.8 demonstrates that the IC and OC performance of the EWMA charts is indeed affected substantially by the estimated IC parameter values in the case when the IC sample size is 100. We can imagine that such an impact would be lessened in cases when the IC sample size gets larger because the estimated values of the IC parameters would get closer to their true values. Jones et al. (2001) studied the performance of the EWMA charts with estimated IC parameters quite systematically, and found that their performance was reliable only when the IC sample size was as large as 2000. In practice, however, it is often difficult to obtain a large IC dataset. Therefore, we still need to find a reliable and effective method to construct EWMA charts when the IC parameters are unknown.

In Subsection 4.5.1, we described the self-starting CUSUM chart that was orig-inally proposed by Hawkins (1987) to handle the case when the IC parameters are unknown. The idea of self-starting CUSUM charts is actually quite general and it can also be applied to other types of control charts, including the EWMA charts. Next, we describe the construction of the self-starting version of the two-sided EWMA chart defined by (5.1) and (5.7). Let the phase II observations be $\{X_1, X_2, \ldots\}$. If the production process is IC at all times, then the observations are i.i.d. with the common distribution $N(\mu_0, \sigma_0^2)$. In such cases, as pointed out in Subsection 4.5.1, the transformed observations $\{Z_1, Z_2, \ldots\}$ are i.i.d. with the common distribution $N(0, 1)$, where Z_n, for $n \geq 3$, are defined by (4.48). Because Z_n is obtained by a strictly increasing transformation from the original observation X_n, detection of mean shifts in the original data $\{X_1, X_2, \ldots\}$ is equivalent to detection of mean shifts in the transformed data $\{Z_1, Z_2, \ldots\}$. By these results, the self-starting two-sided EWMA chart for detecting arbitrary mean shifts can be defined by

$$E_{n,SS} = \lambda Z_n + (1 - \lambda)E_{n-1,SS}, \tag{5.25}$$

where $\lambda \in (0, 1]$ is a weighting parameter, and the chart gives a signal when

$$|E_{n,SS}| > \rho_{SS} \sqrt{\frac{\lambda}{2 - \lambda} [1 - (1 - \lambda)^{2n}]}, \tag{5.26}$$

where $E_{0,SS} = 0$ and $\rho_{SS} > 0$ is a constant chosen to achieve a pre-specified ARL_0 value. In (5.26), we have used the facts that $\{Z_1, Z_2, \ldots\}$ are i.i.d. with the $N(0, 1)$ distribution when the process is IC and that

$$\sigma_{E_{n,SS}}^2 = \frac{\lambda}{2 - \lambda} [1 - (1 - \lambda)^{2n}].$$

If the asymptotic variance of $E_{n,SS}$ is used instead, then the corresponding decision rule of the chart is that it gives a signal when

$$|E_{n,SS}| > \rho_{SS} \sqrt{\frac{\lambda}{2 - \lambda}}. \tag{5.27}$$

Note that, because the transformed data $\{Z_1, Z_2, \ldots\}$ are i.i.d. with the $N(0, 1)$ distribution when the process is IC, the ρ_{SS} values for some commonly used ARL_0 and λ values can be found from Table 5.1.

Example 5.9 *Assume that 10 independent observations are generated from the $N(0, 1)$ distribution and another 30 independent observations are generated from the $N(0.5, 1)$ distribution. These observations are used for simulating a production process whose IC distribution is $N(0, 1)$ and which has a mean shift of size 0.5 at the 11th time point. The 40 observations are shown in Figure 5.13(a), and they are actually the same as those considered in Example 4.12 in Subsection 4.5.1. The first 10 observations $\{X_n, n = 1, 2, \ldots, 10\}$, the computed sample means $\{\overline{X}_n, n = 1, 2, \ldots, 10\}$, the sample standard deviations $\{s_n, n = 2, 3, \ldots, 10\}$, and the transformed observations $\{Z_n, n = 3, 4, \ldots, 10\}$ are presented in Table 4.7. Now, we apply the self-starting two-sided EWMA chart defined by (5.25) and (5.27) to this dataset, in which λ and ARL_0 are chosen to be 0.05 and 200, respectively. Then, by Table 5.1, ρ_{SS} should be chosen to be 2.216. The charting statistic $E_{n,SS}$ is shown in Figure 5.13(b) by the solid curve, and the control limits are shown in the same plot by two horizontal dashed lines. This chart gives a first signal at the 23rd time point. As a comparison, the conventional two-sided EWMA chart defined by (5.1) and (5.7) with the same values of λ, ARL_0, and the two control limits is also shown in the plot by the dotted curve. This chart assumes that the IC mean and variance are known. It gives a first signal at the 23rd time point as well.*

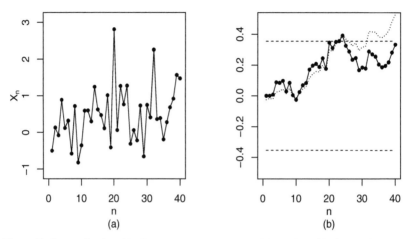

Figure 5.13 *(a) The first 10 observations are generated from the $N(0,1)$ distribution and the remaining 30 observations are generated from the $N(0.5,1)$ distribution. All observations are independent. (b) The self-starting two-sided EWMA chart defined by (5.25) and (5.27) (solid curve) and the conventional two-sided EWMA chart defined by (5.1) and (5.7) (dotted curve). In both charts, λ is chosen to be 0.05 and ARL_0 is chosen to be 200.*

Figure 5.13(b) shows that the self-starting two-sided EWMA chart defined by (5.25) and (5.27) performs reasonably well, without assuming the IC mean and variance to be known. However, this plot also shows that the signal of mean shift from this chart disappears quite fast. Therefore, if we do not react to its first signal of mean shift quickly, a persistent mean shift could be missed, as explained in Subsection 4.5.1 using the expression (4.50). As a comparison, the signal from the conventional two-sided EWMA chart gets stronger and stronger over time.

The self-starting versions of other EWMA charts can be constructed in a similar way. For instance, to construct the self-starting version of the omnibus EWMA schemes with the charting statistics defined in (5.19), $Y_n = [(X_n - \mu_0)/\sigma_0]^2$ can be simply replaced by Z_n^2, where $\{Z_n, n \geq 3\}$ are defined by (4.48). With the EWMA chart defined by (5.20)–(5.23) for detecting process variance shifts, to construct its self-starting version, we can replace σ_0^2 in (5.20) by a reasonable variance estimator constructed from the previous $n-1$ samples, denoted as $\widehat{\sigma}_{0,n-1}^2$. Then, the statistic $s_n^2/\widehat{\sigma}_{0,n-1}^2$ would have an F-distribution after it is properly rescaled, and the control limits of the chart can be chosen accordingly. One possible variance estimator constructed from the first $n-1$ samples is $\widehat{\sigma}_{0,n-1}^2 = (s_1^2 + s_2^2 + \cdots + s_{n-1}^2)/(n-1)$, although some other estimators are also possible.

5.4.2 Adaptive EWMA charts

In Subsection 5.2.1, we gave a general guideline for choosing the weighting parameter λ in the conventional two-sided EWMA chart defined by (5.1) and (5.7). By this guideline, λ should be chosen relatively small if the target shift size is small, and relatively large otherwise. In practice, however, the size of a potential shift is often unknown. In such cases, if we choose a relatively small value for λ, then the resulting EWMA chart would perform poorly in cases when the real shift size is large. Similarly, if λ is chosen relatively large, then the resulting chart would be ineffective for detecting small shifts. Can we propose a data-driven scheme for choosing λ such that the resulting EWMA chart performs reasonably well in a wide range of situations? This is the focus of this subsection.

In Subsection 4.5.2, we discussed a similar problem about the selection of the allowance constant k when implementing a CUSUM chart. In that scenario, theoretical results say that k should be chosen as half of a target shift size δ in order for the CUSUM chart to achieve its optimal performance in detecting that shift. Therefore, as long as we can provide a reasonable estimate of δ using all available observations, selection of k can be solved accordingly. The adaptive CUSUM chart by Sparks (2000) described in Subsection 4.5.2 is based on an EWMA estimate of δ. See the expressions (4.51) and (4.52) for details. For the EWMA charts, however, there is no corresponding theory about the functional relationship between the target shift size δ and the optimal value of λ, although certain researchers have explored such a relationship through numerical studies (e.g., Crowder, 1989; Lucas and Saccucci, 1990). For this reason, the idea of the method by Sparks (2000) is difficult to use here for adaptively choosing λ of an EWMA chart.

Capizzi and Masarotto (2003) provided a solution to the problem of choosing λ adaptively such that the resulting EWMA chart could perform reasonably well in a variety of different situations. Next, we describe their *adaptive exponentially weighted moving average (AEWMA)* control chart in detail. Assume that the IC process distribution is $N(\mu_0, \sigma^2)$, and we are interested in detecting process mean shifts. Then, the charting statistic of the AEWMA chart is defined by

$$A_n = A_{n-1} + \eta(e_n), \tag{5.28}$$

where $A_0 = \mu_0$, $e_n = X_n - A_{n-1}$, and $\eta(\cdot)$ is an increasing *score function*. The chart gives a signal of process mean shift if

$$|A_n - \mu_0| > h \tag{5.29}$$

where $h > 0$ is the control limit chosen to achieve a given ARL_0 value. In (5.28), A_{n-1} can be regarded as a prediction of the process mean based on all previous $n-1$ observations, and e_n is the prediction error. The formula (5.28) updates A_n according to the prediction error. If the prediction error is small, then the update from A_{n-1} to A_n is also small, and the update is large otherwise, which should be intuitively reasonable.

The expression (5.28) can be written in a more familiar form as follows:

$$A_n = w(e_n)X_n + (1 - w(e_n))A_{n-1}, \tag{5.30}$$

where $w(e_n) = \eta(e_n)/e_n$. In (5.30), A_n is a weighted average of the current observation X_n and the charting statistic value at the previous time point A_{n-1}, as in a conventional EWMA chart; but, the weights $w(e_n)$ change over time and they are determined by the prediction errors. In that sense, the EWMA chart (5.28)–(5.29) is adaptive over time to all observed data.

To design the AEWMA chart, the score function $\eta(\cdot)$ should be chosen properly. First, we notice that when $\eta(e_n) = e_n$ (i.e., $\eta(\cdot)$ is an identity function), the AEWMA chart becomes a Shewhart chart. On the other hand, when $\eta(e_n) = \lambda e_n$ with $\lambda \in (0,1)$ a constant, the AEWMA chart is just a conventional EWMA chart. Because Shewhart charts are good for detecting large shifts and conventional EWMA charts with small values of λ are good for detecting small shifts, Capizzi and Masarotto suggested choosing $\eta(e_n)$ between e_n and λe_n. More specifically, they suggested choosing $\eta(e_n)$ such that

(i) $\eta(e_n)$ is strictly increasing in e_n,

(ii) $\eta(-e_n) = -\eta(e_n)$,

(iii) $\eta(e_n)$ is close to λe_n when $|e_n|$ is small, and

(iv) $\eta(e_n)$ is close to e_n when $|e_n|$ is large.

Based on these considerations, they proposed several specific score functions, including the following two:

$$\eta_1(e_n) = \begin{cases} e_n + (1-\lambda)u, & \text{if } e_n < -u \\ \lambda e_n, & \text{if } |e_n| \leq u \\ e_n - (1-\lambda)u, & \text{if } e_n > u \end{cases}$$

and

$$\eta_2(e_n) = \begin{cases} e_n \left[1 - (1-\lambda)(1 - (e_n/u)^2)^2\right], & \text{if } |e_n| \leq u \\ e_n, & \text{otherwise,} \end{cases}$$

where $\lambda \in (0,1]$ and $u \geq 0$ are two constants. When $\lambda = 0.1$, $u = 2,4,6$, and $e_n \in [-10,10]$, $\eta_1(e_n)$, $w_1(e_n) = \eta_1(e_n)/e_n$, $\eta_2(e_n)$, and $w_2(e_n) = \eta_2(e_n)/e_n$ are presented in plots (a)–(d) of Figure 5.14, respectively. From the figure, it can be seen that (i) both $w_1(e_n)$ and $w_2(e_n)$ are between the values of λ and 1 (i.e., the resulting AEWMA chart would behave between the performance of the EWMA and Shewhart charts), (ii) both $w_1(e_n)$ and $w_2(e_n)$ are closer to the value of λ when u is larger (i.e., the AEWMA chart is more like an EWMA chart in such cases), and (iii) for a given value of u, $w_2(e_n)$ is closer to 1, compared to $w_1(e_n)$ (i.e., the AEWMA chart with $w_2(e_n)$ is more like a Shewhart chart, compared to the AEWMA chart with $w_1(e_n)$, if the two charts use the same value of u).

In either $\eta_1(e_n)$ or $\eta_2(e_n)$, there are two parameters λ and u. So, the AEWMA chart using $\eta_1(e_n)$ or $\eta_2(e_n)$ has a total of three parameters h, λ, and u to choose. To this end, Capizzi and Masarotto suggested a numerical algorithm with three steps described below.

Step 1 Choose a desired IC ARL value, denoted as ARL_0, and two shift sizes $\delta_1 < \delta_2$. δ_1 is a small target shift size and δ_2 is a large target shift size. They are chosen such that we hope the AEWMA chart can perform well in detecting shifts in the range $[\delta_1, \delta_2]$.

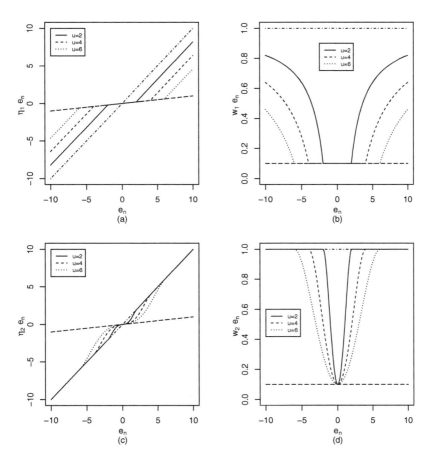

Figure 5.14 *In cases when $\lambda = 0.1$, $u = 2,4,6$, and $e_n \in [-10,10]$, the four quantities $\eta_1(e_n)$, $w_1(e_n) = \eta_1(e_n)/e_n$, $\eta_2(e_n)$, and $w_2(e_n) = \eta_2(e_n)/e_n$ are presented in plots (a)–(d), respectively. In each plot, the dash-dotted and the long-dashed lines denote the corresponding quantities of the Shewhart chart and the conventional EWMA chart, respectively.*

Step 2 Solve the minimization problem

$$\min_{h,\lambda,u} ARL_1(\delta_2;h,\lambda,u)$$

$$\text{subject to} \quad ARL_1(0;h,\lambda,u) = ARL_0,$$

where $ARL_1(\delta;h,\lambda,u)$ denotes the OC ARL value when the shift size is δ and the control chart has parameters h, λ, and u. The solution to the above problem may not be unique. Let (h^*,λ^*,u^*) be one specific solution.

Table 5.6 *"Optimal" parameter values of the AEWMA chart determined by the algorithm suggested by Capizzi and Masarotto (2003) in cases when $ARL_0 = 100$ or 500, $v = 0.05$, $\eta(e_n) = \eta_1(e_n)$ or $\eta_2(e_n)$, and $\sigma = 1$. This table is reproduced from Tables 3 and 4 in Capizzi and Masarotto (2003). In cases when $\sigma \neq 1$, the values of h and u should be multiplied by σ.*

$\eta(e_n)$	μ_1	μ_2	$ARL_0 = 100$			$ARL_0 = 500$		
			h	λ	u	h	λ	u
	0.25	4	0.1471	0.0162	2.7459	0.2017	0.0117	3.0326
	0.50	4	0.3927	0.0614	2.6306	0.4306	0.0398	2.8990
	1.00	4	0.7874	0.1813	2.5752	0.8238	0.1253	2.7765
	0.25	5	0.1457	0.0192	3.3249	0.1835	0.0137	3.4473
$\eta_1(e_n)$	0.50	5	0.3767	0.0670	3.2654	0.3960	0.0437	3.3402
	1.00	5	0.7688	0.1913	3.2907	0.7931	0.1354	3.2587
	0.25	6	0.1515	0.0207	4.2537	0.2091	0.0175	4.2176
	0.50	6	0.3671	0.0657	4.2638	0.4052	0.0474	4.1490
	1.00	6	0.7310	0.1792	7.9825	0.7610	0.1305	4.1534
	0.25	4	0.3542	0.0188	12.6145	0.4866	0.0082	12.5467
	0.50	4	0.4858	0.0463	11.4741	0.5807	0.0256	11.9897
	1.00	4	0.7697	0.1359	10.6114	0.8215	0.0830	11.1408
	0.25	5	0.2065	0.0196	24.0162	0.3381	0.0094	17.2037
$\eta_2(e_n)$	0.50	5	0.3729	0.0520	19.9865	0.4763	0.0296	15.4062
	1.00	5	0.6821	0.1473	20.1147	0.8551	0.1199	13.6702
	0.25	6	0.1430	0.0168	48.7509	0.2283	0.0130	31.1230
	0.50	6	0.3484	0.0577	40.6934	0.4133	0.0394	25.8760
	1.00	6	0.7224	0.1736	47.4411	0.7659	0.1226	25.5065

Step 3 Let $v > 0$ be a pre-specified small constant (e.g., $v = 0.05$). Solve the minimization problem

$$\min_{h,\lambda,u} ARL_1(\delta_1; h, \lambda, u)$$

$$\text{subject to} \quad ARL_1(0; h, \lambda, u) = ARL_0,$$

$$\text{and} \quad ARL_1(\delta_2; h, \lambda, u) \leq (1 + v)ARL_1(\delta_2; h^*, \lambda^*, u^*).$$

Then, the solution to this minimization problem is used by the AEWMA chart.

The AEWMA chart using the parameters determined by the above algorithm would perform reasonably well for detecting both large and small shifts. Source codes of the algorithm in c are available on the web page http://sirio.stat.unipd.it/caewma. When $ARL_0 = 100$ or 500, $v = 0.05$, $\eta(e_n) = \eta_1(e_n)$ or $\eta_2(e_n)$, and $\sigma = 1$, the parameters determined by the algorithm for some pairs of (μ_1, μ_2) are given in Capizzi and Masarotto (2003), which are also presented in Table 5.6. In cases when $\sigma \neq 1$, the values of h and u should be multiplied by σ.

Example 5.10 *Let us consider two datasets. The first one has 100 observations, the first 50 observations are generated from the $N(0,1)$ distribution, the remaining 50 observations are generated from the $N(0.5,1)$ distribution, and all observations are independent of each other. The second dataset is generated in the same way, except* *that its last 50 observations are generated from the $N(2,1)$ distribution. These two datasets are used for simulating two production processes that have mean shifts at the 51st time point with different shift sizes. The first process has a relatively small shift size of 0.5, while the second process has a relatively large shift size of 2.0. We now apply three control charts to both datasets. The first chart is the conventional two-sided EWMA chart defined by (5.1) and (5.7) with $\lambda = 0.01$ and $\rho = 1.973$. The second chart is also the conventional two-sided EWMA chart with $\lambda = 0.5$ and $\rho = 3.071$. The third chart is the AEWMA chart using $\eta_1(e_n)$ with $\lambda = 0.0398, u = 2.8990$, and $h = 0.4306$. By Tables 5.1 and 5.6, all these three charts have the same ARL_0 value of 500. The two datasets are shown by the two plots in the 1st row of Figure 5.15. The charts are shown by the plots in the 2nd, 3rd, and 4th rows of Figure 5.15. From the figure, it can be seen that, for the first dataset, the first EWMA chart gives a first signal around the 80th time point, the second EWMA cannot give any signal, and the AEWMA chart gives a first signal around the 75th time point. In this case, the second EWMA chart does not perform well because it is designed for detecting relatively large mean shifts while the true mean shift in this case is quite small. The first EWMA chart and the AEWMA chart perform reasonably well. For the second dataset, all three charts can give signals because the mean shift in this case is quite large. The first EWMA chart gives a first signal at the 57th time point, the second EWMA chart gives a first signal at the 52nd time point, and the AEWMA chart gives a first signal at the 54th time point. Therefore, the second EWMA chart performs the best in this case and the first EWMA chart performs the worst. As a summary, this example demonstrates that the conventional two-sided EWMA chart cannot be effective in detecting both small and large shifts, while the performance of the AEWMA chart is quite robust to shift sizes.*

5.5 Some Discussions

In the previous sections, we have described various EWMA charts for detecting process mean and/or variance shifts in cases when the IC and OC process distributions are assumed normal. These charts are good for detecting relatively small and persistent shifts. Their performance is generally similar to the performance of the corresponding CUSUM charts, while they are easier to understand and implement. On the other hand, CUSUM charts have certain theoretical optimality properties; the corresponding theory for the EWMA charts is still lacking. For these reasons, both CUSUM and EWMA charts are popularly used in practice.

In the literature, EWMA charts in cases when the distribution of the production process in question follows several other parametric forms have also been proposed. For instance, Pascual (2010) and Zhang and Chen (2004) constructed EWMA charts for monitoring production processes with Weibull distributions. Some authors, including Borror et al. (1998), Gan (1990a), and Weiß (2009), discussed process monitoring using EWMA charts in cases when the process distributions were Poisson. Gan (1990b), Perry and Pignatiello (2005), Sparks et al. (2011), and some others discussed how to monitor processes with binomial or negative binomial distributions using EWMA control charts.

All the EWMA charts discussed in this chapter are for detecting step shifts in the process mean and/or variance. In some applications, when the process in question becomes OC, its mean and/or variance would depart gradually from the IC level(s).

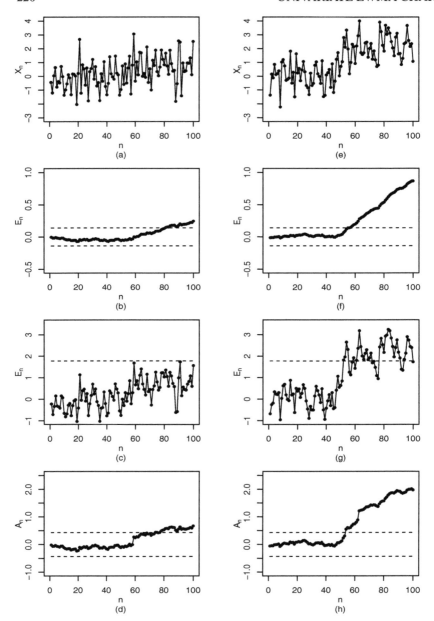

Figure 5.15 *Plot (a) shows a dataset with 100 independent observations, the first 50 of which are generated from the $N(0,1)$ distribution, and the remaining 50 of which are generated from the $N(0.5,1)$ distribution. Plots (b), (c), and (d) show the conventional two-sided EWMA chart defined by (5.1) and (5.7) with $\lambda = 0.01$ and $\rho = 1.973$, the conventional two-sided EWMA chart with $\lambda = 0.5$ and $\rho = 3.071$, and the AEWMA chart using $\eta_1(e_n)$ with $\lambda = 0.0398, u = 2.8990$, and $h = 0.4306$, respectively. All three charts have the same ARL_0 value of 500. Plots (e)–(h) show the corresponding results for another dataset with 100 independent observations, the first 50 of which are generated from the $N(0,1)$ distribution, and the remaining 50 of which are generated from the $N(2,1)$ distribution. Dashed horizontal lines in various plots denote the control limits.*

Such gradual departures are called *drifts*, as discussed in Section 4.7. In the literature, some authors have modified the conventional EWMA charts for detecting various mean/variance drifts. Interested readers are referred to papers including Gan (1991b), Su et al. (2011), Tseng et al. (2007, 2010), and Zou et al. (2009a).

EWMA charts average the current and all previous observations using an exponentially weighted averaging scheme. In the statistical literature, because the observed data are often assumed to have random noise involved, data smoothing through data averaging is the major statistical tool for removing random noise. Many different data averaging techniques have been proposed in the literature, which include kernel smoothing, smoothing splines, wavelet transformations, and so forth (cf., Qiu, 2005, Chapter 2). Among all existing data smoothing techniques, the ones in jump regression analysis (JRA) might be especially relevant to the SPC problem for the reasons explained below. The major goal of JRA is to estimate regression functions with jumps and other singularities (cf., Qiu, 2005). The SPC problem can be regarded as a special JRA problem in the sense that the regression function equals the IC process mean μ_0 when the production process in question is IC, and it jumps to an OC process mean μ_1 at a specific time point τ after the process has a mean shift at τ. There are two major differences between the phase II SPC problem and the JRA problem though. One is that the regression function equals a constant before or after the jump point τ in the SPC problem, while the regression function could be any continuous curves on the two sides of a jump point in the JRA problem. The second difference is that the sample size of a JRA problem is often fixed, while the sample size in the phase II SPC problem keeps increasing (i.e., it is a sequential problem). The second difference would disappear if we are concerned about the phase I SPC problem, because the sample size in phase I SPC is often fixed as well. Because of these substantial differences between the SPC and JRA problems, the methods in the JRA literature may not be efficient if they are applied to the SPC problems directly. However, ideas of certain existing JRA methods should be useful for us to construct efficient SPC procedures. For more information about JRA methods, read Gijbels et al. (2007), Joo and Qiu (2009), Loader (1996), Müller (1992, 2002), Qiu (1994, 1998, 2003, 2004), Qiu et al. (1991), Qiu and Yandell (1998), Wu and Chu (1993), Wu and Zhao (2007), among many others.

5.6 Exercises

5.1 The EWMA charting statistic E_n defined in (5.1) is a weighted average of the current and all previous observations $\{X_n, X_{n-1}, \ldots, X_1\}$. When $n = 20$, list all the weights received by these observations in the following cases:

(i) $\lambda = 1$,

(ii) $\lambda = 0.5$,

(iii) $\lambda = 0.2$, and

(iv) $\lambda = 0.05$.

Summarize your major findings from the weights listed. How would these different weighting schemes affect the properties of E_n?

5.2 For the EWMA charting statistic E_n defined in (5.1), derive the results in (5.2)–(5.5).

5.3 In Example 5.1, assume that we use the control limits defined in (5.7) with $\lambda = 0.1$ and $\rho = 2.703$ when implementing the two-sided EWMA chart. Plot the control chart using the data in that example, and discuss how this change (i.e., the control limits are changed from those in (5.6) to those in (5.7)) would affect the IC and OC performance of the control chart in general.

5.4 Assume that we obtain the following 30 observations from a production process whose IC distribution is assumed to be known $N(10, 2^2)$:

$$13, 12, 10, 10, 9, 7, 11, 12, 7, 10, 11, 10, 11, 10, 9,$$
$$18, 14, 16, 18, 12, 13, 17, 18, 15, 18, 15, 14, 18, 14, 16.$$

Construct different two-sided EWMA charts defined by (5.1) and (5.7) with the following specifications:

(i) $ARL_0 = 200$, $\lambda = 0.1$,

(ii) $ARL_0 = 200$, $\lambda = 0.5$,

(iii) $ARL_0 = 500$, $\lambda = 0.1$,

(iv) $ARL_0 = 500$, $\lambda = 0.5$.

Discuss how the values of ARL_0 and λ would affect the performance of the control chart.

5.5 For the dataset in the previous exercise, consider using the upward EWMA chart defined by (5.8) and (5.9) with $ARL_0 = 200$ and $\lambda = 0.1$. Its ρ_U value can be computed using the function xewma.crit() in the R-package spc in which the argument sided should be set as "one". Compare the performance of this chart with that of the two-sided chart in part (i) of the previous exercise.

5.6 For the two-sided EWMA charts defined by (5.1) and (5.7), compute its ARL_0 values in the following cases, using the function xewma.arl() in the R-package spc:

(i) $\lambda = 0.1$, $\rho = 1$,

(ii) $\lambda = 0.1$, $\rho = 2$,

(iii) $\lambda = 0.5$, $\rho = 1$,

(iv) $\lambda = 0.5$, $\rho = 2$.

Summarize your major findings about the relationship between ARL_0 and (λ, ρ).

5.7 For the two-sided EWMA charts defined by (5.1) and (5.7), compute its ρ values in the following cases, using the function xewma.crit() in the R-package spc:

(i) $ARL_0 = 150$, $\lambda = 0.1$,

(ii) $ARL_0 = 150$, $\lambda = 0.5$,

(iii) $ARL_0 = 450$, $\lambda = 0.1$,

(iv) $ARL_0 = 450$, $\lambda = 0.5$

Summarize your major findings about the relationship between ρ and (ARL_0, λ).

5.8 Figure 5.4 demonstrates the effectiveness of the EWMA chart defined by (5.1) and (5.7) with different values of the weighting parameter λ for detecting process mean shifts of different sizes. Produce a similar plot in cases when the IC process distribution is $N(0,1)$, $ARL_0 = 200$, $\lambda = 0.01, 0.3$, or 0.75, and the shift size δ changes from 0 to 3.0 with a step of 0.1. Summarize your results from the plot.

5.9 Assume that observations from a production process can be adequately described by the following AR(1) model:

$$X_n = 10 + 0.2(X_{n-1} - 10) + e_n, \qquad \text{for } n \geq 1,$$

where $X_0 = 10$, and $\{e_n, n \geq 1\}$ are i.i.d. with the common distribution $N(0,1)$. Design an EWMA chart with the adjusted control limits using the method discussed at the end of Subsection 5.2.2 with $M = 25$, $\lambda = 0.1$, and $\rho = 2.454$. Then, apply the designed EWMA chart to the observed data given below.

9, 10, 10, 11, 11, 10, 10, 10, 9, 11, 10, 9, 10, 11, 11,

16, 16, 15, 17, 17, 17, 16, 18, 18, 17, 16, 16, 15, 15, 16

5.10 Reproduce Figure 5.6 discussed in Subsection 5.2.3.

5.11 At the end of Subsection 5.2.3, it is explained that EWMA charts may not be robust to the normality assumption. To further investigate this issue, consider a simulation with the following several steps:

Step 1 Generate 50 random numbers from the χ_k^2 distribution.

Step 2 Compute the value of the charting statistic E_n of the conventional EWMA chart defined by (5.1) and (5.7) when $n = 50$.

Step 3 Repeat Steps 1 and 2 100 times, and make a histogram of the 100 computed values of E_{50}.

Do the above simulation in the following cases:

(i) $k = 1$, $\lambda = 0.5$,
(ii) $k = 1$, $\lambda = 0.1$,
(iii) $k = 5$, $\lambda = 0.5$,
(iv) $k = 5$, $\lambda = 0.1$.

Summarize your simulation results about the distribution of E_{50} in the different cases considered.

5.12 For the EWMA charting statistic defined in (5.12), verify the results in (5.13)–(5.15).

5.13 Assume that the following 40 observations are obtained from a production process with the IC distribution $N(50, 5^2)$:

33, 55, 51, 57, 57, 56, 49, 45, 55, 50,

50, 49, 50, 55, 57, 50, 47, 39, 51, 55,

52, 54, 36, 37, 43, 57, 52, 49, 51, 47,

40, 35, 36, 51, 39, 44, 35, 54, 38, 47.

Apply the upward EWMA chart for detecting variance shifts, defined by (5.12) and (5.16) with $ARL_0 = 200$ and $\lambda = 0.1$, to this dataset. Summarize your results.

5.14 For each of the two datasets discussed in Example 5.5, apply both the upward and downward EWMA charts for detecting variance shifts defined by (5.12), (5.16), and (5.17). In each chart, use $ARL_0 = 200$ and $\lambda = 0.05$. Summarize your results.

5.15 For the four datasets considered in Example 5.6, apply the upward EWMA chart defined by (5.12) with the upper control limit U in (5.18), $\lambda = 0.3$, and $\rho_U = 3.702$. Generate a figure similar to Figure 5.10. Summarize your results and compare these results with those in Example 5.6.

5.16 For the batch data discussed in Exercise 4.15, apply the EWMA chart defined by (5.20) and (5.22) with $\lambda = 0.1$ and $\rho_U = 2.836$ for detecting possible upward variance shifts. The related production process is assumed to have the IC distribution $N(1, 0.5^2)$. From Example 5.7, the EWMA chart described above has the ARL_0 value of 400. Summarize your results.

5.17 For the batch data discussed in Exercise 4.15, besides the EWMA chart considered in the previous exercise for detecting upward variance shifts, apply the two-sided EWMA chart based on the sample means $\overline{X}_n$ for detecting possible process mean shifts as well. The formulas of its control limits can be found in Example 5.7. In this chart, choose $\lambda = 0.1$ and $\rho = 2.731$ so that its ARL_0 is also 400. Summarize your results from this joint monitoring scheme.

5.18 Reproduce the results presented in the first and last rows of Table 5.5.

5.19 For the data considered in Exercise 4.21, apply the self-starting two-sided EWMA chart defined by (5.25) and (5.27) with $\lambda = 0.2$ and $ARL_0 = 200$. Summarize your results.

5.20 Assume that a production process has the IC distribution $N(0, 1)$, and the first 30 observations obtained for phase II process mean monitoring are listed below.

$$0.0, \; -2.3, \; 0.6, \; -1.0, \; -0.2, \; 0.3, \; 0.5, \; -0.4, \; 0.3, \; -0.5,$$
$$-0.9, \; -0.5, \; 1.0, \; -0.9, \; 1.5, \; 1.4, \; 0.0, \; 2.2, \; 2.7, \; 0.7,$$
$$-0.4, \; -0.8, \; -0.1, \; 0.0, \; 0.3, \; -1.0, \; 2.5, \; 1.3, \; 0.3, \; -0.3.$$

Use the adaptive EWMA chart discussed in Subsection 5.4.2 with the following specifications:

(i) $ARL_0 = 100$, $\eta(e_n) = \eta_1(e_n)$, $(\mu_1, \mu_2) = (0.25, 4)$, and $v = 0.05$,
(ii) $ARL_0 = 100$, $\eta(e_n) = \eta_1(e_n)$, $(\mu_1, \mu_2) = (1.0, 4)$, and $v = 0.05$,
(iii) $ARL_0 = 100$, $\eta(e_n) = \eta_1(e_n)$, $(\mu_1, \mu_2) = (0.25, 6)$, and $v = 0.05$,
(iv) $ARL_0 = 100$, $\eta(e_n) = \eta_2(e_n)$, $(\mu_1, \mu_2) = (0.25, 4)$, and $v = 0.05$,
(v) $ARL_0 = 100$, $\eta(e_n) = \eta_2(e_n)$, $(\mu_1, \mu_2) = (1.0, 4)$, and $v = 0.05$,
(vi) $ARL_0 = 100$, $\eta(e_n) = \eta_2(e_n)$, $(\mu_1, \mu_2) = (0.25, 6)$, and $v = 0.05$.

Summarize your results.

Chapter 6

Univariate Control Charts by Change-Point Detection

6.1 Introduction

From the description in the previous three chapters about the statistical process control (SPC) problem and various SPC control charts, we know that a major goal of SPC is to monitor a production process to make sure that the process runs stably. When the process runs stably, the distribution of its observations satisfies the requirements related to the quality of design and it remains unchanged. This distribution is called the in-control (IC) process distribution in the SPC literature. In cases when the process distribution changes from the IC distribution to another distribution (i.e., the out-of-control (OC) distribution) at an unknown time point, the process becomes OC. An SPC control chart should detect such a shift as soon as possible. From this description of the SPC problem, we can see that the process observations have a common distribution (i.e., the IC distribution) before the shift, and they have another common distribution (i.e., the OC distribution) after the shift. In SPC, our major goal is to detect any distributional shift as quickly as possible, and to locate the shift position accurately as well.

In the statistical literature, there is a research area called *change-point detection (CPD)* that is closely related to the SPC problem described above. In CPD, we are concerned about a sequence of random variables. The distribution of the first part of the random variables is assumed to be the same, the distribution of the remaining part of the random variables is assumed to be the same as well, but the distribution of the first part of the random variables and the one of the remaining random variables are assumed to be different. Then, the specific position in the sequence at which the distribution of the random variables changes from one to the other is called a *change-point*. In CPD, our major goal is to estimate the change-point position. See, for instance, Gombay (2003), Hinkley (1970), Smith (1975), Worsley (1983, 1986), and Yao (1987) for a more detailed description.

In CPD, the sample size is usually fixed. When the two related distributions are assumed to have some parametric forms, the change-point is often estimated by the maximum likelihood estimation method (cf., Subsection 2.7.2). In phase I SPC, the size of a set of observations obtained for SPC analysis is also fixed. So, the CPD methods can be applied to the phase I SPC problem directly. In phase II SPC, the number of process observations increases sequentially over time. In such cases, con-

225

ventional CPD methods cannot be applied to the phase II SPC problem directly. Instead, they should be modified properly for handling cases with sequentially increasing numbers of observations.

In recent years, CPD methodologies have been modified and applied to the SPC problem (cf., Hawkins et al., 2003; Hawkins and Zamba, 2005a,b). Compared to the three types of commonly used SPC control charts (i.e., the Shewhart charts, the CUSUM charts, and the EWMA charts), control charts based on CPD, which are called *CPD charts* in this book, have the advantage that they are efficient for detecting small and persistent shifts and can provide an estimator of the shift position as well at the same time when they give a shift signal. The Shewhart charts, although they can also tell us the shift position when they give a shift signal, are inefficient for detecting small and persistent shifts. On the other hand, the CUSUM and EWMA charts, although they are efficient for detecting small and persistent shifts, usually cannot provide an efficient estimator of the shift position when they give a shift signal. With these control charts, to estimate the shift position, a separate change-point detection procedure is often needed after a shift signal is delivered by them (cf., Samuel and Pignatiello, 2001).

In this chapter, we first describe some fundamental CPD methodologies, and then discuss some recent CPD control charts for detecting mean and/or variance shifts of univariate processes. Our discussion will mainly focus on the case when the IC and OC process distributions are normal. Process monitoring by CPD charts in cases when the processes are multivariate and their distributions are normal will be discussed in Chapter 7. Cases when the process distribution is nonparametric will be discussed in Chapters 8 and 9. Furthermore, similar to the CUSUM and EWMA charts, it would be more economical to use the CPD charts with individual observation data, although they can be easily applied to batch data.

6.2 Univariate Change-Point Detection

In this section, we describe some fundamental CPD methods in univariate cases. These methods should be useful for phase I SPC. They can also be modified properly for phase II SPC, which will be discussed in Section 6.3. Our description in this section is divided into two parts. In Subsection 6.2.1, some CPD methods for detecting a single change-point are described. Cases when multiple change-points are present are discussed in Subsection 6.2.2.

6.2.1 *Detection of a single change-point*

Assume that $\{X_1, X_2, \ldots, X_n\}$ are n independent random variables, $\{X_1, X_2, \ldots, X_r\}$ have a common pdf (or pmf) $f(x; \theta_0)$, and $\{X_{r+1}, X_{r+2}, \ldots, X_n\}$ have another common pdf (or pmf) $f(x; \theta_1)$, where $1 \leq r \leq n-1$ is an unknown integer, $f(x, \theta)$ is a parametric pdf (or pmf) with parameter θ, and $\theta_0 \neq \theta_1$ are two different values of θ. Then, r is the change-point, and the major goal of CPD is to estimate the value of r based on the observations $\{X_1, X_2, \ldots, X_n\}$. In cases when the values of θ_0 and θ_1 are unknown, they also need to be estimated. When θ_0 and θ_1 are the means of $f(x; \theta_0)$

and $f(x;\theta_1)$, respectively, the random variables $\{X_1,X_2,\ldots,X_n\}$ can be described by the model

$$X_i = \begin{cases} \theta_0 + \varepsilon_i, & \text{if } i = 1,2,\ldots,r, \\ \theta_1 + \varepsilon_i, & \text{if } i = r+1, r+2, \ldots, n, \end{cases} \qquad (6.1)$$

where $\{\varepsilon_1, \varepsilon_2, \ldots, \varepsilon_n\}$ is a sequence of i.i.d. random variables with a common pdf (or pmf) $f(x;0)$. Equation (6.1) describes the situation when $\{X_1,X_2,\ldots,X_n\}$ have a mean shift from θ_0 to θ_1 at the unknown change-point r.

In this section, we assume that the parametric pdf (or pmf) $f(x;\theta)$ is given beforehand. In such cases, the likelihood function of the change-point problem is

$$L(r,\theta_0,\theta_1;X_1,X_2,\ldots,X_n) = [\Pi_{i=1}^r f(X_i;\theta_0)] \times [\Pi_{i=r+1}^n f(X_i;\theta_1)].$$

The log-likelihood function is

$$\log\left(L(r,\theta_0,\theta_1;X_1,X_2,\ldots,X_n)\right) = \sum_{i=1}^r \log\left(f(X_i;\theta_0)\right) + \sum_{i=r+1}^n \log\left(f(X_i;\theta_1)\right). \quad (6.2)$$

In the case when both θ_0 and θ_1 are known, the maximum likelihood estimator (MLE) of the change-point r is defined by

$$\hat{r} = \arg \max_{1 \leq r \leq n-1} \log\left(L(r,\theta_0,\theta_1;X_1,X_2,\ldots,X_n)\right). \qquad (6.3)$$

Hinkley (1970) studied the asymptotic distribution of the MLE $\hat{r}$. Worsley (1986) provided a confidence interval for r and a hypothesis testing procedure for testing the existence of the change-point in cases when the pdf (or pmf) $f(x;\theta)$ belonged to the exponential family.

In cases when $\{X_1,X_2,\ldots,X_r\}$ have the normal distribution $N(\mu_0,\sigma^2)$ and $\{X_{r+1},X_{r+2},\ldots,X_n\}$ have the normal distribution $N(\mu_1,\sigma^2)$, where $\mu_0 \neq \mu_1$ are two different constants, the random variables $\{X_1,X_2,\ldots,X_n\}$ have a change-point in their means at r, and their variances are unchanged. In such cases, it can be checked that the log-likelihood function defined in (6.2) becomes

$$\log\left(L(r,\mu_0,\mu_1,\sigma^2;X_1,X_2,\ldots,X_n)\right)$$
$$= -n\log\left(\sqrt{2\pi}\sigma\right) - \frac{1}{2\sigma^2}\left[\sum_{i=1}^r (X_i - \mu_0)^2 + \sum_{i=r+1}^n (X_i - \mu_1)^2\right].$$

Thus, the negative log-likelihood function is proportional to

$$\begin{aligned} \widetilde{S}_r^2 &= \sum_{i=1}^r (X_i - \mu_0)^2 + \sum_{i=r+1}^n (X_i - \mu_1)^2 \\ &= \sum_{i=1}^r \left[(X_i - \mu_0)^2 - (X_i - \mu_1)^2\right] + \sum_{i=1}^n (X_i - \mu_1)^2. \end{aligned} \qquad (6.4)$$

Therefore, the MLE of r defined in (6.3) becomes

$$\hat{r} = \arg \min_{1 \leq r \leq n-1} \widetilde{S}_r^2 = \arg \min_{1 \leq r \leq n-1} \sum_{i=1}^r \left[(X_i - \mu_0)^2 - (X_i - \mu_1)^2\right]. \qquad (6.5)$$

If μ_0 and μ_1 are both unknown, then conditional on r, their MLEs are respectively

$$\overline{X}_r = \frac{1}{r}\sum_{i=1}^{r} X_i, \qquad \overline{X}'_r = \frac{1}{n-r}\sum_{i=r+1}^{n} X_i.$$

In such cases, the MLE of r can be computed by

$$\hat{r} = \arg\min_{1\leq r\leq n-1} \widetilde{S}^2_r = \arg\max_{1\leq r\leq n-1} \frac{r(n-r)}{n}(\overline{X}_r - \overline{X}'_r)^2, \tag{6.6}$$

where

$$
\begin{aligned}
\widetilde{S}^2_r &= \sum_{i=1}^{r}(X_i - \overline{X}_r)^2 + \sum_{i=r+1}^{n}(X_i - \overline{X}'_r)^2 \\
&= \sum_{i=1}^{n}(X_i - \overline{X})^2 - \left[r(\overline{X}_r - \overline{X})^2 + (n-r)(\overline{X}'_r - \overline{X})^2 \right] \\
&= \sum_{i=1}^{n}(X_i - \overline{X})^2 - \frac{n}{r(n-r)}\left[\sum_{i=1}^{r}(X_i - \overline{X})\right]^2 \\
&= \sum_{i=1}^{n}(X_i - \overline{X})^2 - \frac{r(n-r)}{n}\left(\overline{X}_r - \overline{X}'_r\right)^2. \tag{6.7}
\end{aligned}
$$

Example 6.1 *Assume that there are 20 observations, the first 10 of them are gener-ated from the distribution $N(0,1)$, and the remaining 10 observations are generated from the distribution $N(1,1)$. So, the sequence of the 20 observations has a mean shift of size 1 at $r+1 = 11$. One realization of these observations is presented in the second column of Table 6.1. First, we assume that μ_0 and μ_1 are known to be 0 and 1, respectively. Then, the MLE $\hat{r}$ can be computed by (6.5). To this end, the values of $\{S^2_i, i = 1,2,\ldots,19\}$ are computed by (6.4) and presented in the third column of Table 6.1, from which we can see that $\hat{r} = 10$. Now, let us assume that both μ_0 and μ_1 are unknown. In such cases, the MLE $\hat{r}$ can be computed by (6.6). For that purpose, the values of $\{\widetilde{S}^2_i, i = 1,2,\ldots,19\}$ are computed and presented in the fourth column of Table 6.1, from which we also obtain $\hat{r} = 10$. In the latter case, we can compute the MLEs of μ_0 and μ_1 to be*

$$\hat{\mu}_0 = \overline{X}_{\hat{r}} = \frac{1}{\hat{r}}\sum_{i=1}^{\hat{r}} X_i = -0.018, \qquad \hat{\mu}_1 = \overline{X}'_{\hat{r}} = \frac{1}{n-\hat{r}}\sum_{i=\hat{r}+1}^{n} X_i = 1.234.$$

Hawkins (1977) discussed the hypothesis testing problem regarding a loca-tion shift in a sequence of normally distributed independent random variables $\{X_1, X_2, \ldots, X_n\}$, which is discussed below. The null and alternative hypotheses of the problem can be formulated as

$$H_0 : X_i \sim N(\mu_0, \sigma^2), \qquad \text{for } i = 1, 2, \ldots, n,$$

Table 6.1 *The second column presents a sequence of 20 observations, and the third and fourth columns present the corresponding values of $\{\widetilde{S}_i^2, i = 1, 2, \ldots, 19\}$ and $\{\widehat{S}_i^2, i = 1, 2, \ldots, 19\}$, respectively.*

i	X_i	$\widetilde{S}_i^2$	$\widehat{S}_i^2$
1	-0.502	18.398	16.03
2	0.132	17.661	15.929
3	-0.079	16.503	15.3
4	0.887	17.277	16.084
5	0.117	16.511	15.68
6	0.319	16.148	15.494
7	-0.582	13.984	13.873
8	0.715	14.413	14.227
9	-0.825	11.763	11.672
10	-0.36	**10.043**	**9.494**
11	1.09	11.223	10.586
12	1.096	12.415	11.502
13	0.798	13.012	11.616
14	1.74	15.492	13.583
15	1.123	16.739	14.153
16	0.971	17.68	14.349
17	0.611	17.902	13.598
18	1.511	19.924	14.685
19	0.086	19.096	9.639
20	3.31	-	-

versus

$$H_1 : X_i \sim \begin{cases} N(\mu_0, \sigma^2), & \text{for } i = 1, 2, \ldots, r, \\ N(\mu_1, \sigma^2), & \text{for } i = r+1, r+2, \ldots, n, \end{cases}$$

where $1 \leq r \leq n-1$ is an unknown change-point. In cases when σ^2 is known, it can be checked that the well-known likelihood ratio test (LRT) statistic $\Lambda(X_1, X_2, \ldots, X_n)$ (cf., Subsection 2.7.3) has the expression

$$-2\sigma^2 \log\left(\Lambda(X_1, X_2, \ldots, X_n)\right) = S_n^2 - \min_{1 \leq r \leq n-1} \widehat{S}_r^2 = S_n^2 - \widehat{S}_{\hat{r}}^2,$$

where $S_n^2 = \sum_{i=1}^n (X_i - \overline{X})^2$ and $\widehat{S}_r^2$ is defined by (6.7). By the relationship that

$$S_n^2 = \widehat{S}_r^2 + T_r^2,$$

where

$$T_r^2 = \frac{r(n-r)}{n}\left(\overline{X}_r - \overline{X}_r'\right)^2 = \frac{n}{r(n-r)}\left[\sum_{i=1}^r (X_i - \overline{X})\right]^2,$$

we have

$$-2\sigma^2 \log\left(\Lambda(X_1, X_2, \ldots, X_n)\right) = \max_{1 \leq r \leq n-1} T_r^2.$$

Therefore, the LRT test is equivalent to the test using the test statistic

$$U = \max_{1 \le r \le n-1} |T_r|, \tag{6.8}$$

and H_1 can be rejected if the observed value of U is too large. To make formal decisions about the null and alternative hypotheses using U defined in (6.8), we still need to derive its distribution under H_0 (i.e., its null distribution). To this end, Hawkins (1977) derived formulas for the null distributions of both U and $\hat{r}$, which were unfortunately too complicated to use in practice. However, these null distributions can always be approximated well using numerical simulations.

In cases when σ^2 is unknown, it can be checked that the LRT statistic $\Lambda(X_1, X_2, \ldots, X_n)$ is a strictly decreasing function of

$$W = \max_{1 \le r \le n-1} |T_r|/\widehat{\widehat{S}}_r. \tag{6.9}$$

Therefore, W can be used as a test statistic, and H_1 can be rejected if the W value is too large. Because W in (6.9) is a strictly increasing function of

$$V = \max_{1 \le r \le n-1} |T_r|/S_n, \tag{6.10}$$

V can also be used as a test statistic for testing the hypotheses H_0 and H_1. Worsley (1979) derived the null distribution of V in (6.10). Again, the related formulas are complicated. In practice, the null distribution of V can be approximated using numerical simulations instead.

The point estimation and hypothesis testing methods discussed above are for handling the case when the random variables $(X_1, X_2, \ldots, X_n)$ are independent of each other. Kokoszka and Leipus (1998) showed that the MLE $\hat{r}$ defined in (6.6) was also statistically consistent in various cases when $(X_1, X_2, \ldots, X_n)$ are correlated. For related discussions, see Box and Tiao (1965), Gombay (2008), Henderson (1986), Kim (1996), Ling (2007), and the references cited therein.

6.2.2   Detection of multiple change-points

In this subsection, we briefly discuss the case when multiple change-points are possible in the sequence of n independent random variables $\{X_1, X_2, \ldots, X_n\}$. Assume that the sequence consists of k segments, random variables in the first segment $\{X_1, X_2, \ldots, X_{r_1}\}$ have the same pdf (or pmf) $f(x, \theta_1)$, those in the second segment $\{X_{r_1+1}, X_{r_1+2}, \ldots, X_{r_2}\}$ have the same pdf (or pmf) $f(x, \theta_2)$, ..., and those in the k-th segment $\{X_{r_{k-1}+1}, X_{r_{k-1}+2}, \ldots, X_n\}$ have the same pdf (or pmf) $f(x, \theta_k)$, where $f(x, \theta)$ is a pre-specified parametric pdf (or pmf), $(\theta_1, \theta_2, \ldots, \theta_k)$ are k different values of θ, and

$$1 \le r_1 < r_2 < \ldots < r_{k-1} \le n-1$$

are $k-1$ change-points. If $k = 2$, then there is only one change-point and the CPD problem becomes the one discussed in the previous subsection. In this subsection, we focus mainly on cases when at least two change-points are present (i.e., $k > 2$).

In cases when $f(x, \theta)$ is the pdf of a normal distribution and θ is its mean μ, by the CPD model described above, the random variables $\{X_1, X_2, \ldots, X_n\}$ have change-points in their means and their variances are unchanged. In such cases, we can describe $\{X_1, X_2, \ldots, X_n\}$ by

$$
X_i = \begin{cases}
\mu_1 + \varepsilon_i, & \text{if } i = 1, 2, \ldots, r_1 \\
\mu_2 + \varepsilon_i, & \text{if } i = r_1 + 1, r_1 + 2, \ldots, r_2 \\
\vdots & \vdots \\
\mu_k + \varepsilon_i, & \text{if } i = r_{k-1} + 1, r_{k-1} + 2, \ldots, n,
\end{cases}
\tag{6.11}
$$

where $\{\varepsilon_i, i = 1, 2, \ldots, n\}$ are i.i.d. with the common distribution $N(0, \sigma^2)$, and $\mu_1, \mu_2, \ldots, \mu_k$ are k different values of μ. To handle this CPD problem, a classic method is the regression tree approach proposed by Breiman et al. (1984), which uses a recursive binary splitting algorithm to estimate the change-points. This method is computationally fast. But, as pointed out by Hawkins (2001), it requires the change-points to have a hierarchic structure, and therefore it may not yield optimal estimators. Next, we describe the *dynamic programming (DP) algorithm* that was proposed by Hawkins (2001) for estimating multiple change-points.

For $0 \le h < m \le n$, let $Q(h, m)$ be the -2 times the maximized log-likelihood based on the sequence of random variables $\{X_{h+1}, X_{h+2}, \ldots, X_m\}$. Then, we have

$$
\begin{aligned}
& Q(0, n) \\
={} & -2 \max_{\mathbf{r}, \mu} \log \left(L(\mathbf{r}, \mu; X_1, X_2, \ldots, X_n) \right) \\
={} & -2 \max_{\mathbf{r}} \left[\max_{\mu} \log \left(L(\mathbf{r}, \mu; X_1, X_2, \ldots, X_n) \right) \right] \\
={} & \min_{\mathbf{r}} \sum_{j=1}^{k} Q(r_{j-1}, r_j),
\end{aligned}
$$

where $\mathbf{r} = (r_1, r_2, \ldots, r_{k-1})'$, $\mu = (\mu_1, \mu_2, \ldots, \mu_k)'$, $r_0 = 0$, and $r_k = n$. From the above expression, we can notice the so-called "principle of optimality" (cf., Bellman and Dreyfus, 1962), described as follows. To split the sequence of n random variables into k segments in an optimal way, the following recursive algorithm can be used. First, the estimator of the last change-point r_{k-1}, denoted as $\hat{r}_{k-1}$, can be determined from the original sequence $\{X_1, X_2, \ldots, X_n\}$. Then, the estimators of the remaining $k - 2$ change-points can be obtained from the sequence $\{X_1, X_2, \ldots, X_{\hat{r}_{k-1}}\}$ by splitting it into $k - 1$ segments. To determine the estimators $\hat{r}_{k-1}, \hat{r}_{k-2}, \ldots, \hat{r}_1$ using such a recursive algorithm, we can make use of the following results. Let $X^2(j, m)$ be the -2 times the maximized log-likelihood by fitting a j-segment model (i.e., $k = j$ in (6.11)) to the sequence $\{X_1, X_2, \ldots, X_m\}$, where $j \le m$ are two integers. Then, after some constants are omitted, we have the following recursive relationship:

$$
X^2(j, m) = \min_{0 < h < m} \left(X^2(j - 1, h) + Q(h, m) \right),
\tag{6.12}
$$

where $X^2(1, m) = Q(0, m)$. Based on the "principle of optimality" and (6.12),

Hawkins (2001) proposed the DP algorithm to find the estimators $\widehat{r}_{k-1}, \widehat{r}_{k-2}, \ldots, \widehat{r}_1$, which is described in the box below.

Dynamic Programming Algorithm

Assume that the sequence of independent random variables $\{X_1, X_2, \ldots, X_n\}$ satisfies the change-point model (6.11). Then, the change-points $r_1, r_2, \ldots, r_{k-1}$ can be estimated recursively as follows.

Step 1 Compute $X^2(j, m)$, for all $j = 1, 2, \ldots, k-1$ and $m = j, j+1, \ldots, n-1$.

Step 2 For $\ell = 1, 2, \ldots, k-1$, search the minimizer of

$$\min_{k-\ell \leq h < \widehat{r}_{k-\ell+1}} \left(X^2(k-\ell, h) + Q(h, \widehat{r}_{k-\ell+1}) \right),$$

where $\widehat{r}_k = n$. The searched minimizer is defined as the estimator of $r_{k-\ell}$, denoted as $\widehat{r}_{k-\ell}$, and $\mu_{k-\ell+1}$ can be estimated accordingly by the MLE $\widehat{\mu}_{k-\ell+1}$ obtained from $Q(\widehat{r}_{k-\ell}, \widehat{r}_{k-\ell+1})$. After $\widehat{r}_1$ and $\widehat{\mu}_2$ are defined, $\widehat{\mu}_1$ can be defined from $Q(0, \widehat{r}_1)$ by the MLE of μ_1.

Hawkins (2001) proposed the above DP algorithm in cases when $f(x, \theta)$ belongs to the exponential family (cf., (4.29) in Subsection 4.4.1). In cases when $f(x, \theta)$ is the pdf of a normal distribution, then for any $0 \leq h < m \leq n$, we can replace $Q(h, m)$ by

$$W(h, m) = \sum_{i=h+1}^{m} (X_i - \overline{X}_{h,m})^2,$$

where $\overline{X}_{h,m}$ is the sample mean of $\{X_{h+1}, X_{h+2}, \ldots, X_m\}$.

Example 6.2 *Assume that there are $n = 30$ observations, the first 10 of them are generated from the distribution $N(0, 0.5^2)$, the second 10 observations are generated from the distribution $N(1, 0.5^2)$, and the remaining 10 observations are generated from the distribution $N(0, 0.5^2)$. These observations are shown in Figure 6.1(a) by the dark dots. So, there are $k = 3$ segments, and $k - 1 = 2$ change-points at $r_1 = 10$ and $r_2 = 20$ in this dataset. To detect the change-points using the DP algorithm, we first compute*

$$\left\{ X^2(2, h) + Q(h, n), \ h = 2, 3, \ldots, n-1 \right\}.$$

This sequence is shown in Figure 6.1(b) by the solid curve. It can be seen that the minimum of this sequence is reached at $h = 20$. So, we have $\widehat{r}_2 = 20$, and $\widehat{\mu}_3 = \overline{X}_{20,30} = -0.065$. To estimate r_1, we compute the sequence

$$\left\{ X^2(1, h) + Q(h, \widehat{r}_2), \ h = 1, 2, \ldots, \widehat{r}_2 - 1 \right\},$$

which is shown in Figure 6.1(b) by the dashed curve. It can be seen that the minimum of this sequence is reached at $h = 10$. So, we have $\widehat{r}_1 = 10$, and $\widehat{\mu}_2 = \overline{X}_{10,20} = 1.117$. Finally, we have $\widehat{\mu}_1 = \overline{X}_{0,10} = -0.009$.

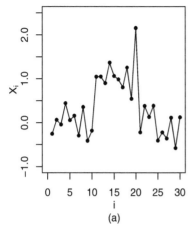
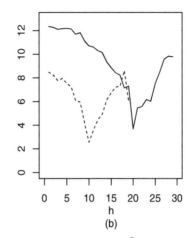

Figure 6.1 *(a) The first 10 observations are generated from the $N(0, 0.5^2)$ distribution, the second 10 observations are generated from the $N(1, 0.5^2)$ distribution, and the remaining 10 observations are generated from the $N(0, 0.5^2)$ distribution. All observations are independent of each other. (b) The sequence $\{X^2(2, h) + Q(h, 30),\ h = 2, 3, \ldots, 29\}$ is shown by the solid curve, and the sequence $\{X^2(1, h) + Q(h, 20),\ h = 1, 2, \ldots, 19\}$ is shown by the dashed curve.*

In the DP algorithm, the number of change-points $k - 1$ is assumed known. In the literature, there is much discussion about estimation of $k - 1$. Theoretically, this is a difficult problem because it is found that most test statistics proposed in the literature for testing hypotheses about $k - 1$ do not have asymptotic distributions (cf., Bhattacharya, 1994; Hinkley, 1970; Hawkins, 1977; Yao, 1987). However, a number of practical approaches for estimating the number and locations of the change-points are available. Most of these approaches treat $k - 1$ as a parameter and estimate it by a model selection approach (e.g., Fearnhead, 2006; Lavielle, 2005; Lavielle and Moulines, 2000; Yao and Au, 1989).

Detection of multiple change-points in time series and other correlated data has also been discussed in the literature. See, for instance, Choi et al. (2008), Lavielle and Ludena (2000), Shao and Zhang (2010), and the references cited therein.

6.3 Control Charts by Change-Point Detection

In this section, we describe some CPD control charts for phase II SPC in cases when process observations are univariate and follow normal distributions. Our description is divided into three parts. Monitoring of process mean is discussed in Subsection 6.3.1, monitoring of process variance is discussed in Subsection 6.3.2, and joint monitoring of both process mean and variance is discussed in Subsection 6.3.3.

6.3.1   Monitoring of the process mean

Assume that the univariate quality characteristic of a production process is X, its observations X_i follow the IC distribution $N(\mu_0, \sigma^2)$ when the process is IC, they follow the OC distribution $N(\mu_1, \sigma^2)$ after the process mean shifts from μ_0 to μ_1 at an unknown time point $r+1$. The major goal of phase II SPC is to detect such a distributional shift as soon as possible. From this description of the phase II SPC problem for monitoring the process mean, it can be seen that it is related to the CPD problem discussed in the previous section. But, the two problems have the following substantial difference. In the CPD problem (6.1), the sample size n is fixed; but, the sample size in the phase II SPC problem keeps increasing until a signal of the mean shift is given. Because of this difference, the CPD methods cannot be applied to the phase II SPC problem directly. Instead, they need to be modified properly.

In cases when the parameters μ_0, μ_1, and σ are all known, it has been demonstrated in the literature that the CUSUM charts have some optimality properties. See Subsection 4.2.4 for a related discussion. In practice, however, these parameters are often unknown. Although the self-starting CUSUMs and the adaptive CUSUMs discussed in Section 4.5 can help solve the problem in such cases, these CUSUM charts would not be optimal any more. Furthermore, most CUSUM charts do not provide a good estimator of the shift time point $\tau = r+1$ at the time when they give a signal of the process distributional shift. In cases when estimation of τ is desirable, then it is often estimated by a separate CPD procedure, after a signal of process distributional shift is given by a CUSUM chart (cf., Pignatiello and Samuel, 2001). In this section, we introduce a CPD control chart that was first proposed by Hawkins et al. (2003) for monitoring the process mean. This chart can handle the case when all parameters μ_0, μ_1, and σ are unknown and it can give an estimator of τ at the same time when it gives a signal of process mean shift.

Assume that n is the current time point, and $\{X_1, X_2, \ldots, X_n\}$ are the observed data up to the current time point. Then, from (6.8) and (6.9), the LRT statistic for testing a possible mean shift at τ is

$$T_{max,n} = \max_{1 \le j \le n-1} \sqrt{\frac{j(n-j)}{n}} \left| \overline{X}_j - \overline{X}_j' \right| \Big/ \widehat{\widetilde{S}}_j, \tag{6.13}$$

where $\widehat{\widetilde{S}}_j$ is defined in (6.7). Clearly, $T_{max,n}$ is exactly the same as W defined in (6.9). We use a different notation here to distinguish the SPC problem discussed in this section from the CPD problem discussed in the previous section. By using $T_{max,n}$, a mean shift can be signaled if

$$T_{max,n} > h_n, \tag{6.14}$$

where $h_n > 0$ is a control limit that may depend on n. From the above description, the charting statistic $T_{max,n}$ is mainly for detecting a change-point in the observations $\{X_1, X_2, \ldots, X_n\}$. In this book, such control charts are called CPD charts.

Example 6.3 For each $1 \le j \le n-1$, let

$$T_{jn} = \sqrt{\frac{j(n-j)}{n}} \left(\overline{X}_j - \overline{X}_j' \right) \Big/ \widehat{\widetilde{S}}_j.$$

Then,

$$T_{max,n} = \max_{1 \le j \le n-1} |T_{jn}|,$$

and it is easy to see that T_{jn} is just the pooled sample t-test statistic defined in (2.36) in Subsection 2.7.3 for testing the equality of two population means (i.e., treating $\{X_1, X_2, \ldots, X_j\}$ and $\{X_{j+1}, X_{j+2}, \ldots, X_n\}$ as two samples from two populations). Under the assumption that the two population means are the same (i.e., there is no mean shift in the current process), each T_{jn} should have the t_{n-2} distribution, which has a mean of 0. If there is a mean shift of size δ at the $(j_0 + 1)$-th time point, then $T_{j_0 n}$ would have a mean of $g(\delta) \ne 0$, where $g(\cdot)$ is a specific function. When j moves away from j_0, the mean of the distribution of T_{jn} would move away from $g(\delta)$ towards 0. Therefore, when the process is IC, the distributions of $|T_{jn}|$, for $1 \le j \le n - 1$, are all the same, and consequently $T_{max,n}$ would take the value of any one of $\{|T_{jn}|, 1 \le j \le n-1\}$ by chance. On the other hand, when the process has a mean shift of size δ at the $(j_0 + 1)$-th time point, the mean of $|T_{j_0 n}|$ would be larger than the means of other variables in the sequence $\{|T_{jn}|, 1 \le j \le n-1\}$; thus, $T_{max,n}$ would most probably take the value of $|T_{j_0 n}|$ which is on average larger than the IC means of $\{|T_{jn}|, 1 \le j \le n-1\}$. Therefore, $T_{max,n}$ is effective in detecting the mean shift. To demonstrate these results, the plot (a) in Figure 6.2 shows 100 observations generated from the $N(0, 1)$ distribution, which can be regarded as the observed data from an IC process with the IC distribution of $N(0, 1)$. The computed values of $\{|T_{jn}|, 1 \le j \le 99\}$ are shown in plot (b), from which it can be seen that these values are similar and there is no obvious pattern among them. The plot (c) shows the same data as those in plot (a), except that the value of each of the last 50 observations was increased by 1. These modified observations can be regarded as the observed data from a process with a mean shift of size $\delta = 1$ which occurred at the 51st time point. The corresponding values of $\{|T_{jn}|, 1 \le j \le 99\}$ are shown in plot (d), from which we can see that this sequence peaks around the true shift time point and it tends to 0 when j moves away from the true shift time point.

From the definition of $T_{max,n}$, the CPD control chart (6.13)–(6.14) can be used only when $n \ge 3$ because the quantities $\widehat{\widetilde{S}}_j$, for $1 \le j \le n - 1$, are 0 or undefined otherwise. Therefore, to use a CPD control chart for phase II process monitoring, we should collect a number of IC observations beforehand. This, however, is not a big constraint of CPD control charts for the following reasons. Recall that conventional Shewhart, CUSUM, and EWMA control charts usually assume that the IC process mean and variance are known. In reality, these IC parameters are actually estimated from IC data collected at the end of a phase I SPC analysis. See Section 3.1 for a related discussion. If the size of the IC data is small, then the accuracy of these IC parameter estimators is not guaranteed. In such cases, the related control charts are unreliable (cf., Example 4.11 in Section 4.5.1). Therefore, in order to have a reliable phase II process monitoring, these conventional control charts usually require a quite large IC dataset. As a comparison, the CPD control chart (6.13)–(6.14) does not assume that the IC process mean and variance are known. Hawkins et al. (2003) have shown that its performance is quite reliable when $n_0 - 1$ IC observations are collected before phase II process monitoring, where n_0 could be as small as 10. In such cases, if these $n_0 - 1$ IC observations are treated as a part of phase II data, then the definition

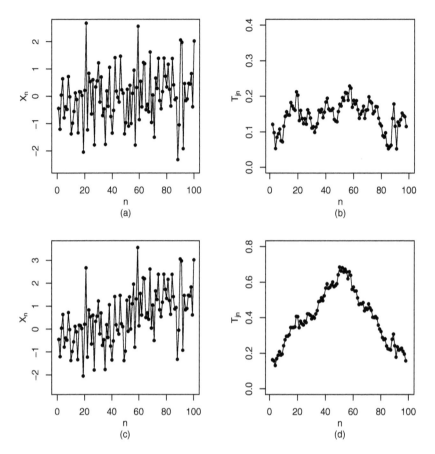

Figure 6.2 *(a) A dataset consisting of 100 observations generated independently from the distribution $N(0,1)$. (b) The computed values of $\{|T_{jn}|, 1 \leq j \leq 99\}$ from the data shown in plot (a). (c) The same dataset as that in plot (a), except that the value of each of the last 50 observations was increased by 1. (d) The computed values of $\{|T_{jn}|, 1 \leq j \leq 99\}$ from the data shown in plot (c).*

of $T_{max,n}$ should be modified to

$$T_{max,n} = \max_{n_0 - 1 \leq j \leq n-1} \sqrt{\frac{j(n-j)}{n}} \left| \overline{X}_j - \overline{X}'_j \right| \Big/ \widehat{\widetilde{S}}_j. \tag{6.15}$$

After a signal is given by the control chart at the time point n, the change-point r can be estimated by the maximizer of the maximization problem (6.13) or (6.15), denoted as $\hat{r}$, and μ_0, μ_1, and σ can be estimated by $\overline{X}_{\hat{r}}, \overline{X}'_{\hat{r}}$, and $\widehat{\widetilde{S}}_{\hat{r}}/\sqrt{n-2}$, respectively. Note that these estimators can be obtained immediately after the signal is given, without any extra computation, which cannot be achieved by the conventional control charts discussed in the previous three chapters.

Next, we discuss the determination of the control limit h_n. In phase I SPC, the sample size n is often fixed. In such cases, for a given significance level $\alpha > 0$, it is reasonable to choose h_n such that, when the process is IC, we have

$$P(T_{max,n} > h_n) \leq \alpha. \qquad (6.16)$$

Namely, the false signal rate is controlled at the level of α. Because the distribution of $T_{max,n}$ is quite complicated, the exact value of h_n satisfying (6.16) is difficult to compute. See a related discussion in Subsection 6.2.1 about the statistics W and V defined in (6.9) and (6.10). In the literature, several approximations to h_n have been proposed. One simple approximation is based on the following Bonferroni inequality

$$
\begin{aligned}
P(T_{\max,n} > h_n) &\leq \sum_{j=1}^{n-1} P\left(\sqrt{\frac{j(n-j)}{n}} \left| \overline{X}_j - \overline{X}'_j \right| \Big/ \widehat{S}_j > h_n \right) \\
&= (n-1)P(|t_{n-2}| > h_n),
\end{aligned}
$$

where t_{n-2} denotes a random variable that has the t distribution with $n - 2$ degrees of freedom. By the above expression, h_n can be chosen the $(1 - \alpha/[2(n-1)])$-th quantile of the t_{n-2} distribution. Other approximations to h_n that satisfy (6.16) are available. See Irvine (1982), Worsley (1979), and Worsley (1982) for related discussions.

In phase II SPC, the observation X_n is collected only in cases when there is no signal of shift at the $(n-1)$-th time point. So, in such cases, a more reasonable way is to choose h_n such that

$$P(T_{max,n} > h_n \,|\, \text{no signal before time } n) \leq \alpha. \qquad (6.17)$$

The expression (6.17) implies that, conditional on the fact that there is no signal before the n-th time point, the chance to have a false signal at the nth time point is not larger than α. In such cases, it is obvious that the ARL_0 value of the corresponding control chart is $1/\alpha$. Because the conditional distribution of $T_{max,n}$ involved in (6.17) is complicated to derive, determination of h_n by (6.17) is difficult analytically. But, we can always use Monte Carlo simulations to calculate the value of h_n that satisfies (6.17). For instance, when $\alpha = 0.05, 0.02, 0.01, 0.005, 0.002$, and 0.001, the corresponding ARL_0 values of the control chart are 20, 50, 100, 200, 500, and 1000, respectively. In such cases, when $n_0 = 10$ and n takes various values between 10 and 200, the computed h_n values of the CPD control chart (6.14)–(6.15) by 16 million replicated simulations are shown in Table 6.2, which is copied from Table 3 in Hawkins et al. (2003). From the table, it can be seen that, for each α value, the value of h_n decreases and then stabilizes when n increases. So, in practice, we can simply set $h_n = h_{200}$ when $n > 200$, and compute the value of h_n by interpolation when $n_0 \leq n \leq 200$ and n is not listed in the table. An alternative approach is to build a numerical relationship between h_n and (α, n) based on some simulation results such as those in Table 6.2. One such numerical relationship given by Hawkins et al. (2003) when $n_0 = 10$ is

$$h_n \approx h^*_\alpha \left(0.677 + 0.019\log(\alpha) + \frac{1 - 0.115\log(\alpha)}{n - 6} \right), \qquad (6.18)$$

Table 6.2 *Control limits h_n when $n_0 = 10$, the sample size n takes various values between 10 and 200, and the conditional false signal rate α equals 0.05, 0.02, 0.01, 0.005, 0.002, and 0.001. This table is copied from Table 3 in Hawkins et al. (2003).*

	α					
n	0.05	0.02	0.01	0.005	0.002	0.001
10	3.662	4.371	4.928	5.511	6.340	7.023
11	3.242	3.908	4.424	4.958	5.697	6.284
12	3.037	3.677	4.167	4.664	5.350	5.890
13	2.909	3.530	3.997	4.468	5.110	5.608
14	2.821	3.424	3.875	4.326	4.931	5.397
15	2.756	3.344	3.780	4.211	4.786	5.229
16	2.704	3.281	3.704	4.121	4.671	5.093
17	2.663	3.228	3.642	4.047	4.576	4.977
18	2.628	3.183	3.587	3.981	4.494	4.885
19	2.599	3.146	3.542	3.926	4.425	4.799
20	2.575	3.115	3.503	3.880	4.367	4.730
22	2.535	3.060	3.437	3.800	4.264	4.610
24	2.504	3.019	3.386	3.736	4.187	4.514
26	2.479	2.985	3.343	3.685	4.119	4.440
28	2.459	2.957	3.308	3.643	4.065	4.375
30	2.440	2.933	3.279	3.609	4.024	4.324
35	2.408	2.888	3.223	3.539	3.937	4.223
40	2.385	2.855	3.184	3.492	3.873	4.147
45	2.368	2.832	3.152	3.454	3.828	4.095
50	2.355	2.811	3.128	3.426	3.791	4.053
60	2.335	2.785	3.094	3.383	3.737	3.989
70	2.324	2.765	3.071	3.355	3.702	3.946
80	2.315	2.752	3.052	3.333	3.677	3.918
90	2.310	2.741	3.040	3.318	3.656	3.895
100	2.302	2.735	3.030	3.307	3.640	3.875
125	2.300	2.717	3.011	3.281	3.611	3.844
150	2.300	2.710	2.997	3.264	3.591	3.821
175	2.300	2.703	2.993	3.257	3.579	3.804
200	2.300	2.700	2.985	3.248	3.570	3.794

where h_{α}^* is a constant depending on α and it equals the computed h_n values when $n = 10$. Namely, it equals 3.662, 4.371, 4.928, 5.511, 6.340, and 7.023, respectively, when α equals 0.05, 0.02, 0.01, 0.005, 0.002, and 0.001.

From (6.15), each time a new observation is obtained, it seems that most quantities on the right-hand side of that equation need to be re-computed. Therefore, the CPD chart (6.14)–(6.15) involves a great amount of computation. However, the involved computation can be greatly reduced if certain quantities can be computed recursively, which is described next. At the current time point n, assume that

$$W_n = \sum_{i=1}^{n} X_i, \qquad S_n^2 = \sum_{i=1}^{n} \left(X_i - \overline{X}_n\right)^2,$$

where $\overline{X}_n$ is the sample mean of the observed data $\{X_1, X_2, \ldots, X_n\}$. Then, W_n and S_n^2 can be computed recursively by the formulas

$$
\begin{aligned}
W_n &= W_{n-1} + X_n, \\
S_n^2 &= S_{n-1}^2 + \left[(n-1)X_n - W_{n-1}\right]^2 / [n(n-1)], \qquad \text{for } n \geq n_0. \quad (6.19)
\end{aligned}
$$

It is obvious that the equation (6.15) can be written as

$$
T_{max,n} = \max_{n_0 - 1 \leq j \leq n-1} |T_{jn}|,
$$

and it can be checked that

$$
T_{jn}^2 = \frac{(n-2)V_{jn}^2}{S_n^2 - V_{jn}^2}, \qquad (6.20)
$$

where

$$
V_{jn}^2 = \frac{(nW_j - jW_n)^2}{nj(n-j)}.
$$

From (6.19) and (6.20), the value of $T_{max,n}$ can be computed easily from the values of $\{W_j, n_0 - 1 \leq j \leq n\}$ and S_n^2, which can be computed recursively by (6.19). Therefore, the computation involved in the CPD control chart (6.14)–(6.15) is manageable.

Example 6.4 *Assume that we collected $n_0 - 1 = 9$ IC observations before phase II process mean monitoring. These IC observations and the first 21 phase II observations are presented in the second column of Table 6.3. Among the total of 30 observations, the first 20 observations are actually generated from the $N(0,1)$ distribution, and the remaining 10 observations are actually generated from the $N(2,1)$ distribution. Therefore, this data simulate a production process that has the IC process distribution $N(0,1)$, and it has a mean shift of size 2 at the 21st time point (note: $r = 20$ in the change-point model (6.1) in this case). The observed data and the computed quantities W_n, S_n^2, $T_{max,n}$, and h_n that are used in the construction of the CPD control chart (6.14)–(6.15) are presented in Table 6.3, where W_n and S_n^2 are computed recursively by (6.19), $T_{max,n}$ is computed from $\{T_{jn}^2\}$ that is defined in (6.20), and h_n is computed by the numerical approximation formula (6.18) when α is fixed at 0.005 (i.e., the ARL$_0$ value of the chart is fixed at $1/0.005 = 200$). From the table, it can be seen that the first signal of mean shift is given at the 24th time point because this is the first time that $T_{max,n} > h_n$. The observed data and the CPD control chart are also shown in the two plots of Figure 6.3. From plot (b), we can see that h_n stabilizes when n increases, and the charting statistic $T_{max,n}$ keeps increasing after the true shift time point.*

Because the CPD control chart (6.14)–(6.15) does not require the IC mean μ_0, the OC mean μ_1, and the process variance σ^2 to be known, and it estimates all these parameters simultaneously when it detects the mean shift, the chart would be more sensitive to a mean shift if the shift occurs later, due to the fact that there would be more information in the observed data about the IC process distribution in such cases. This result has been confirmed by Hawkins et al. (2003) in the numerical example described below. Assume that α takes one of the values

Table 6.3 *The second column presents a sequence of 30 observations, and the third, fourth, fifth, and sixth columns present the corresponding values of $\{W_n, n = 9, 10, \ldots, 30\}$, $\{S_n^2, n = 9, 10, \ldots, 30\}$, $\{T_{max,n}, n = 10, 11, \ldots, 30\}$, and $\{h_n, n = 10, 11, \ldots, 30\}$.*

n	X_n	W_n	S_n^2	$T_{max,n}$	h_n
1	0.550	-	-	-	-
2	-0.842	-	-	-	-
3	0.033	-	-	-	-
4	0.524	-	-	-	-
5	-1.728	-	-	-	-
6	-0.278	-	-	-	-
7	0.361	-	-	-	-
8	-0.591	-	-	-	-
9	0.976	-0.995	12.706	-	-
10	-1.446	-2.440	14.310	1.005	5.393
11	0.295	-2.145	14.575	0.473	4.950
12	0.555	-1.591	15.090	0.721	4.654
13	-0.499	-2.089	15.214	0.471	4.443
14	0.196	-1.893	15.332	0.577	4.285
15	-0.456	-2.349	15.428	0.443	4.162
16	-0.363	-2.712	15.467	0.368	4.063
17	-0.157	-2.869	15.468	0.367	3.982
18	-0.765	-3.634	15.804	0.590	3.915
19	-1.166	-4.800	16.684	1.115	3.858
20	-0.323	-5.123	16.689	0.993	3.810
21	1.650	-3.473	20.150	1.985	3.767
22	1.413	-2.060	22.528	2.637	3.730
23	0.410	-1.650	22.771	2.498	3.698
24	3.690	2.040	36.329	3.707	3.669
25	2.564	4.603	42.227	4.418	3.643
26	4.668	9.271	61.555	5.926	3.620
27	2.357	11.627	65.407	6.112	3.598
28	1.638	13.265	66.812	5.915	3.579
29	2.569	15.834	71.050	6.325	3.562
30	2.029	17.863	73.175	6.475	3.546

$\{0.02, 0.01, 0.005, 0.002\}$. Then, the corresponding ARL_0 value of the CPD control chart (6.14)–(6.15) is one of the values $\{50, 100, 200, 500\}$. We consider a shift of the size $\delta \in \{0, 0.25, 0.5, 1.0, 2.0\}$ occurring at $\tau \in \{10, 25, 50, 100, 250\}$. Then, in various cases considered, the computed ARL_1 values of the CPD control chart using the approximation formula (6.18) for computing the control limit h_n are presented in Table 6.4. From the table, it can be seen that (i) when $\delta = 0$ (i.e., the process is IC), the actual ARL_0 values are close to $1/\alpha$, (ii) when δ is larger, the ARL_1 value tends to be smaller, as expected, and (iii) when $\delta > 0$ and τ is larger (i.e., the mean shift occurs later), the ARL_1 value tends to be smaller as well.

When the IC mean μ_0 and the IC variance σ^2 are unknown, the self-starting CUSUM chart discussed in Subsection 4.5.1 can also be used. It should have a rea-

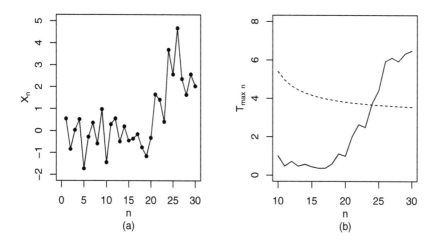

Figure 6.3 *(a) The first 20 observations are generated from the $N(0,1)$ distribution, and the remaining 10 observations are generated from the $N(2,1)$ distribution. All observations are independent of each other. (b) The CPD control chart (6.14)–(6.15) (solid line) with $n_0 - 1 = 9$ and the control limit h_n computed by the numerical approximation formula (6.18) when $\alpha = 0.005$ (dashed line).*

sonably good performance in cases when the shift size δ is known and its allowance constant k is chosen to be $\delta/2$. Hawkins et al. (2003) made a numerical comparison about the performance of such a self-starting CUSUM chart and the CPD control chart (6.14)–(6.15). To make the two charts comparable, the two-sided version of the self-starting CUSUM chart was considered, because the CPD control chart (6.14)–(6.15) can detect arbitrary mean shifts. In the self-starting CUSUM chart, k was chosen to be 0.25, 0.5, or 1.0, and the ARL_0 value was specified to be 100 or 500. In the CPD control chart, α was chosen accordingly to be 0.01 or 0.002. In both charts, it was assumed that the first 9 observations were IC (i.e., $n_0 - 1 = 9$). The computed ARL_1 values of the two charts for detecting a mean shift of size $\delta \in [0,3]$ occurring at $\tau = 10$ or 100 are shown in Figure 6.4. From the figure, it can be seen that the self-starting CUSUM chart performs well only when its allowance constant k is carefully chosen to match half of the shift size (i.e., $\delta/2$) and when τ is reasonably large. Its performance may not be reliable when k and $\delta/2$ are quite different or when the mean shift occurs quite early. As a comparison, the CPD control chart is quite robust to both the shift size and the shift time point.

6.3.2   Monitoring of the process variance

The CPD control chart discussed in the previous subsection is for detecting process mean shifts only. As pointed out in previous chapters (e.g., Subsection 4.3.1), online monitoring of the process variance is also important for ensuring product quality.

Table 6.4 ARL_1 values of the CPD control chart (6.14)–(6.15) when $\alpha \in \{0.02, 0.01, 0.005, 0.002\}$, and a shift of size $\delta \in \{0, 0.25, 0.5, 1.0, 2.0\}$ occurs at $\tau \in \{10, 25, 50, 100, 250\}$.

α	τ	δ 0	0.25	0.5	1.0	2.0
0.02	10	54.5	51.8	43.3	21.3	4.1
	25	54.2	46.7	30.8	9.4	2.4
	50	53.8	42.8	23.3	7.0	2.1
	100	54.7	41.4	19.4	6.0	1.9
	250	58.7	33.4	16.6	6.3	1.8
0.01	10	99.3	95.9	82.4	39.1	5.3
	25	99.4	86.9	55.2	13.4	2.8
	50	99.2	76.3	38.2	8.8	2.3
	100	99.7	69.3	28.5	7.6	2.2
	250	96.9	55.4	23.9	7.0	2.1
0.005	10	195.9	186.2	168.3	84.9	7.0
	25	196.6	174.7	113.9	20.4	3.3
	50	196.3	154.9	68.7	11.2	2.7
	100	196.9	131.3	42.2	9.4	2.5
	250	194.8	99.0	29.7	8.2	2.3
0.002	10	535.8	531.1	492.7	280.9	10.8
	25	538.7	504.3	357.5	43.7	4.3
	50	539.5	457.8	195.4	15.7	3.4
	100	542.8	373.1	77.3	12.3	3.0
	250	546.2	222.9	43.3	10.8	2.8

Therefore, in the CPD framework, Hawkins and Zamba (2005a) proposed a control chart for monitoring the process variance, which is described below.

Assume that the IC distribution of the quality characteristic X is $N(\mu_0, \sigma_0^2)$. After an unknown time point τ, this distribution may shift to the OC distribution $N(\mu_1, \sigma_1^2)$, where $\mu_0 \neq \mu_1$ are the IC and OC process means, $\sigma_0^2 \neq \sigma_1^2$ are the IC and OC process variances, and all these parameters are unknown. Then, the major goal of phase II SPC is to detect such a distributional shift as soon as it occurs. Note that the control charts for process variance monitoring discussed in Sections 3.2, 4.3, and 5.3 can only be used in cases when the IC parameters μ_0 and σ_0^2 are known. Their performance is good only in cases when their procedure parameters are properly chosen to match the unknown OC parameters well, which is challenging to achieve. The self-starting control charts and the adaptive control charts discussed in Sections 4.5 and 5.4 can help overcome these limitations. An alternative approach to overcome these limitations is to adopt the CPD framework when handling the SPC problem, which is discussed below.

The phase II SPC problem described above is closely related to the following CPD problem. Assume that $\{X_1, X_2, \ldots, X_n\}$ is a sequence of independent process observations up to the time point n. When the process is IC, these random variables follow the distribution

$$X_i \sim N(\mu_0, \sigma_0^2), \qquad \text{for } i = 1, 2, \ldots, n.$$

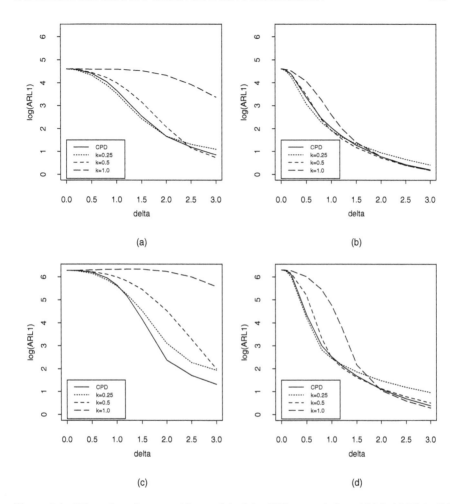

Figure 6.4 *ARL$_1$ values (in natural log scale) of the CPD control chart (6.14)–(6.15) (solid curves) and the self-starting CUSUM chart (dotted, short-dashed, and long-dashed curves) when the process mean shift size δ changes from 0 to 3. (a) α = 0.01 (or ARL$_0$ = 100) and τ = 10. (b) α = 0.01 (or ARL$_0$ = 100) and τ = 100. (c) α = 0.002 (or ARL$_0$ = 500) and τ = 10. (d) α = 0.002 (or ARL$_0$ = 500) and τ = 100.*

When the process has a shift at a certain time point, they follow the distributions

$$X_i \sim \begin{cases} N(\mu_0, \sigma_0^2), & \text{for } i = 1, 2, \dots, r \\ N(\mu_1, \sigma_1^2), & \text{for } i = r+1, r+2, \dots, n, \end{cases}$$

where r is the change-point, and $\tau = r + 1$ is the shift time point. If r is known, then it is easy to check that the MLEs of μ_0, μ_1, σ_0^2, and σ_1^2 are

$$
\begin{aligned}
\widehat{\mu}_0 &= \overline{X}_r, \\
\widehat{\mu}_1 &= \overline{X}'_r, \\
\widehat{\sigma}_0^2 &= \frac{1}{r-1} \sum_{i=1}^{r} (X_i - \overline{X}_r)^2, \\
\widehat{\sigma}_1^2 &= \frac{1}{n-r-1} \sum_{i=r+1}^{n} \left(X_i - \overline{X}'_r \right)^2.
\end{aligned}
$$

Under the assumption that $\sigma_0^2 = \sigma_1^2$, the MLE of the common variance σ^2 is

$$
\widehat{\sigma}^2 = \frac{1}{n-2} \widehat{S}_r^2,
$$

where $\widehat{S}_r^2$ is defined in (6.7). Then, Bartlett (1937, 1955) and Bartlett and Kendall (1946) derived the following likelihood ratio statistic

$$
B_{rn} = \left[(r-1)\log\left(\frac{\widehat{\sigma}^2}{\widehat{\sigma}_0^2}\right) + (n-r-1)\log\left(\frac{\widehat{\sigma}^2}{\widehat{\sigma}_1^2}\right) \right] \Big/ C_{rn}, \tag{6.21}
$$

for testing the hypotheses

$$
H_0 : \sigma_0^2 = \sigma_1^2 \qquad \text{versus} \qquad H_1 : \sigma_0^2 \neq \sigma_1^2, \tag{6.22}
$$

where

$$
C_{rn} = 1 + \left[\frac{1}{r-1} + \frac{1}{n-r-1} - \frac{1}{n-2} \right] \Big/ 3
$$

is the so-called *Bartlett correction factor*. It has been shown that B_{rn} in (6.21) has the null distribution of χ_1^2. Further, it can be written as

$$
\begin{aligned}
B_{rn} &= \frac{1}{C_{rn}} \big[(r-1)\log\left(r-1+(n-r-1)F_{rn}\right) + \\
&\quad (n-r-1)\log\left((r-1)F_{rn}^{-1} + n - r - 1\right) - \\
&\quad (n-2)\log(n-2) \big],
\end{aligned} \tag{6.23}
$$

where

$$
F_{rn} = \frac{\widehat{\sigma}_1^2}{\widehat{\sigma}_0^2}
$$

is the F-test statistic for testing the equality of the two variances. From (6.23), it is clear that the test statistic B_{rn} is for testing the two-sided alternative hypothesis H_1 in (6.22), and it is sensitive to both upward and downward variance shifts. In cases when r is unknown, the MLE of r is defined by

$$
\widehat{r} = \arg \max_{2 \leq r \leq n-2} B_{rn}, \tag{6.24}
$$

and the test statistic for testing the hypothesis in (6.22) would be

$$B_{max,n} = \max_{2 \le r \le n-2} B_{rn}. \tag{6.25}$$

In both (6.24) and (6.25), r is considered in the range from 2 to $n-2$ because B_{rn} is not well defined when $r = 1$ or when $r = n-1$.

From the discussion in the previous subsection, the statistic $B_{max,n}$ defined in (6.25) can also be used for monitoring the process variance in phase II SPC. More specifically, a variance shift is signaled at the n-th time point if

$$B_{max,n} > h_n, \tag{6.26}$$

where $h_n > 0$ is a control limit. Similar to (6.17), the control limit h_n can be chosen such that

$$P\left(B_{max,n} > h_n \mid \text{no signal before time } n\right) \le \alpha, \tag{6.27}$$

where $\alpha \in [0, 1]$ is a given conditional probability. If h_n is chosen by (6.27), then the IC ARL of the CPD control chart (6.25)–(6.26) is $1/\alpha$. From the definition of the charting statistic $B_{max,n}$, this chart can be used for online process monitoring when $n \ge 4$. But, similar to the CPD control chart (6.14)–(6.15), its performance would be more reliable if $n_0 - 1$ IC observations are collected before the online process monitoring. In such cases, for both (6.24) and (6.25), the range of r considered on the right-hand side of both expressions should be changed from $2 \le r \le n-2$ to $n_0 + 1 \le r \le n-2$. When $n_0 - 1 = 9$, α equals 0.05, 0.02, 0.01, 0.005, 0.002, and 0.001, and n changes from 10 to 30, the computed h_n by a Monte Carlo simulation with 5 million replicated simulations are shown in Table 6.5, which is part of Table 1 in Hawkins and Zamba (2005a). From the table, it can be seen that, for a given α value, h_n first decreases and then increases and stabilizes when n gets larger. In practice, Hawkins and Zamba (2005a) proposed the following approach to compute h_n:

- For $n = 10, 11, \ldots, 15$, use the h_n values in Table 6.5.
- For $n > 15$, use the approximation formula

$$h_n \approx \begin{cases} -1.38 - 2.241 \log(\alpha) + \frac{1.61 + 0.691 \log(\alpha)}{\sqrt{n-9}}, & \text{if } \alpha \in [0.001, 0.05) \\ 5 + 0.066 \log(n-9), & \text{if } \alpha = 0.05. \end{cases}$$

For phase II process monitoring, the quantities B_{rn} used in computing the charting statistic $B_{max,n}$ can be computed recursively as follows. For $2 \le r \le n-2$, let

$$\overline{X}_r = \frac{1}{r} \sum_{i=1}^{r} X_i, \qquad \overline{X}'_r = \frac{1}{n-r} \sum_{i=r+1}^{n} X_i,$$

$$\widetilde{S}_r^2 = \sum_{i=1}^{r} \left(X_i - \overline{X}_r\right)^2, \qquad \widetilde{S}_r^{2\prime} = \sum_{i=r+1}^{n} \left(X_i - \overline{X}'_r\right)^2.$$

Then, $W_n = \sum_{k=1}^{n} X_k$ and $S_n^2 = \sum_{k=1}^{n} (X_i - \overline{X}_n)^2$ can be computed recursively by (6.19).

Table 6.5 *Control limits h_n when $n_0 = 10$, the sample size n takes various values between 10 and 30, and the conditional false signal rate α equals 0.05, 0.02, 0.01, 0.005, 0.002, and 0.001.*

			α			
n	0.05	0.02	0.01	0.005	0.002	0.001
10	6.374	8.003	9.229	10.451	12.039	13.238
11	5.651	7.328	8.585	9.840	11.489	12.734
12	5.357	7.077	8.373	9.653	11.357	12.631
13	5.228	6.988	8.312	9.634	11.367	12.672
14	5.173	6.960	8.304	9.658	11.423	12.760
15	5.149	6.960	8.323	9.692	11.469	12.828
16	5.141	6.974	8.357	9.731	11.541	12.885
17	5.145	6.992	8.386	9.776	11.596	12.962
18	5.142	7.010	8.413	9.808	11.651	13.034
19	5.145	7.020	8.434	9.838	11.696	13.070
20	5.150	7.034	8.458	9.875	11.722	13.120
22	5.160	7.064	8.500	9.921	11.788	13.191
24	5.173	7.085	8.529	9.961	11.853	13.297
26	5.184	7.108	8.562	10.000	11.894	13.340
28	5.196	7.125	8.585	10.035	11.947	13.385
30	5.204	7.136	8.610	10.065	11.981	13.408

Furthermore, we have

$$\overline{X}'_r = (W_n - W_r)/(n - r)$$

$$\widetilde{S}^{2'}_r = S^2_n - \widetilde{S}^2_r - \frac{r(n - r)}{n}(\overline{X}_r - \overline{X}'_r)^2$$

$$\hat{\sigma}^2_0 = \widetilde{S}^2_r/(r - 1)$$

$$\hat{\sigma}^2_1 = \widetilde{S}^{2'}_r/(n - r - 1).$$

By (6.21) (or (6.23)), we can compute B_{rn} from $\hat{\sigma}^2_0$, $\hat{\sigma}^2_1$, $\widetilde{S}^2_r$, and $\widetilde{S}^{2'}_r$. Therefore, to compute the values of $\{B_{rn}, n_0 + 1 \leq r \leq n - 2\}$, we only need to maintain the two sequences $\{W_n, n \geq n_0\}$ and $\{S^2_n, n \geq n_0\}$, which can be computed recursively.

Example 6.5 *Assume that we collected $n_0 - 1 = 9$ IC observations before phase II process variance monitoring. These IC observations and the first 21 phase II observations are presented in the second column of Table 6.6. Among the total of 30 observations, the first 20 observations are actually generated from the $N(0, 1)$ distribution, and the remaining 10 observations are actually generated from the $N(0, 2^2)$ distribution. Therefore, these data simulate a production process that has the IC process distribution $N(0, 1)$, and has a variance shift of size $4 - 1 = 3$ at the 21st time point. The observed data and the computed quantities W_n, S^2_n, $B_{max,n}$, and h_n that are used in the construction of the CPD control chart (6.25)–(6.26) are presented in Table 6.6, where W_n and S^2_n are computed recursively by (6.19), $B_{max,n}$ is computed from $\{B_{rn}\}$ that is defined in (6.21), and h_n is computed by the numerical approximation formula given in the previous paragraph. From the table, it can be seen that*

Table 6.6 *The second column presents a sequence of 30 observations, and the third, fourth, fifth, and sixth columns present the corresponding values of $\{W_n, n = 9, 10, \ldots, 30\}$, $\{S_n^2, n = 9, 10, \ldots, 30\}$, $\{B_{max,n}, n = 11, 12, \ldots, 30\}$, and $\{h_n, n = 10, 11, \ldots, 30\}$.*

n	X_n	W_n	S_n^2	$B_{max,n}$	h_n
1	−0.502	-	-	-	-
2	0.132	-	-	-	-
3	−0.079	-	-	-	-
4	0.887	-	-	-	-
5	0.117	-	-	-	-
6	0.319	-	-	-	-
7	−0.582	-	-	-	-
8	0.715	-	-	-	-
9	−0.825	0.180	2.935	-	-
10	−0.360	−0.180	3.065	-	10.451
11	0.090	−0.090	3.075	0.399	9.840
12	0.096	0.007	3.085	6.484	9.653
13	−0.202	−0.195	3.123	2.661	9.634
14	0.740	0.545	3.652	0.585	9.658
15	0.123	0.668	3.659	1.186	9.692
16	−0.029	0.639	3.664	1.850	9.718
17	−0.389	0.250	3.837	1.796	9.768
18	0.511	0.761	4.070	1.666	9.810
19	−0.914	−0.153	4.936	1.136	9.845
20	2.310	2.157	10.042	8.023	9.875
21	−0.876	1.281	10.964	9.685	9.901
22	1.528	2.809	13.018	10.367	9.925
23	0.524	3.333	13.169	9.365	9.945
24	1.547	4.880	15.052	9.371	9.964
25	−1.629	3.251	18.274	12.657	9.981
26	−0.877	2.374	19.249	12.799	9.996
27	−1.440	0.934	21.509	13.653	10.010
28	0.462	1.396	21.685	12.972	10.023
29	−2.315	−0.920	27.086	15.287	10.035
30	0.494	−0.426	27.354	14.762	10.046

the first signal of variance shift is given at the 22nd time point. The observed data and the CPD control chart are also shown in the two plots of Figure 6.5. From plot (b), we can see that h_n stabilizes when n increases, and the charting statistic $B_{max,n}$ is convincingly above the control limit curve after the 25th time point.

6.3.3 Monitoring of both the process mean and variance

Because both the process mean shift and the process variance shift would have an impact on the product quality, we often need to monitor the process mean and variance simultaneously in practice. To this end, one possible approach is to jointly use the two CPD control charts described in the previous two subsections. Alternatively, we can develop a single CPD control chart from the corresponding CPD problem,

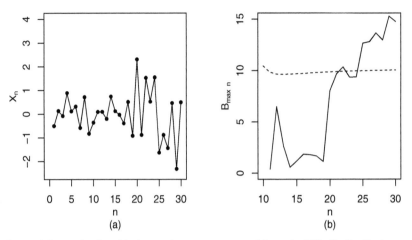

Figure 6.5 *(a) The first 20 observations are generated from the $N(0,1)$ distribution, and the remaining 10 observations are generated from the $N(0,2^2)$ distribution. All observations are independent of each other. (b) The CPD control chart (6.25)–(6.26) (solid line) with $n_0 - 1 = 9$ and the control limit h_n computed by the numerical approximation formula (dashed line).*

as was done in Hawkins and Zamba (2005b). In this subsection, we briefly describe such a CPD control chart that is proposed by Hawkins and Zamba (2005b).

Assume that $\{X_1, X_2, \ldots, X_n\}$ is a sequence of independent process observations collected up to the current time point n, and we are concerned about the hypotheses

$$H_0 : X_i \sim N(\mu_0, \sigma_0^2), \qquad \text{for } i = 1, 2, \ldots, n$$

versus

$$H_1 : X_i \sim \begin{cases} N(\mu_0, \sigma_0^2), & \text{for } i = 1, 2, \ldots, r, \\ N(\mu_1, \sigma_1^2), & \text{for } i = r+1, r+2, \ldots, n, \end{cases}$$

where either $\mu_0 \neq \mu_1$ or $\sigma_0^2 \neq \sigma_1^2$ or both, and $1 \leq r \leq n-1$ is the change-point. If r is known, then the likelihood ratio test statistic for testing the above hypotheses with the Bartlett correction factor (cf., Kendall and Stuart, 1961; Lawley, 1956) is

$$J_{rn} = \left[r \log\left(\frac{\widehat{\sigma}_{H_0}^2}{\widehat{\sigma}_0^2} \right) + (n-r) \log\left(\frac{\widehat{\sigma}_{H_0}^2}{\widehat{\sigma}_1^2} \right) \right] \Big/ C_{rn}^*, \qquad (6.28)$$

where $\widehat{\sigma}_{H_0}^2 = S_n^2/n$ is the MLE of σ^2 when H_0 is assumed to be true, $\widehat{\sigma}_0^2$ and $\widehat{\sigma}_1^2$ are the same as those defined in the previous subsection, and

$$C_{rn}^* = 1 + \frac{11}{12} \left[\frac{1}{r} + \frac{1}{n-r} - \frac{1}{n} \right] + \left[\frac{1}{r^2} + \frac{1}{(n-r)^2} - \frac{1}{n^2} \right].$$

By comparing J_{rn} defined in (6.28) with B_{rn} defined in (6.21), it can be seen that their major difference is in estimating the variance parameter. In (6.21), the estimator

Table 6.7 *Control limits h_n when $n_0 = 10$, the sample size n takes values between 10 and 14, and the conditional false signal rate α equals 0.05, 0.02, 0.01, 0.005, 0.002, and 0.001.*

n	α 0.05	0.02	0.01	0.005	0.002	0.001
10	10.128	12.237	13.795	15.330	17.352	18.840
11	9.213	11.389	12.996	14.556	16.609	18.173
12	8.854	11.083	12.719	14.313	16.397	17.965
13	8.690	10.961	12.631	14.265	16.353	17.950
14	8.616	10.917	12.610	14.249	16.361	17.978

$\widehat{\sigma}^2 = S_r^2/(n-2)$ is used, which is robust to a potential mean shift at the change-point r. As a comparison, in (6.28), the estimator $\widehat{\sigma}_{H_0}^2$ is used, which is affected by a potential mean shift at the change-point r. Because of this difference, J_{rn} is sensitive to both process mean shifts and process variance shifts, while B_{rn} is sensitive to process variance shifts only.

In phase II process monitoring, the change-point r is unknown. In such cases, if $n_0 - 1$ IC observations are collected beforehand, then we can use the charting statistic

$$J_{max,n} = \max_{n_0+1 \leq r \leq n-2} J_{rn}, \tag{6.29}$$

where the range $n_0 + 1 \leq r \leq n-2$ is considered because J_{rn} may not be well defined when $r \leq n_0$ or $r = n-1$. Then, a signal of process mean/variance shift is given when

$$J_{max,n} > h_n, \tag{6.30}$$

where $h_n > 0$ is a control limit that is properly chosen to achieve a given ARL_0 value. Similar to (6.17) and (6.27), for a given conditional probability α, h_n can be chosen such that

$$P\left(J_{max,n} > h_n \,|\, \text{no signal before time } n\right) \leq \alpha.$$

In cases when $n_0 - 1 = 9$ IC observations are collected beforehand, Hawkins and Zamba (2005b) proposed the following numerical approach for computing h_n:

- For $n = 10, 11, \ldots, 14$, use the h_n values in Table 6.7, which are computed using a Monte Carlo simulation with 10 million replicated simulations.

- For $n > 14$, use the approximation formula

$$h_n \approx \begin{cases} 1.58 - 2.52\log(\alpha) + \frac{0.094+0.33\log(\alpha)}{\sqrt{n-9}}, & \text{if } \alpha \in [0.001, 0.05) \\ 8.43 + 0.074\log(n-9), & \text{if } \alpha = 0.05. \end{cases}$$

To compute the charting statistic $J_{max,n}$, we only need to maintain the two sequences $\{W_n, n \geq n_0\}$ and $\{S_n^2, n \geq n_0\}$, which can be computed recursively by (6.19), as in computing the charting statistic $B_{max,n}$ in the previous subsection.

After a signal is given by the CPD control chart (6.29)–(6.30), the following three root causes are possible: (i) the process mean has shifted, (ii) the process variance has shifted, and (iii) both the process mean and the process variance have shifted. To distinguish these three scenarios, Hawkins and Zamba (2005b) suggested performing the two post-signal tests described below.

- Let $\hat{r}$ be the change-point estimator defined by the maximizer of (6.29). Consider the two-sided F-test with the test statistic

$$F = \frac{\hat{\sigma}_0^2}{\hat{\sigma}_1^2},$$

where $\hat{\sigma}_0^2 = S_{\hat{r}}^2/(\hat{r}-1)$ and $\hat{\sigma}_1^2 = S_{\hat{r}}^{2\prime}/(n-\hat{r}-1)$. Then, under the hypothesis that the process variance does not shift on or before the n-th time point, the distribution of this statistic should be close to the $F_{\hat{r}-1,n-\hat{r}-1}$ distribution (cf., Subsection 2.3.4). Therefore, this F-test can be used to conclude whether the process variance has shifted by the time point n.

- Consider the two-sample t-test with the test statistic

$$t = \frac{\overline{X}_{\hat{r}} - \overline{X}_{\hat{r}}'}{\sqrt{\hat{\sigma}_0^2/\hat{r} + \hat{\sigma}_1^2/(n-\hat{r})}}.$$

Then, under the hypothesis that the process mean does not shift on or before the n-th time point, the distribution of this statistic should be close to the t_k distribution (cf., Subsection 2.3.3), where

$$k = \frac{\left[\hat{\sigma}_0^2/\hat{r} + \hat{\sigma}_1^2/(n-\hat{r})\right]^2}{\frac{1}{\hat{r}-1}\left[\hat{\sigma}_0^2/\hat{r}\right]^2 + \frac{1}{n-\hat{r}-1}\left[\hat{\sigma}_1^2/(n-\hat{r})\right]^2}.$$

Therefore, this t-test can be used to conclude whether the process mean has shifted by the time point n.

Regarding the above two tests, the specified null distributions of their test statistics are valid only in cases when the true change-point r is known and $\hat{r}$ in their test statistics is replaced by r. With the extra randomness in $\hat{r}$, the actual null distributions of their test statistics could be quite different from the specified null distributions. Furthermore, if the CPD control chart (6.29)–(6.30) performs well, the time lag from the occurrence of the mean/variance shift to its first signal (i.e., $n-\hat{r}$) is usually short. Consequently, the two test statistics might be quite variable. For these reasons, the above two post-signal tests can only be used as a reference, and their results may not be reliable.

Example 6.6 *In this example, we consider three different datasets. Each dataset has 30 observations, the first 9 of which are regarded as IC observations. In the first dataset, the first 20 observations are generated from the $N(0,1)$ distribution, and the remaining 10 observations are generated from the $N(2,1)$. So, a process mean shift occurs at the 21st time point. The data are shown in Figure 6.6(a). Then, the three CPD control charts described in this and the previous two subsections, with $n_0 - 1 = 9$, $\alpha = 0.005$, and h_n computed by the related approximation formulas, are applied to the dataset, and the three control charts are shown in Figure 6.6(b)–(d).*

From the plots, it can be seen that the CPD control chart (6.14)–(6.15) gives the first signal of process mean shift at the 24th time point, the CPD control chart (6.25)–(6.26) does not give any signal, and the CPD control chart (6.29)–(6.30) gives the first signal of process mean or process variance shift at the 26th time point. For the CPD control chart (6.29)–(6.30), after the signal, we obtain

$$\hat{r} = 20, \qquad \overline{X}_{\hat{r}} = -0.256, \qquad \overline{X}'_{\hat{r}} = 2.399,$$
$$\hat{\sigma}_0^2 = 0.508, \qquad \hat{\sigma}_1^2 = 2.466.$$

Then, the computed value of the F-test statistic is 0.206, its null distribution is $F_{19,5}$, and the corresponding two-sided p-value is about 0.01 (note: the two-sided p-value is defined as $2P(F_{19,5} < 0.206)$). On the other hand, the computed value of the t-test statistic is -4.019, its null distribution is t_6, and the corresponding p-value for a two-sided t-test is 0.007. Based on these two tests, it seems that both the process mean and the process variance have shifted, which is different from the truth that only the process mean shifts in the first dataset.

In the second dataset, the first 20 observations are generated from the $N(0,1)$ distribution, and the remaining 10 observations are generated from the $N(0, 3^2)$. So, a process variance shift occurs at the 21st time point. The data are shown in Figure 6.6(e), and the three CPD control charts in the same setup as that in analyzing the first dataset are shown in Figure 6.6(f)–(h). From the plots, it can be seen that the CPD control chart (6.14)–(6.15) gives signals of process mean shift, but the signals are quite weak. As a comparison, the CPD control charts (6.25)–(6.26) and (6.29)–(6.30) both give convincing signals. For the CPD control chart (6.29)–(6.30), after the first signal at the 24th time point, we obtain

$$\hat{r} = 18, \qquad \overline{X}_{\hat{r}} = 0.042, \qquad \overline{X}'_{\hat{r}} = 0.913,$$
$$\hat{\sigma}_0^2 = 0.226, \qquad \hat{\sigma}_1^2 = 2.830.$$

Then, the computed value of the F-test statistic is 0.080, its null distribution is $F_{17,5}$, and the corresponding p-value is 6.73×10^{-5}. On the other hand, the computed value of the t-test statistic is -1.252, its null distribution is t_5, and the corresponding p-value for a two-sided test is 0.266. Based on these two tests, we conclude that the process mean does not shift but the process variance shifts after the 18th time point, which is close to the truth since the true variance shift happens after the 20th time point.

In the third dataset, the first 20 observations are generated from the $N(0,1)$ distribution, and the remaining 10 observations are generated from the $N(1, 2^2)$. So, both the process mean and the process variance shift at the 21st time point. The data are shown in Figure 6.6(i), and the three CPD control charts in the same setup as that in analyzing the first two datasets are shown in Figure 6.6(j)–(l). From the plots, it can be seen that all three CPD control charts give signals in this case. For the CPD control chart (6.29)–(6.30), after the first signal at the 26th time point, we obtain

$$\hat{r} = 23, \qquad \overline{X}_{\hat{r}} = -0.312, \qquad \overline{X}'_{\hat{r}} = 4.281,$$
$$\hat{\sigma}_0^2 = 0.618, \qquad \hat{\sigma}_1^2 = 4.435.$$

Then, the computed value of the F-test statistic is 0.139, its null distribution is $F_{22,2}$, and the corresponding two-sided p-value is 0.008. On the other hand, the computed *value of the t-test statistic is* -3.744, *its null distribution is* t_2, *and the corresponding p-value for a two-sided test is 0.065. Based on these two tests, we conclude that the process variance shifts and there is marginally significant evidence of a process mean shift in this case.*

6.4   Some Discussions

We have described some recent CPD control charts in this chapter. For a long time, the SPC literature focused mainly on the Shewhart, CUSUM, and EWMA charts. While these control charts are effective for phase I analysis and phase II online process monitoring, their performance depends largely on their design and a good design requires certain information about the IC and OC process distributions. For instance, a conventional CUSUM chart (cf., Chapter 4) assumes that the IC process distribution is completely known, and a distributional shift is in one or more parameters (e.g., the mean, or the variance, or both). In such cases, it has good performance when its allowance constant k (cf., (4.7)) is chosen properly for detecting a target shift. In practice, however, the IC process distribution may not be known, and the target shift is often difficult to specify. Furthermore, a CUSUM chart tuned properly for a given target shift may not perform well when detecting other possible shifts. Although the self-starting control charts and the adaptive control charts (e.g., Section 4.5) can alleviate the impact of these limitations on the practical use of the conventional control charts, the limitations have not been completely eliminated. The CPD control charts provide a new approach to the SPC problem. Two major features of this approach are that (i) it does not require much information about the IC and OC process distributions, and (ii) it can provide reasonably good estimators of the shift time and the IC and OC distribution parameters at the same time as when it gives a signal of process distributional shift. These two properties make the CPD approach appropriate to use in cases when we have little prior information about the IC and OC process distributions and when it is desirable to have the estimators of the shift time and the IC and OC distribution parameters immediately after obtaining a signal of shift.

The CPD methods discussed in Section 6.2 are for analyzing data from retrospective studies in which the data sizes are fixed. These methods can be used directly for phase I process analysis, although besides change-points there are many other issues to handle (i.e., outliers detection) in a typical phase I process analysis. The methods described in that section are traditional, and there are many other CPD methods in the literature. For more recent developments in the CPD area, see Brodsky and Darkhovsky (1993), Chen and Gupta (1997, 2000), Chib (1998), Gustafsson (2000), Herberts and Jensen (2004), Lebarbier (2005), Spokoiny (2009), among many others.

The CPD control charts discussed in Section 6.3 are for phase II process monitoring in which the sample size sequentially increases. Their charting statistics are modified from the test statistics of the corresponding change-point hypothesis testing problems. In their design and implementation, it is critically important to develop proper numerical approximation formulas for choosing the control limits and to formulate the recursive computation of the charting statistics. The control charts discussed in that section are for general cases when all parameters, including the IC

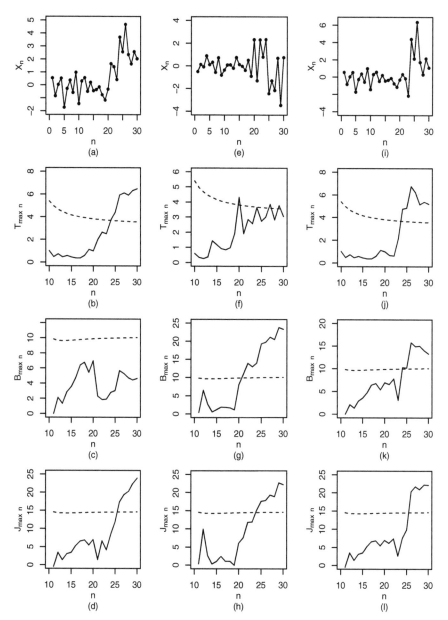

Figure 6.6 *(a) The first 20 observations are generated from the $N(0,1)$ distribution, and the remaining 10 observations are generated from the $N(2,1)$ distribution. All observations are independent of each other. (b)–(d) The CPD control charts (6.14)–(6.15), (6.25)–(6.26), and (6.29)–(6.30) in cases with $n_0 - 1 = 9$, $\alpha = 0.005$, and h_n computed by the related approximation formulas. Plots (e)–(h) and (i)–(l) show the corresponding results when the process distribution shifts at the 21st time point from $N(0,1)$ to $N(0,3^2)$ and from $N(0,1)$ to $N(1,2^2)$, respectively.*

and OC means and variances and the shift time point, are unknown. Various cases when some of these parameters are known are discussed extensively in the literature. See, for instance, Lai (1995, 2001), and Mei (2006). CPD control charts for analyzing correlated data are discussed by Choi et al. (2008), and Perry et al. (2011); CPD control charts for cases when the process distribution is Poisson or other parametric distributions are discussed by Perry and Pignatiello (2008), and Perry et al. (2007).

Although the CPD control charts have their advantages, compared to the conventional control charts, as described above, their computation is still relatively complicated, even though certain quantities involved in their charting statistics can be computed in a recursive way. The main reason for this computational limitation is that the change-point estimator $\hat{r}$ (cf., (6.24)) has to be re-computed each time when a new observation X_n is obtained, which involves a search in a sequence of random variables of length in the order of n. In certain applications, a quick computation is desirable when monitoring a process. For such applications, a proper modification of the conventional CPD control charts is needed. One possible modification is to search within a number of recent observation times only, each time when the change-point estimator is updated after a new observation is obtained. However, this approach has the risk of missing a true shift, especially when only a small number of recent observation times are considered in the search algorithm. Therefore, much future research is needed to balance the computational consideration and the risk of missing a true shift. Also, as discussed in Chapters 4 and 5, self-starting and adaptive CUSUM and EWMA charts can also handle situations in which the IC and OC distributional parameters are unknown. A systematic comparison between this approach and the CPD approach is currently lacking, which requires much future research as well.

6.5 Exercises

6.1 Assume that the sequence of independent random variables $\{X_1, X_2, \ldots, X_n\}$ follows the change-point model below.

$$X_i \sim \begin{cases} N(\mu_0, \sigma^2), & \text{for } i = 1, 2, \ldots, r \\ N(\mu_1, \sigma^2), & \text{for } i = r+1, r+2, \ldots, n, \end{cases}$$

where n is fixed, μ_0, μ_1, and σ^2 are unknown parameters, and $1 \leq r \leq n-1$ is an unknown change-point.

(i) Verify the expression (6.6).

(ii) Verify all equations in the expression (6.7).

6.2 Generate a dataset from the change-point model described in the previous exercise with $n = 100$, $r = 50$, $\mu_0 = 0$, $\mu_1 = 1$, and $\sigma = 1$. Compute the MLEs of μ_0, μ_1, σ, and r.

6.3 In the change-point model described in Exercise 6.1, assume that $n = 100$, $\mu_0 = \mu_1 = 0$, and $\sigma = 1$. Namely, there is actually no change-point in the model. In such cases, it has been demonstrated in the literature that the time points closer to 1 or n would have larger chances to be chosen as the estimated change-point. To verify this result, perform a simulation study with the following steps.

Step 1 Generate a dataset from the change-point model, and compute the values of

$$g(r) = \frac{r(n-r)}{n}(\overline{X}_r - \overline{X}'_r)^2, \qquad \text{for } 1 \le r \le n-1.$$

Step 2 Repeat Step 1 for a total of 100 times, and average $\{g(r), 1 \le r \le n-1\}$ over the 100 replications.

Step 3 Make a histogram (cf., Subsection 2.6.3) of the averaged values of $\{g(r), 1 \le r \le n-1\}$.

By (6.6), the MLE of r is defined by

$$\widehat{r} = \arg \max_{1 \le r \le n-1} g(r).$$

From the histogram, is it true that the time points closer to 1 or n have larger values of $g(r)$?

6.4 In the change-point model described in Exercise 6.1, in cases when σ^2 is unknown, check that the LRT statistic $\Lambda(X_1, X_2, \ldots, X_n)$ discussed in Subsection 6.2.1 is a strictly decreasing function of W defined in (6.9). Also, check that it is a strictly decreasing function of V defined in (6.10).

6.5 Check that the recursive formula (6.12) is valid.

6.6 The following 30 observations are obtained from a production process:

8.308, 7.525, 12.549, 13.127, 6.874, 7.536, 10.430, 9.840, 9.261, 9.601,

7.981, 4.932, 5.968, 4.379, 8.861, 4.562, 7.132, 7.022, 9.164, 8.590, 17.642,

14.252, 14.583, 14.443, 12.573, 12.950, 15.315, 14.700, 14.475, 16.102.

Use the dynamic programming algorithm described in Subsection 6.2.2 to detect possible change-points in the data and compute point estimates of all related distribution parameters.

6.7 Assume that 9 IC observations have been collected for phase II monitoring of a production process. These observations along with the first 21 phase II observations are presented below.

11.528, 8.990, 9.590, 9.648, 16.807, 14.850,

10.708, 7.042, 5.427, 10.107, 0.029, 9.836,

6.807, 12.428, 18.920, 7.535, 11.330, 3.883,

8.874, 16.850, 9.627, 10.887, 21.789, 12.482,

12.741, 13.236, 18.913, 15.113, 16.006, 10.006

Use the CPD control chart (6.14)–(6.15) to monitor the process mean. In the control chart, use $n_0 - 1 = 9$, $\alpha = 0.005$, and the control limit h_n computed by the approximation formula (6.18).

6.8 The approximation formula (6.18) is obtained empirically from simulation results such as those in Table 6.2. It is developed for the convenience of online process monitoring. Please use the results in Table 6.2 to verify that this approximation formula provides a reasonably good approximation to the computed values of h_n in the table.

6.9 Reproduce all the results in Example 6.4.

6.10 Verify the expressions (6.19) and (6.20).

6.11 Show that the expressions (6.21) and (6.23) are equivalent.

6.12 Assume that 9 IC observations have been collected for phase II monitoring of a production process. These observations along with the first 21 phase II observations are presented below.

5.139, 5.097, 5.329, 4.449, 4.890, 4.356, 5.065, 4.883, 4.580, 6.065,

5.525, 4.847, 4.630, 5.036, 5.273, 5.154, 5.226, 5.050, 4.548, 4.857,

1.682, 5.582, 6.180, 2.352, 4.557, 6.713, 6.225, 4.463, 7.486, 3.954

Use the CPD control chart (6.25)–(6.26) to monitor the process variance. In the control chart, use $n_0 - 1 = 9$, $\alpha = 0.005$, and the control limit h_n computed by the approximation formula given in Subsection 6.3.2.

6.13 Reproduce all the results in Example 6.5.

6.14 The following 50 observations are collected from a production process, of which the first 9 observations are known to be IC observations.

4.806, 5.710, 4.329, 4.269, 5.307, 4.441, 5.917, 5.182, 4.899, 5.448,

4.874, 5.090, 4.606, 4.934, 5.688, 4.430, 4.509, 5.577, 5.620, 5.814,

4.426, 5.266, 4.776, 4.115, 5.502, 5.250, 4.871, 5.211, 4.104, 5.237,

7.188, 8.417, 7.354, 8.161, 5.617, 8.435, 3.447, 9.775, 3.864, 8.452,

6.863, 8.866, 9.478, 7.270, 7.462, 5.669, 6.177, 4.383, 7.971, 5.823

Use the CPD control chart (6.29)–(6.30) to monitor both the process mean and the process variance. In the control chart, use $n_0 - 1 = 9$, $\alpha = 0.005$, and the control limit h_n computed by the approximation formula given in Subsection 6.3.3. If the control chart gives a signal, then perform the two post-signal tests discussed in Subsection 6.3.3 to see whether the signal is due to a process mean shift, or a process variance shift, or both.

6.15 Apply the CPD control chart (6.29)–(6.30) with $n_0 - 1 = 9$ and $\alpha = 0.005$ to the data in Exercise 6.7. Compare the results here to the results found in that exercise.

Chapter 7

Multivariate Statistical Process Control

7.1 Introduction

The control charts discussed in Chapters 3–6 are for handling cases when the quality characteristic in question is univariate. In most applications, however, the quality of a product is affected by multiple characteristics of the product. For instance, the quality of a bottle of soft drink may be affected by the weight, amounts of different ingredients, quality of the paper or plastic label on the bottle, and so forth. Regarding the quality of a product with a complicated structure (e.g., cars, airplanes, space shuttles), we can easily list hundreds of quality characteristics that are important to monitor during its production. As described in Chapter 1, quality is a multifaceted concept. Garvin (1987) gave eight dimensions to the definition of quality, including performance, features, reliability, conformance, durability, serviceability, aesthetics, and perceived quality (cf., Section 1.1). Therefore, most statistical process control (SPC) applications have multiple quality characteristics involved. Such SPC problems are often referred to as *multivariate SPC (MSPC)* problems, which are the main focus of this chapter.

In cases when multiple quality characteristics need to be monitored, one natural idea is to use a joint monitoring scheme consisting of a set of univariate SPC control charts, each control chart is for monitoring a single quality characteristic, and the joint monitoring scheme gives a signal of process distributional shift when at least one individual control chart gives a signal. In practice, some people prefer this idea because it only has univariate SPC control charts involved. Thus, it is easy to understand and implement. However, such a joint monitoring scheme has at least two limitations, described briefly below. First, a proper design of the joint monitoring scheme is quite complicated. Each univariate SPC control chart has its own procedure parameters; thus, the joint monitoring scheme would have many parameters to determine before it can be used. For instance, if a univariate CUSUM chart is used for monitoring a single quality characteristic, then the univariate CUSUM chart usually has two parameters involved (i.e., the allowance constant k and the control limit h). Thus, the joint monitoring scheme to monitor p quality characteristics would have $2p$ parameters to determine. Intuitively, different probability distributions of the quality characteristics and the possible correlation among them should be taken into account when we determine the parameters. These issues, along with a relatively large number of procedure parameters involved, make the proper design of the joint monitoring scheme challenging. Second, in the statistical hypothesis testing literature, it has been

demonstrated that, to test the possible difference between two p-dimensional population means, where $p > 1$ is a given integer, it would be more powerful to treat
the two population means as two vectors and compare them by a multivariate testing
procedure (e.g., the Hotelling's T^2 test described in Section 7.2 below) than to utilize
the strategy to use p univariate testing procedures jointly and use the j-th univariate
testing procedure for testing the possible difference between the j-th components
of the two mean vectors, for $1 \le j \le p$ (cf., Johnson and Wichern, 2007, Chapters
5 and 6). Similarly, in the SPC literature, it has been demonstrated that an MSPC
procedure that is constructed properly in the multivariate setup (i.e., treating all observations as p-dimensional vectors and using them for detecting possible shifts in
the p-dimensional distribution of the process) would be more effective than a joint
monitoring scheme consisting of a set of univariate SPC control charts (cf., Crosier,
1988). A main reason for this result is that a vector-based MSPC procedure could
accommodate the possible association among the components of a multivariate process well, while with the joint monitoring scheme based on a set of univariate control
charts it is often difficult to achieve that goal.

In this chapter, we mainly describe some basic MSPC control charts for detecting shifts in the mean and/or covariance matrix of the distribution of a multivariate
production process in cases when the IC and OC process distributions are assumed to
be normal distributions. Cases when the process distributions are multivariate non-
normal will be discussed in Chapter 9. The remaining part of this chapter is organized
as follows. In Sections 7.2–7.5, some fundamental multivariate Shewhart charts, multivariate CUSUM charts, multivariate EWMA charts, and multivariate CPD charts
are discussed, respectively. In Section 7.6, some recent MSPC charts that integrate
the variable selection problem in regression with the MSPC problem are discussed.
Some concluding remarks are given in Section 7.7.

7.2   Multivariate Shewhart Charts

In this section, we describe some Shewhart charts for monitoring multivariate production processes. Our description is divided into two parts. In Subsection 7.2.1,
multivariate normal distributions and their major properties are briefly discussed.
Then, in Subsection 7.2.2, the Hotelling's T^2 statistic and certain related multivariate
Shewhart charts are discussed.

7.2.1   Multivariate normal distributions and some basic properties

In this subsection, we briefly introduce multivariate normal distributions and their
major properties. For a more detailed discussion on this topic, see textbooks such as
Anderson (2003), Eaton (2007), and Johnson and Wichern (2007).

Assume that $\mathbf{X} = (X_1, X_2, \ldots, X_p)'$ is a p-dimensional numerical random vector.
Its cumulative distribution function (cdf) and probability density function (pdf) are
defined by (2.8) and (2.9) in Section 2.2. Its *mean vector*, denoted as $\mu_{\mathbf{X}}$ or $E(\mathbf{X})$, is
defined by

$$\mu_{\mathbf{X}} = (\mu_{X_1}, \mu_{X_2}, \ldots, \mu_{X_p})',$$

and its *covariance matrix*, denoted as $\text{Cov}(\mathbf{X})$, is defined by

$$\text{Cov}(\mathbf{X}) = \text{E}\left[(\mathbf{X} - \mu_\mathbf{X})(\mathbf{X} - \mu_\mathbf{X})'\right],$$

which is a $p \times p$ matrix with the (i, j)-th element equal to $\text{E}[(X_i - \mu_{X_i})(X_j - \mu_{X_j})]$, for $i, j = 1, 2, \ldots, p$. Then, $\text{Cov}(\mathbf{X})$ is a symmetric nonnegative definite matrix. Its j-th diagonal element is $\text{E}(X_j - \mu_{X_j})^2$, which is just the variance of X_j, for $j = 1, 2, \ldots, p$ (cf., (2.5) and (2.7)). Also, its (i, j)-th element is called the *covariance of the two random variables X_i and X_j*, denoted as $\text{Cov}(X_i, X_j)$, which measures the linear association between X_i and X_j. Generally speaking, the linear association between X_i and X_j is stronger if their covariance is larger. However, the value of $\text{Cov}(X_i, X_j)$ depends on the variances of X_i and X_j. To get rid of this dependence, the quantity

$$\frac{\text{Cov}(X_i, X_j)}{\sigma_{X_i} \sigma_{X_j}}, \qquad \text{for } i, j = 1, 2, \ldots, p$$

is defined to be the *correlation coefficient* of the random variables X_i and X_j, denoted as $\text{Cor}(X_i, X_j)$. The $p \times p$ matrix of all pairwise correlation coefficients

$$
\begin{pmatrix}
\dfrac{\text{Cov}(X_1, X_1)}{\sigma_{X_1}^2}, & \dfrac{\text{Cov}(X_1, X_2)}{\sigma_{X_1} \sigma_{X_2}}, & \cdots, & \dfrac{\text{Cov}(X_1, X_p)}{\sigma_{X_1} \sigma_{X_p}} \\[2mm]
\dfrac{\text{Cov}(X_2, X_1)}{\sigma_{X_2} \sigma_{X_1}}, & \dfrac{\text{Cov}(X_2, X_2)}{\sigma_{X_2}^2}, & \cdots, & \dfrac{\text{Cov}(X_2, X_p)}{\sigma_{X_2} \sigma_{X_p}} \\[2mm]
\vdots & \vdots & \ddots & \vdots \\[2mm]
\dfrac{\text{Cov}(X_p, X_1)}{\sigma_{X_p} \sigma_{X_1}}, & \dfrac{\text{Cov}(X_p, X_2)}{\sigma_{X_p} \sigma_{X_2}}, & \cdots, & \dfrac{\text{Cov}(X_p, X_p)}{\sigma_{X_p}^2}
\end{pmatrix}
$$

$$
= \begin{pmatrix}
1, & \text{Cor}(X_1, X_2), & \cdots, & \text{Cor}(X_1, X_p) \\
\text{Cor}(X_2, X_1), & 1, & \cdots, & \text{Cor}(X_2, X_p) \\
\vdots & \vdots & \ddots & \vdots \\
\text{Cor}(X_p, X_1), & \text{Cor}(X_p, X_2), & \cdots, & 1
\end{pmatrix}
$$

is called the *correlation matrix* of $\mathbf{X}$, denoted as $\text{Cor}(\mathbf{X})$. It can be checked that it is always true that

$$-1 \leq \text{Cor}(X_i, X_j) \leq 1, \qquad \text{for } i, j = 1, 2, \ldots, p.$$

The linear association between X_i and X_j is weak when $\text{Cor}(X_i, X_j)$ is close to 0, it is positively strong when $\text{Cor}(X_i, X_j)$ is close to 1, and it is negatively strong when $\text{Cor}(X_i, X_j)$ is close to -1. In cases when $\text{Cor}(X_i, X_j) = 0$, X_i and X_j are said to be *uncorrelated*.

In cases when the pdf of $\mathbf{X}$ has the parametric form

$$f(\mathbf{x}) = \frac{1}{(2\pi)^{p/2} |\Sigma|^{1/2}} \exp\left[-\frac{1}{2}(\mathbf{x} - \mu)'\Sigma^{-1}(\mathbf{x} - \mu)\right], \text{ for } \mathbf{x} = (x_1, x_2, \ldots, x_p)' \in R^p,$$

$$\tag{7.1}$$

where μ is a p-dimensional parameter vector and Σ is a $p \times p$ parameter matrix, the distribution of $\mathbf{X}$ is called a *p-dimensional normal distribution* and $\mathbf{X}$ is called a *p-dimensional normal random vector*. It can be checked that, if $\mathbf{X}$ has a *p*-dimensional normal distribution with the pdf in (7.1), then its mean vector is just μ and its covariance matrix is just Σ. Namely,

$$\mu_{\mathbf{X}} = \mu, \qquad \text{Cov}(\mathbf{X}) = \Sigma.$$

Because the *p*-dimensional normal distribution is uniquely determined by its mean vector and covariance matrix, in cases when $\mathbf{X}$ has the pdf (7.1), it is denoted as $\mathbf{X} \sim N_p(\mu, \Sigma)$. Also, it can be checked that, in cases when $\mathbf{X}$ has a multivariate normal distribution, for any pair (X_i, X_j) of its components with $1 \le i \ne j \le p$, they are uncorrelated if and only if they are independent (cf., (2.10) for the definition of independence).

In cases when $p = 2$, the pdf in (7.1) can be written as

$$f(x_1, x_2) = \frac{1}{2\pi \sigma_{X_1} \sigma_{X_2} \sqrt{1 - \text{Cor}(X_1, X_2)^2}} \times$$

$$\exp\left\{ -\frac{1}{2(1 - \text{Cor}(X_1, X_2)^2)} \left[\left(\frac{x_1 - \mu_{X_1}}{\sigma_{X_1}} \right)^2 - 2\text{Cor}(X_1, X_2) \left(\frac{x_1 - \mu_{X_1}}{\sigma_{X_1}} \right) \left(\frac{x_2 - \mu_{X_2}}{\sigma_{X_2}} \right) + \left(\frac{x_2 - \mu_{X_2}}{\sigma_{X_2}} \right)^2 \right] \right\}.$$

When $\sigma_{X_1} = 1$, $\sigma_{X_2} = 0.5$, and $\text{Cor}(X_1, X_2) = 0$, the pdf function $f(x_1, x_2)$ is shown in Figure 7.1(a). Three contour curves $\{(x_1, x_2) : f(x_1, x_2) = c\}$ with $c = 0.05, 0.1$, and 0.2 are shown in Figure 7.1(b). From both plots, it can be seen that, in cases when X_1 and X_2 are uncorrelated, the surface of $f(x_1, x_2)$ is bell shaped with elliptical contour curves, and the major and minor axes of the contour curves are parallel to the x and y axes. In cases when $\sigma_{X_1} = 1$, $\sigma_{X_2} = 0.5$, and $\text{Cor}(X_1, X_2) = 0.8$ (i.e., X_1 and X_2 are positively correlated), the pdf function $f(x_1, x_2)$ is shown in Figure 7.1(c). Three contour curves $\{(x_1, x_2)) : f(x_1, x_2) = c\}$ with $c = 0.05, 0.1$, and 0.2 are shown in Figure 7.1(d). It can be seen that the surface of $f(x_1, x_2)$ is still bell shaped with elliptical contour curves, but the major and minor axes of the contour curves are no longer parallel to the x and y axes. As a matter of fact, the major axes of the contour curves have a positive slope in such a case, indicating a positive linear relationship between X_1 and X_2.

Now, if $\mathbf{X} \sim N_p(\mu, \Sigma)$, A is any $m \times p$ constant matrix, and $\mathbf{b}$ is any m-dimensional constant vector, then it can be checked that (cf., Johnson and Wichern, 2007, Chapter 4)

$$A\mathbf{X} + \mathbf{b} \sim N_m (A\mu + \mathbf{b}, A\Sigma A'). \tag{7.2}$$

From this property, we have the conclusion that, if $\mathbf{X} \sim N_p(\mu, \Sigma)$ and $\widetilde{\mathbf{X}}$ is a vector of any subset of the components of $\mathbf{X}$, then $\widetilde{\mathbf{X}}$ must have a multivariate normal distribution. Another property of the multivariate normal distribution is described in the box below, which plays an important role in MSPC.

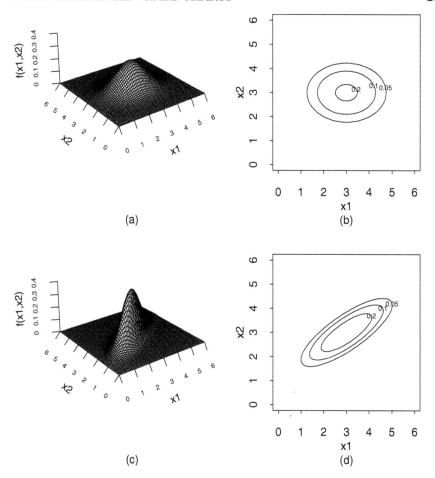

Figure 7.1 *(a) Surface of $f(x_1,x_2)$ when (X_1,X_2) has a bivariate normal distribution, $\sigma_{X_1} = 1$, $\sigma_{X_2} = 0.5$, and $Cor(X_1,X_2) = 0$. (b) Three contour curves $\{(x_1,x_2)) : f(x_1,x_2) = c\}$ of $f(x_1,x_2)$ shown in plot (a) when $c = 0.05, 0.1$, and 0.2. (c) Surface of $f(x_1,x_2)$ when $(X_1,X_2)'$ has a bivariate normal distribution, $\sigma_{X_1} = 1$, $\sigma_{X_2} = 0.5$, and $Cor(X_1,X_2) = 0.8$. (b) Three contour curves $\{(x_1,x_2)) : f(x_1,x_2) = c\}$ of $f(x_1,x_2)$ shown in plot (c) when $c = 0.05, 0.1$, and 0.2.*

Distribution of a Quadratic Form of a Multivariate Normal Random Vector

Assume that $\mathbf{X} \sim N_p(\mu,\Sigma)$ and Σ is a non-singular covariance matrix. Then, the distribution of $(\mathbf{X} - \mu)'\Sigma^{-1}(\mathbf{X} - \mu)$ is χ_p^2.

The property in the above box is easy to verify. Let $\mathbf{Z} = \Sigma^{-1/2}(\mathbf{X} - \mu)$. Then, it can be checked by (7.2) that $\mathbf{Z} \sim N_p(\mathbf{0}, I_{p \times p})$ and $(\mathbf{X} - \mu)'\Sigma^{-1}(\mathbf{X} - \mu) = \mathbf{Z}'\mathbf{Z}$, where $I_{p \times p}$ is the $p \times p$ identity matrix. Let $\mathbf{Z} = (Z_1, Z_2, \ldots, Z_p)'$. Then, $\{Z_1, Z_2, \ldots, Z_p\}$ is a sequence of i.i.d. random variables with the standard normal distribution. By the definition of a chi-square distribution in Subsection 2.3.2, we have $(\mathbf{X} - \mu)'\Sigma^{-1}(\mathbf{X} - \mu) = Z_1^2 + Z_2^2 + \cdots + Z_p^2 \sim \chi_p^2$. Now, $\Sigma^{-1/2}(\mathbf{X} - \mu)$ can be interpreted as a weighted difference between $\mathbf{X}$ and μ, with the components of $\mathbf{X}$ that have larger variances receiving smaller weights. Therefore,

$$d_S^2(\mathbf{X}, \mu) = (\mathbf{X} - \mu)'\Sigma^{-1}(\mathbf{X} - \mu)$$

is often referred to as the squared *statistical distance* from $\mathbf{X}$ to its mean vector μ.

By the result in the above box, if $\mathbf{X} \sim N_p(\mu, \Sigma)$, then $\mathbf{X}$ has a $1 - \alpha$ chance to be contained in the following region:

$$\mathbf{C} = \left\{ \mathbf{x} : \mathbf{x} \in R^p, d_S^2(\mathbf{x}, \mu) \le \chi_{1-\alpha,p}^2 \right\},$$

where $\alpha \in [0, 1]$ is a given significance level, and $\chi_{1-\alpha,p}^2$ is the $(1 - \alpha)$-th quantile of the χ_p^2 distribution. If we ignore the possible association among the components of $\mathbf{X}$ and assume that all components are independent of each other, then we can use the method described below to construct a similar region. First, for the j-th component X_j, it has a $1 - \alpha_j$ chance to be in the region

$$C_j = \left\{ x_j : \mu_j - Z_{1-\alpha_j/2}\sigma_{X_j} \le x_j \le \mu_j + Z_{1-\alpha_j/2}\sigma_{X_j} \right\},$$

where $\alpha_j \in [0, 1]$ is a significance level, and $Z_{1-\alpha_j/2}$ is the $(1 - \alpha_j/2)$-th quantile of the $N(0, 1)$ distribution. To guarantee that there is a $1 - \alpha$ chance that $X_j \in C_j$ for all j, we should choose α_j such that

$$
\begin{aligned}
1 - \alpha \\
= \quad & P\left(\bigcap_{j=1}^{p} (X_j \in C_j) \right) \\
= \quad & (1 - \alpha_1)(1 - \alpha_2) \cdots (1 - \alpha_p).
\end{aligned}
$$

For simplicity, let us assume that $\alpha_1 = \alpha_2 = \cdots = \alpha_p = \alpha'$. Then, we have

$$\alpha' = 1 - (1 - \alpha)^{1/p}.$$

In the case shown in Figure 7.1(c) when $p = 2, \mu_{X_1} = 3, \mu_{X_2} = 3, \sigma_{X_1} = 1, \sigma_{X_2} = 0.5$, and $\mathrm{Cor}(X_1, X_2) = 0.8$, if $\alpha = 0.05$, then $\alpha' = 0.0253$. In such a case, the region $\mathbf{C}$ is shown in Figure 7.2 by the ellipse, and the intervals C_1 and C_2 are shown in the same plot by the dotted and dashed lines, respectively. From the plot, we can see that, although it is guaranteed that $\mathbf{X}$ has a 95% chance to be in the square surrounded by the dotted and dashed lines, it is often too conservative (i.e., the actual probability would be larger than 0.95), compared to the elliptical region.

Assume that there is a p-dimensional population with the population distribution $N_p(\mu, \Sigma)$, and $(\mathbf{X}_1, \mathbf{X}_2, \ldots, \mathbf{X}_n)$ is a simple random sample of size n from the

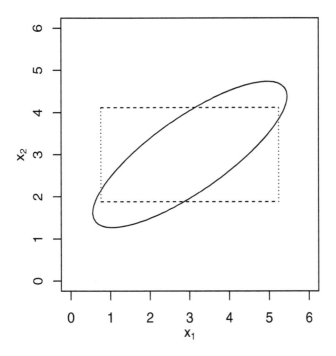

Figure 7.2 *In the case when* $p = 2, \mu_{X_1} = 3, \mu_{X_2} = 3, \sigma_{X_1} = 1, \sigma_{X_2} = 0.5,$ *and* $Cor(X_1, X_2) = 0.8,$ *the elliptical region is constructed from the distribution of* $d_S^2(\mathbf{X}, \mu)$ *directly such that* $\mathbf{X}$ *has a 95% chance to be in the region, and the square has the same property but it is constructed from individual components of* $\mathbf{X}$ *by assuming the components are independent.*

population (cf., Section 2.5). Then, $\mathbf{X}_1, \mathbf{X}_2, \dots, \mathbf{X}_n$ is a sequence of i.i.d. random vectors having the common distribution $N_p(\mu, \Sigma)$. The likelihood function based on the sample (cf., Subsection 2.7.2) is

$$
\begin{aligned}
& L(\mu, \Sigma; \mathbf{X}_1, \mathbf{X}_2, \dots, \mathbf{X}_n) \\
={} & \Pi_{i=1}^n \left\{ \frac{1}{(2\pi)^{p/2}|\Sigma|^{1/2}} \exp\left[-\frac{1}{2}(\mathbf{X}_i - \mu)'\Sigma^{-1}(\mathbf{X}_i - \mu) \right] \right\} \\
={} & \frac{1}{(2\pi)^{np/2}|\Sigma|^{n/2}} \exp\left[-\frac{1}{2}\sum_{i=1}^n (\mathbf{X}_i - \mu)'\Sigma^{-1}(\mathbf{X}_i - \mu) \right].
\end{aligned}
$$

By some algebraic manipulations (cf., Johnson and Wichern, 2007, Section 4.3), we have the results summarized in the box below.

Maximum Likelihood Estimators of μ and Σ

Assume that $(\mathbf{X}_1, \mathbf{X}_2, \ldots, \mathbf{X}_n)$ is a simple random sample from a p-dimensional population with the population distribution $N_p(\mu, \Sigma)$. Then, the maximum likelihood estimators (MLEs) of μ and Σ are

$$\widehat{\mu} = \overline{\mathbf{X}}, \qquad \widehat{\Sigma} = \frac{n-1}{n}\mathbf{S}^2, \tag{7.3}$$

where $\overline{\mathbf{X}} = \frac{1}{n}\sum_{i=1}^{n}\mathbf{X}_i$ is the sample mean and $\mathbf{S}^2 = \frac{1}{n-1}\sum_{i=1}^{n}(\mathbf{X}_i - \overline{\mathbf{X}})(\mathbf{X}_i - \overline{\mathbf{X}})'$ is the sample covariance matrix.

To describe the sampling distribution of $\mathbf{S}^2$ (also the sampling distribution of $\widehat{\Sigma}$ in (7.3)), we need to define the so-called *Wishart distribution* as follows. Let $\mathbf{Z}_1, \mathbf{Z}_2, \ldots, \mathbf{Z}_m$ be a sequence of m i.i.d. random vectors having the common distribution $N_p(\mathbf{0}, \Sigma)$. Then, the distribution of the random matrix $\sum_{i=1}^{m}\mathbf{Z}_i\mathbf{Z}_i'$ is called the Wishart distribution with m degrees of freedom (df) and with the covariance matrix Σ, denoted as $W_{m,\Sigma}$. Then, results about the sampling distributions of $\overline{\mathbf{X}}$ and $\mathbf{S}^2$ are summarized in the box below.

Sampling Distributions of $\overline{\mathbf{X}}$ and $\mathbf{S}^2$

Assume that $(\mathbf{X}_1, \mathbf{X}_2, \ldots, \mathbf{X}_n)$ is a simple random sample from a p-dimensional population with the population distribution $N_p(\mu, \Sigma)$. Then, the sampling distributions of the sample mean $\overline{\mathbf{X}}$ and the sample covariance matrix $\mathbf{S}^2$ have the following properties:

(i) $\overline{\mathbf{X}} \sim N_p(\mu, \Sigma/n)$,

(ii) $(n-1)\mathbf{S}^2 \sim W_{n-1,\Sigma}$, and

(iii) $\overline{\mathbf{X}}$ and $\mathbf{S}^2$ are independent of each other.

In cases when the p-dimensional population distribution is non-normal or unknown but the sample size n is large, the sampling distribution of $\overline{\mathbf{X}}$ would be close to normal, which is the central limit theorem (CLT) summarized in the box below.

Central Limit Theorem

Assume that $(\mathbf{X}_1, \mathbf{X}_2, \ldots, \mathbf{X}_n)$ is a simple random sample from a p-dimensional population, and $\overline{\mathbf{X}}$ is the sample mean. Then, the sampling distribution of $\sqrt{n}(\overline{\mathbf{X}} - \mu)$ converges to $N_p(\mathbf{0}, \Sigma)$, and the sampling distribution of $n(\overline{\mathbf{X}} - \mu)'(\mathbf{S}^2)^{-1}(\overline{\mathbf{X}} - \mu)$ converges to χ_p^2, when n increases.

7.2.2 *Some multivariate Shewhart charts*

Assume that $(\mathbf{X}_1, \mathbf{X}_2, \ldots, \mathbf{X}_n)$ is a simple random sample from a p-dimensional population with the distribution $N_p(\mu, \Sigma)$, and $\overline{\mathbf{X}}$ and $\mathbf{S}^2$ are the sample mean vector and the sample covariance matrix. Then, the random variable

$$T^2 = n(\overline{\mathbf{X}} - \boldsymbol{\mu})' (\mathbf{S}^2)^{-1} (\overline{\mathbf{X}} - \boldsymbol{\mu}) \tag{7.4}$$

is the so-called *Hotelling's T^2 statistic* in the literature, which plays an important role in MSPC. The Hotelling's T^2 statistic was first proposed by Hotelling (1931) as a generalization of the t-test statistic in univariate statistical inference. Recall from Subsection 2.7.1 that, if $(X_1, X_2, \ldots, X_n)$ is a simple random sample from a univariate population with the distribution $N(\mu, \sigma^2)$, then the random variable

$$\frac{\overline{X} - \mu}{s/\sqrt{n}}$$

would have the t_{n-1} distribution (cf., (2.18)). Its square can be written as

$$\left(\frac{\overline{X} - \mu}{s/\sqrt{n}}\right)^2 = n(\overline{X} - \mu)(s^2)^{-1}(\overline{X} - \mu).$$

Obviously, the Hotelling's T^2 statistic defined in (7.4) is the multivariate version of the above random variable. Hotelling (1947) was the first to apply the T^2 statistic to the MSPC problem. His MSPC chart and several variants are described below. For related discussions about the Hotelling's T^2 statistic, its properties, and some control charts constructed from it, see Alt (1985), Fuchs and Kenett (1998), Lowry and Montgomery (1995), Mason and Young (2002), and the references cited therein. In this book, certain quadratic statistics similar to T^2 in (7.4), such as $T_{0,i}^2$ and $T_{1,i}^2$ defined in (7.5) and (7.7) below, are also called the Hotelling's T^2 statistics for simplicity of presentation. Hopefully, this will not cause any confusion.

Let us first discuss the phase I MSPC problem. Assume that $(\mathbf{X}_1, \mathbf{X}_2, \ldots, \mathbf{X}_M)$ is a phase I dataset obtained from a p-dimensional production process. In cases when the process is IC, its distribution is assumed to be $N_p(\boldsymbol{\mu}_0, \Sigma_0)$, and we are interested in detecting any mean shifts in the process. In cases when μ_0 and Σ_0 are known, at the i-th time point, it is natural to consider the charting statistic

$$T_{0,i}^2 = (\mathbf{X}_i - \boldsymbol{\mu}_0)' \Sigma_0^{-1} (\mathbf{X}_i - \boldsymbol{\mu}_0). \tag{7.5}$$

When the process is IC at the i-th time point, it is obvious that

$$T_{0,i}^2 \sim \chi_p^2.$$

Therefore, the Shewhart chart using $T_{0,i}^2$ as its charting statistic would give a signal of mean shift if

$$T_{0,i}^2 > \chi_{1-\alpha,p}^2, \tag{7.6}$$

where $\alpha \in [0, 1]$ is a given significance level, and $\chi_{1-\alpha,p}^2$ is the $(1 - \alpha)$-th quantile of the χ_p^2 distribution. Obviously, $T_{0,i}^2$ is the IC squared statistical distance between $\mathbf{X}_i$ and the IC mean vector μ_0, and the multivariate Shewhart chart (7.5)–(7.6) gives a signal of mean shift when this distance is larger than the critical value $\chi^2_{1-\alpha,p}$. In cases when all observations are independent and the process is IC, we have

$$ARL_0 = \frac{1}{\alpha}.$$

In practice, both μ_0 and Σ_0 are usually unknown. In such cases, they can be estimated by the sample mean vector $\overline{\mathbf{X}}$ and the sample covariance matrix $\mathbf{S}^2$. Then, the resulting charting statistic is

$$T^2_{1,i} = (\mathbf{X}_i - \overline{\mathbf{X}})' (\mathbf{S}^2)^{-1} (\mathbf{X}_i - \overline{\mathbf{X}}). \tag{7.7}$$

When the process is IC and the sample size M is large, the IC distribution of $T^2_{1,i}$ would be close to the χ^2_p distribution because $\overline{\mathbf{X}}$ and $\mathbf{S}^2$ are close to μ_0 and Σ_0, respectively. However, when M is small, the distribution of $T^2_{1,i}$ could be substantially different from χ^2_p, because much randomness is added to $T^2_{1,i}$ by $\overline{\mathbf{X}}$ and $\mathbf{S}^2$. Tracy et al. (1992) studied the IC distribution of $T^2_{1,i}$, and found that the IC distribution of $\frac{M}{(M-1)^2} T^2_{1,i}$ was actually a *beta distribution* with parameters $p/2$ and $(M-p-1)/2$, denoted as $Beta(p/2, (M-p-1)/2)$. The $Beta(a,b)$ distribution is an absolutely continuous distribution with the pdf

$$f(x) = \frac{1}{B(a,b)} x^{a-1} (1-x)^{b-1}, \qquad \text{for } x \in [0,1],$$

where $B(a,b) = \int_0^1 u^{a-1} (1-u)^{b-1}\, du$ is a constant. Then, the Shewhart chart with the charting statistic $T^2_{1,i}$ gives a signal of mean shift in cases when

$$T^2_{1,i} > \frac{(M-1)^2}{M} Beta_{1-\alpha}(p/2, (M-p-1)/2), \tag{7.8}$$

where $\alpha \in [0,1]$ is a given significance level, and $Beta_{1-\alpha}(p/2, (M-p-1)/2)$ is the $(1-\alpha)$-th quantile of the $Beta(p/2, (M-p-1)/2)$ distribution.

Example 7.1 *Table 7.1 contains 30 observations from a 3-dimensional production process for phase I analysis. To apply the Shewhart chart (7.7)–(7.8), we first compute the sample mean vector and sample covariance matrix to be*

$$\overline{\mathbf{X}} = (0.490, 0.464, 0.416)'$$

and

$$\mathbf{S}^2 = \begin{pmatrix} 0.832 & 0.644 & 0.492 \\ 0.644 & 0.943 & 0.794 \\ 0.492 & 0.794 & 1.059 \end{pmatrix}.$$

If we choose $\alpha = 0.005$ (note: $ARL_0 = 1/\alpha = 200$), then the control limit in (7.8) is determined to be

$$\frac{(30-1)^2}{30} Beta_{0.995}(1.5, 13) = 10.773.$$

The corresponding Shewhart chart (7.7)–(7.8) is shown in Figure 7.3 when it is applied to the data in Table 7.1, from which we can see that a signal is given at the 24th time point. Therefore, the root cause of this signal should be investigated.

Table 7.1: *This table contains 30 observations from a 3-dimensional production process.*

i	X_1	X_2	X_3	i	X_1	X_2	X_3
1	−0.224	−0.464	−0.662	16	0.217	−0.260	−0.005
2	−0.082	−0.203	0.682	17	−0.611	−0.048	−0.428
3	−0.436	0.342	−0.174	18	0.473	0.066	0.890
4	0.455	0.715	1.229	19	−1.130	−1.214	−0.064
5	0.437	0.237	−0.377	20	1.379	2.443	2.350
6	0.669	−0.002	0.230	21	0.671	0.534	1.120
7	−0.186	−0.480	−0.907	22	1.467	0.945	0.402
8	0.672	0.384	0.903	23	1.077	0.949	−0.085
9	−0.121	−0.812	−1.279	24	2.126	3.964	3.775
10	−0.683	0.093	−0.436	25	2.237	0.959	0.175
11	0.063	0.193	−0.028	26	0.362	0.344	0.375
12	−1.029	0.503	0.728	27	2.119	1.569	1.031
13	−0.080	−0.238	−0.218	28	1.147	1.003	1.417
14	0.943	0.272	0.831	29	0.667	0.525	1.676
15	0.147	0.655	−0.543	30	1.944	0.935	−0.130

Next, let us discuss the phase II MSPC problem. Assume that $X_1, X_2, \ldots$ are phase II observations obtained from a p-dimensional production process with the IC distribution $N_p(\mu_0, \Sigma_0)$. In the case when the IC distribution is known (i.e., both μ_0 and Σ_0 are known), we can use the Shewhart chart (7.5)–(7.6) for process monitoring. In cases when both μ_0 and Σ_0 are unknown, they need to be estimated from an IC dataset. Assume that an IC sample of size M is available, and μ_0 and Σ_0 are estimated by the sample mean and the sample covariance matrix, respectively, denoted as $\widehat{\mu}_0$ and $\widehat{\Sigma}_0$. Then, the phase II observations $X_1, X_2, \ldots$ are independent of both $\widehat{\mu}_0$ and $\widehat{\Sigma}_0$. At the n-th time point, we can use the charting statistic

$$T_{2,n}^2 = (\mathbf{X}_n - \widehat{\mu}_0)' \left(\widehat{\Sigma}_0\right)^{-1} (\mathbf{X}_n - \widehat{\mu}_0), \tag{7.9}$$

which can be regarded as the estimated value of the squared statistical distance between the observation $\mathbf{X}_n$ and the IC mean vector μ_0 (i.e., $d_S^2(\mathbf{X}_n, \mu_0)$). Tracy et al. (1992) have shown that when the process is IC, we have

$$\frac{(M-p)M}{p(M-1)(M+1)} T_{2,n}^2 \sim F_{p,M-p},$$

where $F_{p,M-p}$ denotes the F distribution with the numerator degrees of freedom p and the denominator degrees of freedom $M - p$ (cf., Subsection 2.3.4). Therefore, the Shewhart chart gives a signal of mean shift at the n-th time point if

$$T_{2,n}^2 > \frac{p(M-1)(M+1)}{(M-p)M} F_{1-\alpha,p,M-p}, \tag{7.10}$$

where $F_{1-\alpha,p,M-p}$ is the $(1-\alpha)$-th quantile of the $F_{p,M-p}$ distribution.

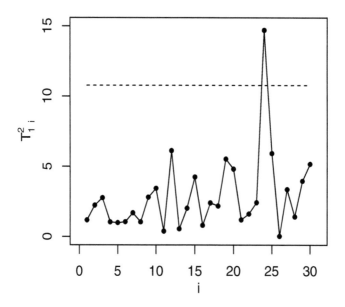

Figure 7.3 *The Shewhart chart (7.7)–(7.8) when it is applied to the data in Table 7.1 and when* $\alpha = 0.005$. *The dashed horizontal line denotes the control limit.*

Example 7.1 (continued) *For the data presented in Table 7.1, assume that the first 20 observations are IC observations, the remaining 10 observations are phase II observations, and we want to use the Shewhart chart (7.9)–(7.10) for phase II monitoring of the process. To this end, we first compute* $\widehat{\mu}_0$ *and* $\widehat{\Sigma}_0$ *from the phase I data to be*

$$\widehat{\mu}_0 = (0.044, 0.109, 0.136)'$$

and

$$\widehat{\Sigma}_0 = \begin{pmatrix} 0.412 & 0.279 & 0.305 \\ 0.279 & 0.532 & 0.451 \\ 0.305 & 0.451 & 0.726 \end{pmatrix}.$$

Next, if we choose $\alpha = 0.005$, *then the control limit is*

$$\frac{3(20-1)(20+1)}{(20-3)20} F_{0.995,3,17} = 21.671.$$

The resulting Shewhart chart (7.9)–(7.10) is shown in Figure 7.4 when it is applied to the 10 phase II observations, from which it can be seen that a signal of mean shift is given at the 4th phase II observation time point.

In the above discussion, we assume that there is only one observation vector at each time point (i.e., the cases with individual observation data). In some applications, we can collect multiple observation vectors at each time point (i.e., the

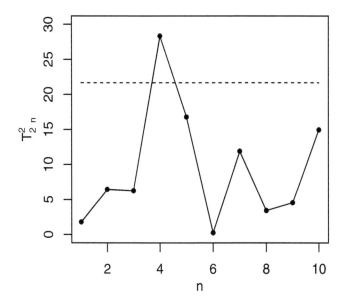

Figure 7.4 *The Shewhart chart (7.9)–(7.10) when it is applied to the last 10 observations presented in Table 7.1. The dashed horizontal line denotes the control limit when $\alpha = 0.005$.*

cases with batch data) for monitoring a production process. Assume that, in a phase I MSPC analysis, the following m observation vectors have been collected from a p-dimensional production process at the i-th time point:

$$\mathbf{X}_{i1}, \mathbf{X}_{i2}, \ldots, \mathbf{X}_{im}, \qquad \text{for } i = 1, 2, \ldots, M,$$

where each $\mathbf{X}_{ij}$ is a p-dimensional observation vector, for $j = 1, 2, \ldots, m$, and m is the batch size. Let

$$\overline{\mathbf{X}}_i = \frac{1}{m} \sum_{j=1}^{m} \mathbf{X}_{ij}, \qquad \text{for } i = 1, 2, \ldots, M,$$

$$\overline{\overline{\mathbf{X}}} = \frac{1}{M} \sum_{i=1}^{M} \overline{\mathbf{X}}_i,$$

$$\mathbf{S}_i^2 = \frac{1}{m-1} \sum_{j=1}^{m} \left(\mathbf{X}_{ij} - \overline{\mathbf{X}}_i \right) \left(\mathbf{X}_{ij} - \overline{\mathbf{X}}_i \right)' \qquad \text{for } i = 1, 2, \ldots, M,$$

$$\overline{\mathbf{S}^2} = \frac{1}{M} \sum_{i=1}^{M} \mathbf{S}_i^2.$$

Then, $\overline{\mathbf{X}}_i$ and $\mathbf{S}_i^2$ are the sample mean and sample covariance matrix of the m observation vectors in the i-th batch, $\overline{\overline{\mathbf{X}}}$ is the grand sample mean of all mM observation vectors, and $\overline{\mathbf{S}^2}$ is the average of $\{\mathbf{S}_i^2, i = 1, 2, \ldots, M\}$. In the case when the IC process distribution $N_p(\mu_0, \Sigma_0)$ is known, we can still use the Shewhart chart (7.5)–(7.6) for process monitoring, except that $T_{0,i}^2$ defined in (7.5) needs to be replaced by

$$\tilde{T}_{0,i}^2 = m \left(\overline{\mathbf{X}}_i - \mu_0\right)' \Sigma_0^{-1} \left(\overline{\mathbf{X}}_i - \mu_0\right). \tag{7.11}$$

In cases when both μ_0 and Σ_0 are unknown, they can be estimated by $\overline{\overline{\mathbf{X}}}$ and $\overline{\mathbf{S}^2}$, respectively. In such cases, the charting statistic $T_{1,i}^2$ in (7.7) can be replaced by

$$\tilde{T}_{1,i}^2 = m \left(\overline{\mathbf{X}}_i - \overline{\overline{\mathbf{X}}}\right)' \left(\overline{\mathbf{S}^2}\right)^{-1} \left(\overline{\mathbf{X}}_i - \overline{\overline{\mathbf{X}}}\right). \tag{7.12}$$

When the process is IC, according to Bersimis et al. (2007) and Mason et al. (2001), we have

$$\frac{Mm - M - p + 1}{p(M-1)(m-1)} \tilde{T}_{1,i}^2 \sim F_{p, Mm-M-p+1}.$$

Therefore, the Shewhart chart gives a signal of mean shift at the i-th time point when

$$\tilde{T}_{1,i}^2 > \frac{p(M-1)(m-1)}{Mm - M - p + 1} F_{1-\alpha, p, Mm-M-p+1}, \tag{7.13}$$

where $\alpha \in [0,1]$ is a given significance level.

The Shewhart chart (7.12)–(7.13) is for phase I MSPC. For phase II MSPC, in cases when batch data with the batch size of m are available and when the IC process distribution $N_p(\mu_0, \Sigma_0)$ is known, we can still use the Shewhart chart (7.5)–(7.6) for process monitoring, except that $T_{0,i}^2$ defined in (7.5) needs to be replaced by $\tilde{T}_{0,i}^2$ defined in (7.11). In cases when both μ_0 and Σ_0 are unknown and they are estimated by $\hat{\mu}_0 = \overline{\overline{\mathbf{X}}}$ and $\hat{\Sigma}_0 = \overline{\mathbf{S}^2}$, respectively, constructed from an IC data of M batches with the batch size m, then we can use the charting statistic

$$\tilde{T}_{2,n}^2 = m(\overline{\mathbf{X}}_n - \hat{\mu}_0)' \left(\hat{\Sigma}_0\right)^{-1} (\overline{\mathbf{X}}_n - \hat{\mu}_0), \tag{7.14}$$

where $\overline{\mathbf{X}}_n$ is the sample mean of the m observations obtained at the n-th time point for phase II process monitoring. In such cases, Mason et al. (2001) showed that the IC distribution of $\tilde{T}_{2,n}^2$ was

$$\frac{Mm - M - p + 1}{p(Mm+1)(m-1)} \tilde{T}_{2,n}^2 \sim F_{p, Mm-M-p+1}.$$

Therefore, the chart gives a signal of mean shift at the n-th time point when

$$\tilde{T}_{2,n}^2 > \frac{p(Mm+1)(m-1)}{Mm - M - p + 1} F_{1-\alpha, p, Mm-M-p+1}, \tag{7.15}$$

where $\alpha \in [0,1]$ is a given significance level.

In the literature, there are also some Shewhart charts proposed for monitoring the process covariance matrix, or for joint monitoring of both the process mean vector and the process covariance matrix. See, for instance, Alt (1985), Aparisi et al. (1999, 2001), Khoo and Quah (2003), Levinson et al. (2002), Tang and Barnett (1996a,b), Yeh and Lin (2002), Yeh et al. (2006), and the references cited therein.

7.3 Multivariate CUSUM Charts

The multivariate Shewhart charts described in the previous section use the observed data at the current time point alone for making decisions about the process performance at the current time point. As pointed out in Section 4.1 about univariate cases, such control charts are effective for detecting relatively large and transient shifts in the process distribution, and are commonly used in phase I SPC analysis. To detect relatively small and persistent shifts in the process distribution, alternative control charts, such as the CUSUM, EWMA, and CPD charts, would be more effective. In this section, we describe some fundamental multivariate CUSUM (MCUSUM) charts in two parts. Those for monitoring process mean vector are described in Subsection 7.3.1, and those for monitoring process covariance matrix are described in Subsection 7.3.2. As in univariate cases, our discussion about these alternative control charts will mainly focus on the phase II MSPC, although they can also be used for the phase I analysis. Also, these alternative control charts are often applied to individual observation data, instead of batch data. See Section 4.1 for a related discussion.

7.3.1   MCUSUM charts for monitoring the process mean

Assume that $\mathbf{X}_1, \mathbf{X}_2, \ldots$ is a sequence of phase II observations obtained from a p-dimensional production process with the IC distribution $N_p(\mu_0, \Sigma_0)$, where μ_0 and Σ_0 are known. To monitor the p-dimensional process, Woodall and Ncube (1985) suggested monitoring the p individual components of the process by a joint monitoring scheme, described below. Without loss of generality, assume that $\mu_0 = \mathbf{0}$ (in cases when $\mu_0 \neq \mathbf{0}$, we can monitor the process using the centered data $\mathbf{X}_1 - \mu_0, \mathbf{X}_2 - \mu_0, \ldots$). For the j-th component with $1 \leq j \leq p$, let us consider a two-sided version of the CUSUM chart whose charting statistics are

$$
\begin{aligned}
C_{n,j}^+ &= \max\left(0, C_{n-1,j}^+ + X_{nj} - k_j\right), \qquad \text{for } n \geq 1 \\
C_{n,j}^- &= \min\left(0, C_{n-1,j}^- + X_{nj} + k_j\right),
\end{aligned}
\tag{7.16}
$$

where $\mathbf{X}_n = (X_{n1}, X_{n2}, \ldots, X_{np})'$ is the n-th observation vector, $C_{0,j}^+ = C_{0,j}^- = 0$, and $k_j > 0$ is an allowance constant. Then, the two-sided CUSUM chart gives a signal of process mean shift in the j-th component at the n-th time point if

$$
C_{n,j}^+ > h_j \qquad \text{or} \qquad C_{n,j}^- < -h_j,
$$

where $h_j > 0$ is a control limit. The joint monitoring scheme suggested by Woodall and Ncube (1985) gives a signal of process mean shift in the p-dimensional process if at least one individual two-sided CUSUM chart gives a signal. Namely, it gives a signal at the n-th time point if

$$
\bigcup_{j=1}^{p} \left(C_{n,j}^+ > h_j \text{ or } C_{n,j}^- < -h_j \right),
\tag{7.17}
$$

where $\bigcup_{j=1}^{p} A_j$ means that at least one of the events $A_1, A_2, \ldots, A_p$ happens.

In cases when the p components of the p-dimensional process are independent, the average run length (ARL) of the joint monitoring scheme (7.16)–(7.17) can be approximated by

$$\left(1 - \frac{1}{ARL}\right) \approx \Pi_{j=1}^{p}\left(1 - \frac{1}{ARL_j}\right)$$

where ARL_j denotes the ARL value of the j-th individual CUSUM chart. From the above expression, in cases when the ARL and ARL_j values are all quite large, the ARL value can be approximated by

$$\frac{1}{ARL} \approx \sum_{j=1}^{p} \frac{1}{ARL_j}. \tag{7.18}$$

For instance, in the case when the IC ARL values of the p individual CUSUM charts, denoted as $\{ARL_{0,j}, j = 1, 2, \ldots, p\}$, are all chosen to be large, the IC ARL value of the joint monitoring scheme, denoted as ARL_0, can be computed by

$$\frac{1}{ARL_0} \approx \sum_{j=1}^{p} \frac{1}{ARL_{0,j}}.$$

In a special case when $ARL_{0,1} = ARL_{0,2} = \cdots = ARL_{0,p}$, we have $ARL_0 \approx ARL_{0,1}/p$.

Example 7.2 *Table 7.2 presents 20 phase II observations obtained from a 3-dimensional production process for online monitoring of the process mean vector. The IC process distribution is assumed to be known $N_3(\mu_0, \Sigma_0)$ where $\mu_0 = (0,0,0)'$ and $\Sigma_0 = I_{3 \times 3}$. Therefore, the three components of the quality characteristic vector $\mathbf{X} = (X_1, X_2, X_3)'$ are independent of each other when the process is IC. The multivariate Shewhart charting statistic*

$$T_{0,n}^2 = (\mathbf{X}_n - \mu_0)' \Sigma_0^{-1} (\mathbf{X}_n - \mu_0)$$

is computed at each observation time point, and the calculated values of $T_{0,n}^2$ are also presented in Table 7.2. When $\alpha = 0.02$ (i.e., the ARL_0 value of the Shewhart chart is $1/0.02 = 50$), the control limit of the Shewhart chart is $\chi^2_{1-0.02,3} = 9.837$ (cf., (7.6)). From the table, we can see that the first signal by the Shewhart chart is given at the 20th observation time point. The 20 observations in Table 7.2 are actually generated from $N_3(\mu_1, \Sigma_0)$ where $\mu_1 = c(1,0,0)'$. Therefore, the multivariate Shewhart chart (7.5)–(7.6) is not quite effective in this example.

Next, we apply the joint monitoring scheme (7.16)–(7.17) to the same dataset. When defining the charting statistics in (7.16), we use $k_1 = k_2 = k_3 = 0.5$. In the decision rule (7.17), we use $h_1 = h_2 = h_3 = 3.892$. So, by Table 4.1 in Subsection 4.2.2, each individual two-sided CUSUM chart in the joint monitoring scheme has an IC ARL value of 150. Consequently, by (7.18), the ARL_0 value of the joint monitoring scheme (7.16)–(7.17) is approximately $150/3 = 50$, which matches the ARL_0 value of the multivariate Shewhart chart (7.5)–(7.6) described above. The observed data and the three individual two-sided CUSUM charts are shown in Figure 7.5, from which we can see that the joint monitoring scheme gives the first signal of process mean shift at the fourth time point, which is much sooner than the first signal given by the multivariate Shewhart chart (7.5)–(7.6).

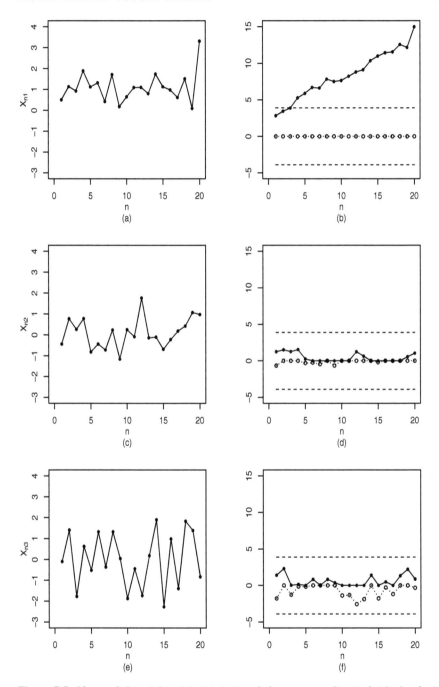

Figure 7.5 *Observed data (plots (a), (c), (e)) and the corresponding individual two-sided CUSUM charts (plots (b), (d), (f)). In each individual CUSUM chart, the upward CUSUM is shown by the dark dots connected by the solid lines, the downward CUSUM is shown by the small circles connected by the dotted lines, and the control limits are shown by the horizontal dashed lines.*

Table 7.2 *This table presents 20 phase II observations from a 3-dimensional production pro-*
cess and the calculated values of the multivariate Shewhart charting statistic $T_{0,n}^2$.

i	X_1	X_2	X_3	$T_{0,n}^2$	i	X_1	X_2	X_3	$T_{0,n}^2$
1	0.498	−0.438	−0.102	0.45	11	1.090	−0.091	−0.447	1.396
2	1.132	0.764	1.403	3.833	12	1.096	1.757	−1.739	7.313
3	0.921	0.262	−1.777	4.074	13	0.798	−0.138	0.179	0.688
4	1.887	0.773	0.623	4.546	14	1.740	−0.111	1.897	6.640
5	1.117	−0.814	−0.522	2.184	15	1.123	−0.690	−2.272	6.900
6	1.319	−0.438	1.322	3.679	16	0.971	−0.222	0.980	1.953
7	0.418	−0.720	−0.363	0.826	17	0.611	0.183	−1.399	2.364
8	1.715	0.231	1.319	4.733	18	1.511	0.417	1.825	5.787
9	0.175	−1.158	0.044	1.373	19	0.086	1.065	1.381	3.050
10	0.640	0.247	−1.879	4.000	20	3.31	0.970	−0.839	12.603

Example 7.2 shows that the joint monitoring scheme (7.16)–(7.17) is much more effective in detecting persistent process mean shifts than the multivariate Shewhart chart (7.5)–(7.6) in that example. Woodall and Ncube (1985) demonstrated that this conclusion was true in general. Also, after the multivariate Shewhart chart (7.5)–(7.6) gives a signal of process mean shift, it is still unknown which components of the mean vector of $\mathbf{X}$ have shifted. With the joint monitoring scheme (7.16)–(7.17), this information is often clear to us. For instance, in the example of Figure 7.5, the upward CUSUM of the first individual CUSUM chart gives the first signal at the fourth time point, and the remaining two individual CUSUM charts do not give any signals before that time point. Based on these individual CUSUM charts, we can conclude that the first component of $\mathbf{X}$ has an upward mean shift on or before the fourth time point, and the remaining two components of $\mathbf{X}$ do not seem shifted by that time point.

In cases when the components of $\mathbf{X}$ are dependent (i.e., Σ_0 is not a diagonal matrix), the result in (7.18) would not be true any more. In such cases, the design of the joint monitoring scheme (7.16)–(7.17) is quite complicated because it has p allowance constants and p control limits to determine. One way to overcome this difficulty is to use the principal components approach, first discussed in the MSPC literature for constructing multivariate Shewhart charts by Jackson (1959), Jackson and Mudholkar (1979), and Jackson (1980). Let Σ_0 have the following matrix decomposition

$$\Sigma_0 = Q'\Lambda Q,$$

where Q is a $p \times p$ orthogonal matrix (i.e., Q has the properties that $QQ' = Q'Q = I_{p \times p}$), and $\Lambda = diag(\lambda_1, \lambda_2, \ldots, \lambda_p)$ is a diagonal matrix. Without loss of generality, let us assume that $\lambda_j > 0$, for all j. Otherwise, some components of $\mathbf{X}$ are redundant (i.e., they can be expressed as linear combinations of the remaining components), and they can be removed from the process monitoring. Then, $\mathbf{X}^* = \Lambda^{-1/2}Q\mathbf{X}$ would have the identity covariance matrix, and thus its p components are independent of each other. In the literature, the components of $Q\mathbf{X}$ are called the principal components of the original random vector $\mathbf{X}$ (cf., Johnson and Wichern, 2007, Chapter 8), and λ_j is just the IC variance of the j-th principal component, for $j = 1, 2, \ldots, p$. Intuitively, $\mathbf{X}^*$ is obtained by first rotating the coordinate system of the data, so that the p components of the rotated data (i.e., $Q\mathbf{X}$) are independent, and then rescaling the rotated data to have the same unit variability on all p axes. Then, instead of monitoring the original data, we can monitor the transformed data

$$\mathbf{X}_n^* = \Lambda^{-1/2} Q \mathbf{X}_n, \qquad \text{for } n \geq 1.$$

It is obvious that the transformed data have a mean shift at a specific time point if and only if the original data have a mean shift at the same time point.

The joint monitoring scheme described above is easy to understand. But, it may not be convenient to use in practice, especially when the dimensionality p is large. In cases when the components of the original quality characteristic vector $\mathbf{X}$ are correlated and its transformed version $\mathbf{X}^*$ is used, its advantage of easy interpretation of the shift signal would disappear. Can we use a single control chart for monitoring a multivariate process? In the literature, there have been many discussions in that direction. Some representative approaches are described below.

Healy (1987) suggested a multivariate CUSUM chart using the general formula (4.20) of the decision interval form of the CUSUM chart (cf., Subsection 4.2.4). Assume that the IC distribution of a p-dimensional production process is $N_p(\mu_0, \Sigma_0)$, and it changes to the OC distribution $N_p(\mu_1, \Sigma_0)$ with $\mu_1 \neq \mu_0$ after an unknown time point. Then, by (4.20), the charting statistic of the CUSUM chart is

$$C_n = \max \left[0, C_{n-1} + \log \left(f_1(\mathbf{X}_n) / f_0(\mathbf{X}_n) \right) \right], \qquad \text{for } n \geq 1,$$

where $C_0 = 0$, and f_0 and f_1 are the pdfs of the IC and OC process distributions. By some routine algebraic manipulations, this statistic is

$$C_n = \max \left[0, C_{n-1} + \mathbf{a}'(\mathbf{X}_n - \mu_0) - 0.5D \right], \tag{7.19}$$

where

$$\mathbf{a}' = \frac{(\mu_1 - \mu_0)' \Sigma_0^{-1}}{D}, \qquad D = \sqrt{(\mu_1 - \mu_0)' \Sigma_0^{-1} (\mu_1 - \mu_0)}.$$

The chart gives a signal of process mean shift when

$$C_n > h \tag{7.20}$$

where $h > 0$ is a control limit. To interpret the meaning of $\mathbf{a}$ and D, let us consider the special case when $\Sigma_0 = I_{p \times p}$. In such cases, it is clear that D is the Euclidean length of $\delta = \mu_1 - \mu_0$, δ is the vector of shift sizes in different process components, and $\mathbf{a}$ is the shift direction. To use the CUSUM chart (7.19)–(7.20), $\mathbf{a}$ should be known. Therefore, this CUSUM chart is actually a univariate control chart, and it aims to detect shifts in the mean of $\mathbf{a}'\mathbf{X}_n$ which is univariate. For this reason, most discussions in Chapter 4 about univariate CUSUM charts can apply to this control chart.

A closely related control chart was proposed by Hawkins (1991), described below. Hawkins (1991) noticed that, in practice, mean shift of a multivariate production process often occurred in a small number of components, although the indices of the shifted components were often unknown. For this reason, instead of monitoring a specific linear combination of $\mathbf{X}$, as done by the CUSUM chart (7.19)–(7.20) described above, we can monitor all individual components of $\mathbf{X}$. Further, when monitoring the individual components of $\mathbf{X}$, their possible association should be taken into account. To this end, Hawkins (1991) considered the following regression relationship between the j-th component of $\mathbf{X}$ and its remaining components:

$$X_j - \mu_{0j} = \sum_{\ell \neq j} \beta_{\ell j}(X_\ell - \mu_{0\ell}) + \varepsilon_j,$$

where $\mathbf{X} = (X_1, X_2, \ldots, X_p)'$, $\mu_0 = (\mu_{01}, \mu_{02}, \ldots, \mu_{0p})'$, $\{\beta_{\ell j}\}$ were regression coefficients, and ε_j was the random error with the distribution $N(0, \tau_{jj})$. By the properties of multivariate normal distributions, if $\mathbf{X} \sim N_p(\mu_0, \Sigma_0)$, then the above regression model is always true, and there are the following relationships:

$$\tau_{jj} = \left(\sigma_0^{jj}\right)^{-1}, \qquad \beta_{\ell j} = -\sigma_0^{\ell j} \tau_{jj}, \text{ for } \ell \neq j,$$

and

$$\tau_{jj} = \sigma_{0,jj} - \sum_{\ell \neq j} \beta_{\ell j} \sigma_{0,\ell j},$$

where $\sigma_{0,\ell j}$ and $\sigma_0^{\ell j}$ are the (ℓ, j)-th elements of Σ_0 and Σ_0^{-1}, respectively. Then, to monitor the j-th component of $\mathbf{X}$, we can use the statistic

$$Z_j = \left[(X_j - \mu_{0j}) - \sum_{\ell \neq j} \beta_{\ell j}(X_\ell - \mu_{0\ell}) \right] \Big/ \tau_{jj}^{1/2}. \qquad (7.21)$$

When the process is IC, then Z_j has the distribution of $N(0,1)$. The statistic Z_j can be regarded as the standardized residual of the regression of $X_j - \mu_{0j}$ on $\{X_\ell - \mu_{0\ell}, \ell \neq j\}$. It has adjusted the variability in $X_j - \mu_{0j}$ due to the possible association between $X_j - \mu_{0j}$ and $\{X_\ell - \mu_{0\ell}, \ell \neq j\}$, and therefore it is expected to be more sensitive to shifts in the mean of X_j, compared to the original variable X_j. Let $\mathbf{Z} = (Z_1, Z_2, \ldots, Z_p)'$. Then, $\mathbf{Z}$ can be computed by

$$\mathbf{Z} = \left[diag(\Sigma_0^{-1}) \right]^{-1/2} \Sigma_0^{-1}(\mathbf{X} - \mu_0), \qquad (7.22)$$

where $diag(\Sigma_0^{-1}) = diag(\sigma_0^{11}, \sigma_0^{22}, \cdots, \sigma_0^{pp})$ is a $p \times p$ diagonal matrix.

It can be checked that Z_j in (7.21) is exactly the same as $\mathbf{a}'\mathbf{X}$ considered in Healy's CUSUM chart (7.19)–(7.20) in cases when $\delta = \mu_1 - \mu_0$ has a non-zero value at the j-th position and 0 elsewhere (i.e., the mean shift is in the j component only). Therefore, CUSUM charts based on Z_j would be optimal for detecting mean shifts in X_j. To monitor the p-dimensional production process, we can use a set of univariate CUSUM charts based on $\{Z_j, j = 1, 2, \ldots, p\}$. To this end, let

$$C_{n,j}^+ = \max\left(0, C_{n-1,j}^+ + Z_{nj} - k\right), \qquad \text{for } n \geq 1$$

$$C_{n,j}^- = \min\left(0, C_{n-1,j}^- + Z_{nj} + k\right), \tag{7.23}$$

where $\mathbf{Z}_n = (Z_{n1}, Z_{n2}, \ldots, Z_{np})'$ is the vector of $\mathbf{Z}$ determined by (7.22) from the n-th observation vector $\mathbf{X}_n$, and $k > 0$ is an allowance constant. Then, the j-th individual CUSUM chart gives a signal of process mean shift in the j-th component of $\mathbf{X}$ if

$$\left(C_{n,j}^+ > h \qquad \text{or} \qquad C_{n,j}^- < -h\right), \tag{7.24}$$

and the joint monitoring scheme gives a signal of process mean shift in $\mathbf{X}$ at the n-th time point if

$$\bigcup_{j=1}^{p} \left(C_{n,j}^+ > h \text{ or } C_{n,j}^- < -h\right),$$

where $h > 0$ is a control limit. Obviously, the above expression is equivalent to

$$C_n = \max_{j=1}^{p} \left[\max\left(C_{n,j}^+, -C_{n,j}^-\right)\right] > h. \tag{7.25}$$

Because Z_{nj} has the IC distribution of $N(0, 1)$, for each j, k and h can be determined to achieve a given ARL_0 value for the individual control charts in the same way as discussed in Chapter 4 about univariate CUSUM charts. For instance, when (k, h) are chosen to be $(0.5, 4.171)$, each individual CUSUM chart (7.24) would have an ARL_0 value of 200 (cf., Table 4.1 in Subsection 4.2.2), and they are good for detecting mean shifts of sizes around ± 1.0 in each component. However, the ARL_0 value of the joint monitoring scheme (7.25) is difficult to determine analytically from the ARL_0 values of the individual CUSUM charts because $\{Z_{nj}, j = 1, 2, \ldots, p\}$ might be associated. One way to determine the ARL_0 value of the joint monitoring scheme is to use a Monte Carlo simulation, as discussed in Subsection 4.2.2.

An alternative approach to construct a multivariate CUSUM chart based on $\mathbf{Z}_n$ is to use the charting statistic

$$\tilde{C}_n = \sum_{j=1}^{p} \left(C_{n,j}^+ + C_{n,j}^-\right)^2, \tag{7.26}$$

and the chart gives a signal of process mean shift if

$$\tilde{C}_n > \tilde{h} \tag{7.27}$$

where $\tilde{h} > 0$ is a control limit chosen to achieve a given ARL_0 value. Again, the value of $\tilde{h}$ can be determined by a Monte Carlo simulation. Between the multivariate CUSUM chart (7.26)–(7.27) and the joint monitoring scheme (7.25), Hawkins (1991) showed by a numerical study that the former was slightly more effective than the latter for detecting mean shifts in $\mathbf{X}$.

Example 7.3 *Consider a 3-dimensional production process with the IC distribution* $N_3(\mu_0, \Sigma_0)$, *where*

$$\mu_0 = (0,0,0)'$$

and

$$\Sigma_0 = \begin{pmatrix} 1.0 & 0.8 & 0.5 \\ 0.8 & 1.0 & 0.8 \\ 0.5 & 0.8 & 1.0 \end{pmatrix}.$$

The first 30 phase II observations are presented in columns 2–4 of Table 7.3. The three components of $\mathbf{X}_n$ *are also shown in Figure 7.6(a)–(c). To monitor the process, we first compute the standardized residuals* $\mathbf{Z}_n$ *by (7.22), and the three components of* $\mathbf{Z}_n$ *are shown in Figure 7.6(d)–(f). Then, we use both control charts (7.25) and (7.26)–(7.27) for process monitoring. In both charts, k and* ARL_0 *are chosen to be 0.5 and 200, respectively, and the control limits h and* $\widetilde{h}$ *are computed by the R codes written by the author to be 5.014 and 42.031. With these parameters, the three individual CUSUM charts defined by (7.23)–(7.24) are shown in Figure 7.6(g)–(i). From the plots, we can see that the first individual CUSUM chart gives the first signal at the 14th time point, and the second and third CUSUM charts both give signals at later times. Therefore, the joint monitoring scheme (7.25) shown in Figure 7.6(j) gives the first signal at the 14th time point as well. The multivariate CUSUM chart (7.26)–(7.27) is shown in Figure 7.6(k), which gives the first signal at the 14th time point too. The production process actually has a mean shift starting from the 11th time point from* μ_0 *to* $\mu_1 = (1,0,0)'$. *From the plots in Figure 7.6, it seems that (i) the mean shift in* X_1 *is more obviously revealed in plot (d), compared to plot (a), and (ii) the mean shift in* X_1 *affects both* Z_1 *and the remaining components of* $\mathbf{Z}$, *although its impact on* Z_1 *is much larger than its impact on the remaining components of* $\mathbf{Z}$ *(cf., plots (g)–(i)). As a matter of fact, the expression (7.22) confirms the second conclusion that the mean shift in* X_1 *would affect all components of* $\mathbf{Z}$ *when the components of* $\mathbf{X}$ *are correlated (i.e., when* Σ_0 *is not a diagonal matrix).*

Besides the above multivariate CUSUM charts, another multivariate CUSUM chart that receives much attention in the literature was proposed by Crosier (1988). Its charting statistic C_n, for $n \geq 1$, is defined as follows. Let

$$\mathbf{U}_n = \begin{cases} \mathbf{0}, & \text{if } Y_n \leq k \\ (\mathbf{U}_{n-1} + \mathbf{X}_n - \mu_0)(1 - k/Y_n), & \text{otherwise}, \end{cases} \quad (7.28)$$

where $\mathbf{U}_0 = \mathbf{0}$, $k > 0$ is an allowance constant, and

$$Y_n = \left[(\mathbf{U}_{n-1} + \mathbf{X}_n - \mu_0)' \Sigma_0^{-1} (\mathbf{U}_{n-1} + \mathbf{X}_n - \mu_0) \right]^{1/2}.$$

Then, the chart gives a signal of process mean shift at the n-th time point when

$$C_n = \left(\mathbf{U}_n' \Sigma_0^{-1} \mathbf{U}_n \right)^{1/2} > h, \quad (7.29)$$

where $h > 0$ is a control limit chosen to reach a pre-specified ARL_0 value. The statistic $\mathbf{U}_n$ is defined using the restarting mechanism of CUSUM charts (cf., a related discussion in Subsection 4.2.1). When Y_n is less than or equal to k, $\mathbf{U}_n$ is set to be $\mathbf{0}$, because

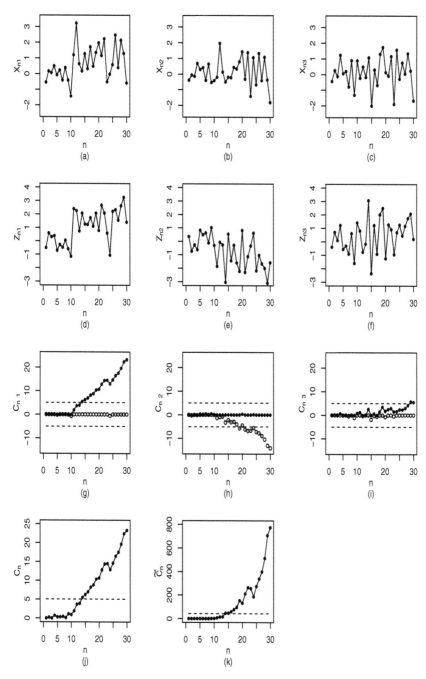

Figure 7.6 *Plots (a)–(c) show the three components of the original observed data* $\mathbf{X}_n$, *plots (d)–(f) show the three components of the standardized residuals* $\mathbf{Z}_n$, *plots (g)–(i) show the three individual CUSUM charts defined by (7.23)–(7.24), and plots (j) and (k) show the multivariate CUSUM charts (7.25) and (7.26)–(7.27). In plots (g)–(k), the horizontal dashed lines denote the control limits. In control charts (7.25) and (7.26)–(7.27), k and* ARL_0 *are chosen to be 0.5 and 200.*

Table 7.3 *This table presents the first 30 phase II observations of a 3-dimensional production process and the calculated values of the charting statistics C_n and $\tilde{C}_n$ defined in (7.25) and (7.26), respectively.*

n	X_1	X_2	X_3	C_n	$\tilde{C}_n$
1	−0.533	−0.385	−0.443	0.003	0.000
2	0.167	−0.052	0.263	0.235	0.103
3	0.068	−0.138	−0.133	0.006	0.000
4	0.492	0.680	1.232	0.709	0.521
5	−0.071	0.306	0.053	0.323	0.151
6	0.229	0.413	0.199	0.305	0.093
7	−0.400	−0.396	−0.789	0.427	0.352
8	0.387	0.640	0.898	0.110	0.012
9	−0.404	−0.529	−1.318	1.105	1.496
10	−1.430	−0.408	0.880	0.907	1.373
11	1.206	−0.194	−0.241	1.902	6.901
12	3.231	1.959	0.492	3.655	14.281
13	0.637	0.140	−0.184	3.897	15.654
14	0.177	−0.497	1.074	5.470	46.898
15	1.270	−0.185	−2.002	6.230	46.872
16	0.322	−0.219	0.302	6.963	58.578
17	1.710	0.434	−0.688	8.201	76.067
18	0.481	0.342	1.306	8.802	95.855
19	1.356	0.798	1.737	10.374	152.797
20	1.972	1.432	0.133	10.657	131.571
21	1.164	−0.310	−0.098	12.817	209.360
22	2.228	1.375	1.160	14.399	262.789
23	−0.511	−1.410	−1.890	14.480	255.437
24	−0.016	1.060	1.565	12.920	185.391
25	0.583	−0.675	−0.287	14.629	272.393
26	2.467	1.322	0.754	16.464	337.168
27	0.399	−0.404	0.037	17.524	396.194
28	2.137	1.089	1.339	19.643	512.923
29	1.304	−0.364	0.234	22.398	705.441
30	−0.575	−1.801	−1.668	23.305	772.824

there is little evidence of process mean shift in such cases. Otherwise, $\mathbf{U}_n$ is set to be $(\mathbf{U}_{n-1} + \mathbf{X}_n - \mu_0)(1 - k/Y_n)$, which shrinks the cumulative sum $\mathbf{U}_{n-1} + \mathbf{X}_n - \mu_0$ by $1 - k/Y_n$ times.

Crosier (1988) studied another multivariate CUSUM chart with the charting statistic

$$\tilde{C}_n = \max\left(0, \tilde{C}_{n-1} + T_n - k\right), \qquad (7.30)$$

where $\tilde{C}_0 \geq 0$ is a pre-specified initial value, $k > 0$ is an allowance constant, and

$$T_n = \left[(\mathbf{X}_n - \mu_0)' \Sigma_0^{-1} (\mathbf{X}_n - \mu_0)\right]^{1/2}.$$

Clearly, T_n is the square-root of the Hotelling's T^2 statistic, and $\tilde{C}_n$ is the conventional

CUSUM charting statistic based on T_n. For this reason, the resulting CUSUM chart is called the CUSUM of T_n, or simply the *COT chart*. This chart gives a signal when

$$\widetilde{C}_n > \widetilde{h}, \tag{7.31}$$

where $\widetilde{h} > 0$ is a control limit chosen to achieve a given ARL_0 value. Based on a numerical study, Crosier (1988) demonstrated that the multivariate CUSUM chart (7.28)–(7.29) was often more effective than the COT chart (7.30)–(7.31).

7.3.2   MCUSUM charts for monitoring the process covariance matrix

All the multivariate CUSUM charts described in the previous subsection can be modified properly for monitoring the process covariance matrix. For instance, in the multivariate CUSUM chart (7.16)–(7.17) by Woodall and Ncube (1985), if its charting statistics $C_{n,j}^+$ and $C_{n,j}^-$ are replaced by the charting statistics for detecting variance shifts in the j-th component of $\mathbf{X}$ (e.g., C_n^+ and C_n^- defined in (4.21) and (4.23)), then the resulting CUSUM chart should be effective for detecting shifts in the process covariance matrix. For a related discussion about variance monitoring by CUSUM charts in univariate cases, see Section 4.3.

Healy (1987) considered a case when the process mean vector would not shift but the process covariance matrix could shift from Σ_0 to $\Sigma_1 = c\Sigma_0$ at an unknown time point, where $c > 0$ was a constant. In such cases, if the p-dimensional process becomes OC, then the variances of the p components of $\mathbf{X}$ change by a same proportion and the correlation between any two individual components does not change. By the general formula (4.20) of the CUSUM charting statistic, Healy (1987) derived the following charting statistic for monitoring the process covariance matrix:

$$C_n = \max\left[0, C_{n-1} + T_n^2 - k\right], \qquad \text{for } n \geq 1, \tag{7.32}$$

where $C_0 \geq 0$ is a pre-specified initial value, $T_n^2 = (\mathbf{X}_n - \mu_0)' \Sigma_0^{-1} (\mathbf{X}_n - \mu_0)$ is the Hotelling's T^2 statistic, and

$$k = \frac{pc\log(c)}{c-1}.$$

This chart gives a signal at the n-th time if

$$C_n > h, \tag{7.33}$$

where $h > 0$ is a control limit chosen to achieve a given ARL_0 value. By the optimality of the CUSUM chart (cf., Subsection 4.2.4), the chart (7.32)–(7.33) should be optimal for detecting the process covariance matrix shift from Σ_0 to $\Sigma_1 = c\Sigma_0$.

By comparing the CUSUM chart (7.32)–(7.33) with the COT chart (7.30)–(7.31), we can see that their only difference is that the Hotelling's T^2 statistic T_n^2 is used in the construction of the former chart while its square-root is used in the construction of the latter chart. Remember that the COT chart is proposed mainly for detecting process mean shifts, while the CUSUM chart (7.32)–(7.33) is derived for detecting process covariance matrix shifts. Therefore, both of them are actually effective for detecting both process mean shifts and process covariance matrix shifts, which is demonstrated in the example below.

Example 7.4 *Assume that a 3-dimensional production process has the IC distribution $N_3(\mu_0, \Sigma_0)$, where*

$$\mu_0 = (0,0,0)'$$

and

$$\Sigma_0 = \begin{pmatrix} 1.0 & 0.8 & 0.5 \\ 0.8 & 1.0 & 0.8 \\ 0.5 & 0.8 & 1.0 \end{pmatrix}.$$

Let us consider the following three scenarios:

(i) *The process has a mean shift only at the 11th time point from μ_0 to $\mu_1 = (1,0,0)'$;*

(ii) *The process has a covariance matrix shift only at the 11th time point from Σ_0 to $\Sigma_1 = 1.1\Sigma_0$; and*

(iii) *The process has a mean shift from μ_0 to $\mu_1 = (1,0,0)'$ and a covariance matrix shift from Σ_0 to $\Sigma_1 = 1.1\Sigma_0$ at the 11th time point.*

In each scenario, 30 observations of the process are generated from the related distribution. In scenario (i), the values of the multivariate Shewhart charting statistic

$$T_{0,n}^2 = (\mathbf{X}_n - \mu_0)'\Sigma_0^{-1}(\mathbf{X}_n - \mu_0), \qquad for\ n = 1,2,\ldots,30$$

are shown in Figure 7.7(a), together with the control limit $\chi_{0.995,3}^2 = 12.838$. This chart has the ARL_0 value of $1/(1-0.995) = 200$. Then, we apply the multivariate CUSUM chart (7.32)–(7.33) and the COT chart (7.30)–(7.31) to the same dataset. In the chart (7.32)–(7.33), k and ARL_0 are chosen to be 4 and 200, respectively, and C_0 is chosen to be 0. By a simulation using an R code written by the author, the control limit h is computed to be 13.062. In the COT chart, k and ARL_0 are chosen to be 1 and 200, respectively, $\tilde{C}_0$ is chosen to be 0, and the control limit $\tilde{h}$ is computed to be 2.713. These two charts are shown in Figure 7.7(b)–(c). From the plots, we can see that the Shewhart chart could not detect the mean shift in this case, the chart (7.32)–(7.33) gives the first signal at the 14th time point, and the COT chart gives the first signal at the 11th time point. The corresponding results for scenarios (ii) and (iii) are shown in the plots in the 2nd and 3rd rows of Figure 7.7. In scenario (ii), the Shewhart chart could not detect the covariance matrix shift, the chart (7.32)–(7.33) gives the first and the only signal at the 24th time point, and the COT chart gives the first signal at the 12th time point. In scenario (iii), the Shewhart chart gives a marginally significant signal at the 12th time point, the chart (7.32)–(7.33) gives the first signal at the 14th time point, and the COT chart gives the first signal at the 11th time point. From this example, it can be seen that the Shewhart chart is not sensitive to either the process mean shift or the process covariance matrix shift in cases when such shifts are small, the multivariate CUSUM chart (7.32)–(7.33) and the COT chart (7.30)–(7.31) are sensitive to both the process mean shift and the process covariance matrix shift, and it seems that the COT chart (7.30)–(7.31) is more effective than the chart (7.32)–(7.33) for detecting either the process mean shift or the process covariance matrix shift or both.

Hawkins (1991) mentioned that the regression-adjusted standardized residuals $\mathbf{Z}$ defined in (7.22) could also be used for monitoring process variance shifts. More specifically, to monitor the variance shift in the j-th component of $\mathbf{X}$, for $1 \leq j \leq p$, we can consider

$$W_{nj} = \left(|Z_{nj}|^{1/2} - 0.822\right)\bigg/0.349, \qquad \text{for } n \geq 1, \qquad (7.34)$$

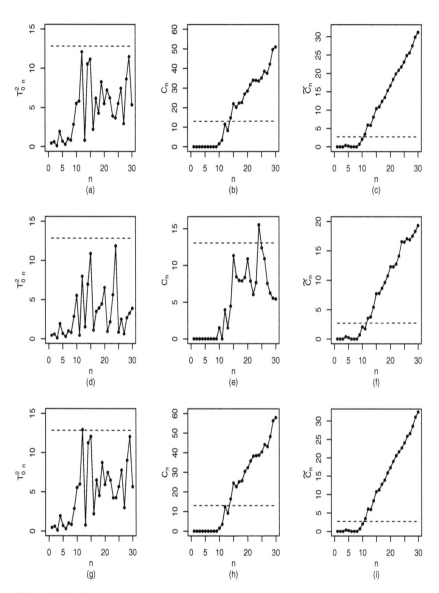

Figure 7.7 *Plots (a)–(c) show the Shewhart chart, the multivariate CUSUM chart (7.32)–(7.33) with k = 4, and the COT chart (7.30)–(7.31) with k = 1, respectively, when they are applied to a dataset of the first 30 observations of a 3-dimensional production process in scenario (i). In each chart, the ARL₀ value is chosen to be 200, and the corresponding control limit is denoted by the horizontal dashed line in the related plot. Results in scenarios (ii) and (iii) are shown in the corresponding plots in the 2nd and 3rd rows, respectively.*

where $\mathbf{Z}_n = (Z_{n1}, Z_{n2}, \ldots, Z_{np})'$ is the n-th regression-adjusted standardized residual vector. It can be checked that the IC distribution of W_{nj} is approximately $N(0,1)$, and its mean would shift upward (downward) when the variance of X_j shifts upward (downward). Define

$$C_{n,j}^+ = \max\left(0, C_{n-1,j}^+ + W_{nj} - k\right), \qquad \text{for } n \geq 1$$

$$C_{n,j}^- = \min\left(0, C_{n-1,j}^- + W_{nj} + k\right), \tag{7.35}$$

where $C_{0,j}^+ = C_{0,j}^- = 0$, and $k > 0$ is an allowance constant. Then, $C_{n,j}^+$ can be used for detecting an upward variance shift in the j-th component of $\mathbf{X}$, and $C_{n,j}^-$ for detecting a downward variance shift in the j-th component. To detect an arbitrary variance shift in any components of $\mathbf{X}$, a joint monitoring scheme can be used, and it gives a signal at the n-th time point if

$$C_n = \max_{j=1}^{p}\left[\max\left(C_{n,j}^+, -C_{n,j}^-\right)\right] > h, \tag{7.36}$$

where $h > 0$ is a control limit chosen to achieve a pre-specified ARL_0 value. An alternative charting statistic for joint monitoring is

$$\tilde{C}_n = \sum_{j=1}^{p}\left(C_{n,j}^+ + C_{n,j}^-\right)^2, \tag{7.37}$$

and the chart gives a signal of process variance shift if

$$\tilde{C}_n > \tilde{h}, \tag{7.38}$$

where $\tilde{h} > 0$ is a control limit chosen to achieve a pre-specified ARL_0 value.

7.4 Multivariate EWMA Charts

In this section, we describe some *multivariate EWMA (MEWMA) charts* for monitoring a p-dimensional quality characteristic vector $\mathbf{X}$ of a production process, where $p \geq 2$ is a given integer. Although MEWMA charts can also be used for phase I SPC analysis, they are mainly used for phase II process monitoring, and the latter is the focus of this section. Our description is divided into two subsections. Those for monitoring the process mean are discussed in Subsection 7.4.1, and those for monitoring the process covariance matrix are discussed in Subsection 7.4.2.

7.4.1   MEWMA charts for monitoring the process mean

The univariate EWMA chart was first generalized to multivariate cases by Lowry et al. (1992) for detecting process mean shifts. Let $\mathbf{X}_1, \mathbf{X}_2, \ldots$ be a sequence of phase II observation vectors obtained from a p-dimensional production process with the IC distribution $N_p(\boldsymbol{\mu}_0, \Sigma_0)$, where $\boldsymbol{\mu}_0$ and Σ_0 are assumed known. Then, a natural generalization of the univariate EWMA charting statistic defined in (5.1) is

$$\mathbf{E}_n = \Lambda(\mathbf{X}_n - \boldsymbol{\mu}_0) + (I_{p \times p} - \Lambda)\mathbf{E}_{n-1}, \tag{7.39}$$

where $\mathbf{E}_0 = \mathbf{0}$, $\Lambda = diag(\lambda_1, \lambda_2, \ldots, \lambda_p)$, and $\lambda_j \in (0, 1]$ is a weighting parameter for the j-th component of $\mathbf{X}$, for $j = 1, 2, \ldots, p$. The corresponding MEWMA chart with the charting statistic $\mathbf{E}_n$ in (7.39) gives a signal of process mean shift at the n-th time point if

$$V_n^2 = \mathbf{E}_n' \Sigma_{\mathbf{E}_n}^{-1} \mathbf{E}_n > h, \qquad (7.40)$$

where $h > 0$ is a control limit chosen to reach a pre-specified ARL_0 value, and $\Sigma_{\mathbf{E}_n}$ is the covariance matrix of $\mathbf{E}_n$.

In practice, if there is no *a priori* reason to weight different components differently, then we can simply choose $\lambda_1 = \lambda_2 = \cdots = \lambda_p = \lambda$. In such cases,

$$\mathbf{E}_n = \lambda\,(\mathbf{X}_n - \mu_0) + (1 - \lambda)\mathbf{E}_{n-1},$$

and by some routine algebraic manipulations, it can be checked from the above expression that

$$\mathbf{E}_n = \lambda \sum_{i=1}^{n} (1 - \lambda)^{n-i} (\mathbf{X}_i - \mu_0).$$

Therefore, when the process is IC up to the time point n, we have

$$
\begin{aligned}
\Sigma_{\mathbf{E}_n} &= \sum_{i=1}^{n} \lambda^2 (1 - \lambda)^{2(n-i)} \Sigma_0 \\
&= \frac{\lambda}{2 - \lambda} \left[1 - (1 - \lambda)^{2n} \right] \Sigma_0, \qquad (7.41)
\end{aligned}
$$

and

$$\mathbf{E}_n \sim N_p\,(\mathbf{0}, \Sigma_{\mathbf{E}_n}).$$

It is obvious that, when n increases, $\Sigma_{\mathbf{E}_n}$ in (7.41) converges to

$$\widetilde{\Sigma}_{0,\lambda} = \frac{\lambda}{2 - \lambda} \Sigma_0. \qquad (7.42)$$

Also, the IC distribution of V_n^2 in (7.40) is χ_p^2. However, because the variables in the sequence $\{V_n^2, n = 1, 2, \ldots\}$ are correlated, the control limit h in (7.40) cannot simply be chosen to be the $(1 - \alpha)$-th quantile $\chi_{1-\alpha,p}^2$ of the χ_p^2 distribution, as we did in (7.6) for the Shewhart charts. Instead, it can be determined by a Monte Carlo simulation (cf., the pseudo code given in Subsection 4.2.2). In cases when the process mean shifts at the time point $\tau \le n$ from μ_0 to μ_1, we have

$$\mathbf{E}_n \sim N_p\left(\lambda \sum_{i=\tau}^{n} (1 - \lambda)^{n-i} (\mu_1 - \mu_0), \Sigma_{\mathbf{E}_n} \right).$$

In such cases, the distribution of V_n^2 is $\chi_p^2(c_n)$ with the noncentrality parameter

$$c_n = \frac{\lambda(2 - \lambda) \left[\sum_{i=\tau}^{n} (1 - \lambda)^{n-i} \right]^2}{1 - (1 - \lambda)^{2n}} (\mu_1 - \mu_0)' \Sigma_0^{-1} (\mu_1 - \mu_0).$$

Table 7.4 ARL_1 values of the chart (7.39)–(7.40) for detecting several mean shifts of various sizes when $p = 3$, $ARL_0 = 200$, and $\lambda = 0.1, 0.2, 0.3, 0.4$, and 0.5.

δ'	$\sqrt{\delta'\Sigma_0^{-1}\delta}$	$\lambda = 0.1$	$\lambda = 0.2$	$\lambda = 0.3$	$\lambda = 0.4$	$\lambda = 0.5$
(0,0,0)	0	200.047	199.967	199.939	199.981	199.914
(0.1614,0,0)	0.292	72.326	90.606	108.387	121.276	134.209
(0.25,0.25,0.25)	0.292	72.408	90.889	108.491	121.850	134.791
(0.25,0.25,0)	0.420	42.199	54.444	68.738	81.786	94.375
(0.25,0,0)	0.452	37.581	48.439	61.813	75.128	87.174
(0.5,0.5,0.5)	0.584	24.768	30.422	39.434	48.901	59.062
(0.5,0.5,0)	0.839	14.418	15.604	19.058	23.430	28.804
(1,1,1)	1.168	9.176	8.807	9.603	11.099	13.227
(1.5,1.5,1.5)	1.752	5.599	4.862	4.758	4.906	5.288
(2,2,2)	2.336	4.102	3.416	3.160	3.070	3.093
(2,2,0)	3.358	2.872	2.324	2.099	1.942	1.827

Therefore, the OC run length distribution of the MEWMA chart (7.39)–(7.40) is affected by the shift size $\delta = \mu_1 - \mu_0$ through $[\delta'\Sigma_0^{-1}\delta]^{1/2}$ which can be interpreted as the statistical distance from δ to $\mathbf{0}$ (cf., Subsection 7.2.1), or the statistical length of δ.

From the above discussion, the IC behavior of the MEWMA chart (7.39)–(7.40) depends only on the parameters p, λ, and ARL_0, and its OC behavior depends on these parameters together with the statistical length of the shift size vector δ. In cases when $p = 3$, $ARL_0 = 200$, and $\lambda = 0.1, 0.2, 0.3, 0.4$, and 0.5, the searched h values of the chart by an R-code written by the author are, respectively,

$$10.820, 11.879, 12.320, 12.559, 12.696.$$

In such cases, the ARL_1 values of the chart for detecting several mean shifts of various different sizes of a 3-dimensional production process with the IC distribution $N_3(\mu_0, \Sigma_0)$ are presented in Table 7.4, where $\mu_0 = (0,0,0)'$ and

$$\Sigma_0 = \begin{pmatrix} 1.0 & 0.8 & 0.5 \\ 0.8 & 1.0 & 0.8 \\ 0.5 & 0.8 & 1.0 \end{pmatrix}.$$

From the table, we can see that (i) the ARL_1 value indeed depends on δ through $[\delta'\Sigma_0^{-1}\delta]^{1/2}$ (cf., lines 2 and 3 of the table in which the values of $[\delta'\Sigma_0^{-1}\delta]^{1/2}$ are about the same and the two sets of ARL_1 values are also about the same), (ii) for a given λ value, the ARL_1 value decreases when $[\delta'\Sigma_0^{-1}\delta]^{1/2}$ increases, and (iii) the chart with a large λ value is good for detecting relatively large shifts and it is good for detecting relatively small shifts when λ is chosen to be small.

Example 7.5 *For the phase II data considered in Example 7.3, let us consider using the MEWMA chart (7.39)–(7.40) for monitoring the process mean. In the chart, we choose $\lambda = 0.2$ and $h = 11.879$ so that its ARL_0 value is 200. The chart is shown in*

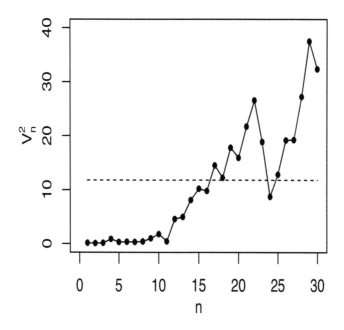

Figure 7.8 *The MEWMA chart (7.39)–(7.40) when it is applied to the data considered in Example 7.3 and when $\lambda = 0.2$ and $ARL_0 = 200$. The dashed horizontal line denotes the control limit.*

Figure 7.8, from which we can see that the first signal of process mean shift is given at the 17th time point.

Hawkins et al. (2007) discussed the cases when the weighting matrix Λ in (7.39) contained non-zero off-diagonal elements. More specifically, they discussed in detail the case when all diagonal elements of Λ were the same to be λ_{on} and all off-diagonal elements of Λ were the same to be λ_{off}. Further, they assumed that $\lambda_{off} = c\lambda_{on}$ with $|c| < 1$, and that the summation of the elements in each column of Λ was the same to be r. The former assumption implies that the j-th component of $\mathbf{X}_n$ has the largest impact on the j-th component of $\mathbf{E}_n$, for each j, and the latter assumption implies that the total weights assigned to each component of $\mathbf{X}_n$ are the same to be r. In such cases, it is easy to check that

$$\lambda_{on} = \frac{r}{1 + (p-1)c}, \qquad \lambda_{off} = \frac{cr}{1 + (p-1)c}.$$

Therefore, the weighting matrix Λ is uniquely determined by the two parameters c and r. Based on the preliminary study in Hawkins et al. (2007), the following two conclusions can be made:

(i) The performance of the MEWMA chart (7.39)–(7.40) can be improved by using the more general weighting matrix Λ described above, compared to its performance when a single weighting parameter λ is used, and

(ii) The ARL performance of the MEWMA chart (7.39)–(7.40) using the more general weighting matrix Λ described above depends on the direction of a mean shift and on the correlation structure of different components of $\mathbf{X}_n$ as well. As a comparison, its ARL performance does not depend on the shift direction, when a single weighting parameter λ is used.

However, the design of the MEWMA chart using the weighting matrix Λ would be much more complicated. For instance, it is still unknown how to choose the parameters c and r properly for detecting a specific process mean shift, partly because the shift direction and the correlation structure among the process components should be taken into account in the design, and the relationship between the ARL and these two factors is unclear yet. Therefore, much future research is required on this topic.

The MEWMA chart (7.39)–(7.40) is based on the assumption that the IC process distribution $N_p(\mu_0, \Sigma_0)$ is known. In reality, both the IC process mean vector μ_0 and the IC process covariance matrix Σ_0 are often unknown. In such cases, certain self-starting multivariate control charts have been proposed in the literature. See, for instance, Hawkins and Maboudou-Tchao (2007), Quesenberry (1997), and Sullivan and Jones (2002). Next, we briefly describe the self-starting MEWMA chart proposed by Sullivan and Jones (2002).

Let $\mathbf{X}_1, \mathbf{X}_2, \ldots$ be a sequence of phase II observation vectors obtained from a p-dimensional production process, and $\overline{\mathbf{X}}_n$ and $\mathbf{S}_n^2$ be the sample mean vector and sample covariance matrix of the first n observation vectors. Then, $\overline{\mathbf{X}}_n$ and $\mathbf{S}_n^2$ can be computed by the recursive formulas below.

$$\overline{\mathbf{X}}_n = \frac{1}{n}\left[(n-1)\overline{\mathbf{X}}_{n-1} + \mathbf{X}_n\right], \text{ for } n \geq 1,$$

$$\mathbf{S}_n^2 = \frac{n-2}{n-1}\mathbf{S}_{n-1}^2 + \frac{1}{n}\left(\mathbf{X}_n - \overline{\mathbf{X}}_{n-1}\right)\left(\mathbf{X}_n - \overline{\mathbf{X}}_{n-1}\right)', \text{ for } n \geq 2, \quad (7.43)$$

where $\overline{\mathbf{X}}_0 = \mathbf{0}$ and $\mathbf{S}_1^2 = \mathbf{0}_{p \times p}$. Further,

$$\mathbf{u}_n = \sqrt{\frac{n-1}{n}}\left(\mathbf{X}_n - \overline{\mathbf{X}}_{n-1}\right) \sim N_p\left(\mathbf{0}, \Sigma_0\right). \quad (7.44)$$

Based on these results, we can consider the following self-starting MEWMA charting statistic

$$\mathbf{E}_{n,SS} = \lambda \mathbf{u}_n + (1-\lambda)\mathbf{E}_{n-1,SS}, \quad (7.45)$$

where $\mathbf{E}_{0,SS} = \mathbf{0}$, and $\lambda \in (0,1]$ is a weighting parameter. The chart gives a signal at the n-th time point if

$$V_{n,SS}^2 = \mathbf{E}_{n,SS}'\widehat{\Sigma}_{\mathbf{E}_{n,SS}}^{-1}\mathbf{E}_{n,SS} > h_{SS}, \quad (7.46)$$

where

$$\widehat{\Sigma}_{\mathbf{E}_{n,SS}} = \frac{\lambda}{2-\lambda}\left[1 - (1-\lambda)^{2n}\right]\mathbf{S}_{n-1}^2$$

is an estimator of $\Sigma_{\mathbf{E}_{n,SS}}$, and $h_{SS} > 0$ is a control limit chosen to achieve a pre-specified ARL_0 value. Note that, in (7.46), we need to compute the inverse of $\widehat{\Sigma}_{\mathbf{E}_{n,SS}}$. When n is small, such an inverse may not exist, or it exists but the matrix $\widehat{\Sigma}_{\mathbf{E}_{n,SS}}$ is close to a singular matrix. To overcome this numerical difficulty, we suggest starting the process monitoring at $n = p+2$, and letting $V^2_{n,SS} = 0$ when $n < p+2$. Namely, to use the self-starting MEWMA chart (7.45)–(7.46), at least $p+1$ IC observation vectors need to be collected beforehand.

To design the self-starting MEWMA chart (7.45)–(7.46), we can determine the value of h_{SS} by a numerical simulation, pretending the IC process distribution is $N_p(\mathbf{0}, I_{p \times p})$. Although the control limit computed in this way is only approximately valid, due to the fact that $\mathbf{u}_n \sim N_p(\mathbf{0}, \Sigma_0)$ and Σ_0 is approximated by the sample covariance matrix $\mathbf{S}^2_n$ in the chart, it is quite reliable based on our numerical studies (cf., Example 7.6 below), especially when ARL_0 is chosen relatively large. Sullivan and Jones (2002) pointed out that the control limit h_{SS} depended on the value of p, besides the values of λ and ARL_0. To get rid of its dependence on p, they suggested the following alternative charting statistic:

$$\widetilde{V}^2_{n,SS} = \sqrt{G^{-1}\left\{F_{p,n-p-1}\left[\frac{n-p-1}{p(n-2)}V^2_{n,SS}\right]\right\}},$$

where G is the cdf of the χ^2_1 distribution, and $F_{p,n-p-1}$ is the cdf of the $F_{p,n-p-1}$ distribution. The IC distribution of $\widetilde{V}^2_{n,SS}$ is the same as the distribution of the square-root of a χ^2_1 distributed random variable, which does not depend on p. Thus, the control limit of $\widetilde{V}^2_{n,SS}$ does not depend on p either.

Example 7.6 *The first 30 observation vectors of a 3-dimensional production process are presented in columns 2–4 of Table 7.5. The three components of $\mathbf{u}_n$ defined by (7.44) are presented in columns 5–7 of the table. Then, the self-starting MEWMA chart (7.45)–(7.46) is applied to this data for online process mean monitoring. In the chart, λ and ARL_0 are chosen to be 0.2 and 200, respectively. By an R-code written by the author, the control limit h_{SS} is computed to be 14.167 and 14.168, respectively, in cases when the IC process distribution is assumed to be $N_p(\mathbf{0}, I_{p \times p})$ and when the true IC distribution $N_p(\mathbf{0}, \Sigma_0)$ is used, where Σ_0 is the same as the one in Example 7.4. The values of the charting statistic $V^2_{n,SS}$ are presented in the last column of Table 7.5. The self-starting MEWMA chart (7.45)–(7.46) is also shown in Figure 7.9. From the figure, it can be seen that the first signal of process mean shift is given at the 25th time point; but, the signal does not last very long. Therefore, we should react to the signal from the chart quickly. Otherwise, the shift could be missed. For a related discussion on this feature of the self-starting control charts, see Subsections 4.5.1 and 5.4.1.*

7.4.2 MEWMA charts for monitoring the process covariance matrix

In this subsection, we discuss some MEWMA charts for monitoring the process covariance matrix. From the discussion in Chapters 4 and 5, it can be seen that

Table 7.5 *This table presents the first 30 phase II observations of a 3-dimensional production process, the three components of $\mathbf{u}_n$ computed by (7.44), and the values of the charting statistic $V_{n,SS}^2$ of the self-starting MEWMA chart (7.45)–(7.46).*

n	X_1	X_2	X_3	u_1	u_2	u_3	$V_{n,SS}^2$
1	−0.224	−0.464	−0.662	0	0	0	0
2	−0.082	−0.203	0.682	0	0	0	0
3	−0.436	0.342	−0.174	−0.231	0.551	−0.150	0
4	0.455	0.715	1.229	0.609	0.713	1.109	0
5	0.437	0.237	−0.377	0.455	0.125	−0.577	4.649
6	0.669	−0.002	0.230	0.583	−0.116	0.083	3.486
7	−0.186	−0.480	−0.907	−0.299	−0.541	−0.982	2.752
8	0.672	0.384	0.903	0.544	0.339	0.842	2.563
9	−0.121	−0.812	−1.279	−0.268	−0.828	−1.315	4.645
10	−0.683	0.093	−0.436	−0.773	0.118	−0.376	1.224
11	0.063	0.193	−0.028	0.012	0.202	0.048	1.019
12	−1.029	0.503	0.728	−1.034	0.482	0.769	4.220
13	−0.080	−0.238	−0.218	−0.039	−0.269	−0.202	1.455
14	0.943	0.272	0.831	0.949	0.242	0.824	0.214
15	0.147	0.655	−0.543	0.114	0.596	−0.561	3.733
16	0.217	−0.260	−0.005	0.175	−0.328	−0.003	0.514
17	−0.611	−0.048	−0.428	−0.639	−0.104	−0.414	0.538
18	0.473	0.066	0.890	0.451	0.014	0.891	0.204
19	−1.130	−1.214	−0.064	−1.133	−1.233	−0.086	8.145
20	1.379	2.443	2.350	1.370	2.394	2.271	5.608
21	0.671	−0.466	0.120	0.613	−0.561	−0.015	2.811
22	1.467	−0.055	−0.598	1.361	−0.134	−0.716	5.203
23	1.077	−0.051	−1.085	0.919	−0.124	−1.161	7.389
24	2.126	2.964	2.775	1.908	2.833	2.667	13.398
25	2.237	−0.041	−0.825	1.938	−0.227	−0.969	17.783
26	0.362	−0.656	−0.625	0.024	−0.821	−0.735	9.438
27	2.119	0.569	0.031	1.746	0.412	−0.063	14.527
28	1.147	0.003	0.417	0.728	−0.158	0.318	11.751
29	0.667	−0.475	0.676	0.232	−0.623	0.561	11.525
30	1.944	−0.065	−1.130	1.479	−0.199	−1.234	14.106

most CUSUM charts can be changed into EWMA charts after proper modifications. By this connection between the two types of charts, the MCUSUM charts discussed in Subsection 7.3.2 for monitoring the process covariance matrix can also be changed into MEWMA charts. For instance, the MCUSUM chart (7.32)–(7.33) can be modified as follows. Let $\mathbf{X}_1, \mathbf{X}_2, \ldots$ be a sequence of phase II observation vectors obtained from a p-dimensional production process with the IC distribution $N_p(\mu_0, \Sigma_0)$, where μ_0 and Σ_0 are assumed known. Then, the Hotelling's T^2 statistic $T_n^2 = (\mathbf{X}_n - \mu_0)' \Sigma_0^{-1} (\mathbf{X}_n - \mu_0)$ would have a χ_p^2 distribution when the process is IC. Instead of using the MCUSUM charting statistic in (7.32), we can consider the

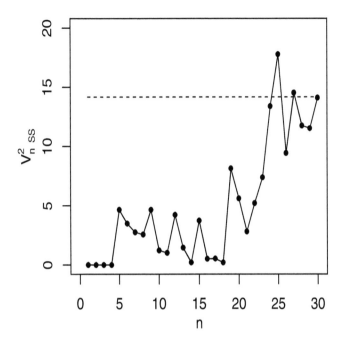

Figure 7.9 *The self-starting MEWMA chart (7.45)–(7.46) when it is applied to the data presented in Table 7.5 and when $\lambda = 0.2$ and $ARL_0 = 200$. The dashed horizontal line denotes the control limit $h_{SS} = 14.167$.*

following MEWMA charting statistic:

$$E_n = \lambda (T_n^2 - p) + (1 - \lambda)E_{n-1}, \qquad \text{for } n \geq 1, \qquad (7.47)$$

where $E_0 = 0$, $\lambda \in (0, 1]$ is a weighting parameter, and the dimensionality p is also the IC mean of T_n^2 (cf., Subsection 2.3.2). The resulting MEWMA chart gives a signal of process covariance matrix shift at the n-th time point if

$$E_n > h \qquad (7.48)$$

where $h > 0$ is a control limit chosen to achieve a pre-specified ARL_0 value, and it can be determined by a Monte Carlo simulation. Similar to the construction of the MEWMA chart (7.47)–(7.48), the COT chart (7.30)–(7.31) and the regression-adjusted MCUSUM chart (7.34)–(7.38) can both be modified properly into the corresponding MEWMA charts.

Hawkins and Maboudou-Tchao (2008) proposed a MEWMA chart specifically for monitoring the process covariance matrix, described as follows. For the IC process covariance matrix Σ_0, assume that its Cholesky decomposition is

$$\Sigma_0 = UU' \qquad (7.49)$$

where U is a $p \times p$ lower triangular matrix. Let $L = U^{-1}$. Then $L\Sigma_0 L' = I_{p \times p}$, and L is a $p \times p$ lower triangular matrix. For the n-th observation vector $\mathbf{X}_n$ with $n \geq 1$, define its transformation

$$\mathbf{Y}_n = L(\mathbf{X}_n - \mu_0). \tag{7.50}$$

Then, it is easy to check that the IC distribution of $\mathbf{Y}_n$ is $N_p(\mathbf{0}, I_{p \times p})$. Thus, its p components are independent of each other, and each of them has the $N(0, 1)$ distribution when the process is IC. As a matter of fact, its j-th component Y_{nj} can be interpreted as the residual of the regression of X_{nj} on $(X_{n1}, X_{n2}, \ldots, X_{n,j-1})$, rescaled to have a unit variance (cf., Hawkins, 1993b). Then, we define

$$\mathbf{E}_n = \lambda \mathbf{Y}_n \mathbf{Y}_n' + (1 - \lambda) \mathbf{E}_{n-1}, \tag{7.51}$$

where $\mathbf{E}_0 = I_{p \times p}$, and $\lambda \in (0, 1]$ is a weighting parameter. The MEWMA chart gives a signal of process covariance matrix shift at the n-th time point if

$$C_n = \text{Tr}(\mathbf{E}_n) - \log |\mathbf{E}_n| - p > h \tag{7.52}$$

where $\text{Tr}(A)$ denotes the trace of a square matrix A (i.e., the summation of all diagonal elements of A), and $h > 0$ is a control limit chosen to achieve a pre-specified ARL_0 value. The charting statistic C_n in (7.52) is derived using the log-likelihood ratio statistic for testing $H_0 : \Sigma = I_{p \times p}$ versus $H_1 : \Sigma \neq I_{p \times p}$ based on $\mathbf{E}_n$, where Σ denotes the true covariance matrix of $\mathbf{Y}_n$.

By Monte Carlo simulations, Hawkins and Maboudou-Tchao (2008) searched for the h values for some commonly used λ and ARL_0 values, part of which are presented in Table 7.6. From the table, it is apparent that the control limit h increases when either one of p, λ, and ARL_0 increases.

Example 7.7 *Consider a 3-dimensional production process with the IC distribution* $N_3(\mu_0, \Sigma_0)$, *where*

$$\mu_0 = (0, 0, 0)'$$

and

$$\Sigma_0 = \begin{pmatrix} 1.0 & 0.8 & 0.5 \\ 0.8 & 1.0 & 0.8 \\ 0.5 & 0.8 & 1.0 \end{pmatrix}.$$

The first 30 phase II observations are presented in columns 2–4 of Table 7.7. Then, by using the function chol() *in the R-package* Matrix, *the Cholesky decomposition of* Σ_0 *(cf., (7.49)) is found to be*

$$\Sigma_0 = \begin{pmatrix} 1.0 & 0 & 0 \\ 0.8 & 0.6 & 0 \\ 0.5 & 0.667 & 0.553 \end{pmatrix} \times \begin{pmatrix} 1.0 & 0.8 & 0.5 \\ 0 & 0.6 & 0.667 \\ 0 & 0 & 0.553 \end{pmatrix}.$$

Therefore, the regression-adjusted standardized residuals $\mathbf{Y}_n$ *can be computed by* (7.50), *where*

$$L = \begin{pmatrix} 1.0 & 0 & 0 \\ -1.333 & 1.667 & 0 \\ 0.704 & -2.010 & 1.809 \end{pmatrix}.$$

Table 7.6 *This table presents the values of the control limit h of the MEWMA chart (7.51)–(7.52) for some commonly used λ and ARL_0 values in cases when $p = 2, 3, 4, 5$. This table is part of Table 1 in Hawkins and Maboudou-Tchao (2008).*

				ARL_0		
p	λ	100	200	250	500	1000
2	0.05	0.182	0.232	0.247	0.295	0.342
	0.10	0.446	0.542	0.573	0.669	0.765
	0.15	0.738	0.883	0.930	1.077	1.228
	0.20	1.051	1.247	1.311	1.515	1.724
	0.25	1.383	1.632	1.714	1.977	2.247
	0.30	1.732	2.037	2.138	2.461	2.794
3	0.05	0.301	0.364	0.383	0.441	0.496
	0.10	0.712	0.831	0.868	0.981	1.093
	0.15	1.163	1.340	1.396	1.569	1.741
	0.20	1.646	1.884	1.960	2.195	2.430
	0.25	2.159	2.459	2.556	2.856	3.157
	0.30	2.701	3.066	3.184	3.550	3.919
4	0.05	0.444	0.521	0.544	0.611	0.673
	0.10	1.028	1.169	1.212	1.342	1.467
	0.15	1.665	1.872	1.937	2.132	2.322
	0.20	2.347	2.623	2.709	2.972	3.231
	0.25	3.072	3.417	3.526	3.860	4.188
	0.30	3.839	4.256	4.388	4.793	5.194
5	0.05	0.612	0.702	0.729	0.805	0.876
	0.10	1.396	1.559	1.608	1.754	1.891
	0.15	2.250	2.486	2.559	2.775	2.984
	0.20	3.163	3.474	3.570	3.861	4.143
	0.25	4.135	4.523	4.643	5.008	5.364
	0.30	5.166	5.631	5.776	6.219	6.652

The computed values of $\mathbf{Y}_n$ are presented in columns 5–7 of Table 7.7. The MEWMA chart (7.51)–(7.52) is then applied to this dataset, in which the λ and ARL_0 values are chosen to be 0.2 and 200, respectively, and thus the control limit h is 1.884 according to Table 7.6. The calculated values of the charting statistic C_n defined in (7.52) are presented in the last column of Table 7.7, and the MEWMA chart is shown in Figure 7.10 as well. From the plot, the chart gives its first signal of process covariance matrix shift at the 14th time point.

In the literature, there is much other discussion about the monitoring of the process covariance matrix. For instance, Huwang et al. (2007), Yeh and Lin (2002), and Yeh et al. (2003, 2004a, 2005) proposed several MEWMA charts for monitoring the process covariance matrix, some of which are based on similar ideas to that of the chart (7.51)–(7.52). Reynolds and Stoumbos (2008) compared a number of joint monitoring schemes for monitoring both the process mean vector and the process covariance matrix. In some of these joint monitoring schemes, both the MEWMA charts and the multivariate Shewhart charts are used.

Table 7.7 *This table presents the first 30 phase II observations of a 3-dimensional production process (columns 2–4), the three components of $\mathbf{Y}_n = (Y_{n1}, Y_{n2}, Y_{n3})'$ computed by (7.50)
(columns 5–7), and the values of the charting statistic C_n of the MEWMA chart (7.51)–(7.52)
(last column) in which λ and ARL_0 are chosen to be 0.2 and 200, respectively.*

n	X_{n1}	X_{n2}	X_{n3}	Y_{n1}	Y_{n2}	Y_{n3}	C_n
1	−0.533	−0.385	−0.443	−0.533	0.070	−0.404	0.053
2	0.167	−0.052	0.263	0.167	−0.309	0.698	0.181
3	0.068	−0.138	−0.133	0.068	−0.322	0.085	0.410
4	0.492	0.680	1.232	0.492	0.476	1.209	0.524
5	−0.071	0.306	0.053	−0.071	0.603	−0.568	0.692
6	0.229	0.413	0.199	0.229	0.383	−0.309	0.959
7	−0.400	−0.396	−0.789	−0.400	−0.127	−0.912	1.215
8	0.387	0.640	0.898	0.387	0.551	0.610	1.454
9	−0.404	−0.529	−1.318	−0.404	−0.343	−1.605	1.698
10	−1.430	−0.408	0.880	−1.430	1.227	1.407	1.060
11	0.363	−0.305	−0.230	0.363	−0.993	0.454	0.772
12	2.157	2.151	−0.149	2.157	0.710	−3.077	1.722
13	−0.574	0.444	−0.337	−0.574	1.505	−1.905	1.583
14	−0.845	−0.914	1.673	−0.845	−0.397	4.269	3.976
15	0.364	0.375	−2.650	0.364	0.140	−5.293	7.629
16	−0.814	−0.212	0.487	−0.814	0.732	0.734	5.938
17	0.737	0.704	−1.118	0.737	0.191	−2.920	6.213
18	−0.814	0.254	1.607	−0.814	1.508	1.825	5.240
19	0.192	0.491	2.024	0.192	0.562	2.810	5.061
20	0.661	1.919	−0.451	0.661	2.316	−4.209	7.267
21	0.368	−0.535	0.020	0.368	−1.383	1.372	5.994
22	1.089	1.324	0.997	1.089	0.756	−0.092	4.162
23	−1.495	−0.984	−1.998	−1.495	0.353	−2.688	3.950
24	−1.826	1.527	1.599	−1.826	4.980	−1.461	6.858
25	−0.266	−0.833	−0.093	−0.266	−1.034	1.318	5.333
26	1.422	1.302	0.476	1.422	0.274	−0.755	3.781
27	−0.632	−0.443	0.214	−0.632	0.104	0.834	2.686
28	1.108	0.820	1.375	1.108	−0.111	1.620	2.048
29	0.601	−0.809	0.508	0.601	−2.150	2.969	3.496
30	−1.387	−1.689	−1.509	−1.387	−0.966	−0.311	2.379

7.5 Multivariate Control Charts by Change-Point Detection

As discussed in Section 6.4, the CUSUM and EWMA charts both require a substantial amount of prior information about the IC and OC process distributions in their
design and construction. As a comparison, the CPD control charts have the advantages that they do not need much prior information about the IC and OC process
distributions for process monitoring and that they can provide an estimator of the
shift time at the same time as they give a signal of process mean or variance shift.
Some of the univariate CPD charts discussed in Chapter 6 have been generalized to
multivariate cases, which are described in this section.

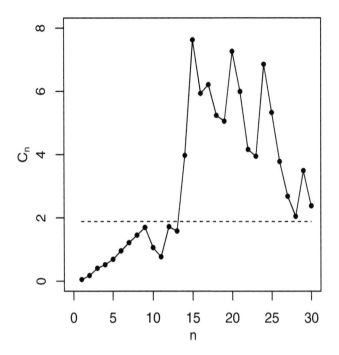

Figure 7.10 *The MEWMA chart (7.51)–(7.52) when it is applied to the data presented in Table 7.7 and when $\lambda = 0.2$ and $ARL_0 = 200$. The dashed horizontal line denotes the control limit.*

Zamba and Hawkins (2006) generalized the univariate CPD chart (6.13)–(6.14) to multivariate cases for online monitoring of the process mean vector. First, let us assume that $\{\mathbf{X}_1, \mathbf{X}_2, \ldots, \mathbf{X}_n\}$ is a sequence of observation vectors obtained from a p-dimensional production process, and the sequence follows the change-point model

$$\mathbf{X}_i \sim \begin{cases} N_p(\mu_0, \Sigma_0), & \text{if } i \leq r \\ N_p(\mu_1, \Sigma_0), & \text{if } i > r, \end{cases}$$

where μ_0 and Σ_0 are the IC process mean vector and the IC process covariance matrix, μ_1 is the OC process mean vector that is assumed different from μ_0, and $1 \leq r \leq n-1$ is the change-point. In this change-point model, the process mean vector has a shift at the time point $\tau = r+1$, but the process covariance matrix is assumed unchanged. In such cases, the Hotelling's T^2 statistic for testing the hypotheses $H_0 : \mu_0 = \mu_1$ versus $H_1 : \mu_0 \neq \mu_1$ is

$$T_r^2 = \frac{r(n-r)}{n} \left(\overline{\mathbf{X}}_{0,r} - \overline{\mathbf{X}}_{r,n} \right)' \left(\widetilde{\mathbf{S}}_r^2 \right)^{-1} \left(\overline{\mathbf{X}}_{0,r} - \overline{\mathbf{X}}_{r,n} \right) \tag{7.53}$$

where

$$\overline{\mathbf{X}}_{j_1,j_2} = \frac{1}{j_2 - j_1} \sum_{i=j_1+1}^{j_2} \mathbf{X}_i, \text{ for any } 0 \le j_1 < j_2 \le n$$

$$\widetilde{\mathbf{S}}_r^2 = \frac{1}{n-2} \left[\sum_{i=1}^{r} (\mathbf{X}_i - \overline{\mathbf{X}}_{0,r})(\mathbf{X}_i - \overline{\mathbf{X}}_{0,r})' + \right.$$

$$\left. \sum_{i=r+1}^{n} (\mathbf{X}_i - \overline{\mathbf{X}}_{r,n})(\mathbf{X}_i - \overline{\mathbf{X}}_{r,n})' \right].$$

Now, let us discuss phase II process monitoring, and assume that $\mathbf{X}_1, \mathbf{X}_2, \dots$ are phase II observation vectors obtained from a p-dimensional production process. To detect a process mean shift at the current time point n, it is natural to consider the charting statistic

$$T_{max,n}^2 = \max_{1 \le r \le n-1} T_r^2, \tag{7.54}$$

where T_r^2 is defined in (7.53). The corresponding CPD chart gives a signal of process mean shift if

$$T_{max,n}^2 > h_n \tag{7.55}$$

where $h_n > 0$ is a control limit chosen to achieve a pre-specified ARL_0 value and it may depend on n. After a signal is given, the shift time τ can be estimated by $\hat{r}+1$, where $\hat{r}$ is the maximizer found in (7.54).

To use the CPD chart (7.54)–(7.55), the control limit h_n should be chosen properly. As discussed in Subsection 6.3.1, one reasonable way is to choose h_n such that, when the process is IC,

$$P\left(T_{max,n}^2 > h_n \,\big|\, \text{no signal before } n\right) \le \alpha,$$

where $\alpha \in [0,1]$ is a pre-specified conditional false alarm rate. If h_n is chosen in this way, then the ARL_0 value of the CPD chart (7.54)–(7.55) would be $1/\alpha$. Based on numerical studies, Zamba and Hawkins (2006) provided the following approximation formulas for computing h_n in cases when $\alpha = 0.001, 0.002,$ and 0.005 and $n > 25$:

$$\log(h_n) = \begin{cases} 3.043 + 0.221p - \frac{p+4}{50}\log(n-25), & \text{if } \alpha = 0.001 \\ 2.897 + 0.226p - \frac{2p+7}{100}\log(n-25), & \text{if } \alpha = 0.002 \\ 2.706 + 0.230p - \frac{p+3}{50}\log(n-25), & \text{if } \alpha = 0.005. \end{cases} \tag{7.56}$$

To simplify the computation involved in the CPD chart (7.54)–(7.55), a recursive algorithm for computing its charting statistic $T_{max,n}^2$ was developed by Zamba and Hawkins (2006), described below. First, for a given n and a given $1 \le r \le n-1$, we can re-write T_r^2 defined in (7.53) as

$$T_r^2 = \frac{(n-2)c_r \mathbf{d}_r' \mathbf{W}_n^{-1} \mathbf{d}_r}{1 - c_r \mathbf{d}_r' \mathbf{W}_n^{-1} \mathbf{d}_r}, \tag{7.57}$$

where $c_r = rn/(n-r)$, $\mathbf{d}_r = \overline{\mathbf{X}}_{0,r} - \overline{\mathbf{X}}_{0,n}$, and

$$\mathbf{W}_n = \sum_{i=1}^{n} (\mathbf{X}_i - \overline{\mathbf{X}}_{0,n})(\mathbf{X}_i - \overline{\mathbf{X}}_{0,n})'.$$

Second, the sample mean $\overline{\mathbf{X}}_{0,n}$ can be computed recursively by

$$\overline{\mathbf{X}}_{0,n} = \overline{\mathbf{X}}_{0,n-1} + \frac{1}{n}(\mathbf{X}_n - \overline{\mathbf{X}}_{0,n-1}). \tag{7.58}$$

Finally, $\mathbf{W}_n$ can be computed recursively by

$$\mathbf{W}_n = \mathbf{W}_{n-1} + \frac{n-1}{n}(\mathbf{X}_n - \overline{\mathbf{X}}_{0,n-1})(\mathbf{X}_n - \overline{\mathbf{X}}_{0,n-1})'.$$

From this formula, we have

$$\mathbf{W}_n^{-1} = \mathbf{W}_{n-1}^{-1} - \frac{(n-1)\mathbf{W}_{n-1}^{-1}(\mathbf{X}_n - \overline{\mathbf{X}}_{0,n-1})(\mathbf{X}_n - \overline{\mathbf{X}}_{0,n-1})'\mathbf{W}_{n-1}^{-1}}{n\left[1 + \frac{n-1}{n}(\mathbf{X}_n - \overline{\mathbf{X}}_{0,n-1})'\mathbf{W}_{n-1}^{-1}(\mathbf{X}_n - \overline{\mathbf{X}}_{0,n-1})\right]}. \tag{7.59}$$

Therefore, T_r^2 can be computed recursively by formulas (7.57)–(7.59).

Example 7.8 *The first 50 observations obtained from a 3-dimensional production process are shown in Figure 7.11(a)–(c). It is assumed that the first 25 of them are IC observations, and that the IC process distribution is $N_3(\mu_0, \Sigma_0)$. But, the IC process mean μ_0 and the IC process covariance matrix Σ_0 are both unknown. We then apply the CPD control chart (7.54)–(7.55) to this data, and start to monitor the process mean vector at the 26th time point. In the chart, α is chosen to be 0.005 (i.e., ARL_0 is chosen $1/0.005 = 200$), h_n is approximated by the third formula in (7.56), and the formulas (7.57)–(7.59) are used for recursively computing the charting statistic $T_{max,n}^2$. The CPD control chart is shown in Figure 7.11(d), from which it can be seen that the first signal of process mean shift is given at the 36th time point. Then, the MLE of the change-point r is obtained from (7.54) to be $\hat{r} = 30$. Therefore, the shift time is estimated to be 31. The IC and OC process mean vectors can then be estimated by the sample mean vectors of the first 30 observations and the following 6 observations, respectively, to be*

$$\hat{\mu}_0 = (-0.027, 0.038, 0.065)', \qquad \hat{\mu}_1 = (1.273, 0.090, -0.030)'.$$

The IC process covariance matrix Σ_0 can be estimated by the pooled sample variance $\widetilde{S}_r^2$ of the first 36 observations, which is computed to be

$$\widetilde{S}_r^2 = \begin{pmatrix} 0.774 & 0.595 & 0.201 \\ 0.595 & 0.856 & 0.586 \\ 0.201 & 0.586 & 0.852 \end{pmatrix}.$$

Zamba and Hawkins (2009) proposed a CPD control chart for monitoring both the process mean vector and the process covariance matrix. Assume that $\mathbf{X}_1, \mathbf{X}_2, \ldots$

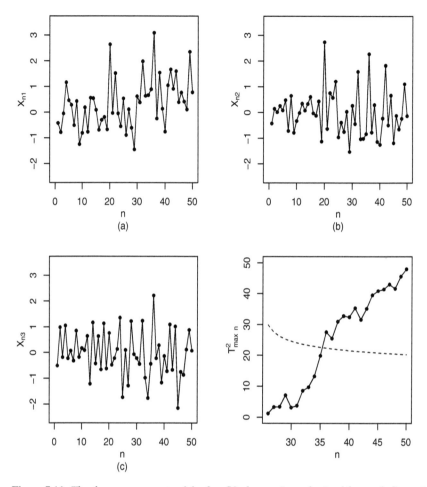

Figure 7.11 *The three components of the first 50 observations obtained from a 3-dimensional production process (plots (a)–(c)) and the CPD control chart (7.54)–(7.55) (plot (d)). In plot (d), the dashed curve denotes the control limit h_n, approximated by the third formula of (7.56) when α is chosen to be 0.005.*

are observation vectors obtained from a p-dimensional production process, and they follow the change-point model

$$\mathbf{X}_i \sim \begin{cases} N_p(\mu_0, \Sigma_0), & \text{if } i \leq r \\ N_p(\mu_1, \Sigma_1), & \text{if } i > r, \end{cases}$$

where μ_0 and Σ_0 are the IC process mean vector and the IC process covariance matrix, μ_1 and Σ_1 are the OC process mean vector and the OC process covariance matrix, and $1 \leq r \leq n-1$ is the change-point. In this change-point model, both the process mean vector and the process covariance matrix may have shifts at the time point $\tau = r + 1$. To use the first n observations to test the hypotheses

$$H_0 : \mu_1 = \mu_0, \Sigma_1 = \Sigma_0 \qquad \text{versus} \qquad H_1 : H_0 \text{ is not true,}$$

under the assumptions that r is given and $r < n$, the LRT statistic is proportional to

$$\Lambda_{r,n} = \frac{|\mathbf{S}_{0,r}^2|^{(r-1)/2}|\mathbf{S}_{r,n}^2|^{(n-r-1)/2}}{|\mathbf{S}_{0,n}^2|^{(n-1)/2}},$$

where $\mathbf{S}_{j_1,j_2}^2$ denotes the sample covariance matrix constructed from the observations $\mathbf{X}_{j_1+1}, \mathbf{X}_{j_1+2}, \ldots, \mathbf{X}_{j_2}$, for any $j_1 < j_2$. Based on this result, Zamba and Hawkins (2009) showed that the IC distribution of the statistic

$$G_{r,n}^2 = -2\log(\Lambda_{r,n})\,[p(p+3)]/C_{r,n}$$

was approximately a chi-square distribution, where

$$
\begin{aligned}
&C_{r,n} \\
&= p\,[\log(2) + (n-1)\log(n-1) - (n-r-1)\log(n-r-1) - (r-1)\log(r-1)] \\
&+ \sum_{j=1}^{p}[(n-1)\psi((n-j)/2) - (r-1)\psi((r-j)/2) - (n-r-1)\psi((n-r-j)/2)]
\end{aligned}
$$

was the Bartlett correction factor. So, Zamba and Hawkins (2009) considered the charting statistic

$$G_{max,n}^2 = \max_{1 \le r \le n-1} G_{r,n}^2. \tag{7.60}$$

The chart gives a signal at the n-th time point if

$$G_{max,n}^2 > h_n, \tag{7.61}$$

where $h_n > 0$ is a control limit, and it can be computed by a Monte Carlo simulation.

7.6 Multivariate Control Charts by LASSO

In MSPC, a multivariate production process is monitored for possible shifts in its mean vector and/or its covariance matrix. In practice, many shifts only involve a small number of unknown process components. If there is a data-driven procedure for figuring out the shifted components, then we can make use of this information in process monitoring, and the resulting MSPC charts could be more efficient. In recent years, certain variable selection procedures, such as the *least absolute shrinkage and selection operator (LASSO)* (cf., Tibshirani, 1996), have been demonstrated to be useful for MSPC. In this section, we briefly discuss how to construct multivariate control charts using LASSO. Our discussion is divided into two parts. In Subsection 7.6.1, the LASSO procedure is briefly described in the context of regression variable selection. Then, a MEWMA chart based on LASSO variable selection is described in Subsection 7.6.2.

7.6.1 LASSO for regression variable selection

Consider the following multiple linear regression model:

$$y_i = \mathbf{Z}_i'\beta + \varepsilon_i, \qquad \text{for } i = 1, 2, \ldots, n,$$

where y, $\mathbf{Z}$, and β are the response variable, vector of predictors, and vector of regression coefficients, respectively, $(\mathbf{Z}_i', y_i)$ is the i-th observation of $(\mathbf{Z}', y)$, and $\{\varepsilon_i\}$ are i.i.d. random errors with the distribution $N(0, \sigma^2)$. In practice, some predictors do not provide much useful information about the response variable, given the remaining predictors; thus, they can be deleted from the model. To find which variables should be deleted, the stepwise and all subset variable selection procedures along with a model selection criterion (e.g., AIC or BIC) are often used (cf., Devore, 2011, Chapter 13). Such model selection procedures are practically useful; but they have several limitations, including the lack of stability (cf., Breiman, 1996), the lack of theoretical properties (cf., Fan and Li, 2001), and the extensive computation (cf., Tibshirani, 1996). To overcome these limitations, some authors suggest using the following penalized least squares method (or equivalently, the penalized likelihood method in the normal error distribution case):

$$\sum_{i=1}^{n} (y_i - \mathbf{Z}_i'\beta)^2 + n \sum_{j=1}^{p} g_{\gamma_j}(|\beta_j|),$$

where β_j denotes the j-th component of β, $\{\gamma_j\}$ are the penalty parameters (also called the regularization parameters), and $\{g_{\gamma_j}\}$ are the penalty functions. When $g_{\gamma_j}(|\beta_j|) = \gamma|\beta_j|$ where γ is a constant parameter, the corresponding penalized least squares method is called LASSO (cf., Tibshirani, 1996). Besides LASSO, another major penalized least squares method is the so-called *smoothly clipped absolute deviation (SCAD)* method (cf., Fan and Li, 2001). Among other good properties, Fan and Li (2001) have shown that, with the proper choice of the penalty functions and regularization parameters, the penalized likelihood estimators of β would perform asymptotically as well as the estimator obtained in cases when the correct submodel was known, which is referred to as the oracle property in the literature. When the penalty functions are chosen to be continuous, the coefficient estimates that correspond to those insignificant predictors would shrink towards 0 as γ increases, and these coefficient estimates could be exactly 0 if γ is chosen sufficiently large. In addition, the penalized likelihood method, especially when LASSO-type penalty functions are used, enjoys efficient computation using the LARS algorithm by Efron (2004). See Zou and Li (2008) for a detailed discussion about the computational issues of LASSO and SCAD.

7.6.2 A LASSO-based MEWMA chart

In this subsection, we describe a MEWMA chart using the LASSO variable selection procedure described in the previous subsection for monitoring the mean vector of a p-dimensional production process, where $p \geq 2$ is a given integer. This MEWMA chart was proposed by Zou and Qiu (2009); thus, a more detailed description can be found in that paper.

From the description about the MSPC control charts in the previous several sections, it can be seen that the fundamental tasks of MSPC control charts for monitoring the process mean vector include (i) determining whether the process mean vector μ has changed from the IC process mean μ_0 up to the current time point n, (ii) detecting the specific time point at which a detected mean shift occurs, and (iii) identifying the shifted components of μ. Let $\mathbf{X}_1, \mathbf{X}_2, \ldots, \mathbf{X}_n$ be the first n observation vectors obtained from a p-dimensional production process with the known IC distribution $N_p(\mu_0, \Sigma_0)$. Then, different methods for accomplishing these tasks usually adopt the following change-point model:

$$\mathbf{X}_i \sim \begin{cases} N_p(\mu_0, \Sigma_0), & \text{for } i = 1, 2, \ldots, r, \\ N_p(\mu_1, \Sigma_0), & \text{for } i = r+1, r+2, \ldots, n, \end{cases}$$

where the IC process covariance matrix Σ_0 is assumed unchanged over time, the process mean vector is assumed changed from μ_0 to μ_1 at $\tau = r+1$, and r is the change-point. Throughout this subsection, we further assume that $\mu_0 = \mathbf{0}$, without loss of generality. Otherwise, we can simply replace $\mathbf{X}_i$ by $\mathbf{X}_i - \mu_0$ in all related descriptions below. Then, a basic strategy for detecting a mean shift is to test hypotheses

$$H_0 : \mu = \mathbf{0} \qquad \text{versus} \qquad H_1 : \mu \neq \mathbf{0}$$

using the likelihood ratio test (LRT) statistic $n\overline{\mathbf{X}}'\Sigma_0^{-1}\overline{\mathbf{X}}$, where μ denotes the true process mean vector, and $\overline{\mathbf{X}} = \frac{1}{n}\sum_{i=1}^n \mathbf{X}_i$ is the sample mean vector. In cases when the shift direction $\mathbf{d}$ with unit length is known, the above alternative hypothesis can be replaced by $H_1' : \mu = \delta\mathbf{d}$, where δ is an unknown constant denoting the shift magnitude in the direction of $\mathbf{d}$. In such cases, Healy (1987) has shown that the LRT statistic and its null distribution are

$$\frac{n\left(\mathbf{d}'\Sigma_0^{-1}\overline{\mathbf{X}}\right)^2}{\mathbf{d}'\Sigma_0^{-1}\mathbf{d}} \stackrel{H_0}{\sim} \chi_1^2. \tag{7.62}$$

The test based on (7.62) should be more efficient than the test based on $n\overline{\mathbf{X}}'\Sigma_0^{-1}\overline{\mathbf{X}}$ in cases when the shift direction $\mathbf{d}$ is known.

In practice, the shift direction $\mathbf{d}$ is usually unknown. Therefore, the procedure by Healy (1987) cannot be used directly in many applications. However, it can be estimated using the idea of the LASSO variable selection as follows. After a constant term is ignored, the penalized likelihood (PL) function of the observed data can be written as

$$PL(\mu) = n\left(\overline{\mathbf{X}} - \mu\right)'\Sigma_0^{-1}\left(\overline{\mathbf{X}} - \mu\right) + n\sum_{j=1}^p g_{\gamma_j}\left(|\mu_j|\right),$$

where $\mu = (\mu_1, \mu_2, \ldots, \mu_p)'$. If the adaptive LASSO (ALASSO) penalty function by Zou (2006) is used, then it becomes

$$PL(\mu) = n\left(\overline{\mathbf{X}} - \mu\right)'\Sigma_0^{-1}\left(\overline{\mathbf{X}} - \mu\right) + n\gamma\sum_{j=1}^p \frac{1}{|\overline{X}_j|}|\mu_j|, \tag{7.63}$$

where $\mathbf{X} = (\overline{X}_1, \overline{X}_2, \ldots, \overline{X}_p)'$. The ALASSO penalty function used in (7.63) is slightly different from the traditional one by Tibshirani (1996) in that the latter uses the same amount of shrinkage for each μ_j while the former determines the amount of shrinkage for μ_j adaptively by the value of $|\overline{X}_j|$. Because of this difference, the coefficient estimator by the traditional LASSO cannot be as efficient as the oracle estimator (cf., Fan and Li, 2001) and its model selection results could be inconsistent in certain cases; but, the ALASSO estimator is asymptotically unbiased and it has certain oracle properties. See Fan and Li (2001), Zhao and Yu (2006), and Zou (2006) for a related discussion. The ALASSO estimator of μ is then defined by

$$\widehat{\mu}_\gamma = \arg\min_\mu PL(\mu).$$

Because of its property that some of its components are exactly zero when γ is properly chosen, the ALASSO estimator $\widehat{\mu}_\gamma$ is an ideal estimator of the shift direction in the MSPC problem. Then, a LASSO-based test statistic can be defined by

$$\widetilde{T}_\gamma = \frac{n\left(\widehat{\mu}_\gamma' \Sigma_0^{-1} \overline{\mathbf{X}}\right)^2}{\widehat{\mu}_\gamma' \Sigma_0^{-1} \widehat{\mu}_\gamma}.$$

Theoretically, $\widehat{\mu}_\gamma \neq \mathbf{0}$ almost surely. For completeness, $\widetilde{T}_\gamma$ can be defined to be any negative number when $\widehat{\mu}_\gamma = \mathbf{0}$. Obviously, $\widetilde{T}_\gamma$ can be regarded as a data-driven version of the test statistic defined in (7.62) by Healy (1987), with the shift direction $\mathbf{d}$ replaced by its estimator $\widehat{\mu}_\gamma$.

To use the above LASSO-based test statistic $\widetilde{T}_\gamma$, the regularization parameter γ should be chosen properly, since it plays an important role in balancing robustness and sensitivity of $\widetilde{T}_\gamma$ to various shifts. From some theoretical results, it is found that selection of γ should depend on the shift size. For detecting a large shift, γ should be chosen to be large, and it should be chosen to be small for detecting a small shift. Further, numerical results show that conventional procedures for choosing regularization parameters, such as the cross-validation (CV), the generalized cross-validation (GCV), AIC, BIC, etc. (cf., Subsection 2.8.5), do not work properly in the current problem, because these parameter selection procedures are mainly for model estimation and they usually do not produce a powerful testing procedure (cf., Bickel and Li, 2006; Hart, 1997). To overcome this difficulty, Zou and Qiu (2009) suggested using the approach proposed by Horowitz and Spokoiny (2001) in the nonparametric testing setup, by combining several values of γ to make the resulting test nearly optimal in various different cases. Let $\Gamma_q = \{\gamma_k, k = 1, \ldots, q\}$ be a set of values of the regularization parameter γ, where q is a pre-specified constant. Then, the modified penalized test statistic is defined by

$$\widetilde{T} = \max_{k=1,\ldots,q} \frac{\widetilde{T}_{\gamma_k} - \mathrm{E}\left(\widetilde{T}_{\gamma_k}\right)}{\sqrt{\mathrm{Var}\left(\widetilde{T}_{\gamma_k}\right)}},$$

where $\mathrm{E}(\widetilde{T}_{\gamma_k})$ and $\mathrm{Var}(\widetilde{T}_{\gamma_k})$ are respectively the mean and variance of $\widetilde{T}_{\gamma_k}$ under H_0.

The pointset Γ_q can be determined as follows. Let us first rewrite the ALASSO-type penalized likelihood function (7.63) as

$$PL(\alpha) = n\left(\overline{\mathbf{X}} - \mathbf{D}\alpha\right)' \Sigma_0^{-1} \left(\overline{\mathbf{X}} - \mathbf{D}\alpha\right) + n\gamma \sum_{j=1}^{p} |\alpha_j|, \qquad (7.64)$$

where $\alpha = (\alpha_1, \alpha_2, \ldots, \alpha_p)'$, $\alpha_j = \mu_j/|\overline{X}_j|$, for $j = 1, 2, \ldots, p$, and $\mathbf{D} = \mathrm{diag}(|\overline{X}_1|, |\overline{X}_2|, \ldots, |\overline{X}_p|)$. This is exactly a LASSO-type penalized likelihood function. According to Zou et al. (2007c), for a given $\overline{\mathbf{X}}$ in (7.64), there is a finite sequence

$$\widetilde{\gamma}_0 > \widetilde{\gamma}_1 > \ldots > \widetilde{\gamma}_K = 0 \qquad (7.65)$$

such that (i) for all $\gamma > \widetilde{\gamma}_0$, $\widehat{\alpha}_\gamma = 0$, where $\widehat{\alpha}_\gamma$ denotes the minimizer of (7.64), and (ii) when γ is in the interval $(\widetilde{\gamma}_{k+1}, \widetilde{\gamma}_k)$, the active set $\mathcal{B}(\gamma) = \{j : \mathrm{sgn}[\widehat{\alpha}_{\gamma,j}] \neq 0\}$ and the sign vector $\mathbf{S}(\gamma) = \{\mathrm{sgn}[\widehat{\alpha}_{\gamma,1}], \ldots, \mathrm{sgn}[\widehat{\alpha}_{\gamma,p}]\}$ are unchanged, where $\widehat{\alpha}_\gamma$ is the j-th component of $\widehat{\alpha}_\gamma$, and sgn is a sign function taking the values of -1, 0, and 1 when its argument is negative, 0, and positive, respectively. These $\widetilde{\gamma}_m$'s are called *transition points* because the active set changes at each $\widetilde{\gamma}_m$ only. The transition points can be determined easily using the LARS algorithm by Efron (2004). Further, according to Efron (2004), the random integer K can be larger than p. Thus, the quantities $\widetilde{\gamma}_{m_k^{\mathrm{last}}}$, for $k = 1, \ldots, q$, can be used for constructing Γ_q, where m_k^{last} is the index of the last $\widetilde{\gamma}$ in the sequence $\{\widetilde{\gamma}_0, \widetilde{\gamma}_1, \ldots, \widetilde{\gamma}_K\}$ defined in (7.65) that the corresponding active set contains exactly k elements. According to Zou et al. (2007c), $\widetilde{T}_{\widetilde{\gamma}_{m_k^{\mathrm{last}}}}$ are well defined, because the "one at a time" condition (cf., Efron, 2004) holds almost everywhere, where "one at a time" means that the two active sets of two consecutive $\widetilde{\gamma}$s differ on at most a single index. After Γ_q is determined in this way, the resulting test statistic becomes

$$\widetilde{T}_L = \max_{k=1,\ldots,q} \frac{\widetilde{T}_{\widetilde{\gamma}_{m_k^{\mathrm{last}}}} - \mathrm{E}\left(\widetilde{T}_{\widetilde{\gamma}_{m_k^{\mathrm{last}}}}\right)}{\sqrt{\mathrm{Var}\left(\widetilde{T}_{\widetilde{\gamma}_{m_k^{\mathrm{last}}}}\right)}}. \qquad (7.66)$$

Note that $\widetilde{T}_{\widetilde{\gamma}_{m_p^{\mathrm{last}}}}$ is equivalent to the Hotelling's T^2 test statistic. Therefore, the test using the test statistic in (7.66) should share certain properties of the Hotelling's T^2 test, which does not take into account any structure of the potential shift, and certain properties of the likelihood ratio tests constructed using some specific structures of the potential shift.

Next, we construct a phase II MEWMA chart using the LASSO-based test statistic $\widetilde{T}_L$ defined in (7.66), and it should be pointed out that other types of MSPC charts (e.g., multivariate CUSUM and Shewhart charts) can also be developed in a similar way. Let $\mathbf{X}_1, \mathbf{X}_2, \ldots$ be phase II observation vectors obtained from a p-dimensional production process. Then, a sequence of MEWMA statistics using a single weighting parameter $\lambda \in (0, 1]$ (cf., (7.39)) is defined by

$$\mathbf{E}_n = \lambda \mathbf{X}_n + (1 - \lambda)\mathbf{E}_{n-1}, \qquad \text{for } n \geq 1, \qquad (7.67)$$

where $\mathbf{E}_0 = \mathbf{0}$. For each $\mathbf{E}_n$, we can compute the q LASSO estimators $\widehat{\mu}_{n,\widetilde{\gamma}_{m_k}^{\text{last}}}$, for $k = 1, 2, \ldots, q$, from the penalized likelihood function

$$(\mathbf{E}_n - \mu)'\Sigma_0^{-1}(\mathbf{E}_n - \mu) + \gamma \sum_{j=1}^{p} \frac{1}{|\mathbf{E}_{n,j}|}|\mu_j|, \tag{7.68}$$

where $\mathbf{E}_n = (\mathbf{E}_{n,1}, \mathbf{E}_{n,2}, \ldots, \mathbf{E}_{n,p})'$. Then, the LASSO-based MEWMA chart, denoted as LEWMA, gives a signal of process mean shift at the n-th time point if

$$Q_n = \max_{k=1,\ldots,q} \frac{W_{n,\widetilde{\gamma}_{m_k}^{\text{last}}} - \mathrm{E}\left(W_{n,\widetilde{\gamma}_{m_k}^{\text{last}}}\right)}{\sqrt{\mathrm{Var}\left(W_{n,\widetilde{\gamma}_{m_k}^{\text{last}}}\right)}} > h, \tag{7.69}$$

where

$$W_{n,\gamma} = \frac{2-\lambda}{\lambda[1-(1-\lambda)^{2n}]} \frac{\left(\mathbf{E}_n'\Sigma_0^{-1}\widehat{\mu}_\gamma\right)^2}{\widehat{\mu}_\gamma'\Sigma_0^{-1}\widehat{\mu}_\gamma},$$

and $h > 0$ is a control limit chosen to achieve a given ARL_0 value. In practice, $(2-\lambda)/\{\lambda[1-(1-\lambda)^{2n}]\}$ can be replaced by its asymptotic form $(2-\lambda)/\lambda$, as discussed in (7.41)–(7.42).

In the LEWMA chart (7.67)–(7.69), selection of λ mainly depends on a target shift, as in conventional EWMA charts (cf., Subsection 5.2.1). It should be chosen large if the target shift is large and small otherwise. Regarding the value of q, if prior information indicating potential shifts in at most r components, with $1 \leq r \leq p$, then the numerical studies show that using $q = r+1$ or $q = r+2$ often provides a satisfactory performance in practice. When such prior information is unavailable, numerical results show that the LEWMA chart with $q = p$ would perform reasonably well. Further, theoretical results show that the quantities $\mathrm{E}(W_{n,\widetilde{\gamma}_{m_k}^{\text{last}}})$ and $\mathrm{Var}(W_{n,\widetilde{\gamma}_{m_k}^{\text{last}}})$ used in (7.69) do not depend on λ and n when the process is IC. Therefore, they can be approximated by the empirical mean and variance of $W_{n,\widetilde{\gamma}_{m_k}^{\text{last}}}$, computed from the simulated IC observation vectors, since $\mathbf{E}_n$ is the same as $\mathbf{X}_n$ when $\lambda = 1$. The control limit h in (7.69) can also be determined by a Monte Carlo simulation.

Example 7.9 *Assume that the IC distribution of a 5-dimensional production process is $N_5(\mathbf{0}, \Sigma_0)$, where $\Sigma_0 = (\sigma_{ij})$ is specified as $\sigma_{ii} = 1$ and $\sigma_{ij} = 0.75^{|i-j|}$, for $i, j = 1, 2, \ldots, 5$. In the LEWMA chart (7.67)–(7.69), q, λ, and ARL_0 are chosen to be 5, 0.2, and 500, respectively. Then, by a simulation with 20,000 replicated runs, its control limit h is computed to be 5.262. Besides this chart, the MEWMA chart (7.39)–(7.40) by Lowry et al. (1992), denoted as MEWMA, and the same chart applied to the regression-adjusted observations $\{\mathbf{Z}_n, n = 1, 2, \ldots\}$ (cf., (7.21) and (7.22)), denoted as REWMA, are also considered. In each of the MEWMA and REWMA charts, a single weighting parameter is used and its value is chosen to be 0.2, as in the LEWMA chart. The control limits of the MEWMA and REWMA charts are computed by simulation to be 18.130 and 3.425, respectively, to reach the ARL_0 value of 500. Then,*

20 different shifts are considered, and a true shift is assumed to occur at $\tau = 26$. The computed OC ARL values of the three charts by a simulation with 20,000 replications are presented in Table 7.8. From the table, we can see that the LEWMA chart may not be the best in a given case, but it is always close to the best. As a comparison, the MEWMA chart does not perform well for detecting the first several shifts in the table while the REWMA chart does not perform well for detecting the last several shifts.

Table 7.8 *OC ARL values of the control charts LEWMA, MEWMA, and REWMA in the case when $p = 5$, $\lambda = 0.2$, and $ARL_0 = 500$. Numbers in parentheses are standard errors. They are shown as ".00" when they are smaller than 0.005.*

Shifts							
X_1	X_2	X_3	X_4	X_5	MEWMA	REWMA	LEWMA
0.91	0.00	0.00	0.00	0.00	17.3 (.09)	14.4 (.07)	14.9 (.07)
0.00	0.36	0.00	0.00	0.00	17.0 (.08)	13.3 (.06)	14.3 (.07)
0.00	0.00	0.48	0.00	0.00	17.3 (.09)	13.8 (.07)	15.0 (.07)
0.00	0.00	0.00	0.34	0.00	17.2 (.09)	13.4 (.06)	14.6 (.07)
0.00	0.00	0.00	0.00	0.46	17.9 (.09)	14.1 (.07)	15.2 (.07)
0.36	0.36	0.00	0.00	0.00	15.0 (.07)	13.5 (.06)	13.7 (.06)
0.54	0.00	0.54	0.00	0.00	12.8 (.06)	12.9 (.06)	12.4 (.05)
0.32	0.00	0.00	0.32	0.00	15.2 (.07)	13.1 (.06)	13.6 (.06)
0.49	0.00	0.00	0.00	0.49	13.2 (.06)	12.6 (.06)	12.5 (.06)
0.00	0.54	0.54	0.00	0.00	8.79 (.03)	12.3 (.06)	8.88 (.03)
0.00	1.60	0.00	1.60	0.00	3.48 (.00)	9.03 (.03)	3.57 (.00)
0.00	0.28	0.00	0.00	0.28	13.0 (.06)	10.5 (.05)	11.3 (.05)
0.00	0.00	0.28	0.28	0.00	13.0 (.06)	10.5 (.05)	11.4 (.05)
0.00	0.00	1.26	0.00	1.26	4.28 (.01)	8.70 (.03)	4.34 (.01)
0.00	0.00	0.00	0.56	0.56	8.60 (.03)	12.2 (.06)	8.70 (.03)
0.01	−0.15	0.07	0.17	−0.09	15.0 (.07)	11.5 (.05)	12.5 (.06)
0.07	−0.13	−0.40	0.19	0.35	10.1 (.04)	12.4 (.06)	10.1 (.04)
0.40	0.63	−0.57	0.47	−0.68	4.74 (.01)	9.28 (.04)	4.94 (.01)
−1.11	0.26	−0.17	0.34	−0.04	11.8 (.05)	14.6 (.07)	12.2 (.05)
2.51	7.11	7.05	7.11	7.08	1.19 (.00)	7.52 (.02)	1.30 (.00)

Assume that the LEWMA chart (7.67)–(7.69) gives a mean shift signal at the n-th observation. Then, estimation of the shift location and identification of the shifted mean components can proceed as follows, which will help engineers to eliminate the root causes of the shift in a timely fashion. With respect to the shift location, it can be determined by a change-point detection approach (e.g., Zamba and Hawkins, 2006). After an estimator of the shift location is obtained, which is denoted as $\hat{\tau}$, we have $n - \hat{\tau} + 1$ OC observations, of which a few mean components have shifted. Among these $n - \hat{\tau} + 1$ observations, a number of them might actually be IC observations, because $\hat{\tau}$ is only an estimator of the true shift location τ. After $\hat{\tau}$ is obtained, the LASSO methodology can be used for specifying the shifted mean components as follows. Let

$$\gamma^* = \arg\min_{\gamma}(n - \hat{\tau} + 1)\left(\overline{\mathbf{X}}_{\hat{\tau}-1,n} - \hat{\mu}_\gamma\right)' \Sigma_0^{-1}\left(\overline{\mathbf{X}}_{\hat{\tau}-1,n} - \hat{\mu}_\gamma\right) + \eta \cdot \hat{df}(\hat{\mu}_\gamma), \quad (7.70)$$

where $\overline{\mathbf{X}}_{\hat{\tau}-1,n} = \sum_{i=\hat{\tau}}^{n} \mathbf{X}_i/(n-\hat{\tau}+1)$, $\widehat{\mu}_{\gamma}$ is the ALASSO estimator of μ from $\overline{\mathbf{X}}_{\hat{\tau}-1,n}$, η is a parameter, and $\widehat{df}(\widehat{\mu}_{\gamma})$ is the number of nonzero coefficients in $\widehat{\mu}_{\gamma}$. Obviously, (7.70) is a model selection criterion (cf., Yang, 2005). If the AIC model selection procedure is used, then $\eta = 2$. It equals $\ln(n-\hat{\tau}+1)$ if the BIC procedure is used. Based on some numerical studies, we found that the diagnostic procedure (7.70) would perform well if the risk inflation criterion (RIC) proposed by George and Foster (1994) was adopted, in which $\eta = 2\log(p)$. Zou et al. (2007c) has shown that $\widehat{\mu}_{\gamma^*}$ is one of $\{\widehat{\mu}_{\tilde{\gamma}_1}, \ldots, \widehat{\mu}_{\tilde{\gamma}_K}\}$. Since some components of $\widehat{\mu}_{\gamma^*}$ are exactly zero, we can simply take its nonzero components as the shifted mean components.

In the literature, there is some related discussion on MSPC using regression variable selection and LASSO. For instance, Wang and Jiang (2009) suggested if a forward search algorithm is used to determine the shifted mean components, the dimensionality of the MSPC problem can then be reduced, and consequently the resulting control charts could be more effective in detecting process mean shifts. Capizzi and Masarotto (2011) suggested a MEWMA chart by integrating the MSPC problem with the variable selection technique using the LARS algorithm. One major difference between this chart and the chart by Zou and Qiu (2009) is that the change-point model in the former approach includes both the "unstructured" scenarios, such as the ones considered by Wang and Jiang (2009) and Zou and Qiu (2009), and the "structured" scenarios, such as shifts occurring at some specific stages of a multistage process or shifts involving parameters of a parametric profile. Besides process mean shifts, the approach by Capizzi and Masarotto (2011) can also detect an increase in the total variability of the process. The LASSO-based MEWMA chart by Li et al. (2013a) is specifically for detecting shifts in the process covariance matrix. Regarding the post-signal diagnostics, Zou et al. (2011) proposed a LASSO-based approach using the BIC variable selection criterion. This approach does not assume the IC process distribution to be known.

7.7   Some Discussions

In the previous sections, we have described some MSPC control charts in cases when multiple quality characteristics of a production process, denoted as a vector $\mathbf{X}$, are concerned, and when the IC and OC distributions of $\mathbf{X}$ are both normal. In most of these control charts, the Hotelling's T^2 statistic plays an important role. The multivariate Shewhart charts described in Section 7.2 are based directly on different versions of the Hotelling's T^2 statistic. Many other types of control charts are usually constructed in two steps. First, a statistic carrying useful information in the observed data about a potential shift in concern is constructed. Such a statistic usually has the same dimension as $\mathbf{X}$. Then, in the second step, this multi-dimensional statistic is transformed to a scalar charting statistic by a quadratic transformation in a similar form to the Hotelling's T^2 statistic. Section 7.2 has provided an explanation about the reason why the Hotelling's T^2 statistic is so important in cases when the distribution of $\mathbf{X}$ is normal. One major reason is that the Hotelling's T^2 statistic provides a good measure of the distance between the observed data and the assumed IC model of the production process.

In this chapter, observation vectors at different time points are assumed independent, which may be invalid in certain applications. The MSPC problem when the observation vectors at different time points are correlated has been discussed by several authors. See, for instance, Jiang (2004), Kramer and Schmid (1997), Pan and Jarrett (2004, 2007), and Vanbrackle and Reynolds (1997). But, this problem has not been completely solved because most of the existing methodologies use some specific time series models to account for the autocorrelation among different observation vectors. However, such time series models may be invalid in certain applications. Many MSPC control charts discussed in this chapter assume that the IC process distribution is known, which is rarely true in practice. As discussed in Section 7.4, a number of self-starting MEWMA charts have been developed to handle the case when the IC process distribution is unknown. But, some questions are still open. For instance, a self-starting MEWMA chart usually requires a number of IC observations to be available before the online process monitoring. If the number of such IC observations is larger, then the self-starting process monitoring would be more reliable, especially in the early stage of the monitoring. But, it might be challenging to collect such IC observations in certain applications. Also, this issue is closely related to the dimensionality p of $\mathbf{X}$. A dozen IC observations might be good enough in 2-dimensional cases; but they would not be enough in 10 or higher dimensional cases. Therefore, much future research is required to address all such issues.

In univariate cases, it has been shown that the CUSUM chart for monitoring process mean shifts has the optimality property when its allowance constant is chosen to be half of a target shift (cf., Subsection 4.2.2). In multivariate cases, similar results about the design of a MSPC chart are still lacking. For instance, in the regression-adjusted multivariate CUSUM chart (7.25), it has been shown that the allowance constant k should be chosen relatively large if the shift size $\delta = \mu_1 - \mu_0$ is large in the sense that its statistical length $\sqrt{\delta' \Sigma_0^{-1} \delta}$ is large. However, the specific relationship between the optimal value of k and the quantity $\sqrt{\delta' \Sigma_0^{-1} \delta}$ is still unknown. For the MEWMA charts, as discussed in Subsection 7.4.1, it is still unclear in which cases we should use a general weighting matrix with non-zero off-diagonal elements, and in which other cases we should use a single weighting parameter. Also, the relationship among the optimal weighting parameters, the shift size δ, and the dimensionality p requires much future research to explore. In the multivariate CPD control charts, the control limits usually depend on the observation times in the early stage of process monitoring, and they stabilize over time. In some cases, numerical formulas are available for approximating the control limits. These formulas often have room for further improvement in their approximation accuracy. The LASSO-based MSPC charts are relatively new. They usually combine a variable selection procedure (e.g., ALASSO, LAR) with a SPC charting technique (e.g., EWMA, CUSUM). Therefore, all design issues related to the multivariate CUSUM, EWMA, and other types of control charts are also relevant to them.

All control charts discussed in this chapter are based on the assumption that the process distribution is multivariate normal, which is rarely valid in practice. Therefore, it is important to investigate their robustness to the normality assumption. If some of them are not robust, then how can they be modified properly so that the modified charts are more robust? One solution to this question is to use the nonparametric control charts discussed in the next two chapters. Another possible solution is to adopt the transformation approach, by which the observed data are first transformed properly so that the transformed data are approximately normally distributed and then the MSPC control charts are applied to the transformed data. So far, we have not found much existing research on this topic. Also, little existing research can be found in handling cases when the distribution of X belongs to a parametric family other than the multivariate normal distribution family (e.g., multivariate t, multivariate Γ), or when only part of the components of X follows a normal distribution.

7.8 Exercises

7.1 Assume that the distribution of the 3-dimensional random vector $X = (X_1, X_2, X_3)'$ is $N_3(\mu, \Sigma)$, where

$$\mu = (1, 2, -1)'$$

and

$$\Sigma = \begin{pmatrix} 1.0 & 0.7 & 0.2 \\ 0.7 & 1.0 & 0.7 \\ 0.2 & 0.7 & 1.0 \end{pmatrix}.$$

Find the distributions of the following random variables or random vectors:

(i) X_2,

(ii) $(X_1, X_2)'$,

(iii) AX, where $A = \begin{pmatrix} 2 & 7 & -9 \\ 5 & -5 & 0 \end{pmatrix}$.

7.2 Assume that the distribution of the 3-dimensional random vector $X = (X_1, X_2, X_3)'$ is $N_3(\mu, \Sigma)$, where

$$\mu = (1, 2, -1)'$$

and

$$\Sigma = \begin{pmatrix} 1.0 & 0.7 & 0 \\ 0.7 & 1.0 & 0 \\ 0 & 0 & 1.0 \end{pmatrix}.$$

For each pair of the random variables below, determine whether they are independent of each other, and provide an explanation.

(i) X_1 and X_2,

(ii) X_1 and X_3,

(iii) $(X_1 + X_2)/2$ and X_3,

(iv) X_1 and $(X_2 + X_3)/2$.

7.3 Make a plot similar to the one in Figure 7.2 in cases when $p = 2, \mu_{X_1} = 3, \mu_{X_2} = 3, \sigma_{X_1} = 1, \sigma_{X_2} = 1$, and $\text{Cor}(X_1, X_2) = 0.9$. Summarize your findings from this plot.

7.4 Assume that $(\mathbf{X}_1, \mathbf{X}_2, \ldots, \mathbf{X}_n)$ is a simple random sample from a p-dimensional population with the distribution $N_p(\mu, \Sigma)$.

(i) Find the distribution of $\overline{\mathbf{X}}$.

(ii) Find the probability $P(d_S^2(\overline{\mathbf{X}}, \mu) > 10)$ when $p = 5$, where

$$d_S^2(\overline{\mathbf{X}}, \mu) = n(\overline{\mathbf{X}} - \mu)' \Sigma^{-1} (\overline{\mathbf{X}} - \mu)$$

is the squared statistical distance from the sample mean vector $\overline{\mathbf{X}}$ to the population mean vector μ.

(iii) The region

$$\{\mu : d_S^2(\overline{\mathbf{X}}, \mu) \leq \chi_{1-\alpha, p}^2\}$$

has a $100(1 - \alpha)\%$ chance to cover the true population mean vector μ, where $\alpha \in [0, 1]$ is a given significance level. This region is often called a $100(1 - \alpha)\%$ *confidence region* for μ. Compute a 95% confidence region for μ in the case when $p = 2$, $\overline{\mathbf{X}} = (5, 5)'$, and $\Sigma = \begin{pmatrix} 1 & 0.5 \\ 0.5 & 1 \end{pmatrix}$. Make a plot of this confidence region.

7.5 Assume that $(\mathbf{X}_1, \mathbf{X}_2, \ldots, \mathbf{X}_n)$ is a simple random sample from a p-dimensional population whose distribution is unknown. Let $\overline{\mathbf{X}}$ and $\mathbf{S}^2$ be the sample mean vector and the sample covariance matrix. In cases when $p = 4$ and $n = 100$, compute the following probabilities:

(i) $P(d_S^2(\overline{\mathbf{X}}, \mu) \leq 9)$, where $d_S^2(\overline{\mathbf{X}}, \mu)$ is defined in the previous exercise,

(ii) $P(n(\overline{\mathbf{X}} - \mu)'(\mathbf{S}^2)^{-1}(\overline{\mathbf{X}} - \mu) > 6)$.

7.6 The first 20 observation vectors in Table 7.1 were actually generated from the distribution $N_3(\mathbf{0}, \Sigma_0)$, where

$$\Sigma_0 = \begin{pmatrix} 1.0 & 0.8 & 0.5 \\ 0.8 & 1.0 & 0.8 \\ 0.5 & 0.8 & 1.0 \end{pmatrix}.$$

Assume that this is the IC process distribution and it is known. Use the multivariate Shewhart chart (7.5)–(7.6) with $\alpha = 0.005$ to monitor the process mean vector by applying it to the 30 observation vectors in Table 7.1. Compare your results with the results in Example 7.1 where the IC process distribution is assumed unknown.

7.7 The data shown in Table 7.9 include the first 50 observation vectors obtained from a sheet metal assembly process. The six quality characteristic variables x_1–x_6 denote the sensor recorded deviation from the nominal thickness (millimeters) at six locations on a car. This dataset was obtained from Table 5.14 in Johnson and Wichern (2007). Assume that the first 30 observation vectors are IC. Use the multivariate Shewhart chart (7.9)–(7.10) with $\alpha = 0.005$ to monitor the remaining 20 observation vectors for detecting a potential shift in the process mean vector. Summarize your results.

Table 7.9 *This table presents the first 50 observation vectors obtained from a sheet metal assembly process.*

i	X_1	X_2	X_3	X_4	X_5	X_6
1	−0.12	0.36	0.40	0.25	1.37	−0.13
2	−0.60	−0.35	0.04	−0.28	−0.25	−0.15
3	−0.13	0.05	0.84	0.61	1.45	0.25
4	−0.46	−0.37	0.30	0.00	−0.12	−0.25
5	−0.46	−0.24	0.37	0.13	0.78	−0.15
6	−0.46	−0.16	0.07	0.10	1.15	−0.18
7	−0.46	−0.24	0.13	0.02	0.26	−0.20
8	−0.13	0.05	−0.01	0.09	−0.15	−0.18
9	−0.31	−0.16	−0.20	0.23	0.65	0.15
10	−0.37	−0.24	0.37	0.21	1.15	0.05
11	−1.08	−0.83	−0.81	0.05	0.21	0.00
12	−0.42	−0.30	0.37	−0.58	0.00	−0.45
13	−0.31	0.10	−0.24	0.24	0.65	0.35
14	−0.14	0.06	0.18	−0.50	1.25	0.05
15	−0.61	−0.35	−0.24	0.75	0.15	−0.20
16	−0.61	−0.30	−0.20	−0.21	−0.50	−0.25
17	−0.84	−0.35	−0.14	−0.22	1.65	−0.05
18	−0.96	−0.85	0.19	−0.18	1.00	−0.08
19	−0.90	−0.34	−0.78	−0.15	0.25	0.25
20	−0.46	0.36	0.24	−0.58	0.15	0.25
21	−0.90	−0.59	0.13	0.13	0.60	−0.08
22	−0.61	−0.50	−0.34	−0.58	0.95	−0.08
23	−0.61	−0.20	−0.58	−0.20	1.10	0.00
24	−0.46	−0.30	−0.10	−0.10	0.75	−0.10
25	−0.60	−0.35	−0.45	0.37	1.18	−0.30
26	−0.60	−0.36	−0.34	−0.11	1.68	−0.32
27	−0.31	0.35	−0.45	−0.10	1.00	−0.25
28	−0.60	−0.25	−0.42	0.28	0.75	0.10
29	−0.31	0.25	−0.34	−0.24	0.65	0.10
30	−0.36	−0.16	0.15	−0.38	1.18	−0.10
31	−0.40	−0.12	−0.48	−0.34	0.30	−0.20
32	−0.60	−0.40	−0.20	0.32	0.50	0.10
33	−0.47	−0.16	−0.34	−0.31	0.85	0.60
34	−0.46	−0.18	0.16	0.01	0.60	0.35
35	−0.44	−0.12	−0.20	−0.48	1.40	0.10
36	−0.90	−0.40	0.75	−0.31	0.60	−0.10
37	−0.50	−0.35	0.84	−0.52	0.35	−0.75
38	−0.38	0.08	0.55	−0.15	0.80	−0.10
39	−0.60	−0.35	−0.35	−0.34	0.60	0.85
40	0.11	0.24	0.15	0.40	0.00	−0.10
41	0.05	0.12	0.85	0.55	1.65	−0.10
42	−0.85	−0.65	0.50	0.35	0.80	−0.21
43	−0.37	−0.10	−0.10	−0.58	1.85	−0.11
44	−0.11	0.24	0.75	−0.10	0.65	−0.10
45	−0.60	−0.24	0.13	0.84	0.85	0.15
46	−0.84	−0.59	0.05	0.61	1.00	0.20
47	−0.46	−0.16	0.37	−0.15	0.68	0.25
48	−0.56	−0.35	−0.10	0.75	0.45	0.20
49	−0.56	−0.16	0.37	−0.25	1.05	0.15
50	−0.25	−0.12	−0.05	−0.20	1.21	0.10

7.8 Assume that the data shown in Table 7.9 are batch data with the batch size of $m = 2$. The first batch includes the first two observation vectors in the table, the second batch includes the next two observation vectors in the table, and so forth. So, there are a total of 25 batches of data in the table. Use the multivariate Shewhart chart (7.14)–(7.15) with $\alpha = 0.005$ to monitor the last 10 batches of observation vectors for detecting potential process mean shift, by assuming the first 15 batches of observation vectors are IC. Compare your results with those obtained in the previous exercise.

7.9 Assume that the 30 observation vectors presented in Table 7.1 are obtained from a 3-dimensional production process with the IC distribution $N_3(\mathbf{0}, \Sigma_0)$, where

$$\Sigma_0 = \begin{pmatrix} 1.0 & 0.8 & 0.5 \\ 0.8 & 1.0 & 0.8 \\ 0.5 & 0.8 & 1.0 \end{pmatrix}.$$

(i) Apply the joint monitoring scheme (7.16)–(7.17) to this data for monitoring the process mean vector. In the joint monitoring scheme, use $k_1 = k_2 = k_3 = 0.5$, and $h_1 = h_2 = h_3 = 4.171$. So, by Table 4.1 in Subsection 4.2.2, each individual two-sided CUSUM chart in the joint monitoring scheme would have the ARL_0 value of 200. Summarize your results.

(ii) Decompose the IC covariance matrix Σ_0 into

$$\Sigma_0 = Q'\Lambda Q,$$

where Q is a $p \times p$ orthogonal matrix, and $\Lambda = diag(\lambda_1, \lambda_2, \ldots, \lambda_p)$ is a diagonal matrix. Then, $\mathbf{X}^* = \Lambda^{-1/2}Q\mathbf{X}$ would have the identity covariance matrix, and thus its p components are independent of each other, as discussed in Subsection 7.3.1 immediately after Example 7.2. Apply the joint monitoring scheme (7.16)–(7.17) to the transformed data

$$\mathbf{X}_n^* = \Lambda^{-1/2}Q\mathbf{X}_n, \qquad \text{for } 1 \le n \le 30.$$

Use the same setup as that in part (i) for the joint monitoring scheme. Summarize your results, and compare the results with those in part (i). (Note: The decomposition of Σ_0 can be accomplished by using the R function eigen().)

7.10 For the data described in the previous exercise, assume that a potential shift is $\mu_1 = (1, 1, 1)'$. Apply the CUSUM chart (7.19)–(7.20) to the data. Design the chart specifically for detecting μ_1 with $ARL_0 = 200$. Summarize your results.

7.11 For the data described in Exercise 7.9, compute the regression-adjusted data by (7.22). Plot the original data and the regression-adjusted data, and discuss their major differences.

7.12 Reproduce the results in Example 7.3.

7.13 For the data shown in Table 7.9, assume that the first 30 observation vectors are IC, and their sample mean and sample covariance matrix can be regarded as the IC process mean and the IC process covariance matrix, respectively. Use the control chart (7.25) to monitor the remaining 20 observation vectors in the table. In the chart, choose $k = 0.25$ and $h = 8,585$, so that the ARL_0 value of each individual two-sided CUSUM chart defined by (7.23)–(7.24) is about 500 (cf., Table 4.1 in Subsection 4.2.2). Summarize your results.

7.14 Apply the CUSUM chart (7.28)–(7.29) to the data described in Example 7.3. In the chart, choose $k = 0.5$ and $h = 6.5$. Summarize your results.

7.15 For the data described in Exercise 7.9, use the multivariate CUSUM chart (7.32)–(7.33) and the COT chart (7.30)–(7.31) to detect possible process co-variance matrix shifts. In the chart (7.32)–(7.33), choose k, ARL_0, and C_0 to be 4, 200, and 0, respectively. By the results in Example 7.4, its control limit h is 13.062. In the COT chart, choose k, ARL_0, and $\widetilde{C}_0$ to be 1, 200, and 0, respectively, and the corresponding control limit $\widetilde{h}$ is 2.713. Summarize your results, and discuss your major findings.

7.16 For the data shown in Table 7.9, assume that the first 30 observation vectors are known to be IC, and we are asked to monitor the remaining 20 observation vectors for possible process mean and process covariance matrix shifts. Use the methods discussed in Section 7.3 to achieve that goal. Describe your monitoring schemes, and summarize your major findings.

7.17 Verify the results in (7.41) and (7.42).

7.18 Reproduce part of the results in Table 7.4 in cases when the shift size $\delta = (0.25, 0.25, 0.25)'$, $(0.25, 0.25, 0)'$, and $(0.25, 0, 0)'$. The sizes of these three shifts seem to decrease from $(0.25, 0.25, 0.25)'$ to $(0.25, 0, 0)'$. Explain why the ARL_1 values of the chart (7.39)–(7.40) also decrease over these three shifts.

7.19 Apply the MEWMA chart (7.39)–(7.40) to the data discussed in Example 7.3. In the chart, use a single weighting parameter λ, and choose the values of λ and ARL_0 to be 0.2 and 200, respectively. Summarize your results.

7.20 Verify the equations in (7.43)

7.21 Apply the self-starting MEWMA chart (7.45)–(7.46) to the data in Table 7.9. In the chart, choose λ and ARL_0 to be 0.2 and 200, respectively. By Example 7.6, the control limit h_{SS} is 14.167 in such cases, if the the IC process distribution is assumed to be $N_p(\mathbf{0}, I_{p \times p})$. Explain that this control limit value is still approximately valid in the current problem. Summarize your major findings from the control chart.

7.22 Consider a 3-dimensional production process with the IC distribution $N_3(\mathbf{0}, \Sigma_0)$ where

$$\Sigma_0 = \begin{pmatrix} 1.0 & 0.7 & 0.3 \\ 0.7 & 1.0 & 0.7 \\ 0.3 & 0.7 & 1.0 \end{pmatrix}.$$

The first 30 phase II observations are presented in Table 7.10. Apply the MEWMA chart (7.51)–(7.52) to this dataset. In the chart, choose λ and ARL_0 to be 0.2 and 250, respectively. Summarize your results.

7.23 Reproduce the results in Example 7.8.

Table 7.10 *This table presents the first 30 phase II observations of a 3-dimensional production process.*

n	X_{n1}	X_{n2}	X_{n3}	n	X_{n1}	X_{n2}	X_{n3}
1	0.526	−0.352	−0.420	16	−0.921	−0.264	0.500
2	0.182	−0.098	0.296	17	0.953	0.604	−1.074
3	0.104	−0.152	−0.135	18	−0.861	0.392	1.783
4	0.397	0.622	1.273	19	0.245	0.887	2.247
5	−0.126	0.350	0.020	20	0.090	1.879	−0.125
6	0.190	0.433	0.155	21	0.287	−0.440	−0.092
7	−0.356	−0.346	−0.816	22	1.390	1.661	1.291
8	0.301	0.616	0.905	23	−1.730	−1.599	−2.279
9	−0.314	−0.451	−1.395	24	−1.691	1.421	2.050
10	−1.606	−0.416	1.128	25	−0.443	−0.866	−0.254
11	0.334	−0.267	−0.314	26	1.756	1.605	0.722
12	2.720	2.439	0.201	27	−0.765	−0.503	0.159
13	−0.516	0.255	−0.256	28	1.302	1.252	1.595
14	−1.123	−0.721	1.627	29	0.469	−0.567	0.368
15	0.533	−0.070	−2.760	30	−1.765	−2.163	−1.899

7.24 Apply the CPD control chart (7.54)–(7.55) to the data presented in Table 7.9 for monitoring the process mean vector. In the data, the first 25 observation vectors are assumed to be IC, and the process monitoring starts at the 26th time point. In the chart, α is chosen to be 0.002, h_n can be approximated by the second formula in (7.56), and the formulas (7.57)–(7.59) can be used for computing the charting statistic $T^2_{max,n}$ recursively. Construct the control chart, and present it in a plot. Do the post-signal analysis (e.g., compute the estimates of the shift time, IC and OC process mean vectors, IC process covariance matrix, etc.), if a signal of process mean shift is given by the chart. Summarize your results.

7.25 For the CPD control chart (7.60)–(7.61), what is the minimum value of n at which the chart can be used? If the chart is used at the n-th time point and n is too small, what are the possible consequences?

7.26 Verify that (7.64) is equivalent to (7.63).

Chapter 8

Univariate Nonparametric Process Control

8.1 Introduction

The SPC control charts discussed in the previous several chapters, for monitoring univariate or multivariate continuous numerical quality characteristics, are all based on the assumption that the related quality characteristics follow normal distributions when the production process in question is IC and after it becomes OC. As pointed out in Subsection 2.3.1, normal distributions play an important role in statistics, because many continuous numerical variables in practice roughly follow normal distributions and much statistical theory is developed for normally distributed random variables. An intuitive explanation about the reason many continuous numerical variables in our daily life roughly follow normal distributions can be given using the central limit theorem (CLT) discussed in Subsection 2.7.1. For instance, a quality characteristic in question (e.g., the lifetime of a machine) is often affected by many different factors, including the quality of the raw material, labor, manufacturing facilities, proper operation in the manufacturing process, and so forth (cf., Figure 1.1 in Section 1.2). So, by the CLT, its distribution would be roughly normal.

In practice, however, there are also many variables whose distributions are substantially different from normal distributions. For instance, economic indices and other nonnegative indices are often skewed to the right. The lifetimes of products can often be described properly by Weibull distributions, which could be substantially different from normal distributions (cf., Subsection 2.3.5). In many cases, it is difficult to find a parametric distribution to describe the distribution of a variable in question. In such cases, various nonparametric methods are available for estimating the variable distribution (cf., Jones et al., 1996; Scott, 1992; Wand and Jones, 1995).

In cases when the normality assumption is invalid, several authors have pointed out that the conventional control charts would be unreliable for process monitoring (cf., Amin et al., 1995; Hackl and Ledolter, 1992; Lucas and Crosier, 1982b; Rocke, 1989). For instance, Qiu and Li (2011a) provided a demonstration example described below. Figure 8.1 shows the actual IC ARL values of the conventional CUSUM chart (4.7)–(4.8) based on the assumption that the IC process distribution is $N(0,1)$, in cases when the allowance constant k of the chart is chosen to be 0.5, the assumed IC ARL value is 500, and the true process distribution is the standardized version (with mean 0 and variance 1) of the chi-square (plot (a)) or t (plot (b)) distribution.

From the plots, it can be seen that the actual IC ARL values of the conventional CUSUM chart are much smaller than the assumed IC ARL value in both cases when the df is small, implying that the related process would be stopped too often by the control chart when it remains IC and consequently a considerable amount of time and resources would be wasted in such cases.

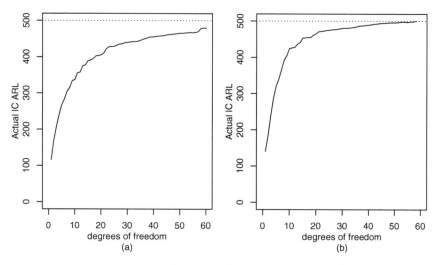

Figure 8.1 *Actual IC ARL values of the conventional CUSUM chart (4.7)–(4.8) in cases when the assumed IC ARL value is 500, and the true process distribution is the standardized version (with mean 0 and variance 1) of the chi-square (plot (a)) or t (plot (b)) distribution with degrees of freedom changing from 1 to 60 in plot (a) and from 3 to 60 in plot (b).*

It has been pointed out that the conventional EWMA charts discussed in Chapter 5 were quite robust to the normality assumption when the weighting parameter λ was chosen to be small (cf., Borror et al., 1999; Testik et al., 2003). As discussed at the end of Subsection 5.2.1, we think that the robustness property of the EWMA charts should be used with care for two main reasons. First, from various numerical results presented in Example 5.2 and in several other sources, it seems that it depends on the magnitude of the difference between the actual IC process distribution and a normal distribution to decide how small the λ value should be chosen in order to have the robustness property, and this magnitude is difficult to measure in practice. Second, if λ is chosen to be small, then the resulting chart would be ineffective in detecting relatively large shifts, as discussed in Subsection 5.2.1 This viewpoint is confirmed recently by an extensive numerical study in Human et al. (2011). Therefore, in cases when the actual IC process distribution is non-normal, we suggest using the nonparametric (or distribution-free) control charts that will be discussed in this and the next chapters, unless we know beforehand that the IC process distribution is quite close to a normal distribution or to another parametric distribution that can be handled more effectively by a parametric control chart.

To handle cases when the normality assumption is invalid, a number of distribution-free or nonparametric control charts have been developed in the literature. See, for instance, Albers and Kallenberg (2004, 2009), Albers et al. (2006), Alloway and Raghavachari (1991), Amin et al. (1995), Amin and Searcy (1991), Amin and Widmaier (1999), Bakir (2004, 2006), Bakir and Reynolds (1979), Chakraborti and Eryilmaz (2007), Chakraborti et al. (2009, 2004), Chakraborti and van der Laan (2000), Chatterjee and Qiu (2009), Hackl and Ledolter (1991), Hawkins and Deng (2010), McDonald (1990), Park and Reynolds (1987), Qiu and Li (2011a,b), Willemain and Runger (1996), Yashchin (1992), and Zou and Tsung (2010) for some related discussions on univariate nonparametric SPC. Chakraborti et al. (2001) gave a quite thorough overview on this topic. In this chapter, we describe some fundamental univariate nonparametric SPC charts. Some important multivariate nonparametric SPC charts will be described in the next chapter. It should be pointed out that, in the literature, people do not always make a clear distinction between the terminologies of "nonparametric control charts" and "distribution-free control charts." In many papers, both terminologies are used to refer to the control charts that can be applied to cases when the process distribution does not have a parametric form. Some "distribution-free control charts" may not be really distribution-free, in the sense that their design may still depend on the process distribution, although they do not require a parametric form of the process distribution.

8.2   Rank-Based Nonparametric Control Charts

As discussed in Section 2.8, in cases when a parametric model is unavailable to describe the process distribution (or, the process distribution is nonparametric), nonparametric statistical methods based on the ranking or ordering information of the observed data can be considered for making inferences about the underlying process distribution. In this section, we describe some fundamental control charts constructed using certain rank-based nonparametric statistical methods for process monitoring in cases when the process distribution is nonparametric. Our description will mainly focus on the monitoring of process mean, although some methods described can also be used for the monitoring of process variance. This section is organized in four parts. Shewhart-type nonparametric control charts are discussed in Subsection 8.2.1, and some nonparametric CUSUM, EWMA, and CPD charts are described in Subsections 8.2.2–8.2.4, respectively.

8.2.1   Nonparametric Shewhart charts

In the literature, a number of nonparametric Shewhart charts have been proposed for both phase I and phase II process monitoring. Some relatively early discussions on this topic can be found in papers, such as Alloway and Raghavachari (1991), Amin et al. (1995), Pappanastos and Adams (1996), Park and Reynolds (1987), Willemain and Runger (1996), and the references cited therein. In this subsection, we first describe the nonparametric Shewhart chart proposed by Bakir (2004), which is based on the Wilcoxon signed-rank test discussed in Subsection 2.8.3.

The nonparametric Shewhart chart by Bakir (2004) works with batch data. Assume that $X_{i1}, X_{i2}, \ldots, X_{im}$ are m observations obtained at the i-th time point, where

$m > 1$ is the batch size. The m observations obtained at the same time point are assumed to be i.i.d.. It is further assumed that the process distribution is symmetric about its median $\widetilde{\mu}$. Then, $\widetilde{\mu}$ is obviously the mean of the process distribution as well. When the process is IC, $\widetilde{\mu} = \widetilde{\mu}_0$. Here, we assume that $\widetilde{\mu}_0$ is known; but, it should be estimated from an IC dataset in practice. Define the charting statistic to be

$$\psi_i = \sum_{j=1}^{m} \text{sign}(X_{ij} - \widetilde{\mu}_0) R_{ij}^+, \tag{8.1}$$

where $\text{sign}(u) = -1, 0, 1$, respectively, when $u < 0, = 0, > 0$,

$$R_{ij}^+ = 1 + \sum_{\ell=1}^{m} I(|X_{i\ell} - \widetilde{\mu}_0| < |X_{ij} - \widetilde{\mu}_0|),$$

and $I(u)$ equals 1 and 0, respectively, when u is "true" and "false". In the above expressions, the value of R_{ij}^+ is actually the rank of $|X_{ij} - \widetilde{\mu}_0|$ in the sequence $|X_{i1} - \widetilde{\mu}_0|, |X_{i2} - \widetilde{\mu}_0|, \ldots, |X_{im} - \widetilde{\mu}_0|$, $\text{sign}(X_{ij} - \widetilde{\mu}_0) R_{ij}^+$ is the conventional Wilcoxon signed-rank of $X_{ij} - \widetilde{\mu}_0$ within the i-th batch of observations, and ψ_i is the sum of all Wilcoxon signed-ranks within the i-th batch (cf., Subsection 2.8.3). Obviously, in cases when the process is IC, positive and negative Wilcoxon signed-ranks are mostly canceled out in ψ_i. Consequently, the absolute value of ψ_i should be small in such cases and ψ_i should be centered at 0. When the process has a positive mean (or median) shift at the i-th time point, however, the positive Wilcoxon signed-ranks would be larger than the negative Wilcoxon signed-ranks in magnitudes within the i-th batch, resulting in a relatively large value of ψ_i. Thus, the upward chart gives a signal of an upward process mean shift if

$$\psi_i \geq U, \tag{8.2}$$

where $U > 0$ is an upper control limit. Similarly, the downward chart gives a signal of a downward process mean shift when $\psi_i \leq L$, and the two-sided chart gives a signal of an arbitrary process mean shift when $\psi_i \leq L$ or $\psi_i \geq U$, where $L < 0$ is a lower control limit. Since the distribution of ψ_i is symmetric about 0 when the process is IC, it is natural to choose $L = -U$. This control chart will be called the *nonparametric signed-rank (NSR) chart* hereafter.

From the definition of ψ_i, it can be checked that

$$\psi_i = 2S_+ - m(m+1)/2,$$

where S_+ is the Wilcoxon signed-rank test statistic described in Subsection 2.8.3. By this connection, the IC distribution of ψ_i can be obtained from the IC distribution of S_+, which has been tabulated in Table 2.7. Then, the false alarm rate α^+ of the upward NSR chart can be computed by

$$\alpha^+ = P(\psi_i \geq U),$$

where the probability is computed using the IC distribution of ψ_i, and the IC ARL value of the upward NSR chart, denoted as ARL_0^+, can be computed by

$$ARL_0^+ = 1/\alpha^+.$$

When $m = 5, 6, 8$, and 10, and U takes various integer values, the values of α^+ and ARL_0^+ have been computed in Bakir (2004), and they are presented in Table 8.1. From the table, it can be seen that, for a given batch size m, the upward NSR chart (8.1)–(8.2) can only reach a limited number of ARL_0 values. This phenomenon is common in cases when the charting statistic can only take a finite number of values, as discussed in Subsection 3.3.1 about the p chart (3.19). For the downward NSR chart, the results in Table 8.1 can also be used after the U values in the table are replaced by the $L = -U$ values. For the two-sided NSR chart, its false alarm rates, denoted as α, are 2 times the values of α^+ presented in the table, and consequently its ARL_0 values are halves of the ARL_0^+ values in the table.

Note that the NSR chart described above can be used for both phase I and phase II SPC. In phase I SPC, the value of $\widetilde{\mu}_0$ is usually unknown. In such cases, it can be replaced by the average of the sample medians (or sample means) of different batches of data obtained for phase I analysis. In phase II SPC, the value of $\widetilde{\mu}_0$ needs to be estimated beforehand from IC data collected at the end of the phase I analysis. In both cases, the symmetry of the process distribution should be verified properly.

Example 8.1 *The data shown in Figure 8.2(a) consist of 30 batches of observations with the batch size of $m = 10$. The first 20 batches of observations are generated from the t_3 distribution, and the remaining 10 batches of observations are generated from the $t_3 + 1$ distribution. So, these data can be regarded as observations from a production process with IC process distribution of t_3, which is symmetric about $\widetilde{\mu}_0 = 0$, and the process has a mean (or median) shift of size 1 at the 21st time point. From the plot of Figure 8.2(a), it seems that the process distribution is symmetric. Now, let us assume that $\widetilde{\mu}_0$ is known to be 0, and we are interested in the online monitoring of the process mean. To this end, we would like to apply the upward NSR chart (8.1)–(8.2) to the observed data, and the values of the charting statistic ψ_i are then computed by the formula (8.1). The upward NSR chart (8.1)–(8.2) is shown in Figure 8.2(b) with the upper control limit of $U = 49$. By Table 8.1, this chart has the ARL_0 value of 204.0816. From the plot, it can be seen that the first signal of the chart is given at the 21st time point. So, the production process should be stopped at that time point for investigating the possible root causes of the detected upward mean shift.*

The NSR chart described above can be modified for various different considerations. For instance, if we are uncertain about the symmetry of the process distribution, then we can consider replacing the Wilcoxon signed-rank test statistic used in the chart by the sign test statistic discussed in Subsection 2.8.3 (cf., Amin et al., 1995). Of course, the resulting control chart will lose certain efficiency in cases when the process distribution is symmetric or close to symmetric. Chakraborti and Eryilmaz (2007) generalized the decision rule (8.2) based on various different runs (cf., their definitions given in Subsection 3.2.1) of the charting statistic ψ_i, and showed

Table 8.1 *This table presents the values of α^+ and ARL_0^+ of the upward NSR chart (8.1)–(8.2) in various cases when $m = 5, 6, 8,$ and 10.*

	$m = 5$			$m = 6$	
U	α^+	ARL_0^+	U	α^+	ARL_0^+
10	0.1563	6.3980	10	0.2188	4.5704
11	0.0938	10.6610	11	0.1563	6.3980
12	0.0938	10.6610	12	0.1563	6.3980
13	0.0625	16.0000	13	0.1094	9.1408
14	0.0625	16.0000	14	0.1094	9.1408
15	0.0313	32.0000	15	0.0781	12.8041
16	0.0000	∞	16	0.0781	12.8041
			17	0.0469	21.3220
			18	0.0469	21.3220
			19	0.0313	31.9489
			20	0.0313	31.9489
			21	0.0156	64.1026
			22	0.0000	∞

	$m = 8$			$m = 10$	
U	α^+	ARL_0^+	U	α^+	ARL_0^+
18	0.1250	8.0000	25	0.1162	8.6059
20	0.0977	10.2354	27	0.0967	10.3413
22	0.0742	13.4771	29	0.0801	12.4844
24	0.0547	18.2815	31	0.0654	15.2905
26	0.0391	25.5755	33	0.0527	18.9753
28	0.0273	36.6300	35	0.0420	23.8095
30	0.0195	51.2821	37	0.0322	31.0559
32	0.0117	85.4701	39	0.0244	40.9836
34	0.0078	128.2051	41	0.0186	53.7634
36	0.0039	256.4103	43	0.0137	72.9927
> 36	0.0000	∞	45	0.0098	102.0408
			47	0.0068	147.0588
			49	0.0049	204.0816
			51	0.0029	344.8276
			53	0.0020	512.0328
			55	0.0010	1024.5900
			> 55	0.0000	∞

that the control charts with the generalized decision rules would be more efficient in various cases.

Next, we discuss the so-called *distribution-free precedence (DFP) control chart* proposed by Chakraborti et al. (2004) for phase II SPC. Assume that a reference sample of size M, denoted as $(Y_1, Y_2, \ldots, Y_M)$, is available beforehand and it is obtained from a production process when it is IC, the IC process distribution is continuous but unknown, and we are interested in the online monitoring of the process mean μ. Let

$$X_{i(1)} \leq X_{i(2)} \leq \cdots \leq X_{i(m)}, \qquad i = 1, 2, \ldots$$

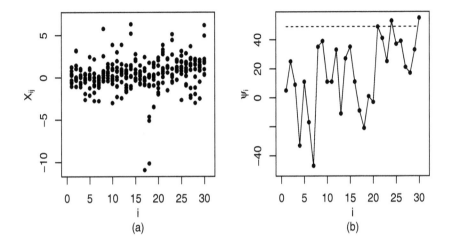

Figure 8.2 *(a) Observed data from a production process with the IC process distribution of t_3, and the process has a median shift at the 21st time point. The data consist of 30 batches of observations with the batch size of $m = 10$. (b) The upward NSR chart (8.1)–(8.2) with the upper control limit of $U = 49$ (dashed horizontal line).*

be m ordered observations obtained at the i-th time point during the online monitoring. Then, this is a case with batch data and the batch size is m. For the j-th ordered observation $X_{i(j)}$ with $1 \leq j \leq m$, we define W_j to be the number of Y-observations in the reference sample that are smaller than or equal to (or, "precede") $X_{i(j)}$. The statistic W_j is called a precedence statistic and a test based on W_j is called a precedence test in the literature (cf., Chakraborti and van der Laan, 1996, 1997; Nelson, 1963, 1993). By some routine combinatorial arguments, the IC distribution of W_j can be derived to be

$$P(W_j = w) = \frac{\dbinom{j+w-1}{w}\dbinom{M+m-j-w}{M-w}}{\dbinom{M+m}{M}}, \qquad w = 0, 1, 2, \dots, M. \quad (8.3)$$

From (8.3), it can be seen that the IC distribution of W_j depends only on M, m, and j, and it does not depend on the IC process distribution. Therefore, it is distribution-free, and consequently any tests or control charts based on W_j would be distribution-free as well.

For simplicity of presentation, let us assume that the batch size $m = 2s + 1$ is an odd integer number, and we are interested in using the sample median $X_{i(s+1)}$ for process monitoring. In such cases, the IC distribution of W_j in (8.3) is obviously symmetric about its center, and therefore it is natural to consider the decision rule that a signal of process mean shift is given when

$$X_{i(s+1)} < L = Y_{(a)} \qquad \text{or} \qquad X_{i(s+1)} > U = Y_{(M-a+1)}, \qquad (8.4)$$

where $1 \leq a < M/2$ is an integer number, and $Y_{(a)}$ and $Y_{(M-a+1)}$ are the a-th and the $(M-a+1)$-th order statistics of the reference sample. Clearly, (8.4) is equivalent to

$$W_{s+1} < a \qquad \text{or} \qquad W_{s+1} > M - a. \qquad (8.5)$$

So, for a given false alarm rate p_0, we can choose the largest a such that

$$
\begin{aligned}
& P(W_{s+1} < a) + P(W_{s+1} > M - a) \\
= \quad & 1 - P(a \leq W_{s+1} \leq M - a) \\
= \quad & \sum_{w=a}^{M-a} \frac{\dbinom{s+w}{w}\dbinom{M+s-w}{M-w}}{\dbinom{M+2s+1}{M}} \\
< \quad & p_0.
\end{aligned}
\qquad (8.6)
$$

For several commonly used p_0, M, and m values, the computed values of the largest a such that (8.6) holds, the actual p_0 values, and the actual ARL_0 value of the DFP chart (8.4)–(8.6) are presented in Table 8.2, which is reorganized from Table 1 in Chakraborti et al. (2004). From the table, it can be seen that the actual p_0 values are generally smaller than the assumed p_0 values, due to the discreteness of the charting statistic W_{s+1}, as mentioned above about the NSR chart (8.1)–(8.2).

For the DFP chart (8.4)–(8.6), its ARL_0 value would be different from the value of $1/p_0$, because the signaling events defined by (8.4) are generally associated at different time points, due to the fact that the same random variables $Y_{(a)}$ and $Y_{(M-a+1)}$ are used in defining all these events. When the reference sample size M is large, $Y_{(a)}$ and $Y_{(M-a+1)}$ would be quite close to the $(p_0/2)$-th and $(1 - p_0/2)$-th quantiles of the IC process distribution, respectively. In such cases, the actual ARL_0 value of the chart should be close to $1/p_0$. However, M must be really large to make this approximation reliable, especially in cases when p_0 is small, which is confirmed by Table 8.2. From the table, it can be seen that the actual ARL_0 values are quite close to $1/p_0$ when $M = 1000$ and $p_0 = 0.01$; but, such ARL_0 values are quite different from $1/p_0$ when $M = 1000$ and $p_0 = 0.0027$. In some applications, it is difficult to obtain a reference sample of a large size. To overcome this difficulty, Chakraborti et al. (2004) derived formulas for the run length distribution and the IC and OC ARL values of the DFP chart (8.4)–(8.6). Based on these formulas, we can compute the value of a such that the ARL_0 value of the chart equals a given value. For several commonly used ARL_0, M, and m values, such a values, along with the actual p_0 values and the actual ARL_0 values, are presented in Table 8.3, which is reorganized from Table 2 in Chakraborti et al. (2004). From the table, it can be seen that the actual and assumed ARL_0 values are generally different, and they are close to each other when M is large, which is similar to the results found in Table 8.2. Between the two approaches for choosing a by specifying either the false alarm rate p_0 or the ARL_0 value, the ARL_0 approach should be preferred, because the p_0 approach is difficult to interpret due to the possible association among signaling events at different time points, while the ARL_0 approach can be interpreted as usual.

Table 8.2 *This table presents the computed values of the largest a such that (8.6) holds (1st line in each entry), the actual p_0 values (2nd line in each entry), and the actual ARL_0 values (3rd line in each entry) of the DFP chart (8.4)–(8.6) for several commonly used p_0, M, and m values.*

p_0	m	s	$M = 50$	$M = 100$	$M = 500$	$M = 1000$
0.0100	5	2	3	7	40	82
			0.0072	0.0086	0.0095	0.0099
			635.7	214.9	114.5	104.6
	11	5	7	15	83	167
			0.0093	0.0085	0.0098	0.0097
			642.2	245.0	113.3	108.4
	25	12	10	23	127	258
			0.0061	0.0080	0.0095	0.0099
			10990.0	510.8	128.3	109.8
0.0050	5	2	2	5	31	64
			0.0030	0.0035	0.0047	0.0049
			5671.0	678.4	242.3	215.1
	11	5	5	13	72	146
			0.0025	0.0045	0.0048	0.0049
			9503.0	574.5	240.9	219.8
	25	12	9	21	118	239
			0.0031	0.0040	0.0049	0.0049
			44750.0	1488.0	261.0	227.5
0.0027	5	2	1	4	25	51
			0.0008	0.0020	0.0025	0.0026
			∞	1550.0	460.2	419.5
	11	5	5	11	64	130
			0.0025	0.0021	0.0026	0.0026
			9503.0	1630.0	456.1	409.8
	25	12	8	19	110	224
			0.0015	0.0018	0.0025	0.0027
			173700.0	5183.0	526.2	430.2

Tables 8.2 and 8.3 are both for the two-sided decision rule (8.4). If an upward or downward decision rule is preferred in a given application, then the value of a can be computed similarly from the IC distribution formula (8.3) of W_j and from the ARL formulas given in Chakraborti et al. (2004).

Example 8.2 *To online monitor the mean of a production process, a reference sample of size $M = 1000$ has been obtained from the production process when it is IC. Figure 8.3(a) shows 30 batches of phase II observations with batch size of $m = 5$. We would like to apply the DFP chart (8.4)–(8.6) to this dataset for the process mean monitoring. In the chart, we choose the assumed ARL_0 value to be 370. Then, by Table 8.3, the value of a should be chosen to be 53. From the reference sample, the lower and upper control limits of the chart (8.4)–(8.6) are found to be*

$$L = Y_{(a)} = -1.0847, \qquad U = Y_{(M-a+1)} = 1.8308.$$

Table 8.3 *This table presents the computed values of a (1st line in each entry), the actual p_0 values (2nd line in each entry), and the actual ARL_0 values (3rd line in each entry) of the DFP chart (8.4)–(8.6) for several commonly used ARL_0, M, and m values.*

ARL_0	m	s	$M = 50$	$M = 100$	$M = 500$	$M = 1000$
370	5	2	6	11	27	53
			0.0055	0.0038	0.0033	0.0029
			359.6	385.5	365.5	374.4
	11	5	14	27	67	132
			0.0063	0.0041	0.0033	0.0029
			367.8	372.4	355.3	377.3
	25	12	24	46	114	227
			0.0111	0.0053	0.0035	0.0030
			316.5	375.7	367.3	377.0
500	5	2	6	10	24	48
			0.0055	0.0029	0.0023	0.0022
			359.6	520.1	520.3	501.9
	11	5	13	26	63	125
			0.0045	0.0034	0.0024	0.0023
			574.5	460.5	497.2	506.9
	25	12	23	45	111	219
			0.0080	0.0044	0.0028	0.0023
			510.8	472.5	480.1	492.2
1000	5	2	5	8	19	38
			0.0035	0.0016	0.0011	0.0011
			678.4	1067.5	1055.8	1005.9
	11	5	11	23	56	110
			0.0021	0.0019	0.0013	0.0011
			940.6	927.8	953.2	1021.7
	25	12	22	42	103	206
			0.0057	0.0024	0.0014	0.0012
			854.6	986.2	1033.6	1004.4

Then, the chart is shown in Figure 8.3(b), from which it can be seen that the first signal of process mean shift is given at the 23rd observation time point. From Figure 8.3(a), it seems that there is an outlier at the 16th time point, which does not have any impact on the charting statistic $X_{i(3)}$ shown in Figure 8.3(b). Therefore, besides its distribution-free property, the DFP chart (8.4)–(8.6) is also robust to a small number of outliers in each batch of the phase II observations.

The decision rule (8.4) has been generalized by Chakraborti et al. (2009) using various different runs. Chakraborti et al. (2009) showed that the runs-type decision rules can enhance the performance of the DFP chart (8.4)–(8.6) in various cases.

8.2.2 Nonparametric CUSUM charts

Since Page (1954) proposed the conventional CUSUM chart under the assumption that the distribution of a production process in question is normal (cf., Section 4.2 for

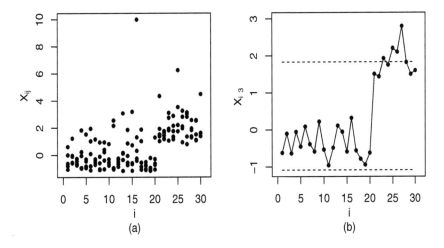

Figure 8.3 *(a) Phase II batch data with the batch size of $m = 5$ obtained from a production process. (b) The DFP chart (8.4)–(8.6) with the assumed ARL_0 value of 370. In the chart, the dashed horizontal lines denote the lower and upper control limits obtained from a reference sample of size $M = 1000$.*

a description), there have been many discussions about its unreliable performance in various cases when the normality assumption is invalid, and a number of nonparametric CUSUM charts have been proposed in the literature. See, for instance, McDonald (1990), Park and Reynolds (1987), Reynolds (1975) for some relatively early discussions. Next, we describe the nonparametric CUSUM chart suggested by Bakir and Reynolds (1979).

Let
$$X_{n1}, X_{n2}, \ldots, X_{nm}, \qquad \text{for } n = 1, 2, \ldots,$$

be m phase II observations obtained at the n-th time point. It is further assumed that the IC process distribution is symmetric about its mean (or median), and the IC mean (or median) is 0. Let

$$\psi_n = \sum_{j=1}^{m} \text{sign}(X_{nj}) R_{nj}^+ \tag{8.7}$$

be the sum of the Wilcoxon signed-ranks within the n-th batch of observations. Then, ψ_n in (8.7) is actually the same as the one defined in (8.1), except that $\tilde{\mu}_0 = 0$ in the current setup. As pointed out in the paragraph containing (8.1) in the previous subsection, ψ_n is centered at 0 when the process is IC at the n-th time point, and its value is lifted upward if there is an upward process mean (or median) shift at that time point. Based on this intuition, Bakir and Reynolds (1979) suggested using the CUSUM charting statistic

$$C_n^+ = \max\left(0, C_{n-1}^+ + \psi_n - k\right), \tag{8.8}$$

where $C_0^+ = 0$ and $k > 0$ was an allowance constant, and the chart gave a signal of upward mean shift if

$$C_n^+ > h, \tag{8.9}$$

where $h > 0$ was a control limit chosen to achieve a pre-specified ARL_0 value. The downward and two-sided versions of this CUSUM chart can be defined similarly. See a related discussion in Subsection 4.2.1. This chart will be called the NSR CUSUM chart hereafter.

The upward NSR CUSUM chart (8.8)–(8.9) has two parameters k and h to choose before the chart can actually be used. Because ψ_n takes integer values, Bakir and Reynolds (1979) suggested choosing k and h to be integer values as well to simplify the theoretical study of the CUSUM chart. By some theoretical studies of the charting statistic C_n^+, Bakir and Reynolds (1979) provided approximate optimal values of k in cases when the process distribution is symmetric, $m = 4, 6$, and 10, and the size of the process mean shift equals $\delta = 0.2\sigma, 0.6\sigma, \sigma, 2\sigma$, and 3σ, where σ is the IC process standard deviation. These approximate optimal values of k are presented in Table 8.4. After k is chosen, the control limit h can be chosen such that a given ARL_0 level is reached. But, because the charting statistic C_n^+ can only take discrete values, not all ARL_0 values can be reached by the chart, as mentioned in the previous subsection. Bakir and Reynolds (1979) demonstrated how to use the Markov chain approach (cf., Subsection 4.6.1) for computing the ARL values of the upward NSR CUSUM chart (8.8)–(8.9). By the Markov chain approach, the ARL_0 values of the chart when the process distribution is assumed symmetric about the IC process mean and when (k, h) take various different values have been computed and tabulated. For instance, in cases when $m = 6$, some computed ARL_0 values are presented in Table 8.5, which is part of Table 2a in Bakir and Reynolds (1979).

Table 8.4 *Approximate optimal values of the allowance constant k of the upward NSR CUSUM chart (8.8)–(8.9) in cases when the process distribution is symmetric, $m = 4, 6$, and 10, and the size of the process mean shift equals $\delta = 0.2\sigma, 0.6\sigma, \sigma, 2\sigma$, and 3σ, where σ is the IC process standard deviation.*

			δ		
m	0.2σ	0.6σ	σ	2σ	3σ
4	1	3	4	5	5
6	2	6	8	10	10
10	6	16	22	27	27

Example 8.3 *Figure 8.4(a) shows 30 batches of phase II observations with the batch size of $m = 6$ obtained from a production process. It is known that the IC process distribution is symmetric about 0, and we are interested in monitoring the process mean by the upward NSR CUSUM chart (8.8)–(8.9). To this end, we choose $k = 8$ and $h = 10$ in the chart. By Table 8.4, this chart is tuned for detecting a mean shift of size 1σ. From Table 8.5, the ARL_0 value of the chart is between 97.04 and 155.36. The chart is shown in Figure 8.4(b), which gives its first signal at the 21st time point. From Figure 8.4(a), it seems that the process mean shifts upward at the 21st time point; but, there are a number of quite large observations before that time point. The chart (8.8)–(8.9) seems quite robust to these scattered large observations.*

Table 8.5 *Computed* ARL_0 *values of the upward NSR CUSUM chart (8.8)–(8.9) in cases when the process distribution is symmetric, $m = 6$, and (k, h) take various different values.*

k	h = 2	h = 4	h = 6	h = 8	h = 10	h = 12	h = 14	h = 16
1	14.22	17.15	20.32	24.49	30.17	36.52	54.41	67.28
3	17.45	21.04	26.00	33.28	42.02	51.80	65.92	80.57
5	20.14	26.94	35.71	46.93	60.11	81.29	104.33	140.63
7	27.43	37.46	51.01	67.54	97.04	130.76	192.07	
9	38.40	53.76	72.98	110.87	155.36	249.32		
11	54.86	75.79	120.17	173.05	301.01			
13	76.80	125.39	183.14	334.79				
15	127.98	190.49	366.46					
17	192.00	378.00						
19	384.00							

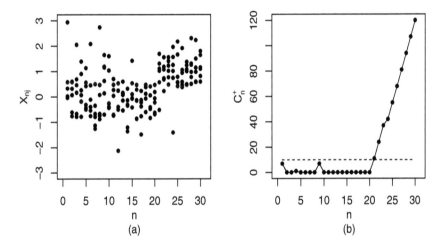

Figure 8.4 *(a) Phase II batch data with the batch size $m = 6$ obtained from a production process. (b) The upward NSR CUSUM chart (8.8)–(8.9) with $k = 8$ and $h = 10$ (denoted by the dashed horizontal line).*

In cases when a reference sample $(Y_1, Y_2, \ldots, Y_M)$ is available beforehand, obtained from a production process when it is IC, Li et al. (2010) proposed a nonparametric CUSUM chart based on the Wilcoxon rank-sum test discussed in Subsection 2.8.3 for phase II process mean monitoring. Let $(X_{n1}, X_{n2}, \ldots, X_{nm})$ be a batch of m observations obtained at the n-th time point of phase II process monitoring, for $n = 1, 2, \ldots$, and W_n be the sum of the ranks of $(X_{n1}, X_{n2}, \ldots, X_{nm})$ in the combined sample

$$(Y_1, Y_2, \ldots, Y_M, X_{n1}, X_{n2}, \ldots, X_{nm}).$$

Then, W_n is the conventional Wilcoxon rank-sum test statistic for comparing the means of the two populations that the two samples $(Y_1, Y_2, \ldots, Y_M)$ and $(X_{n1}, X_{n2}, \ldots, X_{nm})$ represent. In cases when the process is IC at the n-th time point, the upper-tail probabilities of the IC distribution of W_n are presented in Table 2.9 in various cases. In such cases, the IC mean and variance of W_n are

$$\mu_{W_n} = m(m+M+1)/2, \qquad \sigma^2_{W_n} = mM(m+M+1)/12.$$

The value of W_n tends to be larger if the process mean shifts upward at the n-th time point, and smaller if the process mean shifts downward. Therefore, W_n can be used for detecting process mean shifts. Define the charting statistic of the upward CUSUM chart to be

$$C_n^+ = \max\left[0, C_{n-1}^+ + (W_n - m(m+M+1)/2) - k\right], \qquad (8.10)$$

where $C_0^+ = 0$ and $k > 0$ is an allowance constant. The chart gives a signal of upward mean shift if

$$C_n^+ > h, \qquad (8.11)$$

where $h > 0$ is a control limit chosen to achieve a pre-specified ARL_0 value. Similarly, to detect downward process mean shifts, the charting statistic is defined to be

$$C_n^- = \min\left[0, C_{n-1}^- + (W_n - m(m+M+1)/2) + k\right], \qquad (8.12)$$

where $C_0^- = 0$, and the chart gives a signal of downward mean shift if

$$C_n^- < -h. \qquad (8.13)$$

To detect an arbitrary process mean shift, a two-sided CUSUM chart should be used, which gives a signal when either (8.11) or (8.13) or both hold. This chart will be called the nonparametric rank-sum (NRS) CUSUM chart hereafter.

In the NRS CUSUM chart (8.10)–(8.13), if the allowance constant k has been determined, then the control limit h can be chosen to reach a pre-specified ARL_0 value, although not all ARL_0 values can be reached with a certain accuracy, because of the discreteness of the charting statistics C_n^+ and C_n^-. The search of such an h value can be accomplished by a numerical algorithm, such as the one described by the pseudo code discussed in Subsection 4.2.2. For this chart, the relationship between the optimal k and the target shift size δ in the process mean is unclear yet. Intuitively, this relationship depends on the IC process distribution because the relationship between δ and the mean of W_n would depend on the IC process distribution. However, the general guidelines for selecting k, i.e., small k values are good for detecting small shifts and large k values are good for detecting large shifts, should still be valid here. See Subsection 4.2.2 for a related discussion. In all the numerical examples in Li et al. (2010), k is chosen to be $0.5\sigma_{W_n} = 0.5\sqrt{mM(m+M+1)/12}$. In such cases, when $M = 100$, $m = 5$, and $ARL_0 = 500$, the value of h was computed by Li et al. (2010) to be 353 for the two-sided NRS CUSUM chart.

Example 8.4 *To online monitor the mean of a production process, $M = 100$ IC observations have been obtained beforehand and they are shown in Figure 8.5(a) by the dark diamonds before the vertical dotted line at $n = 0$. Then, 30 batches of phase II observations with the batch size of $m = 5$ are collected from the same production* *process, and they are shown in the same plot by the dark dots after the vertical dot-ted line at $n = 0$. The two-sided NRS CUSUM chart (8.10)–(8.13) is then applied to this dataset for monitoring the process mean, in which k and h are chosen to be $0.5\sigma_{W_n} = 0.5\sqrt{mM(m+M+1)/12}$ and 353, respectively. In such cases, the two-sided chart has the ARL_0 value of 500. The charting statistics C_n^+ and C_n^- are shown in Figure 8.5(b) by the dark dots and dark diamonds, respectively, and the upper and lower control limits h and $-h$ are shown by the dashed horizontal lines in the plot. From the plot, the first signal of process mean shift is given at the 23rd time point, and it seems that the mean shift is upward because the signal is from the statistic C_n^+. From Figure 8.5(a), a mean shift can be noticed at $n = 21$.*

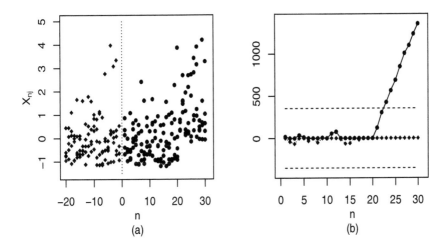

Figure 8.5 (a) An IC dataset of size $M = 100$ (dark diamonds before the vertical dotted line) and a phase II batch dataset with the batch size $m = 5$ obtained from a production process. (b) The two-sided NRS CUSUM chart (8.10)–(8.13) with $k = 0.5\sqrt{mM(m+M+1)/12}$ and $h = 353$. In plot (b), the dark dots and dark diamonds denote the values of C_n^+ and C_n^-, respectively, and the upper and lower control limits are shown by the dashed horizontal lines.

The NSR and NRS nonparametric CUSUM charts described above both use the ranks of process observations for monitoring the mean of a production process. Chat-terjee and Qiu (2009) proposed an alternative approach for constructing nonparamet-ric CUSUM charts, briefly described below. The approach by Chatterjee and Qiu is mainly for phase II process mean monitoring, and it is designed for cases with individual observation data, although it can be adapted easily to cases with batch data and to cases for monitoring process variance and other parameters. Assume that $\{X_n, n \geq 1\}$ are phase II observations obtained from a production process with the IC cdf F_0 that has the mean of 0. Instead of assuming F_0 to be known, we assume that there is an IC dataset available, from which some procedure parameters can be estimated. To monitor the process median, we use the charting statistics of the conventional CUSUM charts described in Section 4.2. Without loss of generality, assume that we are interested in detecting upward mean shifts. In such cases, the charting statistic is defined by (4.7). For readers' convenience, it is written below as well.

$$C_n^+ = \max\left(0, C_{n-1}^+ + (X_n - \mu_0) - k\right), \qquad n \geq 1, \qquad (8.14)$$

where $C_0^+ = 0$ and k is an allowance constant. Instead of using a single control limit to make a decision regarding the process performance at the current time point n, we consider using a sequence of control limits $\{h_j, j \geq 1\}$, depending on the value of the so-called *sprint length* defined by

$$T_n = \begin{cases} 0, & \text{if } C_n^+ = 0 \\ j, & \text{if } C_n^+ \neq 0, C_{n-1}^+ \neq 0, \ldots, C_{n-j+1}^+ \neq 0, C_{n-j}^+ = 0, \text{ for } j = 1, 2, \ldots, n. \end{cases}$$
$$(8.15)$$

The sprint length T_n is the time elapsed since the last time the CUSUM statistic C_n^+ was zero, and it can be easily computed based on the computed values of the charting statistic. At the current time point n, in cases when $T_n = j$, for an integer $1 \leq j \leq n$, a signal of process mean shift is given if

$$C_n^+ > h_j. \qquad (8.16)$$

In practice, we can use a fixed number of control limits $\{h_j, 1 \leq j \leq j_{max}\}$ in the control chart (8.14)–(8.16), where j_{max} is a fixed integer. In such cases, if the observed value of T_n is larger than j_{max}, then $h_{j_{max}}$ is always used in (8.16). Chatterjee and Qiu (2009) proposed a bootstrap algorithm for computing the control limits $\{h_j, 1 \leq j \leq j_{max}\}$, and it was shown that the control chart was effective in most cases considered if j_{max} was chosen in the range 20–50.

From the above description about the control chart (8.14)–(8.16), it can be seen that the chart is nonparametric because it does not require specification of a parametric form for the IC process distribution F_0. Also, it makes decisions based on both T_n and C_n^+, which is intuitively appealing. However, there are still many open questions about this method. For instance, the relationship between the joint distribution of (T_n, C_n^+) and the parameters k and j_{max} is unclear yet. Consequently, more practical guidelines are needed regarding the selection of these two parameters.

8.2.3   Nonparametric EWMA charts

The nonparametric CUSUM charts discussed in the previous subsection can all be modified properly to become nonparametric EWMA charts. For instance, Graham et al. (2011) changed the NSR CUSUM chart (8.8)–(8.9) to a nonparametric EWMA chart, described as follows. Assume that the IC process distribution is symmetric about the IC mean, which is assumed to be 0. Let $(X_{n1}, X_{n2}, \ldots, X_{nm})$ be m observations obtained at the n-th time point when online monitoring the process mean of a production process, for $n = 1, 2, \ldots$, and ψ_n be the sum of the Wilcoxon signed-ranks within these observations, as defined in (8.7). Then, the charting statistic of the nonparametric EWMA chart can be defined by

$$E_n = \lambda \psi_n + (1 - \lambda) E_{n-1}, \qquad \text{for } n \geq 1, \qquad (8.17)$$

where $E_0 = 0$ and $\lambda \in (0, 1]$ is a weighting parameter. From the relation between ψ_n and S_+ that

$$\psi_n = 2S_+ - m(m+1)/2,$$

where S_+ is the Wilcoxon signed-rank test statistic described in Subsection 2.8.3 (note: the sample size n in Subsection 2.8.3 is actually m here), and from the expressions of the IC mean and variance of S_+ given in Subsection 2.8.3, it can be checked that

$$\mu_{\psi_n} = 0, \qquad \sigma_{\psi_n}^2 = m(m+1)(2m+1)/6.$$

Therefore, it is easy to check that the IC mean of E_n is 0, and the IC variance of E_n is

$$\sigma_{E_n}^2 = \left[\frac{m(m+1)(2m+1)}{6}\right]\left[\frac{\lambda}{2-\lambda}\right]\left[1 - (1-\lambda)^{2n}\right].$$

The center line C, the upper control limit U, and the lower control limit L of the nonparametric EWMA chart can then be defined by

$$
\begin{aligned}
U &= \rho\sqrt{\left[\frac{m(m+1)(2m+1)}{6}\right]\left[\frac{\lambda}{2-\lambda}\right]\left[1 - (1-\lambda)^{2n}\right]} \\
C &= 0 \\
L &= -U,
\end{aligned}
\qquad (8.18)
$$

where $\rho > 0$ is a parameter chosen to reach a given ARL_0 value. When n increases, the IC distribution of E_n converges to its "steady-state" distribution with the variance of $[m(m+1)(2m+1)/6][\lambda/(2-\lambda)]$. The "steady-state" control limits can be defined to be

$$
\begin{aligned}
U &= \rho\sqrt{\left[\frac{m(m+1)(2m+1)}{6}\right]\left[\frac{\lambda}{2-\lambda}\right]} \\
C &= 0 \\
L &= -U.
\end{aligned}
\qquad (8.19)
$$

As mentioned in Subsection 5.2.1, the control chart with the control limits in (8.19) would be easier to use, the one with the control limits in (8.18) would be slightly more reliable, and the two versions should have similar performance, especially in cases when the potential mean shift does not occur early in the production process. The chart (8.17)–(8.18) (or (8.17) and (8.19)) will be called the NSR EWMA chart hereafter. The NSR EWMA chart described above is for detecting arbitrary process mean shifts. If only the upward or the downward mean shifts are our concern in a given application, then a one-sided EWMA chart can be defined in a similar way to those in (5.8)–(5.11) given in Subsection 5.2.1.

For the NSR EWMA chart defined by (8.17) and (8.19), Graham et al. (2011) computed the values of ρ for some commonly used values of m, λ, and ARL_0. These values of ρ are presented in Table 8.6. From the table, it can be seen that ρ should be chosen larger when λ is larger or ARL_0 is larger; but, it is quite robust to the value of m in cases when ARL_0 is chosen to be quite large. By the way, the actual ARL_0 values of the chart with the ρ values in the table are all within 10% of the assumed ARL_0 values.

Table 8.6 *Computed values of ρ used by the NSR EWMA chart defined by (8.17) and (8.19), for some commonly used values of m, λ, and ARL_0.*

m	ARL_0	$\lambda = 0.01$	$\lambda = 0.025$	$\lambda = 0.05$	$\lambda = 0.10$	$\lambda = 0.20$
5	370	1.822	2.230	2.481	2.668	2.764
	500	1.975	2.368	2.602	2.775	2.852
10	370	1.821	2.230	2.486	2.684	2.810
	500	1.975	2.367	2.610	2.794	2.905

Example 8.5 *Figure 8.6(a) shows 30 batches of phase II observations with batch size $m = 10$ obtained from a production process. It is known that the IC process distribution is symmetric about 0, and we are interested in monitoring the process mean by the NSR EWMA chart defined by (8.17) and (8.19). To this end, we choose $\lambda = 0.1$ and $ARL_0 = 370$. By Table 8.6, the value of ρ used in (8.19) should be chosen as 2.684. The control chart is shown in Figure 8.6(b), which gives its first signal at the 23rd time point*

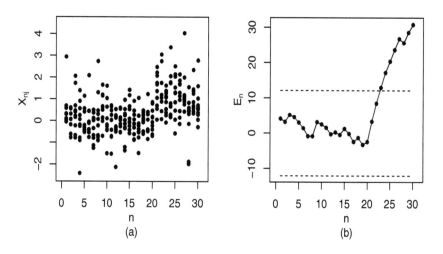

Figure 8.6 *(a) Phase II batch data with the batch size of $m = 10$ obtained from a production process. (b) The NSR EWMA chart defined by (8.17) and (8.19) with $\lambda = 0.1$ and $ARL_0 = 370$. The dashed horizontal lines in plot (b) denote the upper and lower control limits of the control chart.*

For the NRS CUSUM chart (8.10)–(8.13), Li et al. (2010) proposed a companion NRS EWMA chart described as follows. Let W_n be the Wilcoxon rank-sum test statistic defined in the previous subsection for comparing the n-th batch of phase

II observations $(X_{n1}, X_{n2}, \ldots, X_{nm})$ with the reference sample $(Y_1, Y_2, \ldots, Y_M)$. Define the EWMA charting statistic to be

$$E_n = \lambda W_n + (1 - \lambda) E_{n-1}, \qquad \text{for } n \geq 1, \tag{8.20}$$

where $E_0 = \mu_{W_n} = m(m + M + 1)/2$, and $\lambda \in (0, 1]$ is a weighting parameter. Then, the center line C, the upper control limit U, and the lower control limit L of the chart can be defined by

$$U = \frac{m(m + M + 1)}{2} + \rho \sqrt{\left[\frac{mM(m + M + 1)}{6}\right]\left[\frac{\lambda}{2 - \lambda}\right]}$$

$$C = \frac{m(m + M + 1)}{2} \tag{8.21}$$

$$L = \frac{m(m + M + 1)}{2} - \rho \sqrt{\left[\frac{mM(m + M + 1)}{6}\right]\left[\frac{\lambda}{2 - \lambda}\right]},$$

where $\rho > 0$ is a parameter chosen to reach a given ARL_0 value. The chart gives a signal of process mean shift when the charting statistic E_n falls outside the interval $[L, U]$. For a given value of λ, Li et al. (2010) suggested determining the value of ρ by simulation. In cases when $\lambda = 0.1$, $M = 100$, $m = 5$, and $ARL_0 = 500$, Li et al. (2010) computed the value of ρ to be 2.1102. In such cases, the control limits are

$$L = 219.5, \qquad C = 265, \qquad U = 310.5.$$

For the phase II SPC problem considered in Example 8.4, the NRS EWMA chart (8.20)–(8.21) with $\lambda = 0.1$ and $ARL_0 = 500$ is shown in Figure 8.7.

Next, we discuss the nonparametric EWMA chart proposed by Zou and Tsung (2010) based on a nonparametric goodness-of-fit test described below. Assume that we have a simple random sample $(X_1, X_2, \ldots, X_n)$ obtained from a population with the cdf $F(x)$. To test the hypotheses

$$H_0 : F(x) = F_0(x), \text{ for all } x \in R \qquad \text{versus} \qquad H_1 : F(x) \neq F_0(x), \text{ for some } x \in R,$$

where $F_0(x)$ is a given cdf, Zhang (2002) proposed the following likelihood-ratio-based test statistic:

$$\sum_{i=1}^{n} \left\{ \frac{1}{1 - F_n(X_i)} \log\left(\frac{F_n(X_i)}{F_0(X_i)}\right) + \frac{1}{F_n(X_i)} \log\left(\frac{1 - F_n(X_i)}{1 - F_0(X_i)}\right) \right\}, \tag{8.22}$$

where $F_n(x) = \frac{1}{n} \sum_{i=1}^{n} I(X_i \leq x)$ is the empirical cumulative distribution function discussed in Subsection 2.8.1. When H_0 is true, $F_n(x)$ should be close to $F_0(x)$. Consequently, the test statistic in (8.22) should be close to 0. On the other hand, if H_0 is false, the above test statistic should be relatively large. Now, in phase II SPC, assume that $\{X_n, n \geq 1\}$ are observations obtained from a production process for process mean monitoring. Define

$$F_{n,\lambda}(x) = a_{n,\lambda}^{-1} \sum_{i=1}^{n} (1 - \lambda)^{n-i} I(X_i \leq x),$$

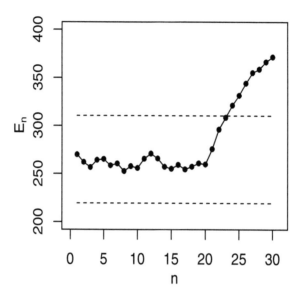

Figure 8.7 *The NRS EWMA chart (8.20)–(8.21) with $\lambda = 0.1$ and $ARL_0 = 500$ when it is applied to the phase II SPC problem considered in Example 8.4.*

where $\lambda \in (0, 1]$ is a weighting parameter, and

$$a_{n,\lambda} = \sum_{i=1}^{n} (1 - \lambda)^{n-i}.$$

Then, $F_{n,\lambda}(x)$ is also an empirical estimator of $F_0(x)$, although observations obtained at time points closer to n would receive more weight in $F_{n,\lambda}(x)$, as in an EWMA charting statistic. After replacing $F_n(x)$ by $F_{n,\lambda}(x)$ and after including the weight $\lambda(1 - \lambda)^{n-i}$ for the i-th term in (8.22), we obtain the following nonparametric EWMA charting statistic

$$
\begin{aligned}
E_n &= \sum_{i=1}^{n} \lambda (1-\lambda)^{n-i} \left\{ \frac{1}{1 - F_{n,\lambda}(X_i)} \log\left(\frac{F_{n,\lambda}(X_i)}{F_0(X_i)} \right) \right. \\
&\quad \left. + \frac{1}{F_{n,\lambda}(X_i)} \log\left(\frac{1 - F_{n,\lambda}(X_i)}{1 - F_0(X_i)} \right) \right\} \\
&= \lambda G_n + (1-\lambda) E_{n-1}, \quad \text{for } n \geq 1,
\end{aligned}
$$

where $E_0 = 0$, and

$$G_n = \frac{1}{1 - F_{n,\lambda}(X_n)} \log\left(\frac{F_{n,\lambda}(X_n)}{F_0(X_n)} \right) + \frac{1}{F_{n,\lambda}(X_n)} \log\left(\frac{1 - F_{n,\lambda}(X_n)}{1 - F_0(X_n)} \right).$$

Obviously, the charting statistic E_n combines the idea of EWMA process monitoring with the nonparametric goodness-of-fit testing procedure by Zhang (2002). It should be sensitive to any distributional shift from $F_0(x)$. However, to use this statistic, we need to specify $F_0(x)$ completely, which is often difficult to achieve in practice. To overcome this limitation, we assume that $F_0(x)$ is unknown, but, there is an IC dataset of size M, denoted as $(X_0, X_{-1}, \ldots, X_{-M+1})$. Then, $F_0(X_n)$ used in G_n can be replaced by its estimator

$$\widehat{F}_0(X_n) = \frac{1}{M+n-1} \sum_{i=-M+1}^{n-1} I(X_i \leq X_n).$$

The resulting charting statistic is

$$E_{n,SS} = \lambda \widetilde{G}_n + (1-\lambda)E_{n-1,SS}, \text{ for } n \geq 1, \tag{8.23}$$

where $E_{0,SS} = 0$, and

$$\widetilde{G}_n = \frac{1}{1-F_{n,\lambda}(X_n)} \log\left(\frac{F_{n,\lambda}(X_n)}{\widehat{F}_0(X_n)}\right) + \frac{1}{F_{n,\lambda}(X_n)} \log\left(\frac{1-F_{n,\lambda}(X_n)}{1-\widehat{F}_0(X_n)}\right).$$

The chart gives a signal of process mean shift at the n-th time point if

$$E_{n,SS} > h_n, \tag{8.24}$$

where $h_n > 0$ is a control limit chosen to achieve a given ARL_0 value. This chart will be called the nonparametric likelihood ratio (NLR) EWMA chart hereafter.

When computing the charting statistic $E_{n,SS}$ in (8.23), there might be numerical difficulties, due to the facts that $F_{n,\lambda}(X_n)$ could equal 1, $\widehat{F}_0(X_n)$ could be 0 or 1, and consequently $\widetilde{G}_n$ used by $E_{n,SS}$ is not defined in such cases. To overcome such numerical difficulties, the following continuity corrections can be considered:

$$F_{n,\lambda}^*(X_n) = \frac{\left[\sum_{i=1}^n (1-\lambda)^{n-i} I(X_i \leq X_n)\right] - 0.5}{a_{n,\lambda}}$$

$$\widehat{F}_0^*(X_n) = \frac{\max\left\{0.5, \left[\sum_{i=-M+1}^{n-1} I(X_i \leq X_n)\right] - 0.5\right\}}{M+n-1}. \tag{8.25}$$

Then, $F_{n,\lambda}(X_n)$ and $\widehat{F}_0(X_n)$ can be replaced by their modified versions $F_{n,\lambda}^*(X_n)$ and $\widehat{F}_0^*(X_n)$, respectively, when computing $\widetilde{G}_n$.

The NLR EWMA chart (8.23)–(8.25) can be regarded as a self-starting version of the chart that uses E_n as its charting statistic. In (8.23), the true IC cdf $F_0(x)$ is estimated by $\widehat{F}_0(x)$, and the estimator is updated every time a new process observation is obtained. It is obvious that both $F_{n,\lambda}(X_n)$ and $\widehat{F}_0(x)$ used in $\widetilde{G}_n$ depend only on the ordering information of the observations $(X_{-M+1}, X_{-M+2}, \ldots, X_0, X_1, X_2, \ldots, X_n)$, which are i.i.d. when the process is IC. Therefore, the IC performance of the chart (8.23)–(8.25) does not depend on the IC cdf $F_0(x)$, and the chart is distribution-free in that sense. Also, from the construction of the charting statistic $E_{n,SS}$, we can see that it aims to detect any distributional shift from the IC distribution $F_0(x)$. Therefore, the chart (8.23)–(8.25) should be able to detect shifts in the location, scale, or other distributional characteristics, which has been confirmed by Zou and Tsung (2010).

By numerical simulations, Zou and Tsung (2010) computed the values of the control limit h_n for some commonly used values of M, λ, and ARL_0. For instance, the computed h_n values in cases when $M = 100$, $\lambda = 0.05$ or 0.1, and $ARL_0 = 370$ or 500 are presented in Table 8.7. This table is modified from a part of Table 2 in Zou and Tsung (2010), after correcting a mistake pointed out by Zou that the values in Table 2 of Zou and Tsung (2010) should be multiplied by λ to be valid control limit values. In the table, only a limited number of n values between 1 and 500 are listed. For a given n value between 1 and 500 that is not listed in the table, the corresponding h_n value can be obtained by linear interpolation, or it can be replaced by the h_n value in the same column of the table whose n value is the first one before the given n. When $n > 500$, we can simply choose $h_n = h_{500}$.

Example 8.6 *The dark diamond points shown before the vertical dotted line in Figure 8.8(a) denote 100 IC observations obtained from a production process for phase II process monitoring. The dark dot points shown in the same plot denote 100 phase II observations collected sequentially from the same production process. To apply the NLR EWMA chart (8.23)–(8.25), we choose $\lambda = 0.05$ and $ARL_0 = 370$. The modifications described in (8.25) are used when computing the values of the charting statistic $E_{n,SS}$. The chart is shown in Figure 8.8(b), from which it can be seen that the first signal is given at the 62nd time point.*

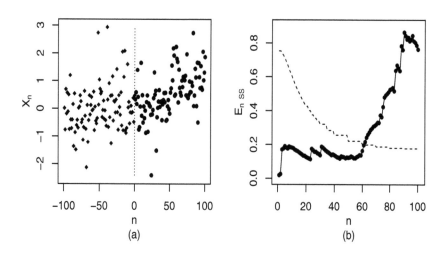

Figure 8.8 *(a) An IC dataset of size $M = 100$ (dark diamond points before the vertical dotted line) and a phase II dataset (dark dot points after the vertical dotted line). (b) The NLR EWMA chart (8.23)–(8.25) with $\lambda = 0.05$ and $ARL_0 = 370$. In plot (b), the dashed line denotes the control limits h_n.*

Table 8.7 *Computed h_n values of the NLR EWMA chart (8.23)–(8.25) in cases when $M = 100$, $\lambda = 0.05$ or 0.1, and $ARL_0 = 370$ or 500.*

	$\lambda = 0.05$		$\lambda = 0.1$	
n	$ARL_0 = 370$	$ARL_0 = 500$	$ARL_0 = 370$	$ARL_0 = 500$
1	0.753	0.803	1.372	1.442
2	0.754	0.799	1.318	1.429
3	0.740	0.793	1.240	1.350
4	0.726	0.788	1.183	1.284
5	0.711	0.772	1.110	1.219
6	0.686	0.744	1.038	1.141
7	0.667	0.726	0.983	1.072
8	0.642	0.701	0.924	1.005
9	0.622	0.681	0.873	0.944
10	0.601	0.657	0.826	0.887
11	0.578	0.633	0.783	0.843
12	0.556	0.610	0.745	0.798
13	0.538	0.586	0.709	0.759
14	0.520	0.568	0.677	0.719
15	0.503	0.548	0.647	0.690
16	0.485	0.528	0.626	0.664
17	0.470	0.510	0.600	0.634
18	0.455	0.493	0.578	0.612
19	0.439	0.478	0.561	0.591
20	0.426	0.462	0.544	0.575
22	0.400	0.432	0.513	0.548
24	0.376	0.408	0.494	0.522
26	0.356	0.384	0.478	0.505
28	0.337	0.362	0.460	0.489
30	0.320	0.344	0.448	0.479
35	0.284	0.303	0.430	0.454
40	0.256	0.274	0.419	0.444
50	0.220	0.235	0.408	0.432
60	0.199	0.212	0.410	0.433
70	0.187	0.198	0.411	0.435
80	0.180	0.193	0.417	0.439
90	0.176	0.188	0.419	0.443
115	0.176	0.188	0.427	0.453
140	0.178	0.191	0.434	0.459
165	0.184	0.194	0.441	0.466
200	0.187	0.199	0.446	0.476
250	0.193	0.207	0.459	0.488
370	0.205	0.219	0.468	0.505
500	0.209	0.227	0.472	0.512

8.2.4 Nonparametric CPD charts

In the literature, some nonparametric control charts have been proposed based on the change-point detection (CPD) framework. As discussed in Chapter 6, CPD control charts have two major advantages, compared to the CUSUM and EWMA control charts. One is that they do not require the specification of certain parameters of the IC process distribution, and the second advantage is that they can provide an estimator of the shift time at the same time as they give a signal of the shift in the process distribution. We describe some fundamental nonparametric CPD charts in this subsection.

First, the CPD chart (6.13)–(6.14) discussed in Subsection 6.3.1 that is based on the normality assumption has been modified to be a nonparametric CPD chart by Hawkins and Deng (2010). Assume that $X_1, X_2, \ldots, X_n$ are independent observations obtained from a production process up to the current time point n, and they follow the change-point model

$$X_i \sim \begin{cases} F_0(x), & \text{if } i = 1, 2, \ldots, r, \\ F_0(x - \delta), & \text{if } i = r+1, r+2, \ldots, n, \end{cases} \quad (8.26)$$

where F_0 is the cdf of the IC process distribution, δ is a location shift size, and $1 \leq r \leq n-1$ is a change-point. To test whether the process has a location shift at or before n, the related hypotheses can be formulated as

$$H_0 : \delta = 0 \qquad \text{versus} \qquad H_1 : \delta \neq 0$$

for some $1 \leq r \leq n-1$, or equivalently, $H_0 : 1 \leq r \leq n-1$ versus $H_1 : r \geq n$ for some $\delta \neq 0$ in (8.26). To test these hypotheses, Pettitt (1979) proposed the following statistic based on the Mann-Whitney two-sample test (cf., Subsection 2.8.3):

$$U_{jn} = \sum_{i=1}^{j} \sum_{i'=j+1}^{n} \text{sign}(X_i - X_{i'}), \qquad \text{for } 1 \leq j \leq n-1, \quad (8.27)$$

where $\text{sign}(u)$ is the sign function defined in (8.1). It is obvious that U_{jn} in (8.27) is 2 times the Mann-Whitney test statistic U defined in Subsection 2.8.3, that is constructed from the two samples $(X_1, X_2, \ldots, X_j)$ and $(X_{j+1}, X_{j+2}, \ldots, X_n)$. By the relationship between the Mann-Whitney test statistic and the Wilcoxon rank-sum test statistic described at the end of Subsection 2.8.3, we have

$$\mu_{U_{jn}} = 0, \qquad \sigma^2_{U_{jn}} = \frac{j(n-j)(n+1)}{3}.$$

Then, the test statistic for testing the above hypotheses is defined by

$$T_{max,n} = \max_{1 \leq j \leq n-1} |T_{jn}|, \quad (8.28)$$

where

$$T_{jn} = \frac{U_{jn}}{\sqrt{j(n-j)(n+1)/3}}.$$

The null hypothesis is rejected and we conclude that there is a location shift at or before n if

$$T_{max,n} > h_n,$$ (8.29)

where $h_n > 0$ is a properly chosen control limit that may depend on n. In cases when the null hypothesis is rejected, the change-point estimator can be defined by

$$\hat{\tau} = \arg \max_{1 \le j \le n-1} |T_{jn}|.$$ (8.30)

From (8.28) and (8.30), it can be seen that $T_{max,n} = |T_{\hat{\tau}n}|$ in such cases.

To use the nonparametric CPD (NCPD) control chart (8.28)–(8.29) for phase II SPC, we need to choose the control limit h_n such that a pre-specified ARL_0 value is reached. Because of the discrete nature of the charting statistic $T_{max,n}$, it is required that n_0 IC observations should be collected before the online process monitoring. Otherwise, a quite large ARL_0 value (or, a quite small false alarm rate) is difficult to achieve within a reasonable accuracy. A related discussion can be found in the paragraph containing (6.15) in Subsection 6.3.1. Hawkins and Deng (2010) found that results of the chart (8.28)–(8.29) were quite reliable when n_0 is chosen to be 14 or larger. When $n_0 = 14$, process monitoring starts at $n = 15$. For some commonly used ARL_0 values, Hawkins and Deng computed the corresponding control limits h_n by simulation for $15 \le n \le 1000$. Some of them are presented in Table 8.8, which is the same as Table 1 in Hawkins and Deng (2010). In the table, not all values of n between 15 and 1000 are listed. For those not in the table, the corresponding h_n values can be obtained by interpolation or by carrying entries forward. For instance, by the approach of carrying entries forward, values of h_{50} in the table can be used as all h_n values for $50 \le n < 60$. The last control limit value in each column of the table (denoted as h_{n*}) can be used as all h_n values for $n > n^*$.

When computing the test statistic $T_{max,n}$, the following recursive formula for computing U_{jn} should be helpful: for any $1 \le j \le n-1$, we have

$$
\begin{aligned}
U_{jn} &= \sum_{i=1}^{j} \sum_{i'=j+1}^{n} \text{sign}(X_i - X_{i'}) \\
&= \sum_{i=1}^{j} \left[\sum_{i'=j+1}^{n-1} \text{sign}(X_i - X_{i'}) + \text{sign}(X_i - X_n) \right] \\
&= U_{j,n-1} + \sum_{i=1}^{j} \text{sign}(X_i - X_n),
\end{aligned}
$$ (8.31)

where $U_{n-1,n-1} = 0$. From (8.31), it can be seen that, after a new observation X_n is obtained, we only need to compare the new observation with all the old observations (i.e., compute $\text{sign}(X_i - X_n)$, for $1 \le j \le n-1$) and then compute $\{U_{jn}, 1 \le j \le n-1\}$ from $\{U_{j,n-1}, 1 \le j \le n-2\}$ by the recursive formula (8.31).

Example 8.7 *Figure 8.9(a) shows the first 100 observations obtained from a production process for phase II process mean monitoring. It is known that the first 14*

Table 8.8 *Computed h_n values of the NCPD control chart (8.28)–(8.29) in cases when $n_0 = 14$, and $ARL_0 = 50, 100, 200, 500, 1000,$ and 2000.*

| n | \multicolumn{6}{c}{ARL_0} |
	50	100	200	500	1000	2000
15	2.700	2.848	2.947	3.069	3.181	3.229
16	2.615	2.767	2.910	3.047	3.142	3.244
17	2.535	2.718	2.862	3.043	3.163	3.247
18	2.535	2.694	2.860	3.034	3.183	3.277
19	2.500	2.695	2.869	3.054	3.186	3.296
20	2.488	2.699	2.851	3.059	3.203	3.311
22	2.468	2.692	2.862	3.082	3.228	3.355
24	2.469	2.676	2.870	3.096	3.249	3.389
26	2.452	2.686	2.875	3.108	3.269	3.415
28	2.455	2.686	2.883	3.121	3.283	3.437
30	2.453	2.684	2.879	3.130	3.297	3.453
35	2.452	2.687	2.894	3.149	3.324	3.487
40	2.447	2.689	2.900	3.162	3.342	3.511
45	2.453	2.690	2.906	3.171	3.356	3.529
50	2.451	2.691	2.908	3.178	3.365	3.542
60	2.452	2.694	2.914	3.188	3.379	3.560
70	2.452	2.694	2.917	3.194	3.388	3.570
80	2.453	2.696	2.918	3.199	3.394	3.579
90	2.452	2.696	2.920	3.200	3.399	3.584
100	2.453	2.697	2.922	3.203	3.402	3.591
125		2.698	2.923	3.206	3.409	3.599
150		2.697	2.924	3.209	3.411	3.603
200		2.699	2.926	3.210	3.415	3.610
250		2.700	2.927	3.212	3.416	3.610
300		2.704	2.926	3.215	3.420	3.616
500			2.927	3.213	3.417	3.612
1000			2.927	3.214	3.418	3.612

observations are IC. So, we would like to use the NCPD control chart (8.28)–(8.29) to monitor the process mean, starting from $n = 15$. In the chart, ARL_0 is chosen to be 200. Its control statistic $T_{max,n}$ is computed by (8.28) and (8.31), and is shown in Figure 8.9(b). From the plot, the first signal of process mean shift is given at $n = 55$. From the figure, it seems that there are a number of outliers in the observed data; but, they do not have a substantial impact on the control chart.

Hawkins and Deng (2010) compared the NCPD control chart (8.28)–(8.29) with the conventional CPD control chart (6.13)–(6.14) in cases when the normality assumption of the process distribution is valid. They found an interesting phenomenon that the former chart outperformed the latter chart in detecting relatively small process mean shifts, and the latter chart was better than the former chart in detecting relatively large process mean shifts only. The second part of this result is intuitively reasonable; but, the first part seems contradictory to our intuition. An explanation given by Hawkins and Deng is that the estimated change-point by the conventional

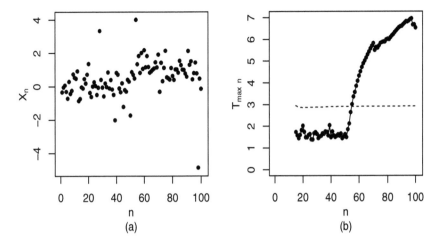

Figure 8.9 *(a) The first 100 observations obtained from a production process, among which the first 14 of them are IC. (b) The NCPD control chart (8.28)–(8.29) starting at $n = 15$ with $ARL_0 = 200$. The dashed line in plot (b) denotes the control limits h_n.*

CPD chart tends to be close to the two ends of the observation time sequence when the process is IC, this tendency is less obvious with the NCPD chart, and consequently the NCPD chart would have a smaller control limit and react faster to relatively small shifts, compared to the conventional CPD chart.

In the literature, there are some other NCPD charts proposed. For instance, Zhou et al. (2009) proposed a NCPD control chart that was closely related to the chart (8.28)–(8.29). Their charting statistic is defined by the maximum of various exponentially weighted moving averages of $\{T_{jn}, 1 \leq j \leq n-1\}$. Ross et al. (2011) proposed two NCPD control charts for monitoring the process standard deviation and for monitoring both the process mean and the process standard deviation, respectively. Ross and Adams (2012) proposed two NCPD control charts for detecting arbitrary shifts in the process distribution. One is based on the Kolmogorov-Smirnov test, and the other is based on the Cramer-von-Mises test (cf., Gibbons and Chakraborti, 2003).

8.3 Nonparametric SPC by Categorical Data Analysis

All the nonparametric control charts described in the previous section are based on the ranking or ordering information of the process observations. In this section, we describe an alternative framework for constructing nonparametric control charts that is based on categorization of numerical process observations. This method has been shown competitive in various cases when process observations are continuous numerical and they do not follow normal distributions. It can also handle cases when process observations are discrete numerical or categorical. Our description here is divided into two parts. The alternative framework is introduced in detail in Subsection 8.3.1, and it is compared to some alternative control charts in Subsection 8.3.2.

8.3.1   Process monitoring by categorizing process observations

When the process distribution is not normal, people traditionally use the ranking or ordering information of process observations for process monitoring. Some non-parametric control charts require multiple observations at each time point (i.e., cases with batch data). Qiu and Li (2011a) proposed an alternative method for handling the univariate SPC problem with an unknown response distribution. By this method, observed data are first categorized, and then certain statistical procedures for categorical data analysis are used for constructing nonparametric SPC charts. Major considerations behind this method are as follows. First, statistical tools for describing and analyzing non-normal numerical data are limited; but, there are many existing statistical tools for handling categorical data (cf., Agresti, 2002). Second, while both data ranking and data categorization could result in a loss of information contained in the observed data, the amount of lost information due to data categorization can be controlled by the number of categories used. The bigger the number of categories used, the less information would be lost. In comparison, the lost information due to ranking is difficult to control. Furthermore, the method by Qiu and Li (2011a) does not require multiple observations at a single time point. It does not require observations to be numerical either. Methods based on ranking are generally difficult to use with non-numerical data. Like some existing nonparametric control charts (cf., Section 8.2), the charts proposed by Qiu and Li (2011a) do not require the assumption that the IC process distribution is known. Instead, we assume that an IC dataset has been collected at the end of a phase I SPC analysis, and the IC dataset can be used for estimating certain IC parameters.

Let $\mathbf{X}_n = (X_{n1}, X_{n2}, \ldots, X_{nm})'$ be m observations obtained at the n-th time point during phase II process monitoring. In cases when $m > 1$, the observed data are batch data with the batch size of m. In this section, the batch size m could be 1. In such cases, the observed data are actually individual observation data. The first step to construct nonparametric control charts based on data categorization is to categorize the observed data as follows. Let

$$A_1 = (-\infty, q_1], \; A_2 = (q_1, q_2], \; \ldots, \; A_p = (q_{p-1}, \infty)$$

be a partition of the real number line R, where $-\infty < q_1 < q_2 < \cdots < q_{p-1} < \infty$ are $p-1$ boundary points of the partitioning intervals. Define

$$Y_{njl} = I(X_{nj} \in A_l), \qquad \text{for } j = 1, 2, \ldots, m, \; l = 1, 2, \ldots, p, \qquad (8.32)$$

and $\mathbf{Y}_{nj} = (Y_{nj1}, Y_{nj2}, \ldots, Y_{njp})'$, where $I(u)$ is the indicator function that equals 1 when u is "True" and 0 otherwise. Then, $\mathbf{Y}_{nj}$ has one and only one component being 1, the index of the component being 1 is the index of the partitioning interval that contains X_{nj}, and this index has a discrete distribution with probabilities

$$f_l = P(X_{nj} \in A_l), \qquad \text{for } l = 1, 2, \ldots, p.$$

So, $\mathbf{Y}_{nj}$ can be regarded as the categorized data, and it records the index of the partitioning interval that contains X_{nj}. For simplicity, $\mathbf{f} = (f_1, f_2, \ldots, f_p)'$ is also called the distribution of $\mathbf{Y}_{nj}$ in this section.

Let $\mathbf{f}^{(0)} = (f_1^{(0)}, f_2^{(0)}, \ldots, f_p^{(0)})'$ be the IC distribution of $\mathbf{Y}_{nj}$, and $\mathbf{f}^{(1)} = (f_1^{(1)}, f_2^{(1)}, \ldots, f_p^{(1)})'$ be its OC distribution. For the distribution of $\mathbf{Y}_{nj}$, it can be checked that, if the support of the IC cdf F_0 of the process distribution contains at least one of the boundary points $\{q_1, q_2, \ldots, q_{p-1}\}$ and $p \geq 2$, then any mean shift in $\mathbf{X}_n$ would result in changes in the distribution $\mathbf{f}$. It should be pointed out that the conditions used in this result are weak. For instance, the supports of most commonly used continuous numerical distributions (e.g., normal, t, χ^2, uniform, exponential distributions) are connected intervals on the number line. For these distributions, any reasonable set of boundary points $\{q_1, q_2, \ldots, q_{p-1}\}$ should have at least one point contained in their supports. Otherwise, there must exist one partitioning interval that contains all observations, which is contradictory to the purpose of categorization. By the above result, under some mild regularity conditions, $\mathbf{f}^{(1)}$ would be different from $\mathbf{f}^{(0)}$ if there is a mean shift in $\mathbf{X}_n$. Therefore, detection of a mean shift in $\mathbf{X}_n$ is equivalent to detection of a change in the distribution of $\mathbf{Y}_{nj}$ in such cases.

To define $\mathbf{Y}_{nj}$ by (8.32), we should choose $\{q_1, q_2, \ldots, q_{p-1}\}$ beforehand. For that purpose, existing research in categorical data analysis demonstrates that it would help detect shifts from $\mathbf{f}^{(0)}$ if they are chosen such that the expected counts of observations under F_0 in the partitioning intervals are roughly the same (cf., Agresti, 2002, Section 1.5). Therefore, we suggest choosing q_l to be the (l/p)-th quantile of F_0, for $l = 1, 2, \ldots, p - 1$. In such cases, $\mathbf{f}^{(0)} = (1/p, 1/p, \ldots, 1/p)'$. In practice, q_l can be estimated by the (l/p)-th sample quantile of the IC dataset.

To test whether the distribution of $\mathbf{Y}_{nj}$ equals its IC distribution $\mathbf{f}^{(0)}$, the Pearson's chi-square test (cf., Subsection 2.8.2) is a natural choice. Let

$$g_{nl} = \sum_{j=1}^{m} Y_{njl}, \qquad \text{for } l = 1, 2, \ldots, p, \; n = 1, 2, \ldots, \tag{8.33}$$

and $\mathbf{g}_n = (g_{n1}, g_{n2}, \ldots, g_{np})'$. Then, g_{nl} in (8.33), for $l = 1, 2, \ldots, p$, denotes the number of original process observations obtained at the n-th time point that fall into the l-th partitioning interval A_l. The Pearson's chi-square test statistic at the n-th time point (cf., (2.43)) is then defined by

$$\widetilde{X}_n^2 = \sum_{l=1}^{p} \frac{\left[g_{nl} - m f_l^{(0)}\right]^2}{m f_l^{(0)}},$$

which measures the discrepancy between the observed counts $\mathbf{g}_n$ and the expected counts $m\mathbf{f}^{(0)}$ in the p partitioning intervals at the n-th time point. However, this statistic uses observations at a single time point; it may not be effective to detect persistent but relatively small changes in the distribution of $\mathbf{Y}_{nj}$ (cf., a related discussion in Section 4.1). To overcome this limitation, Qiu and Li (2011a) suggested the following CUSUM chart, which adopted the structure of the multivariate CUSUM chart (7.28)–(7.29), although the two charts were for two different purposes. Let $\mathbf{U}_0^{obs} = \mathbf{U}_0^{exp} = \mathbf{0}$

be two $p \times 1$ column vectors, and

$$
\begin{cases}
\mathbf{U}_n^{obs} = \mathbf{0}, & \text{if } B_n \le k_P \\
\mathbf{U}_n^{exp} = \mathbf{0}, & \text{if } B_n \le k_P \\
\mathbf{U}_n^{obs} = \left(\mathbf{U}_{n-1}^{obs} + \mathbf{g}_n\right)\left(1 - k_P/B_n\right), & \text{if } B_n > k_P \\
\mathbf{U}_n^{exp} = \left(\mathbf{U}_{n-1}^{exp} + m\mathbf{f}^{(0)}\right)\left(1 - k_P/B_n\right), & \text{if } B_n > k_P,
\end{cases}
$$

where

$$
B_n = \left\{\left(\mathbf{U}_{n-1}^{obs} - \mathbf{U}_{n-1}^{exp}\right) + \left(\mathbf{g}_n - m\mathbf{f}^{(0)}\right)\right\}' \left(\mathrm{diag}(\mathbf{U}_{n-1}^{exp} + m\mathbf{f}^{(0)})\right)^{-1}
$$
$$
\left\{\left(\mathbf{U}_{n-1}^{obs} - \mathbf{U}_{n-1}^{exp}\right) + \left(\mathbf{g}_n - m\mathbf{f}^{(0)}\right)\right\},
$$

$k_P \ge 0$ is the allowance parameter, $\mathrm{diag}(\mathbf{a})$ denotes a diagonal matrix with its diagonal elements equal to the corresponding elements of the vector $\mathbf{a}$, and the superscripts "obs" and "exp" denote observed and expected counts, respectively. Define

$$
C_{n,P} = \left(\mathbf{U}_n^{obs} - \mathbf{U}_n^{exp}\right)' \left(\mathrm{diag}(\mathbf{U}_n^{exp})\right)^{-1} \left(\mathbf{U}_n^{obs} - \mathbf{U}_n^{exp}\right). \tag{8.34}
$$

Then, a mean shift in $\mathbf{X}_n$ is signaled if

$$
C_{n,P} > h_P, \tag{8.35}
$$

where $h_P > 0$ is a control limit chosen to achieve a given IC ARL level. Chart (8.34)–(8.35) is called the nonparametric P-CUSUM chart hereafter, to reflect the fact that it is from the Pearson's chi-square test.

When $k_P = 0$, it is not difficult to check that $\mathbf{U}_n^{obs}$ is a frequency vector with its l-th element denoting the cumulative observed count of observations in the l-th interval A_l as of the time point n, for $l = 1, 2, \ldots, p$, and $\mathbf{U}_n^{exp}$ equals $nm\mathbf{f}^{(0)}$, which is the vector of the corresponding cumulative expected counts. Therefore, in such cases, $C_{n,P}$ is the conventional Pearson's chi-square test statistic that measures the difference between the cumulative observed and expected counts as of the time point n. Further, it can be checked (cf., Qiu and Hawkins, 2001, Appendix C) that

$$
C_{n,P} = \max\left(0, B_n - k_P\right).
$$

Therefore, the charting statistic $C_{n,P}$ is defined in the way that the CUSUM chart can be repeatedly restarted when there is little evidence of distributional shift in $\mathbf{Y}_{nj}$.

For the nonparametric P-CUSUM chart (8.34)–(8.35), the control limit h_P can be determined easily by a numerical algorithm as follows. First, choose an initial value for h_P. Then, compute the IC ARL value of the P-CUSUM chart based on a large number (e.g., 10000) of replicated simulation runs in which the IC multinomial observations $\mathbf{Y}_{nj}$ are sequentially generated from the IC distribution $\mathbf{f}^{(0)} = (1/p, 1/p, \ldots, 1/p)'$. If the computed IC ARL value is smaller than the nominal IC ARL value, then increase the value of h_P. Otherwise, choose a smaller h_P value. The above process is repeated until the nominal IC ARL value is reached within a desired precision. See the pseudo code described in Subsection 4.2.2 for a related discussion, in which the bisection search algorithm is used. From the above description, it can be seen that determination of h_P does not require any information about the IC process distribution F_0. Instead, it only depends on the nominal IC ARL value, the allowance constant k_P, the batch size m, and the number of categories p. In this sense, the nonparametric P-CUSUM chart is distribution-free.

As a remark, due to the fact that the charting statistic $C_{n,P}$ takes discrete values on the positive number line, certain pre-specified nominal IC ARL values cannot be reached within a desired precision. This phenomenon is common when handling discrete data, as discussed in the previous section (cf., Table 8.5). To overcome this limitation, when implementing the nonparametric P-CUSUM chart, we can simply use an IC ARL value that the chart can reach. However, in practice, it is often desirable to use a common IC ARL value (e.g., 200, 370, etc.). In such cases, Qiu and Li (2011a) proposed the following simple but efficient modification of the charting statistic $C_{n,P}$. Let $\mathbf{b}_j(n)$ be a sequence of i.i.d. random vectors generated from the distribution $\mathbf{N}_p(\mathbf{0}, v^2 \mathbf{I}_{p \times p})$, where v^2 is a small positive number and $\mathbf{I}_{p \times p}$ is the $p \times p$ identity matrix. Then, when computing $C_{n,P}$, we can replace $\mathbf{Y}_{nj}$ by $\mathbf{Y}_{nj} + \mathbf{b}_j(n)$. Namely, we add a small random number from the distribution $N(0, v^2)$ to each component of $\mathbf{Y}_{nj}$ to alleviate the discreteness of the charting statistic $C_{n,P}$. Based on our numerical experience, as long as v is chosen to be small (e.g., $v = 0.01$), the OC behavior of the nonparametric P-CUSUM chart is hardly affected by this modification. However, most nominal IC ARL values can be reached within a desired precision after the modification. When $ARL_0 = 200, 300, 500,$ or 1000, $p = 2, 3, 5, 10, 15,$ or 20, $k_P = 0.001, 0.005, 0.01,$ or 0.05, $m = 1,$ or 5, and $M = 500$, the computed h_P values based on 10,000 replicated simulations, as described in this and the previous paragraphs, are presented in Table 8.9. From the table, it can be seen that h_P increases with ARL_0, k_P, and p, and decreases with m.

To use the nonparametric P-CUSUM chart (8.34)–(8.35), besides h_P, we also need to choose k_P and p, and the chart requires an IC data of size M. Based on an extensive simulation study, Qiu and Li (2011a) provided some practical guidelines on choosing k_P, p, and M, which are described below. First, k_P should be chosen as small as 0.05 or 0.01. Second, the chart would perform better when M is larger, and its results do not change much when $M \geq 200$. Third, regarding the selection of p, we have the following practical guidelines. (i) p can be chosen smaller when m is larger. (ii) In cases when we do not have any prior information about the process distribution, then we can choose $p = 10$. In such cases, the P-CUSUM chart should perform reasonably well. (iii) If we know that the process distribution is quite symmetric, or that it is skewed but the potential shift is in the direction of the longer tail, then p can be chosen as small as 5.

From the construction of the nonparametric P-CUSUM chart (8.34)–(8.35), we can see that it should be able to detect any distributional shift of the original observations $\mathbf{X}_n$ that results in a shift in the distribution of the corresponding categorized data $\mathbf{Y}_{nj}$. Therefore, the P-CUSUM chart should have certain ability to detect shifts in variance, or in both mean and variance, of the distribution of the original observations $\mathbf{X}_n$. To demonstrate this, let us consider cases when the IC process distribution

Table 8.9 *Computed h_P values of the nonparametric P-CUSUM chart (8.34)–(8.35) based on 10,000 replications when $ARL_0 = 200, 300, 500,$ or $1000, p = 2, 3, 5, 10, 15,$ or $20, k_P = 0.001, 0.005, 0.01,$ or $0.05, m = 1,$ or $5,$ and $M = 500.$*

				$m = 1$			
k_P	ARL_0	$p = 2$	3	5	10	15	20
	200	4.144	4.429	5.654	10.783	15.777	20.883
0.001	300	4.898	4.614	5.887	11.015	16.230	21.422
	500	5.445	4.833	6.265	11.305	16.570	21.827
	1000	6.402	5.552	6.843	11.960	16.979	22.155
	200	5.089	5.148	6.215	11.180	16.142	21.297
0.005	300	5.848	5.481	6.650	11.531	16.407	21.589
	500	6.497	6.099	7.251	12.006	16.992	22.106
	1000	7.735	7.146	8.056	12.758	17.655	22.677
	200	5.529	5.555	6.665	11.377	16.400	21.448
0.01	300	6.123	6.061	7.209	11.842	16.739	21.856
	500	6.954	6.899	7.929	12.343	17.307	22.312
	1000	8.078	7.979	8.600	13.205	18.056	22.993
	200	5.918	6.860	7.957	12.171	16.981	21.931
0.05	300	6.656	7.685	8.438	12.841	17.550	22.397
	500	7.595	8.492	9.269	13.616	18.181	22.958
	1000	8.559	9.773	10.614	14.461	19.149	23.986
				$m = 5$			
k_P	ARL_0	$p = 2$	3	5	10	15	20
	200	0.899	0.983	1.318	2.430	3.588	4.753
0.001	300	1.080	1.065	1.379	2.496	3.647	4.814
	500	1.265	1.183	1.485	2.598	3.734	4.916
	1000	1.469	1.384	1.636	2.718	3.879	5.054
	200	1.214	1.245	1.512	2.575	3.732	4.912
0.005	300	1.326	1.375	1.618	2.678	3.824	4.987
	500	1.522	1.556	1.759	2.799	3.948	5.115
	1000	1.778	1.815	1.966	2.981	4.119	5.254
	200	1.276	1.389	1.635	2.686	3.826	5.024
0.01	300	1.412	1.530	1.750	2.777	3.909	5.084
	500	1.600	1.725	1.911	2.911	4.032	5.214
	1000	1.889	2.007	2.152	3.100	4.228	5.396
	200	1.348	1.704	1.994	2.927	4.071	5.255
0.05	300	1.488	1.898	2.149	3.047	4.187	5.365
	500	1.718	2.040	2.362	3.220	4.355	5.528
	1000	1.924	2.272	2.628	3.459	4.577	5.745

is the standardized version with mean 0 and standard deviation 1 of χ_1^2 or χ_4^2, and the OC distribution has a (mean, variance) shift from $(0, 1)$ to (μ_1, σ_1^2), where (μ_1, σ_1^2) take the values of $(0, 1.5), (0, 2.0), (0.5, 1.5),$ or $(0.5, 2.0)$. The computed optimal OC ARL values of the P-CUSUM chart (i.e., the minimum OC ARL values when k_P changes) when $M = 500, m = 5, ARL_0 = 500,$ and $p = 5$ are presented in Table 8.10. The small ARL values indicate that the P-CUSUM chart can detect variance shifts, and its computed optimal OC ARL values are much smaller in cases when both the mean and variance shift, compared to cases when only the variance shifts, which is reasonable because the resulting shifts in the distribution of the categorized data Y_{nj} are bigger in the former cases. For a similar reason, the P-CUSUM chart is more sensitive to variance shifts when the IC process distribution is more skewed. However, based on the P-CUSUM chart alone, we cannot distinguish mean shifts from variance shifts, after a signal of shift is given. It requires much future research to design diagnostic procedures for the P-CUSUM chart to make a distinction between the two types of shifts after a signal is given.

Table 8.10 *Computed optimal OC ARL values of the nonparametric P-CUSUM chart (8.34)–(8.35) when $M = 500$, $m = 5$, $ARL_0 = 500$, $p = 5$, the IC observation distribution is the standardized version with mean 0 and standard deviation 1 of χ_1^2 or χ_4^2, and the OC distribution has a (mean, variance) shift from $(0,1)$ to (μ_1, σ_1^2).*

IC Distribution	$(\mu_1, \sigma_1^2)=(0,1.5)$	$(0,2.0)$	$(0.5,1.5)$	$(0.5,2.0)$
χ_1^2	51.90	17.30	8.13	15.87
χ_4^2	279.24	117.69	86.96	79.94

Example 8.8 *To online monitor a production process, 500 IC observations are obtained beforehand. Then, the first 100 phase II process observations are collected for online process monitoring, and they are shown in Figure 8.10(a). To apply the nonparametric P-CUSUM chart (8.34)–(8.35) to the phase II observations, we choose $p = 10$ and $k_P = 0.01$ in the chart, by the empirical guidelines discussed above. The boundary points q_l for phase II observation categorization can be estimated by the l/p-th sample quantiles, for $l = 1,2,\ldots,p-1$, of the IC dataset, and the estimates are*

$$\widehat{q}_1 = -0.946, \quad \widehat{q}_2 = -0.585, \quad \widehat{q}_3 = -0.350,$$
$$\widehat{q}_4 = -0.161, \quad \widehat{q}_5 = 0.005, \quad \widehat{q}_6 = 0.199,$$
$$\widehat{q}_7 = 0.340, \quad \widehat{q}_8 = 0.530, \quad \widehat{q}_9 = 0.941.$$

In such cases, the IC distribution of the categorized data Y_{nj} can be regarded as $\mathbf{f}^{(0)} = (1/p, 1/p, \ldots, 1/p)'$. In the chart, we further choose $ARL_0 = 200$. By Table 8.9, the corresponding control limit is $h_P = 11.377$. The control chart is then constructed and shown in Figure 8.10(b), where the horizontal dashed line denotes the control limit. From the plot, we can see that the first signal of distributional shift is given at $n = 63$.

The above discussion focuses on cases when process observations are continuous numerical. Situations when observations are discrete numerical or categorical can be handled similarly. For instance, when process observations are categorical, the categorization step (cf., equation (8.32)) can be skipped when constructing the nonparametric P-CUSUM chart (8.34)–(8.35). When process observations are discrete numerical and the number of different observation values is quite small, we can use each possible observation value as a single category, and construct the chart as usual. In cases when process observations are discrete numerical but the number of

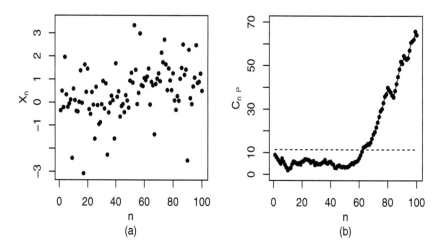

Figure 8.10 *(a) The first 100 phase II observations obtained from a production process for online monitoring. (b) The nonparametric P-CUSUM chart (8.34)–(8.35) with p = 10, k_P = 0.01, and ARL_0 = 200. In the chart, the boundary points q_l are estimated by the l/p-th sample quantiles, for l = 1, 2, ..., p − 1, of an IC data of size M = 500. The dashed horizontal line in plot (b) denotes the control limit h_P.*

different observation values is quite large, proper combination of certain observation values might be necessary when defining the categories.

8.3.2 *Alternative control charts and some comparisons*

In this subsection, we first describe some alternative nonparametric control charts based on observation categorization, and then discuss some numerical results given by Qiu and Li (2011a) about the comparison among the nonparametric P-CUSUM chart (8.34)–(8.35), these alternative nonparametric control charts, and some other competitive control charts. Such numerical results should be helpful for users to choose a proper control chart from many existing charts.

We first introduce the competitive control charts considered in the numerical comparison in Qiu and Li (2011a), including some alternative nonparametric control charts based on observation categorization. To detect changes in the distribution of the categorized data $\mathbf{Y}_{nj}$, besides the Pearson's chi-square test (cf., $\widetilde{X}_n^2$ defined in the previous subsection), another popular test is the likelihood ratio test (cf., Subsection 2.8.2) with the test statistic

$$\widetilde{G}_n^2 = 2 \sum_{l=1}^{p} g_{nl} \log \left(\frac{g_{nl}}{m f_l^{(0)}} \right).$$

Then, a CUSUM chart can be constructed similarly to (8.34)–(8.35) as follows. Let $\widetilde{\mathbf{U}}_n^{obs}$ and $\widetilde{\mathbf{U}}_n^{exp}$ be quantities defined in the same way as $\mathbf{U}_n^{obs}$ and $\mathbf{U}_n^{exp}$ used in (8.34), except that the allowance constant k_P is replaced by another constant k_L, and B_n is replaced by

$$\widetilde{B}_n = 2\left(\widetilde{\mathbf{U}}_{n-1}^{obs} + \mathbf{g}_n\right)' \log\left(\frac{\widetilde{\mathbf{U}}_{n-1}^{obs} + \mathbf{g}_n}{\widetilde{\mathbf{U}}_{n-1}^{exp} + m\mathbf{f}^{(0)}}\right),$$

where $\mathbf{a}/\mathbf{b}$ denotes a vector obtained by componentwise division of the vector $\mathbf{a}$ by the vector $\mathbf{b}$, and similarly $\log(\mathbf{a}/\mathbf{b})$ denotes a vector obtained by applying the logarithm transformation to all components of $\mathbf{a}/\mathbf{b}$. Define

$$C_{n,L} = 2\left(\widetilde{\mathbf{U}}_n^{obs}\right)' \log\left(\frac{\widetilde{\mathbf{U}}_n^{obs}}{m\mathbf{f}^{(0)}}\right). \tag{8.36}$$

Then, a shift in the distribution of $\mathbf{X}_n$ is signaled if

$$C_{n,L} > h_L, \tag{8.37}$$

where the control limit $h_L > 0$ is chosen to achieve a given IC ARL level. Control chart (8.36)–(8.37), which is based on the likelihood ratio test, will be called the nonparametric L-CUSUM chart hereafter.

In the categorical data analysis, there is much discussion about the relationship between the two tests based on $\widetilde{X}_n^2$ and $\widetilde{G}_n^2$ (cf., Agresti, 2002, Section 1.5.5). The two tests are asymptotically equivalent, and the distribution of $\widetilde{X}_n^2$ converges to a chi-square distribution faster than that of $\widetilde{G}_n^2$, when the observed counts in $\mathbf{g}_n$ increase. In SPC, however, asymptotic properties are usually irrelevant because the related control chart should be stopped immediately after a signal of distributional shift is obtained. Therefore, the relationship between the nonparametric P-CUSUM and L-CUSUM charts still needs to be studied.

From the construction of the nonparametric P-CUSUM and L-CUSUM charts, we notice that the partitioning intervals $\{A_1, A_2, \ldots, A_p\}$ are ordered on the number line; but, both the charting statistics $C_{n,P}$ and $C_{n,L}$ ignore such ordering information completely. In the nonparametric statistics literature, a popular test that takes into account the ordering information among observations when testing for a potential change in the observation distribution is the Kolmogorov-Smirnov goodness-of-fit test (cf., Chakravarti et al., 1967). To apply this test here for detecting potential shifts from the IC process distribution F_0, the test statistic can be written as

$$D_n = \max_{1 \leq j \leq m}\left(\widehat{F}_0\left(X_{n(j)}\right) - \frac{j-1}{m}, \frac{j}{m} - \widehat{F}_0\left(X_{n(j)}\right)\right),$$

where $\{X_{n(j)}, j = 1, 2, \ldots, m\}$ are the order statistics of $\{X_{nj}, j = 1, 2, \ldots, m\}$, and $\widehat{F}_0$ is the IC empirical cumulative distribution function (cf., (2.43)) constructed from the IC data. Based on D_n, a CUSUM chart can be constructed as follows. Let $C_{0,K} = 0$, and for $n \geq 1$,

$$C_{n,K} = \max\left[0, C_{n-1,K} + \left(D_n - \mu_D^{(0)}\right) - k_K\right], \tag{8.38}$$

where $\mu_D^{(0)}$ denotes the IC mean of D_n, which can be estimated from the IC data, and k_K is an allowance constant. Then, a mean shift in $\mathbf{X}_n$ is signaled if

$$C_{n,K} > h_K, \tag{8.39}$$

where $h_K > 0$ is a control limit. Chart (8.38)–(8.39) is called the nonparametric K-CUSUM chart hereafter. It should be pointed out that the K-CUSUM chart is constructed from the original observations $\mathbf{X}_n$, instead of from their categorized version $\{Y_{nj}, j = 1, 2, \ldots, m\}$. Although this chart can also be constructed from $\{Y_{nj}, j = 1, 2, \ldots, m\}$, this approach is not recommended due to the reasons that (i) the chart (8.38)–(8.39) based on $\mathbf{X}_n$ is already nonparametric, and (ii) it would unnecessarily lose a certain efficiency in detecting potential mean shifts in $\mathbf{X}_n$ if the categorized data $\{Y_{nj}, j = 1, 2, \ldots, m\}$ were used.

Besides the nonparametric P-CUSUM, L-CUSUM, and K-CUSUM charts described above, some other competitive control charts are also included in the numerical comparison in Qiu and Li (2011a), which are briefly described below. The traditional CUSUM chart described in Section 4.2 is a standard tool for monitoring univariate process mean in practice. Its charting statistics of the two-sided version in cases when the IC process mean is known to be 0 are defined by

$$C_{n,N}^+ = \max\left(0, C_{n-1,N}^+ + \overline{X}_n - k_N\right), \qquad \text{for } n \geq 1$$

$$C_{n,N}^- = \min\left(0, C_{n-1,N}^- + \overline{X}_n + k_N\right), \tag{8.40}$$

where $C_{0,N}^+ = C_{0,N}^- = 0$, k_N is an allowance constant, $\overline{X}_n = \frac{1}{m}\sum_{j=1}^m X_{nj}$, and the subscript "$N$" denotes the fact that this method is based on the normal-distribution assumption. Then, a mean shift in $\mathbf{X}_n$ is signaled if

$$C_{n,N}^+ > h_N^+ \qquad \text{or} \qquad C_{n,N}^- < -h_N^- \tag{8.41}$$

where the control limits $h_N^+, h_N^- > 0$ are chosen to achieve a given IC ARL level. When the true IC process distribution is symmetric (e.g., normal or t), then we can set $h_N^+ = h_N^-$. For skewed IC process distributions, different h_N^+ and h_N^- can be chosen as follows. As demonstrated in Section 3.3 of Hawkins and Olwell (1998), when n is large enough (e.g., $n \geq 50$), the distributions of the charting statistics $C_{n,N}^+$ and $C_{n,N}^-$ become stable (i.e., the steady-state phenomenon discussed in Section 4.2). When the IC process distribution is known, we can determine the steady-state distributions of $C_{n,N}^+$ and $C_{n,N}^-$ by simulation and choose h_N^+ and h_N^- such that $P(u_{n,N}^+ > h_N^+) = P(u_{n,N}^- < -h_N^-)$ and the pre-specified IC ARL value is achieved. When the IC process distribution is unknown but an IC dataset is available, the steady-state distributions of $C_{n,N}^+$ and $C_{n,N}^-$ can be estimated using bootstrap samples drawn from the IC data, as in Chatterjee and Qiu (2009), and then h_N^+ and h_N^- can be determined in a similar way to that in cases when the IC process distribution is known. The chart (8.40)–(8.41) in such cases when the IC process distribution is unknown is called the N-CUSUM chart hereafter.

As mentioned at the end of Subsection 5.2.1, when the process distribution is not normal, Borror et al. (1999) showed that a properly designed EWMA chart was robust to departures from normality. More specifically, the EWMA charting statistic in cases when the IC process mean is known to be 0 is defined by

$$E_n = \lambda \overline{X}_n + (1 - \lambda)E_{n-1},$$

where $E_0 = 0$, and $\lambda \in (0, 1]$ is a weighting parameter. Then, a mean shift in $\mathbf{X}_n$ is signaled if

$$|E_n| \geq h_R,$$

where $h_R > 0$ is a control limit chosen to achieve a pre-specified IC ARL level under the normality assumption. Borror et al. (1999) showed that, when $\lambda = 0.05$, this chart performed reasonably well in various cases when the IC process distribution was actually non-normal. This EWMA chart with $\lambda = 0.05$ is called the R-EWMA chart hereafter, where R stands for "robust."

The upward NSR CUSUM chart (8.8)–(8.9) discussed in Subsection 8.2.2 is for detecting upward process mean shifts. Its two-sided version is also considered in the numerical comparison below. For convenience, the two-sided version is called the S-CUSUM chart hereafter.

Phase II monitoring of categorized (or grouped) data was also discussed in Steiner et al. (1996) and Steiner (1998). In their papers, Steiner and co-authors considered the case when the IC distribution had a known parametric form (e.g., normal), individual observations might not be completely known, and it was known that they belonged to certain given intervals. In such cases, a CUSUM chart based on the likelihood ratio formulation and a corresponding EWMA chart were proposed. Obviously, these charts are different from the nonparametric P-CUSUM, L-CUSUM, and K-CUSUM charts discussed above, in that the former requires a parametric form of the IC process distribution to be given beforehand while the latter does not have this requirement. Next, we briefly describe the CUSUM chart by Steiner et al. (1996). To use this chart, we need to assume that the process distribution has a parametric cdf $F(x, \theta)$, where θ is a parameter. When the process is IC, $\theta = \theta_0$. After the process becomes OC, θ shifts to θ_1. Assume that the data are grouped into s intervals: $(-\infty, t_1], (t_1, t_2], \ldots, (t_{s-1}, \infty)$. For $j = 1, 2, \ldots, s$, let

$$
\begin{aligned}
\pi_j^+(\theta) &= F(t_j, \theta) - F(t_{j-1}, \theta), \\
l_j^+ &= \log\left(\pi_j^+(\theta_1)/\pi_j^+(\theta_0)\right), \\
w_j^+ &= \mathrm{round}\left(Q^+ l_j^+\right),
\end{aligned}
$$

where $t_0 = -\infty$, $t_s = \infty$, round(a) is a function that rounds a to its nearest integer, and $Q^+ = 50/[\max(l_1^+, \ldots, l_s^+) - \min(l_1^+, \ldots, l_s^+)]$. Then, to detect the shift in θ from θ_0 to θ_1, the charting statistic is defined to be

$$T_n^+ = \max\left(0, T_{n-1}^+ + \sum_{j=1}^{s} m_j^+(n) l_j^+\right), \qquad \text{for } n \geq 1,$$

where $T_0^+ = 0$, and $m_j^+(n)$ denotes the number of observations that belong to the interval $(t_{j-1}, t_j]$ at the n-th time point, for $j = 1, 2, \ldots, s$. The chart gives a signal of shift when

$$T_n^+ > h_{St},$$

where $h_{St} > 0$ is a control limit chosen to achieve a given IC ARL value. In the numerical study presented below, we choose F to be the cdf of $N(\theta, 1)$, $\theta_0 = 0$, and $s = 5$. As suggested by Steiner et al. (1996), $\{t_j, j = 1, 2, \ldots, s\}$ are chosen to be the optimal CUSUM gauge points listed in Table 2 of their paper that are scaled by θ_1. The control chart using T_n^+ is one-sided; but, all other control charts considered in the numerical comparison are two-sided. To make the chart comparable with its peers, we modify it to a two-sided control chart as follows. For $j = 1, 2, \ldots, s$, let

$$\pi_j^-(\theta) = F(-t_{j-1}, \theta) - F(-t_j, \theta),$$
$$l_j^- = \log\left(\pi_j^-(-\theta_1)/\pi_j^-(\theta_0)\right),$$
$$w_j^- = \text{round}\left(Q^- l_j^-\right),$$

and

$$T_n^- = \min\left(0, T_{n-1}^- - \sum_{j=1}^{s} m_j^-(n) l_j^-\right), \qquad \text{for } n \geq 1,$$

where $T_0^- = 0$, $Q^- = 50/[\max(l_1^-, \ldots, l_s^-) - \min(l_1^-, \ldots, l_s^-)]$, and $m_j^-(n)$ denotes the number of observations that belong to the interval $(-t_j, -t_{j-1}]$ at the n-th time point, for $j = 1, 2, \ldots, s$. Then, the two-sided control chart gives a signal of shift if

$$T_n^+ > h_{St} \quad \text{or} \quad T_n^- < -h_{St},$$

where h_{St} is chosen to achieve a given IC ARL value. This chart is called the St-CUSUM chart hereafter.

Next, we present some numerical comparison results about the control charts P-CUSUM, L-CUSUM, K-CUSUM, N-CUSUM, R-EWMA, S-CUSUM, and St-CUSUM. In the comparison, the IC distribution is chosen to be the standardized version with mean 0 and standard deviation 1 of one of the following four distributions: $N(0, 1)$, t_4, χ_1^2, and χ_4^2. The t_4 distribution represents symmetric distributions with heavy tails, and the χ_1^2 and χ_4^2 distributions represent skewed distributions with different skewness. It is also assumed that the pre-specified IC ARL value is 500, and the batch size of phase II observations at each time point is $m = 5$. We fix $p = 5$ in the P-CUSUM and L-CUSUM charts.

First, we compute the actual IC ARL values of the seven control charts, based on 10,000 replicated simulations, for each of the four actual IC process distributions described above. For the charts P-CUSUM, L-CUSUM, K-CUSUM, N-CUSUM, and S-CUSUM, their control limits are determined based on 500 IC observations when their allowance constants are chosen to be 0.1. The control limit of the chart R-EWMA is chosen when $\lambda = 0.05$ and the IC distribution is assumed normal. For the chart St-CUSUM, it is assumed that the IC process distribution is normal in all cases, and the OC mean $\theta_1 = 0.6$. The actual IC ARL values are shown in Table 8.11. From the table, it can be seen that the actual IC ARL values of the charts P-CUSUM, L-CUSUM, and K-CUSUM are close to 500 in all cases, as expected. The actual IC ARL values of the charts N-CUSUM and R-EWMA are quite different from 500 when the actual IC process distribution is not normal. For the chart S-CUSUM, its actual IC ARL values are moderately different from 500 due to the discreteness of its charting statistic, as described in the previous section. Because the chart St-CUSUM is constructed using a likelihood based on the normal distribution assumption, its actual IC ARL value is close to 500 when the actual IC process distribution is $N(0, 1)$; but, its actual IC ARL values are far from 500 in the other three cases.

Table 8.11 *The actual IC ARL values and their standard errors (in parentheses) of the seven control charts when the nominal IC ARL values are fixed at 500 and the actual IC process distribution is the standardized version of $N(0, 1)$, t_4, χ_1^2, and χ_4^2.*

	$N(0,1)$	t_4	χ_1^2	χ_4^2
P-CUSUM	501.9 (5.51)	503.3 (5.55)	504.8 (5.46)	501.1 (5.45)
L-CUSUM	498.1 (4.96)	495.3 (5.02)	504.2 (5.13)	497.4 (5.06)
K-CUSUM	499.0 (4.48)	499.7 (4.47)	504.4 (4.61)	496.4 (4.46)
N-CUSUM	498.9 (4.35)	156.0 (1.13)	321.4 (2.63)	371.5 (3.27)
R-EWMA	502.2 (6.24)	578.2 (20.87)	605.5 (8.65)	533.7 (6.57)
S-CUSUM	497.3 (5.22)	532.2 (5.57)	544.5 (5.88)	518.3 (5.63)
St-CUSUM	499.5 (4.87)	3037.7 (27.33)	9316.6 (20.46)	1862.2 (18.08)

Next, we compare the OC performance of the related control charts in cases when the IC sample size $M = 500$. In order to make the comparison more meaningful, we intentionally adjust the parameters of the charts N-CUSUM, R-EWMA, and St-CUSUM so that their actual IC ARL values equal 500 in all cases considered. For the chart S-CUSUM, we use the same modification procedure as the one described in the previous subsection for the P-CUSUM chart to overcome the difficulty caused by the discreteness of its charting statistic, and the actual IC ARL value of its modified version can reach 500 in all cases considered. In this study, 10 mean shifts ranging from -1.0 to 1.0 with a step of 0.2 are considered, representing small, medium, and large shifts. Due to the fact that different control charts have different parameters (e.g., k_P in the chart P-CUSUM, k_N in the chart N-CUSUM, and λ in the chart R-EWMA), and that the performance of different charts may not be comparable if their parameters are pre-specified, we compare the optimal performance of all the charts when detecting each shift, by selecting their parameters to minimize the OC ARL values for detecting each individual shift, while their IC ARL values are all fixed at 500. Based on 10,000 replications, the optimal OC ARL values of the related control charts are shown in Figures 8.11 and 8.12. The results are shown in two figures to better demonstrate the difference among different control charts. The optimal OC ARL values of the P-CUSUM chart are plotted in both figures for the convenience of comparison. When reading the plots in these figures, readers are reminded that the scale on the y-axis is in natural logarithm, to better demonstrate the difference among different control charts in detecting relatively large shifts.

From Figures 8.11(a) and 8.12(a), we can see that when the normality assumption is valid, the N-CUSUM chart is the best, as expected. The St-CUSUM and P-CUSUM charts lose a certain power in detecting mean shifts because of data grouping in the St-CUSUM chart and the data categorization in the P-CUSUM chart. The R-EWMA and S-CUSUM charts perform well for detecting small shifts only.

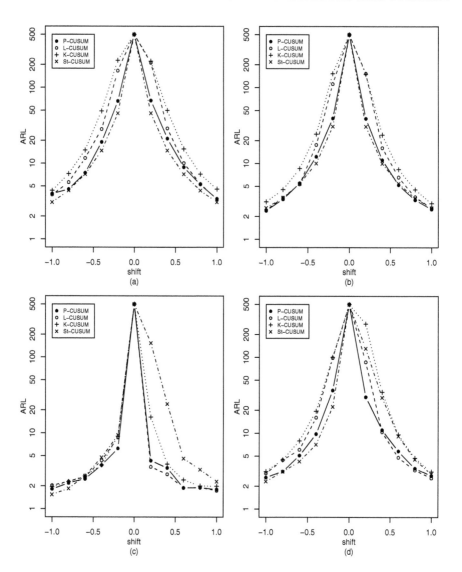

Figure 8.11 *Optimal OC ARL values of four control charts when the IC ARL is 500, M = 500, m = 5, and the actual IC process distribution is the standardized version of $N(0,1)$ (plot (a)), t_4 (plot (b)), χ_1^2 (plot (c)), and χ_4^2 (plot (d)). Scale on the y-axis is in natural logarithm.*

It seems that the K-CUSUM and L-CUSUM charts are not appropriate to use in this scenario. In the scenario of Figures 8.11(b) and 8.12(b), the IC process distribution is t_4, which is symmetric with heavy tails. In such cases, results are quite similar to those in the $N(0,1)$ case, with two exceptions: (i) the P-CUSUM chart performs the best or close to the best among all control charts in most cases of this scenario,

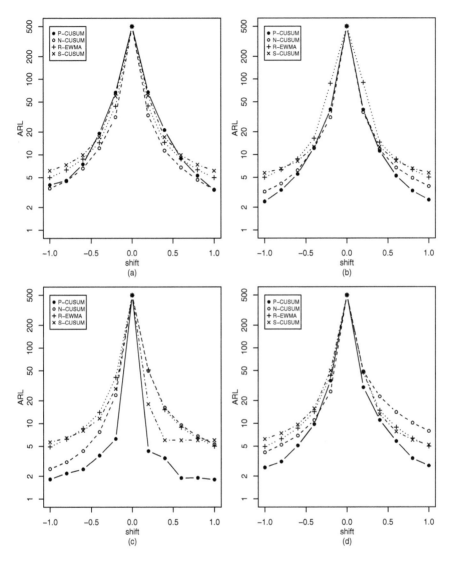

Figure 8.12 *Optimal OC ARL values of four control charts when the IC ARL is 500, M = 500, m = 5, and the actual IC process distribution is the standardized version of N(0, 1) (plot (a)), t_4 (plot (b)), χ_1^2 (plot (c)), and χ_4^2 (plot (d)). Scale on the y-axis is in natural logarithm.*

and (ii) the N-CUSUM chart is not among the best any more when the shift is large, although it still performs well when the shift is small. When the actual IC distribution is skewed (cf., Figures 8.11(c)–(d) and 8.12(c)–(d)), the charts P-CUSUM and L-CUSUM perform well. The chart St-CUSUM performs well only when detecting downward shifts. The K-CUSUM chart performs well only when the skewness is large. The other two charts do not perform well in such cases.

In Qiu and Li (2011a) and its supplementary file, there are corresponding numerical results in cases when $M = 200$ and in cases when procedure parameters of the control charts are chosen to be the ones that minimize the OC ARL values for detecting the shift of size 0.6 and these parameters are used in all other cases. These extra results suggest the same relative performance of the seven control charts as observed in Figures 8.11 and 8.12.

From the numerical results described above, we can make the following conclusions. (i) The control charts constructed based on the normality assumption (e.g., the N-CUSUM, R-CUSUM, and St-CUSUM charts) would not be reliable to use in cases when the normality assumption is invalid, because their actual IC ARL values could be substantially different from the nominal IC ARL value. Even when they are adjusted such that their actual IC ARL values equal the nominal IC ARL value, their OC performance could be poor, especially when the process distribution is quite different from a normal distribution. (ii) The nonparametric control charts based on the ranking information of process observations (e.g., the K-CUSUM and S-CUSUM charts) are reliable to use, but they are not quite effective in various different cases. (iii) The control charts based on observation categorization (e.g., the P-CUSUM and L-CUSUM charts) are reliable to use, and they are quite effective as well in different cases considered above. Between the P-CUSUM and L-CUSUM charts, it seems that the P-CUSUM chart is generally more effective.

8.4   Some Discussions

In this chapter, we have discussed some nonparametric control charts for monitoring univariate production processes in cases when the process distribution cannot be described properly by a parametric form or when no parametric form of the process distribution is available. From Figure 8.12(a), it can be seen that, in cases when the assumed parametric form of the process distribution is valid, the parametric control charts that makes use of the assumed parametric form would be more effective than the nonparametric control charts. However, in practice, the assumed parametric form of the process distribution (e.g., a normal distribution) is often invalid, or no parametric form is available. See, for instance, the aluminum smelter example discussed in Qiu and Hawkins (2001) and the exchange rate example discussed in Qiu and Li (2011a). In cases when the assumed parametric form is invalid, SPC results from the related parametric control charts may not be reliable, and the results could even be misleading because the actual false alarm rates of the parametric charts could be substantially larger or smaller than the assumed false alarm rate (cf., Table 8.11). As mentioned in Section 8.1, in cases when the actual false alarm rate of a control chart is larger than the assumed false alarm rate (or, the actual IC ARL value is smaller than the assumed IC ARL value), much labor and other resources would be wasted because the production process would be stopped too often unnecessarily. On the other hand, if the actual false alarm rate of a control chart is smaller than the assumed false alarm rate (or, the actual IC ARL value is larger than the assumed

IC ARL value), then the chart cannot give signals of process distributional shift in a timely manner. A direct consequence could be that many defective products are manufactured by the production process without notice. Therefore, in cases when no parametric form of the process distribution is available or when no parametric form is validated properly beforehand, nonparametric control charts should be considered.

In Sections 8.2 and 8.3, two different types of nonparametric control charts have been described. One is based on the ranking or ordering information in the observed data, and the other is based on observation categorization. Obviously, both types of methods would lose information in the observed data. But the methods based on observation categorization seem to be more efficient in process monitoring (cf., Figures 8.11 and 8.12) because they have used certain information in the observation magnitudes. The larger the number of partitioning intervals p, the more magnitude information they use. Of course, the boundary points of the partitioning intervals need to be estimated from the observed data in practice. When p is chosen to be large, the estimation error could affect the control chart performance. So there is a trade-off between these two considerations. In Subsection 8.3.1, we have described some practical guidelines on choosing p and some other procedure parameters, from which p can be chosen to be 10 or smaller, depending on the batch size and the shape of the process distribution. But much future research is needed to further explore the relationship between the control chart performance and the values of p, M, the shape of the process distribution, and so forth.

Most nonparametric control charts described in this chapter are for phase II process monitoring. In the literature, there is only a limited discussion on phase I process analysis in cases when the process distribution does not have any parametric form (cf. Graham et al., 2010; Jones-Farmer et al., 2009). Also, many existing nonparametric control charts assume that certain parameters (e.g., the median) of the IC process distribution are known, or that there is an IC dataset available for phase II SPC. In practice, some extra research effort is needed to estimate these parameters or collect the IC dataset properly before phase II online process monitoring. Furthermore, existing research on nonparametric SPC focuses mainly on detection of location shifts in the process distribution, although a few nonparametric control charts can also detect scale shifts or arbitrary shifts in the process distribution (e.g., the NLR EWMA chart in Subsection 8.2.3 and the two NCPD charts by Ross et al. (2011) that are mentioned in Subsection 8.2.4). Much future research is needed to study the monitoring of process variance and the joint monitoring of process mean and variance.

The major limitation of the nonparametric control charts discussed in Section 8.2 is that they only use the ranking or ordering information in the process observations for process monitoring and much useful information in the original process observations has been ignored. This limitation is partially overcome by the nonparametric control charts discussed in Section 8.3 that are based on observation categorization. However, the useful information in the process observations has not been used sufficiently yet by the latter type of nonparametric control charts. For instance, the partitioning intervals $\{A_1, A_2, \ldots, A_p\}$ used in the P-CUSUM and L-CUSUM charts are ordered on the number line. This kind of ordering information is ignored completely in the construction of the two control charts. The charts do not use any quantitative measures of the intervals (e.g., their center locations and their lengths) either. Therefore, these charts still have much room for improvement.

8.5   Exercises

8.1 The following 11 numbers constitute the i-th batch of observations obtained from a production line for process monitoring, where i is given.

  3.718, 0.695, 1.094, 1.175, 2.888, −0.234, −0.231, 0.581, 0.307, −0.193, 2.343

Assume that the IC process distribution is symmetric and the IC process median is $\tilde{\mu}_0 = 1$.

(i) Compute the value of ψ_i defined in (8.1).

(ii) Compute the value of the Wilcoxon signed-rank test statistic S_+ described in Subsection 2.8.3.

(iii) Verify the equation $\psi_i = 2S_+ - m(m+1)/2$, where $m = 11$ is the batch size.

8.2 Perform a Monte Carlo simulation study to verify the following results about the upward NSR chart (8.1)–(8.2) given in Table 8.1:

(i) When $m = 10$ and $U = 47$, ARL_0^+ is about 147.

(ii) When $m = 10$ and $U = 49$, ARL_0^+ is about 204.

Then, in cases when m is fixed at 10, compute the ARL_0^+ values of the chart when U changes from 47 to 49. Summarize your findings about the computed ARL_0^+ values.

8.3 In the simulation study described in the previous exercise, you can assume that the IC process distribution is $N(0, 1)$, and generate all IC process observations from this distribution. You can also assume that the IC process distribution is t_5. Verify by a simulation study that the IC performance of the upward NSR chart (8.1)–(8.2) is about the same in these two scenarios (i.e., the computed ARL_0^+ values are about the same in the two scenarios). Therefore, the chart is indeed distribution-free.

8.4 The upward NSR chart (8.1)–(8.2) requires the assumption that the IC process distribution is symmetric about its median $\tilde{\mu}_0$. Discuss the possible consequences if the chart is used in cases when this assumption is invalid.

8.5 The data shown in Table 8.12 are the first 30 batches of phase II observations with the batch size $m = 10$, obtained from a production process for process mean monitoring. Assume that the IC process distribution is symmetric about the IC median of $\tilde{\mu}_0 = 0$. Monitor the process mean using the upward NSR chart (8.1)–(8.2) with $U = 51$ (the ARL_0 value of the chart is about 345 according to Table 8.1). Summarize your results.

8.6 The following 50 observations constitute a reference sample of size $M = 50$ obtained from a production process when it is IC:

  −0.79, −0.95, −0.34, 2.87, −1.91, −0.84, 0.32, −0.35, −1.08, 0.25,
  −0.11, −0.65, 0.43, −0.30, −1.29, −2.45, −0.09, 0.15, −1.89, −0.07,
  0.57, −0.29, 0.07, −0.22, −0.66, −0.8, 0.12, −0.15, 0.85, 0.15,

Table 8.12 *The first 30 batches of phase II observations with the batch size* $m = 10$ *obtained from a production process for process monitoring.*

i	X_{i1}	X_{i2}	X_{i3}	X_{i4}	X_{i5}	X_{i6}	X_{i7}	X_{i8}	X_{i9}	X_{i10}
1	−0.33	−0.37	0.72	0.16	−0.60	−0.18	0.25	0.46	−0.31	0.71
2	−0.04	−0.61	−0.10	0.83	0.02	−0.19	0.60	0.83	1.82	1.85
3	0.07	0.01	−0.33	0.18	0.02	0.61	−0.63	0.52	0.14	−0.45
4	−0.30	0.27	0.20	−0.17	−0.17	−0.01	−1.51	−1.10	−0.31	−0.11
5	−0.71	0.04	−1.21	−0.09	0.55	0.18	1.42	0.22	0.64	−0.08
6	0.29	−0.49	−0.20	0.01	0.28	−0.24	−0.56	1.26	−1.59	−1.02
7	−0.42	−0.06	−0.29	0.08	0.01	−0.67	−0.41	−0.53	−1.60	−0.17
8	−0.24	3.34	0.41	0.46	−0.01	1.47	0.17	−0.19	0.03	0.32
9	0.74	0.41	0.30	0.15	−0.22	0.66	0.35	0.94	−0.03	−0.22
10	0.53	−0.05	−1.74	0.91	−0.41	0.62	−0.15	−0.19	0.11	2.24
11	0.48	−0.58	−0.13	−1.31	1.13	0.42	−0.59	1.73	0.69	0.04
12	0.92	0.19	−0.29	0.43	0.30	0.48	−1.39	0.57	0.43	2.24
13	−0.85	−0.02	−0.51	−0.68	0.51	−0.97	0.84	0.28	0.41	0.05
14	−0.73	0.85	3.00	0.12	−0.57	−0.59	3.65	0.07	0.59	0.75
15	−0.03	−0.10	0.34	1.14	−0.19	0.56	1.26	−0.16	0.27	1.60
16	−0.10	0.40	0.83	−0.44	0.43	−0.38	0.23	0.11	−0.41	0.21
17	0.46	−0.17	−0.17	−0.50	−0.22	1.40	−6.32	−0.01	0.64	−0.45
18	0.19	−0.49	1.02	−0.58	−5.86	−2.67	−2.94	0.10	0.20	0.02
19	0.72	−2.02	0.06	0.11	−0.53	0.91	0.20	−0.13	1.02	−1.99
20	1.36	0.83	1.18	−0.36	−1.13	−1.01	0.27	−0.43	−0.35	−0.72
21	0.47	2.33	0.72	1.07	0.99	2.17	2.65	2.02	0.08	0.86
22	1.22	1.73	1.09	1.18	1.68	1.18	−0.44	1.15	1.71	0.77
23	0.59	0.47	0.62	0.09	0.16	1.99	0.14	3.38	1.03	2.99
24	1.10	0.92	1.51	0.85	1.57	0.97	1.65	1.15	1.53	0.28
25	1.16	1.05	0.67	−1.11	1.53	0.89	1.92	2.23	0.88	1.49
26	0.82	0.95	−0.11	1.45	−0.04	1.07	1.35	0.54	1.49	2.08
27	−0.29	2.56	−0.80	1.25	1.28	1.36	1.67	1.76	−0.96	0.95
28	1.63	−1.31	0.99	1.69	1.30	−0.70	0.48	−0.26	1.35	1.58
29	1.12	1.22	−1.01	0.94	1.20	0.14	1.49	1.29	0.90	1.64
30	1.42	1.50	0.65	1.45	3.28	1.69	4.00	1.54	0.62	0.99

0.21, 1.30, −0.06, 0.62, 0.73, 0.62, −0.06, 0.08, 0.26, −0.23,
0.62, −0.50, −0.64, 0.48, 0.67, 0.39, −0.01, −0.24, −0.64, 0.13

Compute the values of the precedence statistic W_3 (cf., its definition in the paragraph containing the expression (8.3)) for the following batches of phase II observations with the batch size $m = 5$ obtained from the same production process:

(i) $(1.23, -0.47, 0.01, -0.10, 1.25)$,

(ii) $(-0.71, -0.27, -0.18, 0.14, 0.88)$,

(iii) $(-0.05, -0.42, -1.42, 0.05, 1.39)$.

8.7 Assume that each row in Table 8.12 contains two batches of phase II observations from a production process. So, the first batch of observations is $(-0.33, -0.37, 0.72, 0.16, -0.60)$, the second batch is $(-0.18, 0.25, 0.46,$

−0.31, 0.71), and so forth. Use the DFP chart (8.4)–(8.6) and the reference sample given in Exercise 8.6 to monitor the 60 batches of phase II observations presented in Table 8.12 for possible process mean shift. In the chart, choose $ARL_0 = 370$. Summarize your results.

8.8 Assume that $\{(X_{i1}, X_{i2}, X_{i3}, X_{i4}, X_{i5}, X_{i6}), i = 1, 2, \ldots, 30\}$ shown in Table 8.12 are the first 30 batches of observations with the batch size of $m = 6$ obtained from a production process for phase II process monitoring. Use the upward NSR CUSUM chart (8.8)–(8.9) with $k = 9$ and $h = 12$ to detect potential upward process mean shift. Summarize your results.

8.9 For the Wilcoxon rank-sum test statistic W_n discussed in the paragraph below Figure 8.4, verify the following results:

$$\mu_{W_n} = m(m + M + 1)/2, \qquad \sigma_{W_n}^2 = mM(m + M + 1)/12.$$

8.10 The following $M = 100$ observations constitute a reference sample obtained from a production process when it is IC. The first 60 batches of phase II observations with the batch size of $m = 5$ from the same production process are those described in Exercise 8.7. Use the two-sided NRS CUSUM chart (8.10)–(8.13) with $k = 0.5\sigma_{W_n} = 0.5\sqrt{mM(m + M + 1)/12}$ and $h = 353$ to online monitor the process mean. Summarize your results.

-0.13, 0.77, −1.01, 0.29, 2.49, 0.64, −0.08, 0.32, 2.1, −0.91,
−0.22, 0.54, −1.16, −0.24, −0.21, 0.95, 0.71, 0.68, −1.22, 0.68,
−2.61, 0.24, −1.88, −1.37, −1.32, −0.33, −2.58, 0.28, −1.31, 0.00,
−0.6, −0.8, 0.7, −0.06, 0.48, −0.33, 0.42, 0.34, −0.54, −0.02,
1.04, −0.08, −0.47, 0.04, 1.04, −0.12, 0.39, 0.00, −0.52, 3.81,
−0.53, −0.89, −0.06, 0.23, −0.03, −0.51, 0.26, −0.26, 0.21, 1.52,
0.05, 0.39, 0.21, 0.8, −0.03, 0.11, −1.00, 0.91, 0.07, 0.00,
0.71, −0.72, −0.13, −1.10, −0.11, −1.38, 1.40, −2.50, −0.37, −1.53,
−0.36, −0.93, −2.92, −0.40, −0.18, 2.09, 0.44, 3.27, 0.13, 0.85,
−0.05, −0.01, 0.44, 0.39, 0.24, −0.04, 0.76, 2.68, 0.11, −0.31

8.11 For the 100 observations shown in the previous exercise, compute the values of the charting statistic C_n^+ defined in (8.14) and the values of the sprint length T_n defined in (8.15). In (8.14), use $\mu_0 = 0$ and $k = 0.5$.

8.12 Apply the NSR EWMA chart defined by (8.17) and (8.19) with $\lambda = 0.1$ and $ARL_0 = 500$ to the phase II observations described in Exercise 8.7. Summarize your results.

8.13 Apply the NRS EWMA chart defined by (8.20) and (8.21) with $\lambda = 0.1$ and $ARL_0 = 500$ to the observed data described in Exercise 8.10. Summarize your results.

8.14 Apply the NLR EWMA chart (8.23)–(8.25) with $\lambda = 0.1$ and $ARL_0 = 500$ to the observed data described in Exercise 8.10. Summarize your results, and compare the results here with those obtained in the previous exercise.

8.15 The first 100 phase II observations obtained from a production process are given below, among which the first 14 observations are known to be IC. Use the NCPD control chart (8.28)–(8.29) with $ARL_0 = 500$ to monitor the process mean, starting from $n = 15$. If a signal of process mean shift is given, then give an estimate of the shift position as well. Summarize your results.

-0.39, -0.05, 0.08, -0.37, -0.79, 0.36, -0.52, -0.30, 0.92, 0.50,
0.56, 0.95, -0.91, -0.86, -0.04, -0.12, 0.51, 0.24, 0.83, 1.60,
-0.44, -0.70, 0.02, -0.20, -0.91, -0.72, 0.00, 1.94, 0.48, -1.35,
0.62, 0.93, -0.11, 0.44, -0.22, -0.58, -1.85, 1.01, 0.85, -0.13,
-0.39, 0.36, 0.80, 2.49, 0.39, 0.82, -0.16, 0.07, 0.13, 0.20,
1.96, 1.20, 0.81, 0.89, 1.01, 1.09, 1.55, 1.17, 2.11, -0.29,
1.55, 0.18, 1.13, 2.34, 0.55, 0.44, 0.32, 1.13, 0.59, 0.28,
1.03, 1.03, 0.80, 1.69, 1.34, 1.01, 1.00, -0.23, 0.42, 0.46,
1.23, 0.57, 2.02, 1.33, 1.57, 0.37, 0.76, 1.51, 0.72, -2.86,
0.41, -0.25, 0.78, 0.78 , 1.67, 0.98, 1.77, 2.12, 0.40, 1.22

8.16 The (l/p)-th sample quantiles, for $p = 10$ and $l = 1, 2, \ldots, 9$, of an IC dataset of the size $M = 500$ obtained from a production process are listed below.

-1.172, -0.708, -0.454, -0.201, -0.040, 0.193, 0.397, 0.617, 1.088

Use this IC dataset and the nonparametric P-CUSUM chart (8.34)–(8.35) with $p = 10$, $k_P = 0.01$, and $ARL_0 = 200$ to monitor the first 100 phase II observations presented in the previous exercise for possible process mean shifts. Summarize your results.

8.17 In Table 8.11, the IC performance of seven control charts is compared by their actual IC ARL values. Please comment on the possible consequences if they are used in the three cases considered in the table when the actual IC process distribution is not normal.

Chapter 9

Multivariate Nonparametric Process Control

9.1 Introduction

The nonparametric control charts discussed in the previous chapter are for cases when the quality characteristic in question is univariate. In practice, however, the quality of a product is usually described by multiple quality characteristic variables, as discussed at the beginning of Chapter 7. That is, most applications require multivariate, instead of univariate, SPC methods. In multivariate cases, the assumption that the multiple quality characteristic variables in question follow a normal distribution is difficult to satisfy, because this assumption implies that all individual quality characteristic variables follow normal distributions and all subsets of them follow joint normal distributions (cf., (7.2) and the related discussion in Subsection 7.2.1), which are difficult to satisfy in applications. Therefore, it is important to develop multivariate SPC control charts that are appropriate to use in cases when the normality assumption is invalid.

The multivariate SPC control charts discussed in Chapter 7 are appropriate to use only in cases when the normality assumption is valid. Their results are unreliable or even misleading in cases when the normality assumption is violated, as demonstrated by the following example originally discussed in Qiu and Hawkins (2001). Assume that there are four quality characteristic variables involved in an SPC application, and they are assumed to follow the normal distribution $N(\mathbf{0}, I_4)$ when the related production process is IC. However, the four quality characteristic variables are actually independent of each other and each follows the standardized version with mean 0 and variance 1 of the $Poisson(\widetilde{\lambda})$ distribution (cf., Subsection 2.4.5). Let us consider using the multivariate CUSUM chart (7.28)–(7.29) for phase II process monitoring, in which the allowance constant k and the control limit h are chosen to be 1.0 and 4.503, respectively, so that its nominal IC ARL value equals 200 under the normality assumption. In cases when $\widetilde{\lambda} = 0.01, 0.1, 0.5, 1.0, 1.5, 2.0$, and 5.0, its actual IC ARL values are presented in Figure 9.1(a) by the small diamonds. It can be seen that these actual IC ARL values differ substantially from the nominal IC ARL value, especially in cases when $\widetilde{\lambda}$ is small. When $\widetilde{\lambda}$ gets larger, the standardized version of the $Poisson(\widetilde{\lambda})$ distribution becomes closer to the standard normal distribution $N(0, 1)$. Consequently, the actual IC ARL values of the control chart get closer to its nominal

IC ARL value, as seen in the plot. One idea to overcome this limitation of the multi-variate CUSUM chart, caused mainly by the invalidity of the normality assumption, is to transform the true IC process distribution to a normal distribution. A commonly used transformation for the Poisson distribution is the square root transformation. The corresponding actual IC ARL values of the chart, after the observations are first square-root transformed and then standardized to have mean 0 and variance 1, are presented in Figure 9.1(b). Again, the actual IC ARL values are far away from the nominal IC ARL value in most situations considered in this plot.

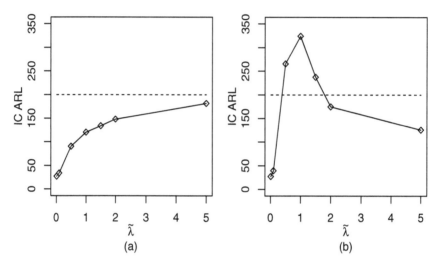

Figure 9.1 (a) The small diamonds denote the actual IC ARL values of the multivariate CUSUM chart (7.28)–(7.29) in cases when there are four quality characteristic variables involved in an SPC application, the variables are independent of each other, and each has the standardized version with mean 0 and variance 1 of the Poisson($\tilde{\lambda}$) distribution where $\tilde{\lambda} = 0.01, 0.1, 0.5, 1.0, 1.5, 2.0,$ and 5.0. (b) The small diamonds denote the actual IC ARL values of the chart after the original observations are square-root transformed and standardized to have mean 0 and variance 1. In the chart (7.28)–(7.29), the control limit h and the allowance constant k are chosen to be 4.503 and 1.0, respectively, such that its nominal IC ARL equals 200 under the normality assumption. The dashed line in each plot denotes the nominal IC ARL value of 200.

Stoumbos and Sullivan (2002) studied the robustness of the conventional multi-variate EWMA charts (cf., Section 7.4) to the normality assumption, and they found that such charts were quite robust as long as the smoothing parameters used in the charts were chosen to be small. However, their numerical results showed that it de-pended on the true process distribution to decide how small the smoothing param-eters should be chosen. This conclusion is also confirmed by the results shown in Figure 9.2, which were originally given in Qiu (2008). In the plot, we consider cases when three quality characteristic variables are involved, their joint IC distribution is assumed to be $N(\mathbf{0}, I_3)$, and they are actually independent of each other and each of them has the standardized version with mean 0 and variance 1 of the χ_r^2 distribution where $r = 1, 5, 10, 20$, and 50. Both the multivariate CUSUM chart (7.28)–(7.29) and the multivariate EWMA chart (7.39)–(7.40) are considered here. In the multivariate CUSUM chart, k and h are chosen to be 1.0 and 3.786, respectively. In the multivariate EWMA chart, a single weighting parameter λ is used, and (λ, h) are chosen to be $(0.05, 9.603)$ or $(0.2, 11.956)$. In such cases, all three charts have nominal IC ARL values of 200. The plot shows their actual IC ARL values in various cases considered. It can be seen that the actual IC ARL values of the multivariate CUSUM chart are quite far away from their nominal IC ARL value, which is consistent with the results shown in Figure 9.1(a). For the multivariate EWMA chart, when $\lambda = 0.2$, its actual IC ARL values are quite close to the nominal IC ARL value only in cases when r is large. When $\lambda = 0.05$, its actual IC ARL values are quite close to the nominal IC ARL value in cases when $r \geq 5$. When $r = 1$, it seems that λ should be chosen smaller than 0.05. In practice, the difference between the true IC process distribution and a normal distribution is often difficult to measure. Also, when the smoothing parameter in the multivariate EWMA chart is chosen to be small, the chart would be ineffective in detecting relatively large shifts, as demonstrated in Table 7.4. Therefore, the robustness property of multivariate EWMA charts should be used with care in practice, as discussed in univariate cases in Section 8.1.

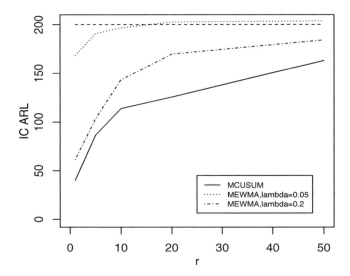

Figure 9.2 *Actual IC ARL values of the multivariate CUSUM chart (7.28)–(7.29), denoted as MCUSUM, and the multivariate EWMA chart (7.39)–(7.40), denoted as MEWMA, in cases when there are three quality characteristic variables involved in an SPC application, the variables are independent of each other, and each has the standardized version with mean 0 and variance 1 of the χ^2_r distribution where $r = 1, 5, 10, 20$, and 50. In the multivariate CUSUM chart, (k, h) are chosen to be $(1.0, 3.786)$ (solid curve). In the multivariate EWMA chart, (λ, h) are chosen to be $(0.05, 9.603)$ (dotted curve) or $(0.2, 11.956)$ (dot-dashed curve). All charts have nominal IC ARL values of 200 (dashed horizontal line) under the normality assumption.*

In cases when the multiple quality characteristic variables are not normally distributed, one natural idea is to transform them properly so that the tranformed variables would be roughly normally distributed. The results shown in Figure 9.1(b) were obtained by this idea. In the statistical literature, in cases when a multivariate dataset is not normally distributed, existing tools for describing such a dataset, or transforming it to a normally distributed dataset, are limited (cf., Eaton, 1983; Fang et al., 1990; Johnson and Wichern, 2007). Therefore, this idea is challenging to accomplish in practice.

To handle multivariate SPC in cases when the normality assumption is invalid, there have been some nonparametric or distribution-free control charts proposed in the literature. See, for instance, Boone and Chakraborti (2012), Liu (1995), Liu et al. (2004), Qiu (2008), Qiu and Hawkins (2001, 2003), Sun and Tsung (2003), Zou and Tsung (2011), Zou et al. (2012), and the references cited therein. In this chapter, we discuss some of them. As in univariate cases, we do not always make a clear distinction between the words "nonparametric multivariate control charts" and "distribution-free multivariate control charts" in this chapter. See Section 8.1 for a related discussion.

9.2 Rank-Based Multivariate Nonparametric Control Charts

In this section, we describe some multivariate nonparametric control charts based on the ranking or ordering information of the observed data. Unlike univariate cases in which the ranking information refers to the order among observations at different time points, in multivariate cases, the order could be among observations at different time points, which is called *longitudinal ranking* or longitudinal ordering in this chapter. It can also refer to the order among different components of an observation vector at a given time point, which is called *cross-component ranking* or cross-component ordering in this chapter. When computing the longitudinal ranking, there are several different ways to consider as well. Some fundamental multivariate nonparametric control charts based on different types of ranking or ordering information of the observed data will be discussed in two parts of this section. Those based on different types of longitudinal ranking are described in Subsection 9.2.1, and those based on cross-component ranking are described in Subsection 9.2.2.

9.2.1   *Control charts based on longitudinal ranking*

Let us first describe two recent multivariate nonparametric Shewhart charts proposed by Boone and Chakraborti (2012) that are based on componentwise signs and componentwise signed-ranks. These two Shewhart charts are mainly for phase II SPC, although they can also be modified properly for phase I process analysis. Assume that there are p quality characteristic variables involved in a production process, which are denoted as a p-dimensional vector $\mathbf{X}$, and at the n-th time point of phase II SPC, we have collected a sample of m i.i.d. p-dimensional observations

$$\mathbf{X}_{n1}, \mathbf{X}_{n2}, \ldots, \mathbf{X}_{nm}, \qquad \text{for } n = 1, 2, \ldots,$$

where $\mathbf{X}_{nj} = (X_{nj1}, X_{nj2}, \ldots, X_{njp})'$, for $j = 1, 2, \ldots, m$. Then, for the l-th component, with $l = 1, 2, \ldots, p$, we can define the componentwise sign statistic

$$\xi_{nl} = \sum_{j=1}^{m} \text{sign}\left(X_{njl} - \tilde{\mu}_{0l}\right), \tag{9.1}$$

where $\text{sign}(\cdot)$ is the sign function defined in (8.1), and $\tilde{\mu}_0 = (\tilde{\mu}_{01}, \tilde{\mu}_{02}, \ldots, \tilde{\mu}_{0p})'$ is the p-dimensional vector of the IC process distribution that is assumed known. Intuitively, in cases when the process is IC, ξ_{nl} is about 0, because the numbers of positive and negative values among $\{X_{njl} - \tilde{\mu}_{0l}, j = 1, 2, \ldots, m\}$ are about the same. In cases when one or more components of $\mathbf{X}$ have location shifts at n, the corresponding components of $\xi_n = (\xi_{n1}, \xi_{n2}, \ldots, \xi_{np})'$ would move away from 0. Therefore, the sign statistic vector ξ_n can be used for detecting process location shifts.

The covariance matrix of ξ_n can be estimated by $\widehat{\mathbf{V}} = (\widehat{v}_{l_1 l_2})$ where $l_1, l_2 = 1, 2, \ldots, p$,

$$\widehat{v}_{l_1 l_2} = \begin{cases} m, & \text{when } l_1 = l_2 \\ \sum_{j=1}^{m} \text{sign}\left(X_{njl_1} - \tilde{\mu}_{0l_1}\right) \text{sign}\left(X_{njl_2} - \tilde{\mu}_{0l_2}\right), & \text{when } l_1 \neq l_2. \end{cases}$$

Then, the Shewhart charting statistic can be defined by

$$T_S^2 = \xi_n' \widehat{\mathbf{V}}^{-1} \xi_n. \tag{9.2}$$

The chart gives a signal of process location shift when

$$T_S^2 > h_S, \tag{9.3}$$

where $h_S > 0$ is a control limit chosen to achieve a given ARL_0 value. The chart (9.2)–(9.3) will be called the *multivariate nonparametric sign (MNS) chart* hereafter.

In cases when the process is IC, it can be shown that T_S^2 converges in distribution to χ_p^2 as m increases (cf., Hettmansperger, 2006). Therefore, when m is large, for a given false alarm rate α (or equivalently, a given $ARL_0 = 1/\alpha$), the control limit h_S can be chosen to be

$$h_S = \chi_{1-\alpha,p}^2, \tag{9.4}$$

where $\chi_{1-\alpha,p}^2$ is the $(1-\alpha)$-th quantile of the χ_p^2 distribution. Boone and Chakraborti (2012) have shown that formula (9.4) provides a reliable control limit only in cases when the batch size m is reasonably large (e.g., $m \geq 50$). In cases when m is small, this formula often provides a conservative control limit. Namely, the control limit computed by (9.4) is often larger than the control limit that achieves a given ARL_0 value, or the actual false alarm rate (actual ARL_0 value) using the control limit by (9.4) is often smaller (larger) than the nominal false alarm rate (nominal ARL_0 value). In such cases, a more reliable control limit can be computed by a bootstrap algorithm from a reasonably large IC dataset, as discussed in Chatterjee and Qiu (2009).

The second multivariate nonparametric Shewhart chart proposed by Boone and Chakraborti (2012) is based on the componentwise Wilcoxon signed-rank statistics. For the l-th component, with $l = 1, 2, \ldots, p$, let

$$\psi_{nl} = \sum_{j=1}^{m} \text{sign}\left(X_{njl} - \tilde{\mu}_{0l}\right) R_{njl}^+, \tag{9.5}$$

where R^+_{njl} is the rank of $|X_{njl} - \tilde{\mu}_{0l}|$ among $\{|X_{njl} - \tilde{\mu}_{0l}|, j = 1,2,\ldots,m\}$. Then, clearly, ψ_{nl} is the sum of all Wilcoxon signed-ranks for the l-th component within the n-th batch of observed data (cf., the paragraph containing (8.1) in Subsection 8.2.1 for a related discussion). Let $\psi_n = (\psi_{n1}, \psi_{n2}, \ldots, \psi_{np})'$. Then, when the IC process distribution is diagonally symmetric in the sense that the distributions of $\mathbf{X} - \tilde{\mu}_0$ and $\tilde{\mu}_0 - \mathbf{X}$ are the same when the process is IC, the marginal distributions of X_{njl}, for given n and j and for $l = 1,2,\ldots,p$, are all symmetric about their medians $\tilde{\mu}_{0l}$. In such cases, ψ_n would be centered at $\mathbf{0}$, and the covariance matrix of ψ_n can be estimated by $\widehat{\mathbf{W}} = (\widehat{w}_{l_1 l_2})$, where $l_1, l_2 = 1,2,\ldots,p$,

$$
\widehat{w}_{l_1 l_2} = \begin{cases} \frac{m(m+1)(2m+1)}{6}, & \text{when } l_1 = l_2 \\ \sum_{j=1}^{m} \text{sign}\left(X_{njl_1} - \tilde{\mu}_{0l_1}\right) R^+_{njl_1} \text{sign}\left(X_{njl_2} - \tilde{\mu}_{0l_2}\right) R^+_{njl_2}, & \text{when } l_1 \neq l_2. \end{cases}
$$

When one or more components of $\mathbf{X}$ have location shifts at n, the corresponding components of ψ_n would move away from 0. Therefore, ψ_n can be used for detecting process location shifts. The Shewhart charting statistic is then defined by

$$
T^2_{SR} = \psi'_n \widehat{\mathbf{W}}^{-1} \psi_n. \tag{9.6}
$$

The chart gives a signal of process location shift when

$$
T^2_{SR} > h_{SR}, \tag{9.7}
$$

where $h_{SR} > 0$ is a control limit chosen to achieve a given ARL_0 value. The chart (9.6)–(9.7) will be called the *multivariate nonparametric signed-rank (MNSR) chart* hereafter.

By Hettmansperger (2006), when m increases, T^2_{SR} converges in distribution to χ^2_p. Therefore, when m is large enough, the control limit h_{SR} can be chosen to be

$$
h_{SR} = \chi^2_{1-\alpha,p}, \tag{9.8}
$$

where α is a given false alarm rate. Similar to h_S in (9.4), Boone and Chakraborti (2012) have shown that formula (9.8) provides a reliable control limit only in cases when m is reasonably large (e.g., $m \geq 50$). In cases when m is small, the formula often provides a conservative control limit. In such cases, a more reliable control limit can be computed by a bootstrap algorithm from a reasonably large IC dataset.

Between the MNS chart (9.2)–(9.3) and the MNSR chart (9.6)–(9.7), the former is more general in the sense that it does not require the IC process distribution to be diagonally symmetric while the latter does. However, in cases when the diagonal symmetry assumption is valid, Boone and Chakraborti (2012) have shown that the MNSR chart would generally be more efficient.

Example 9.1 *Assume that there are $p = 2$ quality characteristic variables involved in a production process. For phase II process mean monitoring, a sample of size $m = 50$ is obtained at each time point. The two components of the observed 2-dimensional data obtained at the first 20 time points are shown in Figure 9.3(a)–(b). From the* *plots, it seems that the first quality characteristic variable has a mean shift starting from the 11th time point, and the second variable does not have any mean shift. We then apply the MNS chart (9.2)–(9.3) and the MNSR chart (9.6)–(9.7) to this dataset. In both charts, α is chosen to be 0.005 (i.e., $ARL_0 = 200$), and their control limits are chosen by (9.4) and (9.8) to both be $\chi^2_{0.995,2} = 10.597$. The two charts are shown in Figure 9.3(c)–(d). From the plots, both charts give their first signals at the 11th time point. The signal by the MNSR chart seems more convincing because its charting statistic takes larger values when $n \geq 11$.*

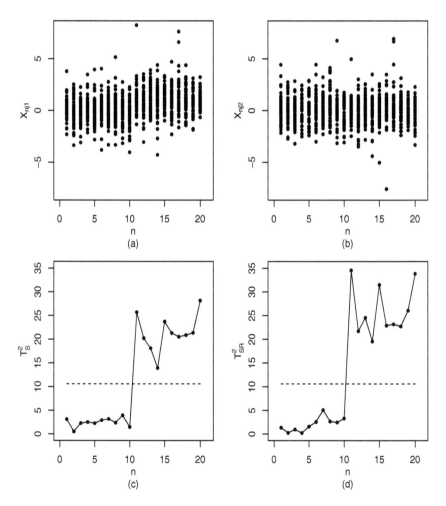

Figure 9.3 *(a) First component of the observed 2-dimensional batch data with batch size $m = 50$ obtained at the first 20 time points. (b) Second component of the observed data. (c) MNS chart (9.2)–(9.3) when it is applied to the observed data. (d) MNSR chart (9.6)–(9.7) when it is applied to the observed data. In both charts, α is chosen to be 0.005, and the control limit is chosen to be $\chi^2_{0.995,2} = 10.597$ (dashed horizontal lines in plots (c) and (d)).*

Both the MNS chart (9.2)–(9.3) and the MNSR chart (9.6)–(9.7) are Shewhart charts. They are good at detecting large and isolated shifts, but ineffective in detecting small and persistent shifts, based on the discussion in previous chapters. It is possible to modify them properly to become multivariate nonparametric CUSUM, EWMA, or CPD charts, as discussed in univariate cases in Section 8.2. For instance, to modify the MNS chart (9.2)–(9.3) to become a multivariate nonparametric EWMA chart, we can define

$$\mathbf{E}_n = \lambda \xi_n + (1-\lambda)\mathbf{E}_{n-1}, \qquad \text{for } n \geq 1,$$

where $\mathbf{E}_0 = \mathbf{0}$, and $\lambda \in (0,1]$ is a weighting parameter. Then, the multivariate non-parametric EWMA chart gives a signal of process mean shift if

$$\mathbf{E}_n' \Sigma_{\mathbf{E}_n}^{-1} \mathbf{E}_n > h, \tag{9.9}$$

where $h > 0$ in (9.9) is a control limit chosen to achieve a given ARL_0 value.

In the MNS chart (9.2)–(9.3) and the MNSR chart (9.6)–(9.7), the component-wise sign statistics ξ_{nl} defined in (9.1) and the componentwise signed-rank statistics ψ_{nl} defined in (9.5) are used. Zou and Tsung (2011) proposed a multivariate EWMA chart based on the so-called *spatial sign* function defined as

$$U(\mathbf{X}) = \begin{cases} \frac{\mathbf{X}-\mu_0}{\|\mathbf{X}-\mu_0\|}, & \text{when } \mathbf{X} \neq \mu_0 \\ \mathbf{0}, & \text{when } \mathbf{X} = \mu_0, \end{cases} \tag{9.10}$$

where μ_0 is the IC mean of $\mathbf{X}$, and $\|\cdot\|$ is the Euclidean length. Obviously, the spatial sign of $\mathbf{X}$ is just the direction of $\mathbf{X}$ from its center μ_0 with unit Euclidean length. Zou et al. (2012) proposed another multivariate EWMA chart based on the so-called *spatial ranks*. For a sample of p-dimensional observations $(\mathbf{X}_1, \mathbf{X}_2, \ldots, \mathbf{X}_n)$, the spatial rank of $\mathbf{X}_i$, for $i = 1, 2, \ldots, n$, is defined as

$$\mathbf{r}_i = \frac{1}{n} \sum_{j=1}^{n} U(\mathbf{X}_i - \mathbf{X}_j). \tag{9.11}$$

In cases when $p = 1$, $\mathbf{r}_i = [2R_i - (n+1)]/n$, where R_i is the conventional rank of $\mathbf{X}_i$ in the sample. Because $\mu_{R_i} = (n+1)/2$, $[2R_i - (n+1)]/n = \frac{2}{n}[R_i - (n+1)/2]$ can be regarded as $2/n$ times the centered rank of $\mathbf{X}_i$. The spatial rank of $\mathbf{X}_i$ is a generalization of the centered rank in p-dimensional cases. For related discussions on spatial signs, spatial ranks, and their statistical properties, see Oja (2010), Oja and Randles (2004), and the references cited therein. Because the two multivariate EWMA charts based on spatial signs and spatial ranks mentioned above are quite complicated to present, they are not described in detail here.

Liu (1995) proposed several multivariate control charts based on the concept of *data depth*. Let $\mathbf{Y}_1, \mathbf{Y}_2, \ldots, \mathbf{Y}_M$ be M observations generated from a p-dimensional distribution F_0. The concept of data depth tries to order the observations from the most central point to the most outlying point with respect to the distribution F_0, using a specific definition of data depth. In the literature, there are several different definitions of data depth, including the Mahalanobis depth, the half-space depth, the convex hull peeling depth, the Oja depth, the simplicial depth, the majority depth, the likelihood depth, and so forth. Next, we describe two of them. For a quite complete description of all these definitions, see Liu et al. (1999) and Liu and Singh (1993).

The Mahalanobis depth of a p-dimensional point $\mathbf{y}$ with respect to the distribution F_0 is defined as

$$D_M(\mathbf{y};F_0) = \frac{1}{1+(\mathbf{y}-\mu_0)'\Sigma_0^{-1}(\mathbf{y}-\mu_0)}, \qquad (9.12)$$

where μ_0 and Σ_0 are the mean vector and covariance matrix of F_0. Clearly, the term $(\mathbf{y}-\mu_0)'\Sigma_0^{-1}(\mathbf{y}-\mu_0)$ on the right-hand side of (9.12) is the squared statistical distance (or, the squared Mahalanobis distance) from $\mathbf{y}$ to the center μ_0 of F_0. The Mahalanobis depth $D_M(\mathbf{y};F_0)$ is defined as a decreasing function of this squared statistical distance in (9.12). So, the Mahalanobis depth of $\mathbf{y}$ with respect to F_0 would be larger if $\mathbf{y}$ is closer to the center of F_0 in terms of the Mahalanobis distance, and vice versa. In cases when F_0 is unknown, but we have a sample $(\mathbf{Y}_1, \mathbf{Y}_2, \ldots, \mathbf{Y}_M)$ from F_0, we can define the empirical Mahalanobis depth of $\mathbf{y}$ as

$$D_M(\mathbf{y};F_{0M}) = \frac{1}{1+(\mathbf{y}-\overline{\mathbf{Y}})'\left(\mathbf{S}_{\mathbf{Y}}^2\right)^{-1}(\mathbf{y}-\overline{\mathbf{Y}})}, \qquad (9.13)$$

where $\overline{\mathbf{Y}}$ and $\mathbf{S}_{\mathbf{Y}}^2$ are the sample mean vector and the sample covariance matrix of the sample $(\mathbf{Y}_1, \mathbf{Y}_2, \ldots, \mathbf{Y}_M)$, and F_{0M} is the empirical estimator of F_0.

The simplicial depth of $\mathbf{y}$ with respect to F_0 (cf., Liu, 1990) is defined as

$$D_S(\mathbf{y};F_0) = P\left(\mathbf{y} \in S(\mathbf{Y}_1, \mathbf{Y}_2, \ldots, \mathbf{Y}_{p+1})\right), \qquad (9.14)$$

where the probability $P(\cdot)$ is under F_0, $(\mathbf{Y}_1, \mathbf{Y}_2, \ldots, \mathbf{Y}_{p+1})$ is a simple random sample of size $p+1$ from F_0, and $S(\mathbf{Y}_1, \mathbf{Y}_2, \ldots, \mathbf{Y}_{p+1})$ is an open simplex with vertices at $\mathbf{Y}_1, \mathbf{Y}_2, \ldots, \mathbf{Y}_{p+1}$. Intuitively, in cases when $\mathbf{y}$ is closer to the center of the distribution F_0, $D_S(\mathbf{y};F_0)$ in (9.14) would be larger because the random simplex would have a larger chance to contain $\mathbf{y}$. So, $D_S(\mathbf{y};F_0)$ measures how central (or, deep) the point $\mathbf{y}$ is with respect to F_0. In cases when F_0 is unknown but there is a sample $(\mathbf{Y}_1, \mathbf{Y}_2, \ldots, \mathbf{Y}_M)$ from F_0, the sample simplicial depth of $\mathbf{y}$ is defined as

$$D_S(\mathbf{y};F_{0M}) = \frac{1}{\binom{M}{p+1}} \sum I\left(\mathbf{y} \in S(\mathbf{Y}_{i_1}, \mathbf{Y}_{i_2}, \ldots, \mathbf{Y}_{i_{p+1}})\right), \qquad (9.15)$$

where $(\mathbf{Y}_{i_1}, \mathbf{Y}_{i_2}, \ldots, \mathbf{Y}_{i_{p+1}})$ is a subset of $(\mathbf{Y}_1, \mathbf{Y}_2, \ldots, \mathbf{Y}_M)$, and the summation on the right-hand side of (9.15) is over all possible subsets.

From the above description, it seems that the Mahalanobis depth is easier to compute and perceive. So, we will describe the three control charts by Liu (1995) using this definition of data depth, although the charts can be constructed in the same way using the simplicial depth. Now, for phase II SPC, assume that $(\mathbf{Y}_1, \mathbf{Y}_2, \ldots, \mathbf{Y}_M)$ is a reference sample obtained beforehand from a production process when it is IC, and F_0 is the IC process distribution. Let $\mathbf{Y}$ be a p-dimensional random vector with the distribution F_0, and for a given point $\mathbf{y} \in R^p$, define

$$r(\mathbf{y};F_0) = P\left(D_M(\mathbf{Y};F_0) \le D_M(\mathbf{y};F_0)\right), \qquad (9.16)$$

where the probability $P(\cdot)$ is under F_0. Then, $r(\mathbf{y}; F_0)$ in (9.16) can be regarded as the cdf of $D_M(\mathbf{Y}; F_0)$ in terms of $D_M(\mathbf{y}; F_0)$. In cases when F_0 is unknown but the reference sample is available, the empirical version of $r(\mathbf{y}; F_0)$ is defined as

$$r(\mathbf{y}; F_{0M}) = \frac{1}{M} \sum_{i=1}^{M} I(D_M(\mathbf{Y}_i; F_{0M}) \leq D_M(\mathbf{y}; F_{0M})). \qquad (9.17)$$

Let $\mathbf{X}_n$ be a phase II observation obtained at the n-th time point. Then, if $\mathbf{X}_n$ is IC (i.e., $\mathbf{X}_n \sim F_0$), Liu and Singh (1993) showed that (i) $r(\mathbf{X}_n; F_0) \sim Uniform(0,1)$ (cf., Exercise 2.1 for a description), and (ii) the distribution of $r(\mathbf{X}_n; F_{0M})$ converged to $Uniform(0,1)$ when M increased. Therefore, to monitor $\mathbf{X}_n$, for $n \geq 1$, if the value of $r(\mathbf{X}_n; F_{0M})$ is too small (i.e., $D_M(\mathbf{X}_n; F_{0M})$ is too small or $\mathbf{X}_n$ is too outlying with respect to the empirical distribution F_{0M}), a signal of distributional shift should be given. The corresponding Shewhart chart, called the r chart in Liu (1995), then gives a signal when

$$r(\mathbf{X}_n; F_{0M}) < \alpha, \qquad (9.18)$$

where $\alpha \in (0,1)$ is a given false alarm rate. For the r chart (9.18), it is often helpful to draw a center line at $C = 1/2$, because the charting statistic $r(\mathbf{X}_n; F_{0M})$ should wave around this line when the process is IC.

The r chart (9.18) can be modified as follows to become a CUSUM-type control chart. Let $\mathbf{X}_1, \mathbf{X}_2, \ldots, \mathbf{X}_n$ be the first n observations obtained during phase II SPC. Then, when F_0 is known,

$$S_n(F_0) = \sum_{i=1}^{n} \left[r(\mathbf{X}_i; F_0) - \frac{1}{2} \right]$$

provides a measure of the cumulative difference between $r(\mathbf{X}_i; F_0)$ and its IC mean $1/2$. In cases when F_0 is unknown, but we have a reference sample $(\mathbf{Y}_1, \mathbf{Y}_2, \ldots, \mathbf{Y}_M)$, the above measure can be estimated by its empirical version

$$S_n(F_{0M}) = \sum_{i=1}^{n} \left[r(\mathbf{X}_i; F_{0M}) - \frac{1}{2} \right],$$

where $r(\mathbf{X}_i; F_{0M})$ is the empirical version of $r(\mathbf{X}_i; F_0)$, defined in (9.17). Liu (1995) showed that (i) $S_n(F_0)$ converged in distribution to $N(0, n/12)$ when n increased, and (ii) $S_n(F_{0M})$ converged in distribution to $N(0, n^2[(1/M) + (1/n)]/12)$ when $\min(M,n)$ increased. Therefore, we can use the charting statistic

$$S_n^*(F_{0M}) = \frac{S_n(F_{0M})}{n\sqrt{[(1/M) + (1/n)]/12}}, \qquad (9.19)$$

and the chart gives a signal of process distributional shift if

$$S_n^*(F_{0M}) < -Z_{1-\alpha}, \qquad (9.20)$$

where $Z_{1-\alpha}$ is the $(1-\alpha)$-th quantile of the standard normal distribution, and $\alpha \in (0,1)$ is a given false alarm rate. This chart was called the S chart in Liu (1995).

Example 9.2 *Assume that there are* $p = 2$ *quality characteristic variables involved in a production process. For phase II process monitoring, a reference sample of size* $M = 100$ *is obtained beforehand, and the two components of these observations are shown in Figure 9.4(a)–(b), respectively, by the dark diamond points before the vertical dotted line at* $n = 0$. *Then, the first 20 phase II observations of the production process are obtained whose two components are shown in Figure 9.4(a)–(b), respectively, by the dark dot points after the vertical dotted line at* $n = 0$. *The r chart (9.18) and the S chart (9.19)–(9.20) are then applied to this data. In both charts,* α *is chosen to be 0.05, and the empirical Mahalanobis depth defined in (9.13) is used. The two charts are shown in Figure 9.4(c)–(d), respectively. From the plots, it can be seen that the r chart gives the first signal at* $n = 14$, *and the S chart gives its first signal at* $n = 18$.

Both the r chart (9.18) and the S chart (9.19)–(9.20) are for cases with individual observation data (i.e., there is only one observation vector at each time point). In cases when the observed data are batch data, Liu (1995) suggested a so-called Q *chart* described below. Let $\mathbf{Y} \sim F_0$ and $\mathbf{X} \sim F_1$ be two p-dimensional random vectors, where F_0 and F_1 are two p-dimensional distributions. Define

$$
\begin{aligned}
Q(F_0, F_1) &= P(D_M(\mathbf{Y}; F_0) \leq D_M(\mathbf{X}; F_0)) \\
&= E[r(\mathbf{X}; F_0)],
\end{aligned} \tag{9.21}
$$

where the probability $P(\cdot)$ in the first equation is under the joint distribution of $(\mathbf{Y}, \mathbf{X})$, and the expectation $E(\cdot)$ in the second equation is with respect to the distribution F_1 of $\mathbf{X}$. From the second equation of (9.21), it can be seen that $Q(F_0, F_1)$ provides a measure about the difference between F_0 and F_1. If they are the same, then $Q(F_0, F_1) = 1/2$, because $r(\mathbf{X}; F_0) \sim Uniform(0, 1)$ in such cases, as mentioned earlier. If $Q(F_0, F_1)$ is too small, then it is an indication that on average $\mathbf{X}$ is far away from the center of F_0. Now, for phase II SPC, assume that $(\mathbf{Y}_1, \mathbf{Y}_2, \ldots, \mathbf{Y}_M)$ is a reference sample obtained from a production process when it is IC, F_0 is the IC process distribution, and at the n-th time point during phase II SPC, we have the following batch data with batch size of m:

$$
\mathbf{X}_{n1}, \mathbf{X}_{n2}, \ldots, \mathbf{X}_{nm}, \qquad \text{for } n \geq 1.
$$

Then,

$$
Q(F_0, F_{1m,n}) = \frac{1}{m} \sum_{i=1}^{m} r(\mathbf{X}_{ni}; F_0)
$$

measures the difference between F_0 and the empirical distribution $F_{1m,n}$ of the observed data at the n-th time point. In cases when F_0 is unknown, $Q(F_0, F_{1m,n})$ can be estimated by

$$
Q(F_{0M}, F_{1m,n}) = \frac{1}{m} \sum_{i=1}^{m} r(\mathbf{X}_{ni}; F_{0M}), \tag{9.22}
$$

where F_{0M} is the empirical distribution of the reference sample. Liu and Singh (1993) showed that (i) $[Q(F_0, F_{1m,n}) - 1/2]$ converged in distribution to $N(0, 1/(12m))$ as m increased, and (ii) $[Q(F_{0M}, F_{1m,n}) - 1/2]$ converged in distribution to $N(0, (1/M +$

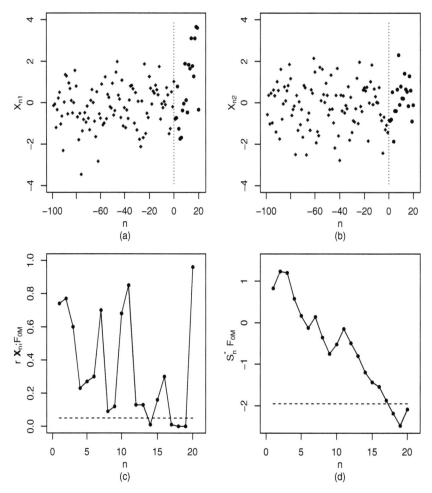

Figure 9.4 *(a) First component of a reference sample of size $M = 100$ (dark diamond points before the vertical dotted line) and a phase II dataset (dark dot points after the vertical dotted line). (b) Second component of the reference sample and the phase II dataset. (c) r chart (9.18). (d) S chart (9.19)–(9.20). In both charts, α is chosen to be 0.05, and the empirical Mahalanobis depth defined in (9.13) is used.*

$1/m)/12$) as $\min(M, m)$ increased. Based on these results and the corresponding results in cases when m is small, the Q chart by Liu (1995) gives a signal of process distributional shift when

$$Q(F_{0M}, F_{1m,n}) < \begin{cases} 0.5 - Z_{1-\alpha}\sqrt{(1/M + 1/m)/12}, & \text{when } m \geq 5 \\ (n!\alpha)^{1/n}/n, & \text{when } m < 5. \end{cases} \quad (9.23)$$

Example 9.3 *Assume that there are $p = 2$ quality characteristic variables involved in a production process. For phase II process monitoring, a reference sample of size*

$M = 100$ is obtained beforehand, and the two components of these observations are shown in Figure 9.5(a)–(b), respectively, by the dark diamond points before the vertical dotted line at $n = 0$. Then, the first 20 batches of phase II observations of the production process with batch size of $m = 10$ are obtained whose two components are shown in Figure 9.5(a)–(b), respectively, by the dark dot points after the vertical dotted line at $n = 0$. The Q chart (9.23) is then applied to this data, in which α is chosen to be 0.05, and the empirical Mahalanobis depth defined in (9.13) is used. The chart is shown in Figure 9.5(c). From the plot, it can be seen that the Q chart gives its first signal at $n = 11$.

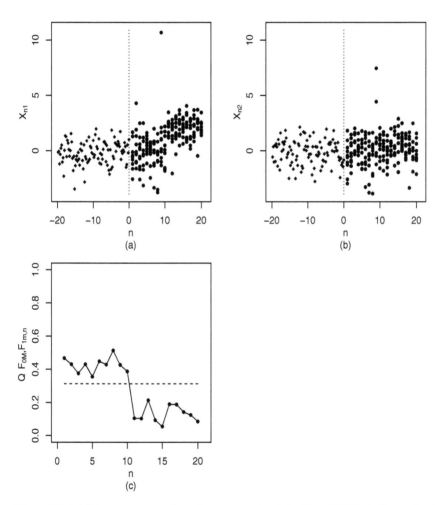

Figure 9.5 *(a) First component of a reference sample of size $M = 100$ (dark diamond points before the vertical dotted line) and a phase II dataset (dark dot points after the vertical dotted line). (b) Second component of the reference sample and the phase II dataset. (c) Q chart (9.23), in which α is chosen to be 0.05, and the empirical Mahalanobis depth defined in (9.13) is used.*

9.2.2   Control charts based on cross-component ranking

The multivariate nonparametric control charts described in the previous subsection are all based on the longitudinal ranking or ordering information of observations at different time points, although the definitions of longitudinal ranking or ordering might be different in different control charts. For multivariate SPC, another strategy is to use the ranking or ordering information across different components of observations at individual time points, and try to detect any change in such cross-component ranking patterns. Qiu and Hawkins (2001, 2003) proposed multivariate nonparametric SPC charts based on this strategy for detecting process mean shifts in phase II SPC, which are described in this subsection.

Assume that $\mathbf{X}_n = (X_{n1}, X_{n2}, \ldots, X_{np})'$ is an observation of a p-dimensional quality characteristics vector $\mathbf{X}$ at the n-th time point during phase II SPC. Without loss of generality, we further assume that the IC mean vector μ_0 is $\mathbf{0}$. In practice, μ_0 needs to be estimated from an IC dataset, and then it should be subtracted from all phase II observations in order for us to use the control charts described below. Let $\mu = (\mu_1, \mu_2, \ldots, \mu_p)'$ be the process mean vector at a given time point. Then, the null hypothesis involved in the multivariate SPC problem is

$$H_0 : \mu_1 = \mu_2 = \cdots = \mu_p = 0. \tag{9.24}$$

Apparently, H_0 in (9.24) is equivalent to the combination of the following two hypotheses:

$$H_0^{(1)} : \mu_1 = \mu_2 = \cdots = \mu_p \tag{9.25}$$

and

$$H_0^{(2)} : \sum_{j=1}^{p} \mu_j = 0. \tag{9.26}$$

The hypothesis $H_0^{(1)}$ is related to the order of magnitudes of $\{\mu_j, j = 1, 2, \ldots, p\}$. When $H_0^{(1)}$ is true, $H_0^{(2)}$ is related to the magnitudes of $\{\mu_j, j = 1, 2, \cdots, p\}$. In cases when μ has a fixed non-zero Euclidean length, violation of H_0 implies violation of either $H_0^{(1)}$ or $H_0^{(2)}$ or both. Furthermore, if μ does not satisfy H_0, then the closer it is to $H_0^{(1)}$ (it is therefore more difficult to detect by a control chart designed for detecting violation of $H_0^{(1)}$), the farther it is from $H_0^{(2)}$ and consequently it is easier to detect by a control chart designed for detecting violation of $H_0^{(2)}$, and vice versa. Hence, a combination of two CUSUM procedures designed for detecting violation of $H_0^{(1)}$ and $H_0^{(2)}$, respectively, will be able to detect all kinds of shifts in the mean vector of the process.

Process mean shifts that violate $H_0^{(2)}$ can be detected by a univariate control chart based on $\sum_{j=1}^{p} X_{nj}$. Next, we introduce a multivariate CUSUM chart for detecting process mean shifts that violate $H_0^{(1)}$. As mentioned above, hypothesis $H_0^{(1)}$ defined in (9.25) is related to the order of magnitudes of $\{\mu_j, j = 1, 2, \cdots, p\}$. It is therefore natural to expect that some rank-based procedures can detect mean shifts that violate

$H_0^{(1)}$. Let

$$\mathbf{A}_n = (A_{n1}, A_{n2}, \ldots, A_{np})'$$

be the antirank vector of $\mathbf{X}_n$. That is, $\mathbf{A}_n$ is a permutation of $(1, 2, \ldots, p)'$ such that $X_{nA_{n1}} \leq X_{nA_{n2}} \leq \ldots \leq X_{nA_{np}}$ are the order statistics of $\{X_{nj}, j = 1, 2, \ldots, p\}$.

Let us first consider A_{n1}. For $1 \leq j \leq p$, define

$$\xi_{n1,j} = \begin{cases} 1, & \text{if } A_{n1} = j \\ 0, & \text{otherwise,} \end{cases} \tag{9.27}$$

and $\xi_{n1} = (\xi_{n1,1}, \xi_{n1,2}, \ldots, \xi_{n1,p})'$. That is, $\xi_{n1,j}$ is defined as an indicator of the event that the j-th component takes the smallest value among p observation components at the time point n. Under $H_0^{(1)}$, assume that $E(\xi_{n1,j}) = g_{1,j}$, for $j = 1, 2, \ldots, p$. Then, the probability distribution of A_{n1} is $\mathbf{g}_1 = (g_{1,1}, g_{1,2}, \ldots, g_{1,p})'$. After a shift in the process mean vector μ, the corresponding distribution of A_{n1} is denoted by $\mathbf{g}_1^* = (g_{1,1}^*, g_{1,2}^*, \ldots, g_{1,p}^*)'$. Qiu and Hawkins (2001) showed that, under some regularity conditions, a process mean shift that violated $H_0^{(1)}$ would result in a shift in the distribution of A_{n1} (i.e., $\mathbf{g}_1$ and $\mathbf{g}_1^*$ were different). Therefore, testing $H_0^{(1)}$ is equivalent to testing

$$H_0^{(1)*} : \text{ the distribution of } A_{n1} \text{ is } \{g_{1,j}, j = 1, 2, \ldots, p\}. \tag{9.28}$$

The hypothesis testing problem (9.28) is similar to the testing problem discussed in Subsection 8.3.1 about the distribution of the categorized data (cf., (8.32), (8.33), and the related discussion). Qiu and Hawkins (2001) proposed a multivariate CUSUM procedure, similar to (8.34)–(8.35), for detecting shifts that violated $H_0^{(1)*}$. Let $\mathbf{U}_0^{obs} = \mathbf{U}_0^{exp} = \mathbf{0}$ be two $p \times 1$ vectors, and

$$\begin{cases} \mathbf{U}_n^{obs} = \mathbf{0}, & \text{if } B_n \leq k_1 \\ \mathbf{U}_n^{exp} = \mathbf{0}, & \\ \mathbf{U}_n^{obs} = \left(\mathbf{U}_{n-1}^{obs} + \xi_{n1}\right)(1 - k_1/B_n), & \text{if } B_n > k_1 \\ \mathbf{U}_n^{exp} = \left(\mathbf{U}_{n-1}^{exp} + \mathbf{g}_1\right)(1 - k_1/B_n), & \end{cases}$$

where

$$B_n = \left\{\left(\mathbf{U}_{n-1}^{obs} - \mathbf{U}_{n-1}^{exp}\right) + (\xi_{n1} - \mathbf{g}_1)\right\}' \left(\text{diag}(\mathbf{U}_{n-1}^{exp} + \mathbf{g}_1)\right)^{-1} \\ \left\{\left(\mathbf{U}_{n-1}^{obs} - \mathbf{U}_{n-1}^{exp}\right) + (\xi_{n1} - \mathbf{g}_1)\right\},$$

$k_1 \geq 0$ is the allowance constant, $\text{diag}(\mathbf{a})$ denotes a diagonal matrix with its diagonal elements equal to the corresponding elements of the vector $\mathbf{a}$, and the superscripts "obs" and "exp" denote observed and expected counts, respectively. Define

$$C_{n1} = \left(\mathbf{U}_n^{obs} - \mathbf{U}_n^{exp}\right)' \left(\text{diag}(\mathbf{U}_n^{exp})\right)^{-1} \left(\mathbf{U}_n^{obs} - \mathbf{U}_n^{exp}\right). \tag{9.29}$$

Then, a shift violating $H_0^{(1)*}$ is signaled if

$$C_{n1} > h_1, \tag{9.30}$$

where $h_1 > 0$ is a control limit chosen to achieve a given ARL_0 level. Chart (9.29)–(9.30) is called *multivariate antirank CUSUM (MA-CUSUM) chart* hereafter. Regarding the allowance constant k_1, Qiu and Hawkins (2001) showed that it should be chosen in the interval

$$\left[0, \max_{\ell=1}^{p} \frac{\sum_{j \neq \ell} g_{1,j}}{g_{1,\ell}} \right].$$

If k_1 is chosen outside this interval, then the MA-CUSUM chart will be restarted at each time point and consequently the given ARL_0 level can not be achieved.

It is apparent that both the IC and OC properties of the MA-CUSUM chart (9.29)–(9.30) are determined uniquely by the distribution of A_{n1} for fixed k_1 and h_1. The null distribution $\mathbf{g}_1$ of A_{n1} could be determined from an IC data by computing the relative frequencies of $(A_{n1} = j)$, for $j = 1, 2, \ldots, p$. If the joint distribution of the original observation $\mathbf{X}$ is known under H_0, then $\mathbf{g}_1$ could also be computed algebraically or numerically from this distribution. As long as $\mathbf{g}_1$ is determined, the IC property of the chart could be obtained by simulating a series of i.i.d. multinomial random variables $\{A_{n1}, n \geq 1\}$. Therefore, the MA-CUSUM chart (9.29)–(9.30) does not require a parametric form of the distribution of $\mathbf{X}$, and it is indeed nonparametric.

When all or $p - 1$ components of $\mathbf{X}$ are continuous, the chance of ties among the p components is negligible for all practical purposes. Therefore A_{n1} is well defined. When two or more components are discrete and these discrete components can take the same values, however, ties among the p components are possible. For instance, assume that $p = 3$ and an observation takes the value of $(-1, -1, 0)'$ at the time point n. Then X_{n1} and X_{n2} are tied, and the value of A_{n1} is not well defined. To overcome this difficulty, there are several existing proposals in the literature (e.g., Gibbons and Chakraborti, 2003). One such proposal is to randomly assign 1 or 2 to A_{n1} with a probability 0.5 for each number. Unfortunately, results from such a random mechanism are not reproducible. Qiu and Hawkins (2001) suggested modifying the definition of $\xi_{n1,j}$ in (9.27) as follows, to overcome the difficulty caused by ties. Suppose that $X_{nj_1}, X_{nj_2}, \ldots, X_{nj_k}$ form a tie and their values reach the minimum among all p components at the time point n. Then, we define

$$\xi_{n1,j} = \begin{cases} 1/k, & \text{if } j \in \{j_1, j_2, \ldots, j_k\} \\ 0, & \text{otherwise.} \end{cases}$$

By using this definition, no information about the antiranks is lost and the results are reproducible.

The MA-CUSUM chart (9.29)–(9.30) is based on one component of the antirank vector. It is possible to construct a similar CUSUM chart based on several components of the antirank vector. Let $(A_{nj_1}, A_{nj_2}, \ldots, A_{nj_q})'$ be any q components of the antirank vector with $j_1 < j_2 < \cdots < j_q$. Assume that the distribution of $(A_{nj_1}, A_{nj_2}, \ldots, A_{nj_q})'$ when $H_0^{(1)}$ is true has been determined in the sample space:

$$S(j_1, j_2, \ldots, j_q)$$
$$= \{(i_1, i_2, \ldots, i_q) : i_1, i_2, \ldots, i_q \text{ are } q \text{ different integers in } (1, 2, \ldots, p)\}.$$

The sample space $S(j_1, j_2, \ldots, j_q)$ has $P_{q,p} = p(p-1)\cdots(p-q+1)$ elements. Similar to (9.27), we can define a $P_{q,p}$-dimensional random vector $\xi_{nj_1 j_2 \cdots j_q}$ with its j-th component equal to 1 when $(A_{nj_1}, A_{nj_2}, \ldots, A_{nj_q})'$ takes the value of the j-th element in $S(j_1, j_2, \ldots, j_q)$ and 0 otherwise. Then, a multivariate CUSUM chart could be constructed similarly to (9.29)–(9.30) with ξ_{n1} replaced by $\xi_{nj_1 j_2 \ldots j_q}$ and p by $P_{q,p}$.

As mentioned above, the MA-CUSUM chart (9.29)–(9.30) can detect shifts in all directions except the one in which the components of the mean vector are all the same but not zero. For many applications, this chart is adequate since this situation is unlikely. If we do care about this situation, however, it can be detected easily by a univariate CUSUM based on

$$SX_n = \frac{\sum_{j=1}^{p} X_{nj}}{\sum_{j_1=1}^{p} \sum_{j_2=1}^{p} \sigma_{j_1, j_2}},$$

where σ_{j_1, j_2} is the (j_1, j_2)-th element of Σ_X. Define $C_0^+ = C_0^- = 0$, and

$$\begin{cases} C_n^+ = \max(0, C_{n-1}^+ + SX_n - k_2) \\ C_n^- = \min(0, C_{n-1}^- + SX_n + k_2), \end{cases} \tag{9.31}$$

where $k_2 > 0$ is an allowance constant. Then, this univariate CUSUM chart signals a shift when

$$C_n^+ > h_2 \qquad \text{or} \qquad C_n^- < -h_2, \tag{9.32}$$

where $h_2 > 0$ is a control limit. Obviously, this CUSUM chart is the two-sided version of the conventional CUSUM chart discussed in Subsection 4.2.1 (cf., (4.7)–(4.10)). Then, a joint CUSUM scheme for detecting all kinds of shift in the process mean vector can be obtained by combining (9.29)–(9.30) and (9.31)–(9.32), which signals a shift whenever (9.30) or (9.32) holds.

One question that may occur is why we use the antirank rather than the more familiar rank of each component within an observation vector. If we were interested in a univariate control chart for a particular component of $\mathbf{X}$, then the rank of that component within the vector might indeed be a relevant summand. However, this problem is not common in multivariate SPC. Instead, we are faced with the detection of shifts in some unknown component (or perhaps components). For this problem, the first antirank is particularly effective in detecting a downward shift in a single arbitrary and unknown component since this component will start to figure prominently in the first antirank. Similarly, the last antirank is attractive for detecting an upward shift in the mean of one of the components. If we do not know the direction of a shift, then the chart based on both the first and the last antirank will be effective (cf., the numerical results in Example 9.5 below). As a comparison, ranks do not have these properties.

Example 9.4 *We illustrate the MA-CUSUM chart with a dataset from an aluminum smelter. The dataset contains 5 variables: the content of SiO_2, Fe_2O_3, MgO, CaO, and Al_2O_3 (labeled as X_1, X_2, X_3, X_4, and X_5 below) in the charge. All these measures are relevant to the operation of the smelter. Stability of the alumina level and calcium oxide level is desirable. The silica, ferric oxide, and magnesium oxide levels are affected by the raw materials and are potential covariates to be taken into account in a fully-fledged multivariate scheme. The five variables are substantially correlated –*

both cross-correlated with each other, and autocorrelated. The dataset contains 189 vectors. We use the first half (95 vectors) as an IC dataset to calibrate the related control charts and the second half to test. A calibration sample of this size is smaller than one would like to fully determine the IC distribution, but suffices to illustrate the use of the method in a real-world setting.

Looking first at the autocorrelation, we model each variable by the following k-th order autoregression model using the R function ar.yw(): *for* $j = 1, 2, 3, 4, 5$,

$$X_j(n) - \mu_j = \alpha_1(X_j(n-1) - \mu_j) + \cdots + \alpha_k(X_j(n-k) - \mu_j) + \varepsilon_j(n), \qquad (9.33)$$

where $X_j(n)$ denotes the j-th variable at the n-th time point, μ_j is its mean, and $\{\varepsilon_j(n), n \geq 1\}$ is an i.i.d. sequence of random noise with mean zero and variance $\sigma_{\varepsilon_j}^2$. The default Akaike's Information Criterion (AIC) is used to determine the value of k. The results are summarized in Table 9.1.

Table 9.1 *Results from the autoregression modeling of the five variables in the aluminum smelter example.*

variable	μ	k	$\alpha_1, \ldots, \alpha_k$
X_1	0.63	3	0.07,0.12,0.28
X_2	24.81	2	0.30,0.24
X_3	12.97	1	0.55
X_4	4.14	1	0.54
X_5	57.86	1	0.32

The results in Table 9.1 are then used to pre-whiten the original data into residual vectors ($\varepsilon_j(n)$ in (9.33)) with cross-component correlation, but with their autocorrelations removed. The correlation matrix of the five residual vectors is

$$\begin{pmatrix} 1.0000 & 0.2287 & -0.1056 & -0.0277 & -0.3062 \\ 0.2287 & 1.0000 & -0.5702 & -0.0632 & 0.0676 \\ -0.1056 & -0.5702 & 1.0000 & -0.3362 & 0.2205 \\ -0.0277 & -0.0632 & -0.3362 & 1.0000 & -0.2826 \\ -0.3062 & 0.0676 & 0.2205 & -0.2826 & 1.0000 \end{pmatrix}.$$

Apparently the residuals are substantially cross-correlated (e.g., between the second and third components). For simplicity of presentation, the five components of the residual vectors are still denoted as X_1, X_2, X_3, X_4, and X_5.

Figures 9.6 and 9.7 show the time series plots and density plots of the residuals. As Figure 9.7 makes abundantly clear, the individual residual components are not normally distributed (e.g., the density of X_1 has a long right tail), and so the residual vector cannot be multivariate normal. It is therefore inadvisable to apply conventional multivariate control charts under the normality assumption, such as those described in Chapter 7, to these data. We use the IC dataset (i.e., the first half of the residual vectors) to estimate the IC distribution of the antirank vector, and then apply the MA-CUSUM charts to the test dataset.

Visually, Figure 9.6 does not suggest step changes in the residuals over the course of the dataset, so it is interesting to see what multivariate methods can indicate. Figure 9.8(a)–(b) show the MA-CUSUM charts of the first antirank, and of the first-and-last combined antiranks of the residual vectors along with their control limits. In both cases, k_1 is fixed at 1 and h_1 is computed to be 11.85 and 34.89, respectively,

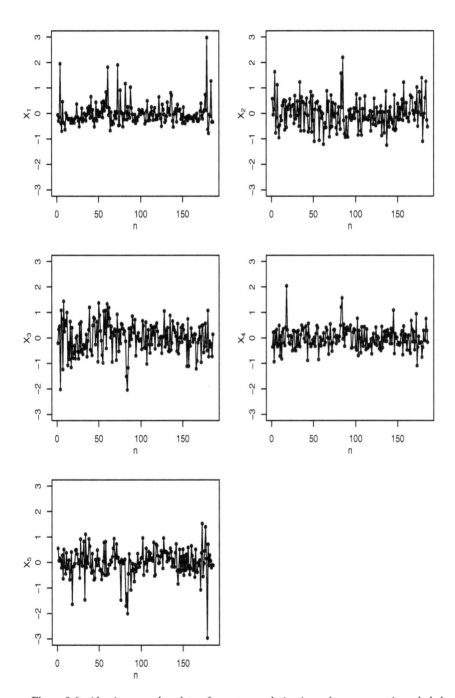

Figure 9.6: *Aluminum smelter data after autocorrelation in each component is excluded.*

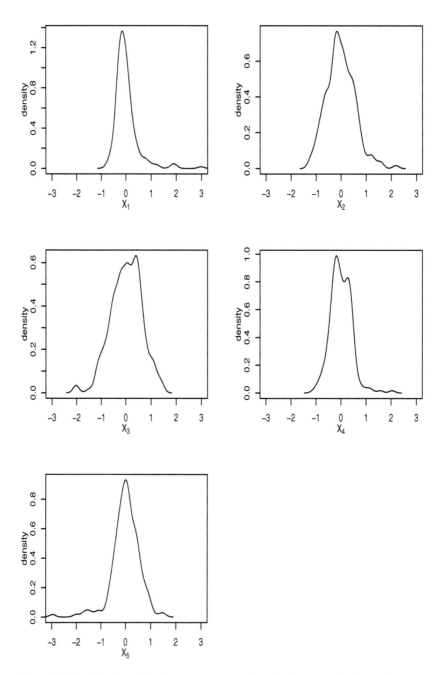

Figure 9.7 *Density plots of the five components of the aluminum smelter data after autocorrelation in each component is excluded.*

such that the ARL_0 value is 200. The first MA-CUSUM chart gives a signal of process distributional shift briefly around the time 153. The second MA-CUSUM chart, however, gives a signal almost immediately into the test dataset, and the charting statistic remains well above the control limit for most part of the test dataset. This result therefore shows a marked change, which seems to have occurred at or near the midpoint of the data set.

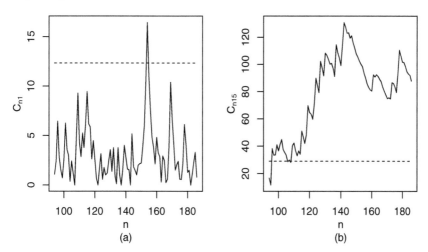

Figure 9.8 (a) MA-CUSUM chart (denoted as C_{n1}) based on the first antirank. (b) MA-CUSUM chart (denoted as C_{n15}) based on both the first and the last antiranks. The horizontal dashed lines in both plots indicate the control limits of the charts such that their ARL_0 values are both 200.

Example 9.5 The MA-CUSUM chart (9.29)–(9.30) is based on the first antirank of the quality characteristic vector $\mathbf{X}$. As mentioned above, the first and last antiranks are more sensitive to process distributional shifts, compared to the remaining antiranks. Consequently, CUSUM charts based on them should be more effective for detecting the shifts, which is demonstrated in this example. Suppose $p = 4$ and the IC joint distribution of $\mathbf{X}$ is $N_4(\mathbf{0}, I_{4 \times 4})$. In all charts considered in this example, their ARL_0 values are fixed at 200, and their allowance constants k_1 are chosen to be 1. In Table 9.2, AR1 denotes the MA-CUSUM chart based on the first antirank, AR12 denotes the MA-CUSUM chart based on the first and second antiranks, and the other notations initiated with "AR" are similarly defined. The control limit values are 6.842 and 15.6887 for AR1–AR4 and AR12–AR34, respectively. Because the four components of $\mathbf{X}$ are "exchangeable" (i.e., their joint distribution remains the same when their positions are switched) when the process is IC, the IC distributions of A_{nj}, for $j = 1, 2, 3, 4$, and (A_{nj_1}, A_{nj_2}), for $j_1, j_2 = 1, 2, 3, 4$ and $j_1 \neq j_2$, are all uniform (e.g., the IC distributions of A_{n1} is $(g_{1,1} = 1/4, g_{1,2} = 1/4, g_{1,3} = 1/4, g_{1,4} = 1/4)$). Assume that there is process mean shift at the initial time point and the shifted process mean is $\mu_1 = (\mu_{11}, \mu_{12}, \mu_{13}, \mu_{14})'$. The numbers in Table 9.2 are the OC ARL values and their standard errors of various control charts for detecting several shifts, computed based on 10,000 replicated simulations. Without loss of generality, in this example, all mean components after the shift are assumed to be non-positive and each mean vector has at least one component equal to 0. (For a general mean vector, *it can be transformed to the mean vector satisfying these conditions by subtracting its maximum component from each component of the vector and this transformation will not affect any property of the antirank vector.)*

We first check the performance of AR1–AR4 which are based on a single antirank component. The first shift considered is $\mu_1' = (-4, 0, 0, 0)$. In this case, there is only one component with a mean shift, and the probability for this component to reach the minimum of the four components is large; consequently, AR1 performs the best. In the second case, shifts occur in two mean components, with one shift larger in magnitude than the other one; still AR1 performs the best. In the third situation, two components have the same shifts; both AR1 and AR4 perform slightly better than the other two. Each of the last four cases has shifts in three mean components. It can be seen that at least one of AR1 and AR4 is better than the other two procedures. This example shows that AR1 and AR4 are indeed more sensitive to a shift than AR2 and AR3 in general. But, we still need to choose between AR1 and AR4. This can be decided by the following rough guideline: use AR1 if the number of components of μ_1 with the smallest or close to the smallest magnitudes is expected to be less than the number of components of μ_1 with the largest or close to the largest magnitudes, and vice versa.

Next, we check the performance of AR12, AR13, ..., AR34, all of which are based on two antirank components. Overall, the OC ARL values of these procedures are smaller than those of AR1–AR4 because they use more antirank information for shift detection. In cases when AR1 performs well (e.g., when $\mu_1' = (-4, 0, 0, 0)$), the charts with the first antirank involved, namely AR12, AR13, and AR14, also perform reasonably well. This phenomenon is also true for AR4. Therefore, we can expect that AR14 works reasonably well in all cases, which is confirmed by this example. We can also expect that MA-CUSUM charts based on more than two antirank components will be even more sensitive to shifts. But the charts themselves will become more complicated, especially in cases when p is large. Therefore, we recommend using more antirank components in cases when p is small, and to use fewer antirank components otherwise. In most cases, the chart based on the first and last antirank components is highly recommended.

Next, we present some numerical results for testing both $H_0^{(1)}$ in (9.25) and $H_0^{(2)}$ in (9.26), by combining the MA-CUSUM chart (9.29)–(9.30) with the univariate CUSUM chart (9.31)–(9.32). As mentioned before, these two charts are designed for two different shift detection purposes, and thus they do not need to be used together in practice. Here, we just want to show that they can support each other in the sense that when one chart has difficulty in detecting a shift, the other one can detect the same shift well. In Table 9.3, the IC process distribution is still assumed to be $N_4(\mathbf{0}, I_4)$. At the initial time point, the process mean vector shifts to μ_1, which takes six possible vectors with unit Euclidean length. The first three mean vectors are ordered by the decreasing Euclidean distance from μ_1 to $(-0.5, -0.5, -0.5, -0.5)$ and by the decreasing numbers of 0 components. The last three vectors are similarly ordered except that the mean components of each vector have the same absolute magnitude of 0.5. In the CUSUM charts, both k_1 and k_2 are fixed at 1. The control limits h_1 and h_2 are searched with a step of 0.0001 such that the ARL_0 value is 200, and the OC ARL values reach the minimum (denoted as $h_{1,min}$ and $h_{2,min}$, respectively, in the table). The minimum OC ARL values and the standard errors of the joint monitoring scheme, computed based on 10,000 replications, are listed in the column labeled by $CUSUM_{12}$. The corresponding OC ARL values and their standard errors of the two individual procedures with $(k_1 = 1, h_1 = h_{1,min})$ and $(k_2 = 1, h_2 = h_{2,min})$ are pre-

Table 9.2 *Comparison of the MA-CUSUM charts based on different antiranks of the p components of* $\mathbf{X}$, *where* $p = 4$ *and* $\mathbf{X}$ *has the IC joint distribution* $N_4(\mathbf{0}, I_4)$. *The* ARL_0 *value of each chart is fixed at 200 and* $k_1 = 1$. *"AR1" denotes the MA-CUSUM chart based on the first antirank, "AR12" denotes the MA-CUSUM chart based on the first and the second antiranks, and the other notations initiated with "AR" are similarly defined. The numbers in this table are the OC ARL values and their standard errors (in parentheses) based on 10,000 replicated simulations.*

| | $(\mu_{11}, \mu_{12}, \mu_{13}, \mu_{14})'$ | | | |
	$(-4,0,0,0)$	$(-4,-2,0,0)$	$(-4,-4,0,0)$	$(-4,-4,-2,0)$
AR1	4.06 (0.00)	4.97 (0.02)	17.48 (0.11)	20.13 (0.15)
AR2	79.40 (0.77)	7.54 (0.04)	18.44 (0.13)	32.52 (0.30)
AR3	76.73 (0.75)	32.92 (0.30)	18.59 (0.13)	7.55 (0.05)
AR4	78.01 (0.76)	20.52 (0.16)	17.72 (0.12)	4.98 (0.02)
AR12	4.09 (0.02)	2.92 (0.01)	3.03 (0.01)	3.87 (0.02)
AR13	4.12 (0.02)	4.29 (0.02)	5.58 (0.03)	4.70 (0.03)
AR14	4.11 (0.02)	3.67 (0.02)	5.45 (0.03)	3.65 (0.02)
AR23	11.51 (0.09)	4.71 (0.03)	5.58 (0.03)	4.70 (0.03)
AR24	11.49 (0.09)	4.70 (0.03)	5.57 (0.03)	4.27 (0.02)
AR34	11.14 (0.09)	3.90 (0.02)	3.04 (0.01)	2.90 (0.01)

| | $(\mu_{11}, \mu_{12}, \mu_{13}, \mu_{14})'$ | | |
	$(-4,-2,-2,0)$	$(-4,-3,-1,0)$	$(-4,-4,-4,0)$
AR1	5.82 (0.03)	8.54 (0.05)	79.60 (0.79)
AR2	35.17 (0.33)	12.14 (0.09)	78.71 (0.77)
AR3	35.38 (0.33)	12.19 (0.09)	79.34 (0.78)
AR4	5.86 (0.03)	8.65 (0.05)	4.06 (0.00)
AR12	4.12 (0.02)	3.26 (0.02)	11.06 (0.08)
AR13	4.85 (0.03)	4.83 (0.03)	11.41 (0.09)
AR14	3.22 (0.02)	4.01 (0.02)	4.07 (0.02)
AR23	5.36 (0.03)	4.95 (0.03)	11.43 (0.09)
AR24	4.85 (0.03)	4.83 (0.03)	4.11 (0.02)
AR34	4.10 (0.02)	3.28 (0.02)	4.07 (0.02)

sented in the columns labeled by $CUSUM_1$ and $CUSUM_2$, respectively. The values of $(g_{1,1}^*, g_{1,2}^*, g_{1,3}^*, g_{1,4}^*)$, $Q = \sum_{j=1}^{p}(g_{1,j}^* - g_{1,j})^2/g_{1,j}$, and $\sum_{j=1}^{4}\mu_{1j}$ are given in Table 9.4 for reference. It can be seen from the tables that (i) at least one of the two charts performs reasonably well for a shift with a fixed Euclidean length, and (ii) the joint monitoring scheme signals each shift reasonably fast.

From the above description, the MA-CUSUM chart (9.29)–(9.30) can detect all process mean shifts except the ones with equal components (i.e., $\mu_1 = (\mu_{11}, \mu_{12}, \ldots, \mu_{1p})'$ with $\mu_{11} = \mu_{12} = \cdots = \mu_{1p}$). Such shifts are called *equal-component shifts* hereafter. Qiu and Hawkins (2003) proposed a simple but effective modification to overcome this limitation, described below.

The main reason the MA-CUSUM chart (9.29)–(9.30) cannot detect equal-component shifts is that such shifts would not change the distribution of the antiranks because they would not change the order of the components of $\mathbf{X}$. In order to detect all possible shifts by a control chart based on the antirank vector alone, the

Table 9.3 *The numbers in the column labeled by CUSUM$_{12}$ denote the minimum OC ARL values and their standard errors (in parentheses) of the joint monitoring scheme (9.29)–(9.30) and (9.31)–(9.32), computed based on 10,000 replicated simulations. $h_{1,min}$ and $h_{2,min}$ are the corresponding control limits in the case when k_1 and k_2 are fixed at 1, and the ARL$_0$ values are fixed at 200. The corresponding OC ARL values and their standard errors (in parentheses) of the two individual charts with $(k_1 = 1, h_1 = h_{1,min})$ and $(k_2 = 1, h_2 = h_{2,min})$ are presented in the columns labeled by CUSUM$_1$ and CUSUM$_2$, respectively.*

$(\mu_{11}, \mu_{12}, \mu_{13}, \mu_{14})$	$h_{1,min}$	$h_{2,min}$	CUSUM$_1$	CUSUM$_2$	CUSUM$_{12}$
$(-1,0,0,0)$	6.99	3.70	25.00 (0.23)	254.65 (2.46)	23.78 (0.21)
$(-\frac{1}{\sqrt{2}}, -\frac{1}{\sqrt{2}}, 0, 0)$	10.13	2.21	399.39 (3.94)	25.94 (0.23)	24.51 (0.22)
$(-\frac{1}{\sqrt{3}}, -\frac{1}{\sqrt{3}}, -\frac{1}{\sqrt{3}}, 0)$	12.55	2.19	3297.65 (27.10)	16.28 (0.14)	16.18 (0.13)
$(-0.5,-0.5,0.5,0.5)$	6.85	4.77	46.72 (0.46)	4742.12 (56.72)	46.72 (0.46)
$(-0.5,-0.5,-0.5,0.5)$	11.25	2.20	1262.59 (12.73)	49.45 (0.46)	48.00 (0.44)
$(-0.5,-0.5,-0.5,-0.5)$	12.70	2.19	4314.73 (37.31)	11.49 (0.09)	11.49 (0.09)

Table 9.4 *The OC distribution of $A_1(i)$, $Q = \sum_{j=1}^{4}(g_{1,j}^* - g_{1,j})^2/g_{1,j}$, and $\sum_{j=1}^{4}\mu_{1j}$ for several shifts considered in Table 9.3.*

$(\mu_{11}, \mu_{12}, \mu_{13}, \mu_{14})$	$(g_{1,1}^*, g_{1,2}^*, g_{1,3}^*, g_{1,4}^*)$	Q	$\sum_{j=1}^{4}\mu_{1j}$
$(-1,0,0,0)$	$(0.55,0.15,0.15,0.15)$	0.48	-1
$(-\frac{1}{\sqrt{2}}, -\frac{1}{\sqrt{2}}, 0, 0)$	$(0.37,0.36,0.14,0.13)$	0.22	$-\sqrt{2}$
$(-\frac{1}{\sqrt{3}}, -\frac{1}{\sqrt{3}}, -\frac{1}{\sqrt{3}}, 0)$	$(0.29,0.29,0.29,0.13)$	0.08	$-\sqrt{3}$
$(-0.5,-0.5,0.5,0.5)$	$(0.41,0.40,0.09,0.10)$	0.38	0
$(-0.5,-0.5,-0.5,0.5)$	$(0.31,0.31,0.31,0.07)$	0.18	-1
$(-0.5,-0.5,-0.5,-0.5)$	$(0.25,0.25,0.25,0.25)$	0	-2

definition of the antirank vector needs to be modified. The modified antirank vector should be sensitive to shifts that violate either $H_0^{(1)}$ or $H_0^{(2)}$ or both. Based on this observation, we define $\tilde{\mathbf{A}}_n = (\tilde{A}_{n,1}, \tilde{A}_{n,2}, \ldots, \tilde{A}_{n,p}, \tilde{A}_{n,p+1})'$ as the antirank vector of $\mathbf{Y}_n = (X_{n1}, X_{n2}, \ldots, X_{np}, 0)'$, a combination of the observation vector $\mathbf{X}_n$ and the common IC mean of 0. Therefore, $\tilde{\mathbf{A}}_n$ is a permutation of $(1, 2, \cdots, p, p+1)'$ such that $Y_{n,\tilde{A}_{n,1}} \leq Y_{n,\tilde{A}_{n,2}} \leq \cdots \leq Y_{n,\tilde{A}_{n,p+1}}$ are the order statistics of the components of $\mathbf{Y}_n$. Clearly, the antirank vector $\tilde{\mathbf{A}}_n$ contains the order information of $\{\mu_1, \mu_2, \ldots, \mu_p, 0\}$. It is sensitive to the ordering of $\{\mu_j, j = 1, 2, \cdots, p\}$ and to the ordering between $\{\mu_j, j = 1, 2, \cdots, p\}$ and 0 as well. Therefore, a CUSUM chart based on this antirank vector has the potential to detect all possible shifts. The *modified MA-CUSUM (MMA-CUSUM) chart* can then be constructed in exactly the same way as the chart (9.29)–(9.30), except that all quantities related to the antirank vector $\mathbf{A}_n$ should be replaced by the corresponding quantities of the modified antirank vector $\tilde{\mathbf{A}}_n$.

Example 9.6 *Assume that $p = 4$ and the IC joint distribution of $\mathbf{X}$ is $N(\mathbf{0}, I_4)$. In such a case, the IC distribution of $\tilde{A}_{n,1}$ is Multinomial$(1, 0.2344, 0.2344, 0.2344, 0.2344, 0.0624)$ (cf., Subsection 2.4.2 for its definition). We compare the MMA-CUSUM chart based on $\tilde{A}_{n,1}$ with the MA-CUSUM chart based on A_{n1} in this example.*

The ARL_0 values of both charts are fixed at 200. The allowance constant k_1 in both charts is chosen to be 0.5. In such cases, the control limits are computed to be 12.488 and 8.029 for the MMA-CUSUM and MA-CUSUM charts, respectively. Remember that the MMA-CUSUM chart depends only on the distribution of the first modified antirank $\widetilde{A}_{n,1}$. So, its control limit can be searched by a simulation in which the CUSUM is constructed from a series of i.i.d. random vectors with the Multinomial$(1, 0.2344, 0.2344, 0.2344, 0.2344, 0.0624)$ distribution. It should be much faster to search the control limit value in this way, compared to the way to search its value based on the CUSUM constructed from the original observations.

We first study the scenario with equal-component shifts, assuming that the mean vector has shifted to $(a, a, a, a)'$, with a taking the values of 0, 0.2, 0.4, 0.6, 0.8, and 1. Figure 9.9(a) shows the ARL's given by 10,000 simulations. The MA-CUSUM chart, as expected, has no ability whatsoever to detect such shifts. The MMA-CUSUM chart, however, is quite effective in cases when a is above 0.4. This is because the mean shift substantially increases the probability that the four components of the data are all positive in such cases. Next, we consider an intermediate scenario where the mean vector shifts to $(a, 1, 1, 1)'$ (the case when $a = 1$ reduces to the equal-component shift scenario). The corresponding results in this scenario are shown in Figure 9.9(b). The MA-CUSUM chart is ineffective when a is close to 1, as expected, but it performs quite well when a is near zero. This is because the situation when $a = 0$ is essentially the same, with respect to the MA-CUSUM chart, as a downward shift in μ_1 and all other mean components remaining unchanged. The MMA-CUSUM chart, however, has a fairly consistent and much better performance for all a values. Figure 9.9(c) shows a situation in which the first component of the mean vector shifts, but in an upward direction, while Figure 9.9(d) shows the situation for which the MA-CUSUM chart is designed, a downward shift in just the first component of the mean vector. In the first case, the MMA-CUSUM chart retains a small superiority. In the second case, it is inferior to the MA-CUSUM chart, though the difference is substantial only when the shift size is relatively small.

From the above example, we can have the following conclusions:

(a) The MMA-CUSUM chart can detect equal-component shifts,

(b) When the shifts are off but close to the equal-component direction, the MMA-CUSUM chart often outperforms the MA-CUSUM chart, especially in cases when the shifts are far away from the origin, and

(c) In cases when only a small number of mean components have shifts, the benefit to use the MMA-CUSUM chart is small. Only in special circumstances does the MMA-CUSUM chart give substantially worse performance than the MA-CUSUM chart.

9.3 Multivariate Nonparametric SPC by Log-Linear Modeling

As mentioned in Section 9.1, in cases when a p-dimensional random vector $\mathbf{X}$ does not follow a multivariate normal distribution, existing statistical tools for describing its distribution are limited. A major difficulty in this task is in describing the possible association among the components of $\mathbf{X}$. In cases when all the components of $\mathbf{X}$ are categorical, in the statistical literature there are some well-developed methods for describing the association among the components of $\mathbf{X}$ (cf., Agresti, 2002). A major

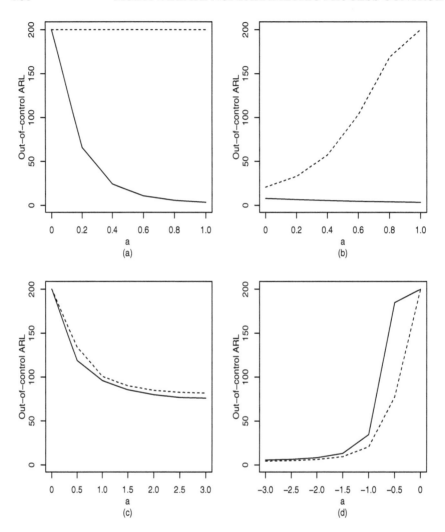

Figure 9.9 *The OC ARL values of the MMA-CUSUM (solid lines) and MA-CUSUM charts (dashed lines). The ARL$_0$ values of the two charts are fixed at 200. The allowance constant k_1 and the control limit h_1 are 0.5 and 12.488 in the MMA-CUSUM chart, and 0.5 and 8.029 in the MA-CUSUM chart. (a) $\mu_1 = (a,a,a,a)'$ and a varies among 0, 0.2, 0.4, 0.6, 0.8, and 1; (b) $\mu_1 = (a,1,1,1)'$ and a varies among 0, 0.2, 0.4, 0.6, 0.8, and 1; (c) $\mu_1 = (a,0,0,0)'$ and a varies among 0, 0.5, 1.0, 1.5, 2.0, 2.5, and 3; (d) $\mu_1 = (a,0,0,0)'$ and a varies among -3.0, -2.5, -2.0, -1.5, -1.0, -0.5, and 0.*

tool is the so-called *log-linear modeling*. This fact opens a door for solving the multivariate nonparametric SPC problem. Qiu (2008) developed a general framework for constructing multivariate nonparametric SPC charts, by first categorizing the observed multivariate data and then applying the log-linear modeling approach to the categorized data. This method is described in this section in two parts. In Subsection

9.3.1, the log-linear modeling approach is described in detail. Then, in Subsection 9.3.2, the multivariate nonparametric SPC chart by Qiu (2008) is discussed.

9.3.1 Analyzing categorical data by log-linear modeling

Assume that Y_1 and Y_2 are two categorical variables, Y_1 has s levels (or categories), denoted as $c_1, c_2, \ldots, c_s$, and Y_2 has t levels, denoted as $d_1, d_2, \ldots, d_t$. Then, observations of the 2-dimensional vector (Y_1, Y_2) could look like $(c_2, d_1), (c_1, d_3)$, and so forth. Assume that a dataset of (Y_1, Y_2) contains n observations. Then, this dataset can be summarized by Table 9.5 below.

Table 9.5: *An $s \times t$ 2-way contingency table.*

	d_1	d_2	$\cdots$	d_t	
c_1	n_{11}	n_{12}	$\cdots$	n_{1t}	n_{1+}
c_2	n_{21}	n_{22}	$\cdots$	n_{2t}	n_{2+}
$\vdots$	$\vdots$	$\vdots$	$\ddots$	$\vdots$	$\vdots$
c_s	n_{s1}	n_{s2}	$\cdots$	n_{st}	n_{s+}
	n_{+1}	n_{+2}	$\cdots$	n_{+t}	n_{++}

In Table 9.5, the column on the left lists all the levels of Y_1, the row on the top lists all the levels of Y_2, the (j_1, j_2)-th entry $n_{j_1 j_2}$ denotes the count of observations taking the value $(Y_1 = c_{j_1}, Y_2 = d_{j_2})$, the j_1-th element of the column on the right, $n_{j_1+} = \sum_{j_2=1}^{t} n_{j_1 j_2}$, is the j_1-th row total, the j_2-th element of the row at the bottom, $n_{+j_2} = \sum_{j_1=1}^{s} n_{j_1 j_2}$, is the j_2-th column total, and n_{++} at the lower-right corner is the grand total (i.e., $n_{++} = \sum_{j_1=1}^{s} n_{j_1+} = \sum_{j_2=1}^{t} n_{+j_2} = n$), where $j_1 = 1, 2, \ldots, s$ and $j_2 = 1, 2, \ldots, t$. In this table, the two categorical variables Y_1 and Y_2 classify all observations into $s \times t$ entries. So, they are also called *classifiers*. Since this table has two classifiers involved, it is called a $s \times t$ *2-way contingency table*. A general p-way contingency table, with $p > 2$, can be defined in a similar way. Because a contingency table provides a neat summary of the original data, it is easy to make, and it contains all useful information in the original categorical data, it is a popular descriptive tool in categorical data analysis.

For $j_1 = 1, 2, \ldots, s$ and $j_2 = 1, 2, \ldots, t$, let

$$\begin{aligned}
\pi_{j_1 j_2} &= P\left(Y_1 = c_{j_1}, Y_2 = d_{j_2}\right), \\
\pi_{j_1+} &= P\left(Y_1 = c_{j_1}\right), \\
\pi_{+j_2} &= P\left(Y_2 = d_{j_2}\right).
\end{aligned}$$

Then, $\{\pi_{j_1 j_2}, j_1 = 1, 2, \ldots, s, j_2 = 1, 2, \ldots, t\}$ defines the joint distribution of (Y_1, Y_2), $\{\pi_{j_1+}, j_1 = 1, 2, \ldots, s\}$ is the marginal distribution of Y_1, and $\{\pi_{+j_2}, j_2 = 1, 2, \ldots, t\}$ is the marginal distribution of Y_2. Define $\mu_{j_1 j_2} = \mathrm{E}(n_{j_1 j_2}) = n\pi_{j_1 j_2}$ to be the expected count of the (j_1, j_2)-th entry, for $j_1 = 1, 2, \ldots, s$ and $j_2 = 1, 2, \ldots, t$. Then, in cases when Y_1 and Y_2 are independent, we have $\pi_{j_1 j_2} = \pi_{j_1+} \pi_{+j_2}$, for all j_1 and j_2, and consequently,

$$\mu_{j_1 j_2} = n\pi_{j_1+} \pi_{+j_2}.$$

After taking log-transformation on both sides of the above expression, and denoting $\log(n)$, $\log(\pi_{j_1+})$ and $\log(\pi_{+j_2})$ by μ, $\lambda_{j_1}^{Y_1}$ and $\lambda_{j_2}^{Y_2}$, respectively, we have

$$\log(\mu_{j_1 j_2}) = \mu + \lambda_{j_1}^{Y_1} + \lambda_{j_2}^{Y_2}, \qquad \text{for } j_1 = 1, 2, \ldots, s, j_2 = 1, 2, \ldots, t, \qquad (9.34)$$

where μ is a constant term, $\{\lambda_{j_1}^{Y_1}\}$ and $\{\lambda_{j_2}^{Y_2}\}$ are parameters satisfying the conditions that

$$\sum_{j_1=1}^{s} \lambda_{j_1}^{Y_1} = 0, \qquad \sum_{j_2=1}^{t} \lambda_{j_2}^{Y_2} = 0.$$

In model (9.34), the term $\lambda_{j_1}^{Y_1}$ reflects the impact of Y_1 on $\mu_{j_1 j_2}$, and the term $\lambda_{j_2}^{Y_2}$ reflects the impact of Y_2 on $\mu_{j_1 j_2}$. They are often called the *main effect* terms of Y_1 and Y_2, respectively (cf., Weisberg, 2005, Section 6.3). In this model, the two main effect terms are assumed additive, reflecting the model assumption that Y_1 and Y_2 are independent. This model is thus called the *log-linear model of independence*.

In cases when Y_1 and Y_2 are associated, we can consider the following *saturated log-linear model*: for $j_1 = 1, 2, \ldots, s$ and $j_2 = 1, 2, \ldots, t$,

$$\log(\mu_{j_1 j_2}) = \mu + \lambda_{j_1}^{Y_1} + \lambda_{j_2}^{Y_2} + \lambda_{j_1 j_2}^{Y_1 Y_2}, \qquad (9.35)$$

where $\{\lambda_{j_1 j_2}^{Y_1 Y_2}\}$ are parameters accounting for the interactive effect of Y_1 and Y_2 on $\mu_{j_1 j_2}$. Therefore, the term $\lambda_{j_1 j_2}^{Y_1 Y_2}$ is called the *2-way interaction term*. Because the main effect term $\lambda_{j_1}^{Y_1}$ has $s - 1$ non-redundant parameters, and the other main effect term $\lambda_{j_2}^{Y_2}$ has $t - 1$ non-redundant parameters, by the analysis of variance or the linear regression theory (cf., Agresti, 2002; Weisberg, 2005), the interaction term allows a maximum of $(s-1)(t-1)$ parameters. To this end, besides the constraints on the two main effect terms described above, we often put the following constraint on the interaction term:

$$\sum_{j_1=1}^{s} \lambda_{j_1 j_2}^{Y_1 Y_2} = 0, \qquad \text{for each } j_2,$$

$$\sum_{j_2=1}^{t} \lambda_{j_1 j_2}^{Y_1 Y_2} = 0, \qquad \text{for each } j_1.$$

Under all these constraints, the model (9.35) has

$$1 + (s-1) + (t-1) + (s-1)(t-1) = st$$

non-redundant parameters, which is the same as the number of observed counts (or, the number of cells) in the contingency table. That is the reason this model is called the saturated model because it already contains the maximal number of non-redundant parameters.

The log-linear models described above can be generalized to cases with $p > 2$ categorical variables. For instance, when $p = 3$ and the three categorical variables $Y_1, Y_2,$

and Y_3 have s_1, s_2, and s_3 categories, respectively, the saturated log-linear model can be defined as follows. For $j_1 = 1, 2, \ldots, s_1$, $j_2 = 1, 2, \ldots, s_2$, and $j_3 = 1, 2, \ldots, s_3$,

$$\log(\mu_{j_1 j_2 j_3}) = \mu + \lambda_{j_1}^{Y_1} + \lambda_{j_2}^{Y_2} + \lambda_{j_3}^{Y_3} + \lambda_{j_1 j_2}^{Y_1 Y_2} + \lambda_{j_1 j_3}^{Y_1 Y_3} + \lambda_{j_2 j_3}^{Y_2 Y_3} + \lambda_{j_1 j_2 j_3}^{Y_1 Y_2 Y_3}, \qquad (9.36)$$

where μ is a constant term, $\lambda_{j_1}^{Y_1}, \lambda_{j_2}^{Y_2}$, and $\lambda_{j_3}^{Y_3}$ are the main effects of Y_1, Y_2, and Y_3, respectively, $\lambda_{j_1 j_2}^{Y_1 Y_2}, \lambda_{j_1 j_3}^{Y_1 Y_3}$, and $\lambda_{j_2 j_3}^{Y_2 Y_3}$ are the 2-way interaction terms, and $\lambda_{j_1 j_2 j_3}^{Y_1 Y_2 Y_3}$ is the 3-way interaction term. To make all parameters estimable, they should be subject to certain constraints. One set of such constraints is as follows:

$$\sum_{j_1} \lambda_{j_1}^{Y_1} = \sum_{j_2} \lambda_{j_2}^{Y_2} = \sum_{j_3} \lambda_{j_3}^{Y_3} = 0;$$

$$\sum_{j_1} \lambda_{j_1 j_2}^{Y_1 Y_2} = \sum_{j_2} \lambda_{j_1 j_2}^{Y_1 Y_2} = 0; \; \sum_{j_1} \lambda_{j_1 j_3}^{Y_1 Y_3} = \sum_{j_3} \lambda_{j_1 j_3}^{Y_1 Y_3} = 0; \; \sum_{j_2} \lambda_{j_2 j_3}^{Y_2 Y_3} = \sum_{j_3} \lambda_{j_2 j_3}^{Y_2 Y_3} = 0;$$

$$\sum_{j_1} \lambda_{j_1 j_2 j_3}^{Y_1 Y_2 Y_3} = \sum_{j_2} \lambda_{j_1 j_2 j_3}^{Y_1 Y_2 Y_3} = \sum_{j_3} \lambda_{j_1 j_2 j_3}^{Y_1 Y_2 Y_3} = 0,$$

where $\sum_{j_1} \lambda_{j_1 j_2}^{Y_1 Y_2} = 0$ means $\sum_{j_1=1}^{s_1} \lambda_{j_1 j_2}^{Y_1 Y_2} = 0$, for all j_2, and the other summations are defined similarly.

Model (9.36) can be denoted by $(Y_1 Y_2 Y_3)$, which lists the highest-order term in the model for each variable. If the 3-way interaction term $\lambda_{j_1 j_2 j_3}^{Y_1 Y_2 Y_3}$ is not in the model, then it can be checked that the conditional association between any two variables of Y_1, Y_2, and Y_3 is the same at all levels of the remaining variable (cf. Agresti, 2002, Chapter 8). In other words, the association between any two variables is homogeneous across the different levels of the third variable. This model is called the *homogeneous association model*, and is denoted by $(Y_1 Y_2, Y_1 Y_3, Y_2 Y_3)$. Similarly, we use $(Y_1 Y_2, Y_1 Y_3)$ to denote the model with the two 2-way interaction terms $\lambda_{j_1 j_2}^{Y_1 Y_2}$ and $\lambda_{j_1 j_3}^{Y_1 Y_3}$ included, and with the other 2-way interaction term $\lambda_{j_2 j_3}^{Y_2 Y_3}$ and the 3-way interaction term $\lambda_{j_1 j_2 j_3}^{Y_1 Y_2 Y_3}$ excluded. That model implies that Y_2 and Y_3 are independent conditional on Y_1, the association between Y_1 and Y_2 is homogeneous across the different levels of Y_3, and the association between Y_1 and Y_3 is also homogeneous across the different levels of Y_2. So, model (9.36) and its variants can describe all kinds of possible associations among Y_1, Y_2, and Y_3, by including appropriate 2-way and 3-way interaction terms.

To estimate a log-linear model, we usually need to specify a probability model for the related contingency table (e.g., the observed count of the (j_1, j_2, j_3)-th entry of a 3-way table has the $Poisson(\mu_{j_1 j_2 j_3})$ distribution). Then, the model can be estimated by the maximum likelihood estimation method (cf., Subsection 2.7.2). However, the likelihood function of the table is often quite complicated, and a numerical algorithm, such as the Newton-Raphson algorithm, is needed. In the software package R, a log-linear model can be estimated easily using the function glm().

There are some standard procedures for model selection and model goodness-of-fit testing. For instance, if M_1 denotes the saturated model $(Y_1 Y_2 Y_3)$, M_0 denotes the homogeneous association model $(Y_1 Y_2, Y_1 Y_3, Y_2 Y_3)$, and we want to know whether M_0 is more appropriate to use (or, equivalently, whether the 3-way interaction term in M_1 can be removed from the model). For this purpose, we can use the following likelihood ratio test statistic:

$$G^2(M_0 | M_1) = -2 \log\left(\frac{\ell_{M_0}}{\ell_{M_1}}\right),$$

where ℓ_{M_0} and ℓ_{M_1} denote the likelihood functions of M_0 and M_1, respectively. Theory on categorical data analysis shows that $G^2(M_0|M_1)$ has the asymptotic distribution of $\chi^2_{(s_1-1)(s_2-1)(s_3-1)}$ (cf. Agresti, 2002, Chapter 8). So, its observed value can be compared to the $(1-\alpha)$-th quantile of the $\chi^2_{(s_1-1)(s_2-1)(s_3-1)}$ distribution for making decisions, where α is a given significance level. Note that this test is based on the asymptotic distribution of $G^2(M_0|M_1)$. So, we should use it with caution when the sample size is small, although it has been shown in the literature that it is still quite reliable with fairly sparse tables (cf., Haberman, 1977). There are several different ways to do model selection using the above model comparison approach, including the backward, forward, and stepwise procedures. By the backward model selection, in the above example with $p = 3$, we first consider removing the 3-way interaction term from the saturated model, as described above. If the likelihood ratio test using $G^2(M_0|M_1)$ suggests that the 3-way interaction term cannot be removed, then stop. Otherwise, consider removing a 2-way interaction term. This process continues until no terms can be removed. In each step of the backward model selection, only one term is considered to be removed and we always consider removing the highest-order term in the current model. If there are several highest-order terms in the current model, then the one with the least significant (or the smallest) $G^2(M_0|M_1)$ value is considered to be removed. After a final model is determined, its goodness-of-fit can be tested using the Pearson's chi-square test or the likelihood ratio G^2 test, as discussed in Subsection 2.8.2.

9.3.2   Nonparametric SPC by log-linear modeling

In this subsection, we introduce the general framework for constructing multivariate nonparametric control charts using the log-linear modeling approach that was first proposed by Qiu (2008). At the end of phase I analysis of a production process, assume that all bugs are fixed and we are left with an IC dataset

$$\{\mathbf{X}_i = (X_{i1}, X_{i2}, \ldots, X_{ip})', i = 1, 2, \ldots, M\},$$

where M is a fixed sample size. This sequence of observations is assumed to be i.i.d. with a common cdf $F_0(\mathbf{x})$, which is also the IC process distribution. Let the IC median of X_{ij} be $\widetilde{\mu}_j$, for $j = 1, 2, \ldots, p$, which can be estimated from the IC data. We then define

$$Y_{ij} = I(X_{ij} > \widetilde{\mu}_j), \qquad \text{for } j = 1, 2, \ldots, p, \tag{9.37}$$

and $\mathbf{Y}_i = (Y_{i1}, Y_{i2}, \ldots, Y_{ip})'$, where $I(a)$ is an indicator function that equals 1 if a is "true" and 0 otherwise. So equation (9.37) transforms the p original quality characteristic variables to p binary variables. Note that, in equation (9.37), the IC median $\widetilde{\mu}_j$ can be replaced by the more general r-th quantile of the IC distribution of X_{ij}, with r being any real number in $(0, 1)$. We consider using the median, which is the 0.5th quantile, because the resulting joint distribution of $\mathbf{Y}_i$ could be more efficiently estimated by the log-linear modeling approach discussed in the previous subsection, due to the fact that relatively fewer cell counts of the contingency table formed by the components of $\mathbf{Y}_i$ would be small in such a case and consequently the model selection and the model goodness-of-fit test would be more reliable.

Of course, we lose information by transforming $\mathbf{X}_i$ to $\mathbf{Y}_i$. But it is not difficult to check that the distribution of $\mathbf{Y}_i$ would be changed by any shift in the median vector $\widetilde{\mu} = (\widetilde{\mu}_1, \widetilde{\mu}_2, \ldots, \widetilde{\mu}_p)'$ of the process, as long as the IC process distribution $F_0(\mathbf{x})$ has a positive probability to take values in any neighborhood of its IC median vector. Therefore, if we are interested in detecting shifts in a location parameter vector (e.g., the median vector $\widetilde{\mu}$), then $\mathbf{Y}_i$ is appropriate to use. If we are also interested in detecting other changes from $F_0(\mathbf{x})$ (e.g., changes in the covariance matrix of $F_0(\mathbf{x})$), then $\mathbf{Y}_i$ needs to be modified. In the latter case, one possible approach is to transform each component of $\mathbf{X}_i$ to a categorical variable with more than two categories, as was done in univariate cases discussed in Section 8.3.

For the categorized data $\mathbf{Y}_i$, for $i = 1, 2, \ldots, M$, we can choose a log-linear model using a model selection procedure described in the previous subsection for analyzing the related $2 \times 2 \times \cdots \times 2$ p-way contingency table. After the final model is determined, the expected cell count $\mu_{j_1 j_2 \cdots j_p}$ of the $(j_1, j_2, \ldots, j_p)$-th entry can be computed from the estimated log-linear model, for $j_1, j_2, \ldots, j_p = 1, 2$, and the IC joint distribution of $Y_1, Y_2, \ldots, Y_p$ can be estimated by

$$\left\{ f^{(0)}_{j_1 j_2 \cdots j_p} = \mu_{j_1 j_2 \cdots j_p} / n, \ j_1, j_2, \ldots, j_p = 1, 2 \right\}.$$

It should be noticed that if the final model is the saturated model (e.g. (9.36)), then $\mu_{j_1 j_2 \cdots j_p} = O_{j_1 j_2 \cdots j_p}$, for all $j_1, j_2 \cdots j_p$, where $O_{j_1 j_2 \cdots j_p}$ denotes the observed count of the $(j_1, j_2, \ldots, j_p)$-th entry. In such a case, $\mu_{j_1 j_2 \cdots j_p} / n$ is just the ordinary relative frequency of the $(j_1, j_2, \ldots, j_p)$-th entry. When the components of $\mathbf{Y}$ have some association structure, the selected log-linear model (e.g., the homogeneous association model) would be simpler than the saturated model to reflect this structure. In such cases, the variability of $\mu_{j_1 j_2 \cdots j_p} / n$ computed from the selected log-linear model would be smaller than the variability of the corresponding relative frequency $O_{j_1 j_2 \cdots j_p} / n$, because the selected log-linear model has fewer parameters than the saturated model. In other words, $\mu_{j_1 j_2 \cdots j_p} / n$ is "smoother" than $O_{j_1 j_2 \cdots j_p} / n$. In that sense, the log-linear modeling is a smoothing process, and the degree of smoothing depends on the association structure of the components of $\mathbf{Y}$. Since both the log-linear estimator $\mu_{j_1 j_2 \cdots j_p} / n$ and the relative frequency estimator $O_{j_1 j_2 \cdots j_p} / n$ are unbiased in cases when the selected model holds, the log-linear estimator

$$\left\{ f^{(0)}_{j_1 j_2 \cdots j_p} = \mu_{j_1 j_2 \cdots j_p} / n, \ j_1, j_2, \ldots, j_p = 1, 2 \right\}$$

would provide a better estimator for the joint distribution of $\mathbf{Y}$ in such cases.

Now, let us discuss phase II SPC and assume that we are interested in monitoring the process median vector $\widetilde{\mu}$. Let $\mathbf{X}_n$ be the original phase II observation vector of the process at the n-th time point, and $\mathbf{Y}_n = (Y_{n1}, Y_{n2}, \ldots, Y_{np})'$ be its categorized version defined by (9.37). For $j_1, j_2, \ldots, j_p = 1, 2$, define

$$g_{n j_1 j_2 \cdots j_p} = I(Y_{n1} = j_1 - 1, Y_{n2} = j_2 - 1, \ldots, Y_{np}(i) = j_p - 1),$$

and $\mathbf{g}_n$ is a vector of all $g_{n j_1 j_2 \cdots j_p}$ values. Then, $\mathbf{g}_n$ is a vector of observed counts at time point n, and $\{ f^{(0)}_{j_1 j_2 \cdots j_p}, j_1, j_2, \ldots, j_p = 1, 2 \}$ are the expected counts. Let $\mathbf{f}^{(0)}$

be the vector of all $f_{j_1 j_2 \cdots j_p}^{(0)}$. Then, a CUSUM chart can be constructed similarly to the chart (8.34)–(8.35) in univariate cases, as follows. Let $\mathbf{U}_0^{obs} = \mathbf{U}_0^{exp} = \mathbf{0}$ be two 2^p-dimensional vectors, and

$$
\begin{cases}
\mathbf{U}_n^{obs} = \mathbf{0}, & \text{if } B_n \leq k \\
\mathbf{U}_n^{exp} = \mathbf{0}, & \\
\mathbf{U}_n^{obs} = \left(\mathbf{U}_{n-1}^{obs} + \mathbf{g}_n\right)(1 - k/B_n), & \text{if } B_n > k \\
\mathbf{U}_n^{exp} = \left(\mathbf{U}_{n-1}^{exp} + \mathbf{f}^{(0)}\right)(1 - k/B_n),
\end{cases}
$$

where

$$
B_n = \left\{\left(\mathbf{U}_{n-1}^{obs} - \mathbf{U}_{n-1}^{exp}\right) + \left(\mathbf{g}_n - \mathbf{f}^{(0)}\right)\right\}' \left(\mathrm{diag}(\mathbf{U}_{n-1}^{exp} + \mathbf{f}^{(0)})\right)^{-1} \\
\left\{\left(\mathbf{U}_{n-1}^{obs} - \mathbf{U}_{n-1}^{exp}\right) + \left(\mathbf{g}_n - \mathbf{f}^{(0)}\right)\right\},
$$

$k \geq 0$ is the allowance constant, $\mathrm{diag}(\mathbf{a})$ denotes a diagonal matrix with its diagonal elements equal to the corresponding elements of the vector $\mathbf{a}$, and the superscripts "obs" and "exp" denote observed and expected counts, respectively. Define

$$
C_n = \left(\mathbf{U}_n^{obs} - \mathbf{U}_n^{exp}\right)' (\mathrm{diag}(\mathbf{U}_n^{exp}))^{-1} \left(\mathbf{U}_n^{obs} - \mathbf{U}_n^{exp}\right). \tag{9.38}
$$

Then, a median shift in $\mathbf{X}_n$ is signaled if

$$
C_n > h, \tag{9.39}
$$

where $h > 0$ is a control limit chosen to achieve a given IC ARL level. Chart (9.38)–(9.39) is called the *multivariate log-linear CUSUM (MLL-CUSUM) chart* hereafter, to reflect the fact that it is based on the log-linear model.

For a given value of ARL_0 and a given k, the value of h of the MLL-CUSUM chart can be searched in a range $[0, U_h]$ by the algorithm described in the box below, where U_h is an upper bound satisfying the condition that the actual ARL_0 value of the chart is larger than the given ARL_0 value when $h = U_h$.

Iterative Algorithm for Selecting h

Step 1 In the i-th iteration, h is searched in the range $[L_h^{(i)}, U_h^{(i)}]$. When $i = 1$, $L_h^{(1)} = 0$ and $U_h^{(1)} = U_h$.

Step 2 A series of random vectors from the multinomial distribution with probability parameters $\{f_{j_1,j_2,\ldots,j_p}^{(0)}, j_1, j_2, \ldots, j_p = 1, 2\}$ are generated by a random number generator.

Step 3 This series of random vectors are used in place of $g(n)$ in (9.38) and the run length distribution is obtained by running the chart (9.38)–(9.39) with $h = h^{(i)} := (L_h^{(i)} + U_h^{(i)})/2$ a number of times (e.g., 10,000 times). The IC ARL value $ARL_0^{(i)}$ is then computed by averaging all the run lengths obtained.

Step 4 If $|ARL_0^{(i)} - ARL_0| < \varepsilon_1$, where $\varepsilon_1 > 0$ is a pre-specified threshold value and ARL_0 is a given ARL_0 value, then the algorithm stops, and the searched value of h is $h^{(i)}$. Otherwise, define

$$L_h^{(i+1)} = h^{(i)} \text{ and } U_h^{(i+1)} = U_h^{(i)}, \text{ if } ARL_0^{(i)} < ARL_0;$$

$$L_h^{(i+1)} = L_h^{(i)} \text{ and } U_h^{(i+1)} = h^{(i)}, \text{ if } ARL_0^{(i)} > ARL_0;$$

$$\text{and } h^{(i+1)} = (L_h^{(i+1)} + U_h^{(i+1)})/2.$$

Step 5 If $|h^{(i+1)} - h^{(i)}| < \varepsilon_2$, where $\varepsilon_2 > 0$ is another pre-specified threshold value, then the algorithm stops, and the searched value of h is $h^{(i)}$. In such a case, a message should be printed, to remind the user of the actual IC ARL value. If $|h^{(i+1)} - h^{(i)}| \geq \varepsilon_2$, then the algorithm executes the next iteration.

Based on our experience, the above algorithm usually stops at Step 4. But, occasionally it can happen that it stops at Step 5, especially when ε_1 is chosen relatively small and ε_2 is chosen relatively large. In such cases, users are reminded by the algorithm that the assumed IC ARL value is not reached within the specified range by the chart (9.38)–(9.39) using the searched value of h; its actual IC ARL value is also printed. We have several options if this situation happens. The first option is to decrease the value of ε_2 and run the iterative algorithm again. But, it can happen that the given ARL_0 value cannot be reached in any accuracy, no matter how small ε_2 is, because of the discreteness of the charting statistic, as discussed in Chapter 8 (cf., Tables 8.1 and 8.5). Then, the second option is to use the actual ARL_0 value of the chart as a new given ARL_0 value. If the new given ARL_0 value is difficult to interpret and the original given ARL_0 value is preferred, then the third option is to adopt the modification procedure discussed in Subsection 8.3.1 about the P-CUSUM chart (8.34)–(8.35) (cf., the third paragraph below expression (8.35)).

The allowance constant k should be chosen in the interval

$$\left[0, \max_{j_1,j_2,\ldots,j_p=1,2} \frac{1 - f_{j_1,j_2,\ldots,j_p}^{(0)}}{f_{j_1,j_2,\ldots,j_p}^{(0)}}\right).$$

Otherwise, the MLL-CUSUM chart (9.38)–(9.39) will restart at each time point when the process is IC. Consequently, the specified IC ARL property can never be achieved. If we have a target shift in the mean or median vector of the process distribution, then for a given ARL_0 value, the optimal value of k can be searched for in a range $[0, U_k]$ by an iterative algorithm described in the box below, where U_k is an upper bound.

Iterative Algorithm for Selecting k

Step 1 In the i-th iteration, k is searched for in the range $[L_k^{(i)}, U_k^{(i)}]$. When $i = 1$, $L_k^{(1)} = 0$ and $U_k^{(1)} = U_k$. Divide $[L_k^{(i)}, U_k^{(i)}]$ into m equally spaced subintervals, where m is pre-specified (e.g., $m = 10$). Then k is searched for among all the end points $\{k_j^{(i)} = L_k^{(i)} + (U_k^{(i)} - L_k^{(i)}) * j/m, j = 0, 1, \ldots, m\}$ of these subintervals.

Step 2 When $k = k_j^{(i)}$, for any $j = 0, 1, \ldots, m$, search for the corresponding h value by the previous iterative algorithm, such that the actual ARL_0 value equals the given ARL_0 value.

Step 3 For the target shift, compute the OC distribution of $\mathbf{Y}_n$, denoted by $\{f_{j_1,j_2,\ldots,j_p}^{(1)}, j_1, j_2, \ldots, j_p = 1, 2\}$, by the log-linear modeling procedure discussed in Subsection 9.3.1 and by using the IC data after adjusting for the location shift.

Step 4 Generate a series of random vectors from the multinomial distribution with probability parameters $\{f_{j_1,j_2,\ldots,j_p}^{(1)}, j_1, j_2, \ldots, j_p = 1, 2\}$, and use this series of random vectors in place of $\mathbf{g}(n)$ when computing C_n in (9.38).

Step 5 For each j, compute the OC ARL value, denoted as $ARL_{1,j}^{(i)}$, by running the chart (9.38)–(9.39) with $k = k_j^{(i)}$ a number of times. The minimizer of $\{ARL_{1,j}^{(i)}, j = 0, 1, \ldots, m\}$ is denoted as $J^{(i)}$.

Step 6 If $(U_k^{(i)} - L_k^{(i)})/m < \varepsilon_3$, where $\varepsilon_3 > 0$ is a pre-specified threshold value, then the algorithm stops, and the searched value of k is $k_{J^{(i)}}^{(i)}$. Otherwise, let $L_k^{(i+1)} = \max(0, k_{J^{(i)}-1}^{(i)})$ and $U_k^{(i+1)} = \min(k_{J^{(i)}+1}^{(i)}, U_k)$, and the algorithm executes the next iteration.

In the above two algorithms, the upper bound U_k can be simply chosen as $\max_{j_1,j_2,\ldots,j_p=1,2}(1 - f_{j_1,j_2,\ldots,j_p}^{(0)})/f_{j_1,j_2,\ldots,j_p}^{(0)}$ if that number is well defined and not too big. Selection of the upper bound U_h should not be difficult. For a given ARL_0 value and a given k, we could try a large number (e.g., 50 or 100) for U_h, and then run the chart (9.38)–(9.39) to make sure that its actual ARL_0 value when $h = U_h$ is larger than the given ARL_0 value. Since both algorithms converge reasonably fast, accurate selection of U_k and U_h is not essential to their convergence speed, which makes the selection of U_k and U_h much easier. That is, the two upper bounds can be chosen relatively large, without sacrificing much convergence speed of the two algorithms. The values of $\varepsilon_1, \varepsilon_2$, and ε_3 are related to the accuracy requirements for the solutions. For example, if we require that the actual ARL_0 value equals the given ARL_0 value up to the third digit after the decimal point, then we can choose $\varepsilon_1 = 0.5 \times 10^{-3}$.

Example 9.7 *In the aluminum smelter example discussed in Example 9.4, let us consider the three quality characteristic variables SiO_2, MgO, and Al_2O_3 for multivariate SPC. Again, the autocorrelation within each variable is excluded beforehand, using the autoregression model (9.33), and the resulting residuals are shown in the three plots in the first column of Figure 9.6. For simplicity of presentation, the residuals of the three variables are denoted as $\varepsilon_1, \varepsilon_2$, and ε_3. Both the Pearson's chi-square test and the Kolmogorov-Smirnov test (cf., Subsection 2.8.2) conclude that ε_1 and ε_3 in this dataset are not normally distributed (Pearson's chi-square test: p-value equals 0 for both ε_1 and ε_3. Kolmogorov-Smirnov test: p-value equals 0 for ε_1, and it equals 0.022 for ε_3). So the joint distribution of $\mathbf{X} = (\varepsilon_1, \varepsilon_2, \varepsilon_3)'$ can not be normal because a joint normal distribution implies that all marginal distributions are normal.*

We then use the first 95 residual vectors as an IC dataset and the remaining for testing. From the IC dataset, the selected log-linear model is $(Y_1, Y_2 Y_3)$, the estimated IC distribution $\{f^{(0)}_{j_1, j_2, j_3}, j_1, j_2, j_3 = 1, 2\}$ of $\mathbf{Y}$ by this selected model is calculated to be $(0.1053, 0.1474, 0.1158, 0.1368, 0.1895, 0.0632, 0.0947, 0.1474)$, where $\mathbf{Y}$ is the categorized version of $\mathbf{X}$. By the first iterative algorithm described above, the control limit value h is searched and found to be 10.793, for $k = 0.1$ and $ARL_0 = 200$. Then, the MLL-CUSUM chart (9.38)–(9.39) is used for detecting shifts in the test data, and the chart is shown in Figure 9.10. It can be seen that there is convincing evidence for a shift occurring right at the beginning of the test data, which is consistent with the results shown in Figure 9.8(b).

Example 9.8 *In this example, we compare the numerical performance of the MLL-CUSUM chart (9.38)–(9.39) with certain competing control charts in cases when $p = 3$ and the IC distribution of $\mathbf{X} = (X_1, X_2, X_3)'$ is the normalized version with mean $\mathbf{0}$ and covariance matrix $I_{3 \times 3}$ of the distribution specified below*

$$X_1, X_2, X_3 \text{ are i.i.d. with the common distribution } \chi_1^2.$$

When using the chart (9.38)–(9.39), the IC distribution of $\mathbf{Y}$ is not assumed known. Instead, it is estimated from an IC dataset. To this end, we randomly generate 100 such IC datasets, each of which has sample size of 100. Then, from each IC dataset, the estimated IC distribution of $\mathbf{Y}$ is computed, using the log-linear modeling approach, and the control limit value h is searched for by the iterative algorithm discussed above, when k and ARL_0 are fixed at 1.0 and 200, respectively. In the searching algorithm, we choose $U_h = 30, \varepsilon_1 = .01$, and $\varepsilon_2 = 10^{-5}$, and the search is based on 10,000 replications. Then, the IC dataset and the corresponding control limit value giving the median actual IC ARL value is used in all cases considered in this example. Suppose that there is a shift of size $(a, 0, 0)$ in the median vector of the process distribution, the shift starts at the 100th phase II observation time, and a changes its value between 0 and 0.4 with a step of 0.04. The specific starting time

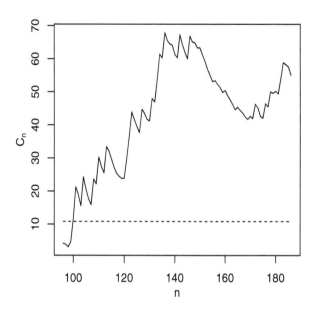

Figure 9.10 *MLL-CUSUM chart (9.38)–(9.39) with $k = 0.1$ and $ARL_0 = 200$ when it is applied to the residuals of the quality characteristic variables $SiO_2, MgO,$ and Al_2O_3 in the aluminum smelter example. The horizontal dashed line denotes the control limit.*

of the shift (i.e., the 100th time point) is used as an approximation to the "steady-state" start, after which the distribution of the charting statistic C_n approaches a "steady-state distribution" that does not depend on n (cf., Subsection 4.2.2 for a related discussion). Then, the OC ARL values of the MLL-CUSUM chart are shown in Figure 9.11(a) by the solid line.

   As pointed out in Section 9.1, the conventional multivariate control charts based on the normality assumption would be inappropriate to use in cases when the normality assumption is violated. To further demonstrate this, the MCUSUM chart (7.28)–(7.29) is used here in two different ways. First, the chart is used in the conventional way that the IC process distribution is assumed to be known $N(\mathbf{0}, I_{3\times3})$. Second, the chart is used under the assumption that the IC process distribution is $N(\mu, I_{3\times3})$, but μ should be estimated from an IC dataset of size 100. The two different versions of the MCUSUM chart are denoted as MCUSUM1 and MCUSUM2, respectively. Their computed OC ARL values are shown in Figure 9.11(a) by the dashed and dotted lines. From Figure 9.11(a), it can be seen that among the three charts MLL-CUSUM, MCUSUM1, and MCUSUM2, only the MLL-CUSUM chart has its actual ARL_0 value equal to the assumed ARL_0 value of 200. The actual IC ARL values of the other two charts are far from the assumed ARL_0 value, which is consistent with the results shown in Figures 9.1 and 9.2. Figure 9.11(a) also shows that the OC ARL values of MLL-CUSUM are small, compared to charts MCUSUM1 and MCUSUM2, when the shift is reasonably large. Considering the fact that the IC ARL of MLL-CUSUM is much larger than the IC ARL values of MCUSUM1 and MCUSUM2, this is an *endorsement of the former chart. To further investigate this issue, let us consider the larger shift* $(a, 0.4, 0.4)$, *where a changes its value between 0 and 0.4 with a step of 0.04. The corresponding OC ARL values of the three charts are shown in Figure 9.11(b). It can be seen that MLL-CUSUM is consistently better than MCUSUM1 and MCUSUM2 in such cases.*

The corresponding results of the MEWMA chart (7.39)–(7.40) in cases when a single weighting parameter λ *is used are shown in Figure 9.11(c)–(d) by the dashed and dotted curves when* $\lambda = 0.05$ *and* $\lambda = 0.2$, *respectively. In the plots, the two different versions of the MEWMA chart are denoted as MEWMA1 and MEWMA2, respectively. For convenience of comparison, the OC ARL values of MLL-CUSUM are shown in these plots again by the solid curves. From the plots, it can be seen that: (i) when* λ *is smaller (i.e.,* $\lambda = 0.05$), *the MEWMA chart is more robust to the normality assumption, as observed in Figure 9.2, because its actual* ARL_0 *value is closer to the assumed value 200, (ii) in such cases, its ability to detect possible shifts is also weaker, compared to the MEWMA chart when* λ *is larger, and (iii) MLL-CUSUM performs better when the process is IC and when the process becomes OC with a quite large shift (e.g., shifts* $(a, 0, 0)$ *when* $a > 0.34$ *in plot (c) and shifts* $(a, 0.4, 0.4)$ *for all a values in plot (d)).*

Next, we compare MLL-CUSUM with the MA-CUSUM chart (9.29)–(9.30), which does not require the normality assumption. We still consider the shift $(a, 0, 0)$, *with a changing its value between 0 and 0.4 with a step of 0.04. Since this shift is upward, the MA-CUSUM chart based on the third antirank is preferred here, among all MA-CUSUM charts based on a single antirank. To use this chart, the IC distribution of the third antirank should be specified. To this end, there are two possibilities. One is that the IC process distribution is assumed known, and the IC distribution of the third antirank can be determined accordingly. In practice, the IC process distribution is often unknown. In such cases, the IC distribution of the third antirank needs to be estimated from an IC dataset. In the first scenario, the control limit value* h_1 *of the chart is searched and found to be 4.300 such that its* ARL_0 *value equals 200 when* $k = 1$. *Results under this condition are labeled by MA-CUSUM1. If the IC distribution is estimated from the same IC dataset used above in estimating the IC distribution of* $\mathbf{Y}$, *then the control limit value* h_1 *is searched and found to be 4.430, to reach the* ARL_0 *value 200 when* $k = 1$. *Results in this case are labeled by MA-CUSUM2. In these two cases, the OC ARL values of the MA-CUSUM chart are shown in Figure 9.11(e) by the dashed and dotted curves, respectively. For convenience in comparison, the OC ARL values of MLL-CUSUM are also presented in this plot by the solid curve. It can be seen that the actual* ARL_0 *value of MA-CUSUM2 is also quite different from the assumed* ARL_0 *value. But the difference is not so large, compared to charts MCUSUM1, MCUSUM2, and MEWMA2. The chart MA-CUSUM1 is indeed appropriate to use here, since its actual* ARL_0 *value is about the same as the assumed* ARL_0 *value. But, in most cases, the MLL-CUSUM chart outperforms the MA-CUSUM1 chart, and the difference is quite large when the shift gets large. The corresponding results when the shift is* $(a, 0.4, 0.4)$ *are shown in Figure 9.11(f), where a changes its value between 0 and 0.4 with a step of 0.04, as before. This shift is closer to the "equal-component" direction discussed in Subsection 9.2.2, in which all components of the shift are the same, when a is closer to 0.4. It is expected that performance of the MA-CUSUM chart would get worse when a is closer to 0.4, because the OC distribution of the third antirank is closer to its IC distribution in such cases, which is demonstrated in Figure 9.11(f). From the plot, it can*

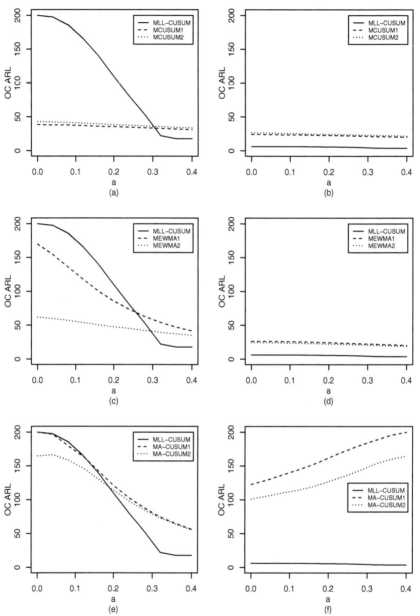

Figure 9.11 *(a) OC ARL values of charts MLL-CUSUM (solid curve), MCUSUM1 (dashed curve), and MCUSUM2 (dotted curve). The shift is assumed to be* $(a, 0, 0)$ *in the median vector of the process distribution starting at the 100th time point. (b) Corresponding results of plot (a) when the shift is* $(a, 0.4, 0.4)$. *(c) OC ARL values of charts MLL-CUSUM (solid curve), MEWMA1 (dashed curve), and MEWMA2 (dotted curve), when there is a shift* $(a, 0, 0)$ *in the median vector starting at the 100th time point. (d) Corresponding results of plot (c) when the shift is* $(a, 0.4, 0.4)$. *(e) OC ARL values of charts MLL-CUSUM (solid curve), MA-CUSUM1 (dashed curve), and MA-CUSUM2 (dotted curve), when there is a shift* $(a, 0, 0)$ *in the median vector starting at the 100th time point. (f) Corresponding results of plot (e) when the shift is* $(a, 0.4, 0.4)$.

be seen that the MLL-CUSUM chart performs much better than both MA-CUSUM1 and MA-CUSUM2 in this case.

From Example 9.8, it can be seen that the MLL-CUSUM chart is competitive, compared to its peers. However, it still has much room for improvement. For instance, in (9.37), each component of the original observation vector **X** is categorized into two categories. We believe that it would be more effective in detecting shifts in the process distribution, if each component is categorized into more than two categories, although the log-linear modeling would become more complex in that case. Also, the MLL-CUSUM chart can handle cases when all components of **X** are categorical or cases when some components of **X** are categorical. In such cases, we only need to categorize the numerical components before using the control chart.

9.4   Some Discussions

The multivariate nonparametric SPC problem is important because most SPC applications are multivariate, the multivariate process distributions are often non-normal, and we do not know much about proper ways to describe the distribution of a multivariate non-normal dataset. In this chapter, some existing multivariate nonparametric SPC charts have been described. They either use the ordering or ranking information in the observed data, or categorize the observed data first and then use the methods of categorical data analysis for process monitoring. The rank-based methods can only handle numerical data. The methods based on categorization can handle both numerical and categorical data.

The existing rank-based methods use either the longitudinal ranking information among the observed data at consecutive observation times or the cross-component ranking information among the different components of observation vectors. At least three types of longitudinal ranking have been discussed in the literature: componentwise longitudinal ranking, spatial longitudinal ranking, and longitudinal ranking by data depth. The componentwise longitudinal ranking treats individual components of the observation vectors separately when computing the longitudinal ranks, while the spatial longitudinal ranking and the longitudinal ranking by data depth consider all components jointly when computing the longitudinal ranks. So, the first type of longitudinal ranking depends only on the marginal distributions of individual components of the quality characteristic vector **X** while the second and third types depend on the joint distribution of **X**.

The MNS chart (9.2)–(9.3) and the MNSR chart (9.6)–(9.7) are both based on componentwise longitudinal ranking. Boone and Chakraborti (2012) have shown that the formulas (9.4) and (9.8) that are based on asymptotic distributions usually provide conservative control limits for them. It requires much future research to provide more appropriate control limits based on the finite-sample distributions of the related charting statistics or based on an IC dataset. Furthermore, these two charts are Shewhart charts. They are good at detecting large shifts, and may not be effective in detecting small and persistent shifts. As discussed in the paragraph containing (9.9), it is not difficult to modify these two charts to become multivariate nonparametric CUSUM, EWMA, or CPD charts. But, the design of these more effective charts for detecting small and persistent shifts still requires much research effort. It seems that spatial ranks defined in (9.11) are more sensitive to potential process distributional shifts, compared to the longitudinal ranking by data depth, because the former carries information about both the magnitude and the direction of a potential shift while the latter has information mainly about the magnitude. Theoretical and numerical comparison of these two types of longitudinal ranking and their applications in the multivariate nonparametric SPC problem, which is lacking in the current SPC literature, should be useful. Stoumbos and Jones (2000) studied the design and properties of a control chart based on data depth, and found that the chart was reliable and effective only in cases when a large reference sample was available.

The MA-CUSUM chart (9.29)–(9.30) is constructed based on the cross-component ranking of individual observation vectors. One major benefit of this type of control chart is that it is relatively simple to compute the cross-component ranks, compared to the computation of the longitudinal ranks, because the number of components p is usually a fixed small integer number (e.g., 5, or 15), but the number of observation times could be very large. In (9.29)–(9.30), antiranks instead of regular ranks are used because antiranks have the following "dimension reduction" property. If we know that a potential location shift can only occur in a single but unknown component of $\mathbf{X}$ and the shift is downward, then it suffices to use the first antirank for process monitoring. If such a shift is upward, then the last antirank is sufficient to use. As a comparison, all regular ranks should be used in such cases because we do not know which components in $\mathbf{X}$ have shifts. With the MA-CUSUM chart, there are still many questions to answer. For instance, in which cases should we use the first or last antirank? In which other cases should we use both the first and the last antiranks? Will the chart be more efficient in certain cases if more than two antiranks are used? Much future research is needed to provide practical guidelines that address such questions.

The multivariate nonparametric SPC charts based on log-linear modeling are promising because the log-linear modeling approach can describe the possible association among different components of the categorized data properly. In my opinion, the essential difficulty with the multivariate nonparametric SPC problem is that we do not know how to describe and estimate the possible association among different components of the original observed data when they do not follow a normal distribution. Although data categorization would result in information loss, the amount of lost information can be reduced if more categories are used in data categorization. With more categories, the related contingency table of the categorized observed data would be large and the model selection and model estimation in the log-linear modeling would become more complex. To handle this issue, the LASSO-based SPC method described in Subsection 7.6.2 should be helpful.

9.5   Exercises

9.1 Figure 9.2 shows that the actual IC ARL values of the multivariate CUSUM chart (7.28)–(7.29) and the multivariate EWMA chart (7.39)–(7.40) could be substantially different from the assumed IC ARL value of 200 in cases when there are three quality characteristic variables involved in an SPC application, the variables are independent of each other, and each has the standardized version with mean 0 and variance 1 of the χ_r^2 distribution where $r = 1, 5, 10, 20$, and 50. Make a similar plot in the same setup, except that each of the three quality characteristic variables has the standardized version with mean 0 and variance 1 of the t_r distribution where $r = 3, 5, 10, 20$, and 50.

9.2 Reproduce the results in Example 9.1.

9.3 Assume that the 20 observation vectors presented in Table 9.6 are obtained from a 2-dimensional production process with the IC distribution $N_2(\mu_0, \Sigma)$, where $\mu_0 = (0.5, 0.8)'$ and

$$\Sigma = \begin{pmatrix} 1.0 & 0.5 \\ 0.5 & 1.0 \end{pmatrix}.$$

The first 10 observation vectors are IC and the remaining 10 observation vectors have a mean shift from μ_0 to $\mu_1 = (1.0, 1.0)'$.

Table 9.6 *This table presents the first 20 observation vectors obtained from a 2-dimensional production process.*

i	X_{i1}	X_{i2}	i	X_{i1}	X_{i2}
1	0.110	0.320	11	0.575	0.666
2	0.662	0.866	12	2.540	0.783
3	0.331	0.832	13	1.158	1.296
4	1.638	1.198	14	1.614	1.725
5	0.663	0.840	15	−0.050	0.640
6	0.761	1.091	16	0.509	0.731
7	−0.198	0.491	17	0.468	0.285
8	1.374	1.163	18	1.409	0.991
9	−0.672	0.542	19	0.530	−0.535
10	1.343	−0.667	20	1.699	0.729

(i) Compute the spatial signs (cf., (9.10)) of all 20 observation vectors, and present them in a plot.

(ii) Compute the spatial ranks (cf., (9.11)) of all 20 observation vectors, and present them in a plot.

(iii) Based on the two plots made in parts (i) and (ii), do you think spatial signs and spatial ranks can be used for detecting process mean shifts? Why?

9.4 For the first 10 observation vectors in Table 9.6, compute their values of the Mahalanobis depth and the empirical Mahalanobis depth. When computing the empirical Mahalanobis depth, treat the first 10 observation vectors in Table 9.6 as a sample of size $M = 10$ from the IC process distribution $N_2(\mu_0, \Sigma)$.

9.5 For the data presented in Table 9.6, construct the r chart (9.18) and the S chart (9.19)–(9.20), using $\alpha = 0.01$ and the first 10 observation vectors in Table 9.6 as a sample of size $M = 10$ from the IC process distribution $N_2(\mu_0, \Sigma)$. Compare the performance of the two charts.

9.6 Reproduce the results in Example 9.3.

9.7 The following 10 vectors are observations from a 5-dimensional production process:

$$(-0.502, -0.582, -0.202, -0.914, -0.814),$$
$$(0.132, 0.715, 0.740, 2.310, -0.438),$$
$$(-0.079, -0.825, 0.123, -0.438, -0.720),$$
$$(0.887, -0.360, -0.029, 0.764, 0.231),$$
$$(0.117, 0.090, -0.389, 0.262, -1.158),$$
$$(0.319, 0.096, 0.511, 0.773, 0.247),$$
$$(-2.091, -0.690, 1.065, -1.777, 0.637),$$
$$(-0.243, -0.222, 0.970, 0.623, 2.319),$$
$$(-2.138, 0.183, -0.102, -0.522, 1.044),$$
$$(-2.111, 0.417, 1.403, 1.322, -0.879).$$

The first six vectors are actually generated from a distribution with mean $\mathbf{0}$ and
covariance matrix $I_{5 \times 5}$, and the remaining four vectors are generated from the
same distribution but with a different mean of $(-2, 0, 0, 0, 1)'$. Therefore, there
is a mean shift from $\mu_0 = \mathbf{0}$ to $\mu_1 = (-2, 0, 0, 0, 1)'$ at the 7th time point.

(i)   Compute the five cross-component ranks for each observation vector.

(ii) Compute the five cross-component antiranks for each observation vector.

(iii) Comment on the useful information in the five ranks and the five antiranks
about the mean shift.

9.8 Apply the two MA-CUSUM charts based on the first antirank and on the first-
and-last antiranks to the data considered in the previous exercise for detecting
process mean shifts. In both charts, choose $k_1 = 1$ and $ARL_0 = 200$. By Example 9.4, the control limits of the two charts are 11.85 and 34.89, respectively.
Also, in this exercise, it is assumed that the five quality characteristic variables
are exchangeable when the process is IC in the sense that the IC process distribution would not be changed if the positions of the five variables are switched.
Summarize your results.

9.9 The following 10 vectors are observations from a 5-dimensional production process:

$$(0.020, -1.316, -0.259, 1.008, -1.378),$$
$$(-0.172, -0.340, 0.922, 0.451, -0.349),$$
$$(-1.510, -1.791, 0.816, -0.657, -0.757),$$
$$(-0.498, -0.213, 0.074, -1.818, -0.725),$$
$$(0.318, 1.189, -1.031, -0.728, -0.110),$$
$$(0.341, 0.661, -0.171, -1.852, -0.222),$$
$$(3.430, -0.603, -1.199, 2.895, 2.000),$$
$$(3.171, 1.541, 0.537, 2.235, -0.266),$$
$$(1.387, 0.506, 0.365, 1.563, 1.291),$$
$$(2.258, 1.466, 1.544, 0.230, -1.289).$$

The first six vectors are actually generated from a distribution with mean $\mathbf{0}$ and
covariance matrix $I_{5 \times 5}$, and the remaining four vectors are generated from the

same distribution but with a different mean of $(1,1,1,1,1)'$. Therefore, there is an equal-component mean shift from $\mu_0 = \mathbf{0}$ to $\mu_1 = (1,1,1,1,1)'$ at the 7th time point.

(i) Compute the first antirank value A_{n1} for each observation vector.

(ii) Compute the modified first antirank value $\tilde{A}_{n,1}$ (cf., its definition in the paragraph immediately before Example 9.6) for each observation vector.

(iii) Comment on the useful information in A_{n1} and $\tilde{A}_{n,1}$ about the mean shift.

9.10 Assume that Y_1, Y_2, and Y_3 are the three categorical variables involved in a research problem. They have s_1, s_2, and s_3 categories, respectively. Write out the log-linear models with the following notations:

$$(Y_1Y_2, Y_1Y_3), \quad (Y_1Y_2, Y_3), \quad (Y_1, Y_2, Y_3).$$

Describe the association among the three variables in each model.

9.11 In the SPC problem considered in Exercise 9.9, assume that the IC process median vector is $\tilde{\mu} = (0,0,0,0,0)'$. For the 10 observation vectors presented in that exercise, compute their categorized observation vectors using (9.37).

9.12 Reproduce the results in Example 9.7.

9.13 Table 9.7 presents the first 40 observation vectors obtained from a 2-dimensional production process. Assume that the first 30 observation vectors are IC, and the IC process median vector is known to be $\tilde{\mu} = (0,0)'$. Categorize these 30 observation vectors using (9.37), and estimate the IC distribution of the categorized data using the log-linear modeling discussed in Subsection 9.3.1.

Table 9.7 *This table presents the first 40 observation vectors obtained from a 2-dimensional production process.*

i	X_{i1}	X_{i2}	i	X_{i1}	X_{i2}
1	−0.502	−0.091	21	−0.438	−0.447
2	0.132	1.757	22	0.764	−1.739
3	−0.079	−0.138	23	0.262	0.179
4	0.887	−0.111	24	0.773	1.897
5	0.117	−0.690	25	−0.814	−2.272
6	0.319	−0.222	26	−0.438	0.980
7	−0.582	0.183	27	−0.720	−1.399
8	0.715	0.417	28	0.231	1.825
9	−0.825	1.065	29	−1.158	1.381
10	−0.360	0.970	30	0.247	−0.839
11	0.090	−0.102	31	0.738	1.449
12	0.096	1.403	32	0.931	−0.064
13	−0.202	−1.777	33	0.621	−0.162
14	0.740	0.623	34	3.582	2.649
15	0.123	−0.522	35	1.130	−1.062
16	−0.029	1.322	36	0.287	1.013
17	−0.389	−0.363	37	1.638	−0.088
18	0.511	1.319	38	1.202	1.271
19	−0.914	0.044	39	0.930	2.008
20	2.310	−1.879	40	0.908	−1.074

9.14 For the data presented in Table 9.7, design the MLL-CUSUM chart (9.38)–(9.39), using $k = 1$, $ARL_0 = 200$, and the estimated IC distribution of the categorized data obtained in the previous exercise. Apply the resulting chart to the last 10 observation vectors in the table for detecting possible process mean shifts. Summarize your results.

Chapter 10

Profile Monitoring

10.1 Introduction

In the SPC problems discussed in the previous chapters, the quality of a product is characterized by one or more variables. Each sampled product has a single observation of the quality characteristic variable, and we monitor the related production process using all available observations of the products sampled over time. In some applications, however, the quality of a product is characterized by the functional relationship between a response variable and one or more explanatory variables. In these applications, for each sampled product, one observes a set of data points of these variables that can be represented by a curve/surface (or a profile), and the major purpose of SPC is to check the stability of this relationship over time based on observed profile data. This SPC problem is called *profile monitoring* in the literature, which is the main topic of this chapter.

The profile monitoring problem has broad applications. For instance, Kang and Albin (2000) discussed two examples of profile monitoring. The first example was about an artificial sweetener, called aspartame. In this example, they were concerned about the amount of aspartame that could be dissolved per liter of water at different temperatures. In order to evaluate the quality of manufactured aspartame, for each sampled product, we needed to collect observations of the amount of aspartame dissolved per liter of water at a given set of temperatures, which formed a profile. The second example was about the relationship between the measured pressure and the level of gas flow controlled by a mass flow controller that was used for etching away photoresist on chips manufactured by a semiconductor production process. Many other interesting applications of profile monitoring were discussed by papers including Jin and Shi (1999), Qiu et al. (2010), Walker and Wright (2002), and Zhou and Jin (2005).

Due to its broad applications, the profile monitoring problem has received much attention from statisticians recently. Early research on this topic assumes that the mean profile is linear when the process is IC, and various profile monitoring charts have been proposed for detecting possible shifts in certain parameters of the linear IC profile. See, for instance, Jensen et al. (2008), Kang and Albin (2000), Kim et al. (2003), Mahmoud et al. (2007), Mahmoud and Woodall (2004), Stover and Brill (1998), Wang and Tsung (2005), and Zou et al. (2006). Linear profile monitoring techniques have been generalized to various cases with parametric nonlinear profiles by some authors, including Ding et al. (2006), Jensen and Birch (2009), Williams et al. (2007a,b), and Zou et al. (2007a). In certain applications, it is difficult to justify the legitimacy of a specific parametric model for describing the IC profile data. Therefore, some nonparametric profile monitoring techniques have been proposed in the literature (cf., Qiu and Zou, 2010; Qiu et al., 2010; Wei et al., 2012; Zou et al., 2009b, 2008). In this chapter, we will describe some representative parametric and nonparametric profile monitoring techniques. For some related discussions, see Noorossana et al. (2011), Woodall (2007), Woodall et al. (2004), and the references cited therein.

10.2 Parametric Profile Monitoring

In some applications, a parametric model is available to describe IC profile data, based on some scientific knowledge or our past experience about the related production process. In such cases, parametric profile monitoring techniques can be considered. In this section, we describe some parametric profile monitoring charts in two parts. Those based on linear regression modeling are described in Subsection 10.2.1, and those based on nonlinear regression modeling are described in Subsection 10.2.2.

10.2.1 Linear profile monitoring

In a profile monitoring problem, assume that we are concerned about the relationship between a response variable y and a predictor x. To monitor such a relationship, some products need to be randomly selected over time from the manufactured products. For the j-th selected product with $j = 1, 2, \ldots$, it is assumed that n observations of the pair (x, y), denoted as $\{(x_i, y_{ij}), i = 1, 2, \ldots, n\}$, are obtained and they follow the linear regression model

$$y_{ij} = a_j + b_j x_i + \varepsilon_{ij}, \qquad \text{for } i = 1, 2, \ldots, n, \ j = 1, 2, \ldots, \qquad (10.1)$$

where a_j and b_j are the intercept and slope of the true regression line $y = a_j + b_j x$, and $\{\varepsilon_{ij}, i = 1, 2, \ldots, n\}$ are i.i.d. random errors with the common distribution $N(0, \sigma_j^2)$. When the j-th product is IC, it is further assumed that $a_j = a_0$, $b_j = b_0$, and $\sigma_j^2 = \sigma_0^2$, where a_0, b_0, and σ_0^2 are the IC values of the intercept, slope, and error variance of the regression model (10.1). The resulting model is called the *IC linear regression model* in this chapter. In (10.1), it has been assumed that (i) the design points $\{x_i, i = 1, 2, \ldots, n\}$ are deterministic and unchanged from one profile to another, and (ii) the sample size n is the same for all profiles. Furthermore, it is conventionally assumed that between-profile observations are independent of each other (i.e., observations of different profiles are independent). Then, the *linear profile monitoring (LPM)* problem is mainly for monitoring the related production process to make sure that profiles of its products follow the IC linear regression model.

Let us first discuss phase II LPM, in which the IC parameters a_0, b_0, and σ_0^2 are assumed to be known (as a matter of fact, they need to be estimated from an IC dataset, as discussed in Subsection 4.5.1). In such cases, to monitor the j-th sampled product, it is natural to first estimate the model (10.1) based on the observations in the j-th observed profile, and then compare the estimated model with the IC linear regression model. This is also the basic idea of an approach proposed by Kang and Albin (2000). By the discussion in Subsection 2.7.2, the least squares (LS) estimators of b_j and a_j are

$$\widehat{b}_j = \frac{\sum_{i=1}^{n}(x_i-\bar{x})(y_{ij}-\bar{y}_j)}{\sum_{i=1}^{n}(x_i-\bar{x})^2}$$

$$\widehat{a}_j = \bar{y}_j - \widehat{b}_j\bar{x}, \tag{10.2}$$

where $\bar{x}$ and $\bar{y}_j$ are the sample means of $\{x_i, i=1,2,\ldots,n\}$ and $\{y_{ij}, i=1,2,\ldots,n\}$, respectively. The commonly used estimator of σ_j^2 is

$$\widehat{\sigma}_j^2 = \frac{1}{n-2}\sum_{i=1}^{n}\left[y_{ij}-\left(\widehat{a}_j+\widehat{b}_jx_i\right)\right]^2. \tag{10.3}$$

When the j-th product is IC and the four conventional model assumptions used in linear regression analysis (cf., the related discussion in the paragraph containing (2.22) in Subsection 2.7.2) are valid, these estimators have the properties that

$$\begin{pmatrix} \widehat{a}_j \\ \widehat{b}_j \end{pmatrix} \sim N\left(\begin{pmatrix} a_0 \\ b_0 \end{pmatrix}, \sigma_0^2\begin{pmatrix} \frac{1}{n}+\frac{\bar{x}^2}{sxx}, & -\frac{\bar{x}}{sxx} \\ -\frac{\bar{x}}{sxx}, & \frac{1}{sxx} \end{pmatrix}\right), \tag{10.4}$$

and

$$\frac{(n-2)\widehat{\sigma}_j^2}{\sigma_0^2} \sim \chi_{n-2}^2, \tag{10.5}$$

where $sxx = \sum_{i=1}^{n}(x_i-\bar{x})^2$. Therefore, to monitor the j-th product, we can use the Shewhart charting statistic

$$T_j^2 = (\widehat{\mathbf{d}}_j-\mathbf{d}_0)'\Sigma_{\widehat{\mathbf{d}}_j}^{-1}(\widehat{\mathbf{d}}_j-\mathbf{d}_0), \tag{10.6}$$

where

$$\widehat{\mathbf{d}}_j = \begin{pmatrix} \widehat{a}_j \\ \widehat{b}_j \end{pmatrix}, \quad \mathbf{d}_0 = \begin{pmatrix} a_0 \\ b_0 \end{pmatrix}, \quad \Sigma_{\widehat{\mathbf{d}}_j} = \sigma_0^2\begin{pmatrix} \frac{1}{n}+\frac{\bar{x}^2}{sxx}, & -\frac{\bar{x}}{sxx} \\ -\frac{\bar{x}}{sxx}, & \frac{1}{sxx} \end{pmatrix}.$$

By (10.4), the IC distribution of T_j^2 is χ_2^2. So, a reasonable decision rule of the chart is that it gives a signal when

$$T_j^2 > \chi_{1-\alpha,2}^2, \tag{10.7}$$

where $\alpha \in (0,1)$ is a given significance level, and $\chi_{1-\alpha,2}^2$ is the $(1-\alpha)$-th quantile of the χ_2^2 distribution. Obviously, the IC ARL of the Shewhart chart (10.6)–(10.7) is $ARL_0 = 1/\alpha$ (cf., (3.8) in Subsection 3.2.1).

Example 10.1 *For a production process, assume that its profiles are known to be linear and the IC linear regression model is*

$$y_{ij} = 2+x_i+\varepsilon_{ij}, \quad for\ i=1,2,\ldots,n,\ j=1,2,\ldots,$$

where $\{x_i, i = 1, 2, \ldots, n\}$ are design points in the design interval $[0, 1]$, and $\{\varepsilon_{ij}, i = 1, 2, \ldots, n, j = 1, 2, \ldots\}$ are i.i.d. random errors with the common distribution $N(0, 0.5^2)$. The first 20 observed profiles of the process obtained during phase II SPC are presented in Table 10.1, where the design points are chosen to be 0.2, 0.4, 0.6, 0.8, and 1.0. For each observed profile, the model (10.1) is fitted, and the estimates $\widehat{a}_j, \widehat{b}_j,$ and $\widehat{\sigma}_j^2$ are shown in Figure 10.1(a)–(c). From the plots, it seems that both the intercept and slope shift at the 11th time point, and the error variance seems quite stable. Then, we apply the Shewhart chart (10.6)–(10.7) to the 20 observed profiles, in which α is chosen to be 0.005. The chart is shown in Figure 10.1(d). From the chart, it can be seen that the first signal is given at the 11th time point. Therefore, the production process should be stopped at that time point to investigate the possible root causes of the detected profile shift.

Table 10.1 *The first 20 observed profiles of a production process obtained during phase II SPC. The design points of all profiles are 0.2, 0.4, 0.6, 0.8, and 1.0.*

j	y_{1j}	y_{2j}	y_{3j}	y_{4j}	y_{5j}
1	2.209	2.308	1.914	2.500	3.147
2	2.395	1.796	2.418	1.987	2.872
3	2.751	2.778	2.481	3.294	3.371
4	2.245	1.923	2.502	3.263	3.241
5	1.902	1.307	2.263	1.740	2.367
6	2.013	2.056	2.164	2.749	2.873
7	1.273	2.361	3.084	2.892	2.310
8	1.482	2.581	1.720	2.638	2.674
9	2.743	2.019	2.186	3.217	2.516
10	2.186	2.516	2.449	2.461	3.328
11	2.900	3.033	3.984	4.469	3.753
12	3.493	2.749	3.566	3.077	3.645
13	2.481	2.972	2.885	3.570	4.033
14	3.708	3.568	3.059	3.681	2.877
15	3.290	2.485	2.776	3.291	2.755
16	3.686	2.460	3.085	2.874	4.261
17	3.396	3.089	3.656	3.758	3.720
18	3.179	3.530	4.410	2.808	3.463
19	2.892	3.104	3.335	3.978	3.797
20	2.390	2.397	3.746	3.474	4.114

Besides the Shewhart chart (10.6)–(10.7), Kang and Albin (2000) also proposed an R chart (cf., (3.7)) for detecting shift in error variance. As discussed in previous chapters, we know that Shewhart charts are effective in detecting large and isolated shifts, but ineffective in detecting small and persistent shifts. To overcome this limitation of the Shewhart charts, Kim et al. (2003) proposed three univariate EWMA charts for detecting shifts in the intercept, slope, and error variance, respectively, which are described below. First, let us rewrite model (10.1) as

$$y_{ij} = a_j^* + b_j(x_i - \bar{x}) + \varepsilon_{ij}, \qquad \text{for } i = 1, 2, \ldots, n, j = 1, 2, \ldots, \qquad (10.8)$$

where $a_j^* = a_j + b_j \bar{x}$. Obviously, the new intercept a_j^* is just the mean value of the

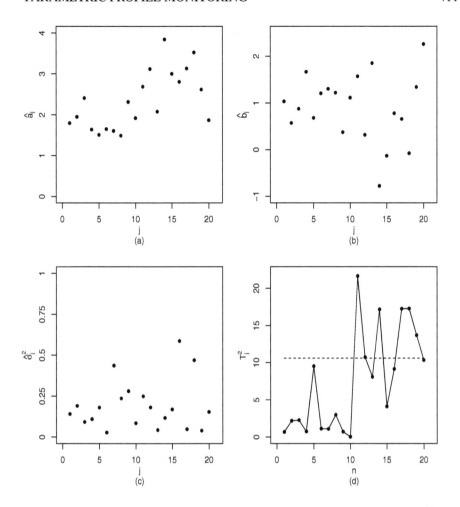

Figure 10.1 *(a) Estimated intercepts $\{\widehat{a}_j, j = 1, 2, \ldots, 20\}$. (b) Estimated slopes $\{\widehat{b}_j, j = 1, 2, \ldots, 20\}$. (c) Estimated error variances $\{\widehat{\sigma}_j^2, j = 1, 2, \ldots, 20\}$. (d) Shewhart chart (10.6)–(10.7) with $\alpha = 0.005$. The dashed horizontal line in (d) denotes the control limit $\chi_{1-\alpha,2}^2$.*

response y_{ij} of the j-th profile when $x_i = \bar{x}$. In the literature, model (10.8) is often called the centered linear regression model because the new design points $\{x_i^* = x_i - \bar{x}, i = 1, 2, \ldots, n\}$ are centered at 0. For model (10.8), it is easy to check that the LS estimator of a_j^* is $\widehat{a}_j^* = \bar{y}_j$, where $\bar{y}_j$ is the sample mean of $\{y_{ij}, i = 1, 2, \ldots, n\}$. In such cases, we can also check that the estimators $\widehat{a}_j^*, \widehat{b}_j$, and $\widehat{\sigma}_j^2$ are independent of each other. Therefore, to monitor the whole profile without losing any information, we can simply monitor these three quantities separately.

When the process is IC, it is easy to check that, for each $j \geq 1$,

$$\widehat{a}_j^* \sim N(a_0 + b_0 \bar{x}, \sigma_0^2/n).$$

To monitor the sequence $\{\widehat{a}_j^*, j = 1, 2, \ldots\}$, Kim et al. (2003) proposed using the following EWMA charting statistic:

$$E_{j,I} = \lambda_I \widehat{a}_j^* + (1 - \lambda_I)E_{j-1,I}, \tag{10.9}$$

where $E_{0,I} = a_0 + b_0 \bar{x}$, and $\lambda_I \in (0,1]$ is a weighting parameter. The upper control limit U, the center line C, and the lower control limit L of the corresponding EWMA chart are

$$
\begin{aligned}
U &= a_0 + b_0 \bar{x} + \rho_I \sigma_0 \sqrt{\frac{\lambda_I}{(2 - \lambda_I)n}} \\
C &= a_0 + b_0 \bar{x} \\
L &= a_0 + b_0 \bar{x} - \rho_I \sigma_0 \sqrt{\frac{\lambda_I}{(2 - \lambda_I)n}},
\end{aligned}
\tag{10.10}
$$

where $\rho_I > 0$ is a parameter chosen to reach a given ARL_0 value.

For the estimator $\widehat{b}_j$, its IC distribution can be derived from (10.4) to be

$$\widehat{b}_j \sim N(b_0, \sigma_0^2/sxx).$$

To monitor the sequence $\{\widehat{b}_j, j = 1, 2, \ldots\}$, we can use the following EWMA charting statistic

$$E_{j,S} = \lambda_S \widehat{b}_j + (1 - \lambda_S)E_{j-1,S}, \tag{10.11}$$

where $E_{0,S} = b_0$, and $\lambda_S \in (0,1]$ is a weighting parameter. The control limits and the center line of the corresponding EWMA chart are

$$
\begin{aligned}
U &= b_0 + \rho_S \sigma_0 \sqrt{\frac{\lambda_S}{(2 - \lambda_S)sxx}} \\
C &= b_0 \\
L &= b_0 - \rho_S \sigma_0 \sqrt{\frac{\lambda_S}{(2 - \lambda_S)sxx}},
\end{aligned}
\tag{10.12}
$$

where $\rho_S > 0$ is a parameter chosen to reach a given ARL_0 value.

The IC distribution of $\widehat{\sigma}_j^2$ is given in (10.5), which is a chi-square distribution that is skewed to the right. To detect an upward shift in the error variance, we can use the upward EWMA chart originally proposed by Crowder and Hamilton (1992) with the charting statistic

$$E_{j,V}^+ = \max(0, \lambda_V \log(\widehat{\sigma}_j^2/\sigma_0^2) + (1 - \lambda_V)E_{j-1,V}), \qquad \text{for } j \geq 1, \tag{10.13}$$

where $E_{0,V}^+ = 0$ and $\lambda_V \in (0,1]$ is a weighting parameter. The main purpose to use the log transformation in (10.13) is to make the distribution of $\widehat{\sigma}_j^2/\sigma_0^2$ more symmetric. See a related discussion at the end of Subsection 5.3.1. The chart gives a signal when

$$E_{j,V}^+ > U = \rho_V \sigma_W^* \sqrt{\frac{\lambda_V}{2 - \lambda_V}}, \tag{10.14}$$

where $\rho_V > 0$ is a parameter chosen to reach a given ARL_0 value, and

$$\sigma_W^* = \sqrt{\frac{2}{n-2} + \frac{2}{(n-2)^2} + \frac{4}{3(n-2)^3} - \frac{16}{15(n-2)^5}}.$$

An EWMA chart for detecting downward variance shift can be constructed in a similar way, although we are mainly concerned about upward variance shifts in most applications, as discussed in Subsection 4.3.1.

Example 10.1 (continued) *For the phase II LPM problem discussed in Example 10.1, let us consider using the three EWMA charts discussed above for monitoring the intercept, slope, and error variance. In the three charts, λ_I, λ_S, and λ_V are all chosen to be 0.2. From Kim et al. (2003), if we choose $\rho_I = 3.0156$, $\rho_S = 3.0109$, and $\rho_V = 1.3723$ in the three charts, then their ARL_0 values are all around 584.5, and the joint monitoring scheme would have an ARL_0 value of about 200. The three charts are shown in Figure 10.2, from which it can be seen that the first signal of the joint monitoring scheme is given at the 13th time point, and it seems that the shift is in the intercept of the regression line. Compared to the Shewhart chart (10.6)–(10.7) shown in Figure 10.1(d), it can be seen that the joint monitoring scheme presented in Figure 10.2 has two benefits. (i) After giving a signal, it can tell us which parameters among the intercept, slope, and error variance of the IC linear regression model have shifted. (ii) The signals in Figure 10.2(a) seem more convincing, compared to the signals in Figure 10.1(d).*

Next, we discuss the phase I LPM problem, in which the IC parameters a_0, b_0, and σ_0^2 are all unknown and they need to be estimated from an observed dataset. Assume that we have m observed profiles with the observations $\{y_{ij}, i = 1, 2, \ldots, n, j = 1, 2, \ldots, m\}$. For the j-th profile with $j = 1, 2, \ldots, m$, its observations are assumed to follow the model (10.1). When the j-th profile is IC, it is further assumed that $a_j = a_0$, $b_j = b_0$, and $\sigma_j^2 = \sigma_0^2$. For phase I LPM, Kang and Albin (2000) proposed the following estimators of a_0, b_0, and σ_0^2:

$$\widehat{a}_0 = \frac{1}{m} \sum_{j=1}^m \widehat{a}_j, \qquad \widehat{b}_0 = \frac{1}{m} \sum_{j=1}^m \widehat{b}_j, \qquad \widehat{\sigma}_0^2 = \frac{1}{m} \sum_{j=1}^m \widehat{\sigma}_j^2, \tag{10.15}$$

where $\widehat{a}_j, \widehat{b}_j$, and $\widehat{\sigma}_j^2$ are defined in (10.2) and (10.3). Instead of using the model (10.1), Mahmoud and Woodall (2004) suggested using the centered model (10.8), due to the property discussed above that the estimators $\widehat{a}_j^*, \widehat{b}_j$, and $\widehat{\sigma}_j^2$ are independent

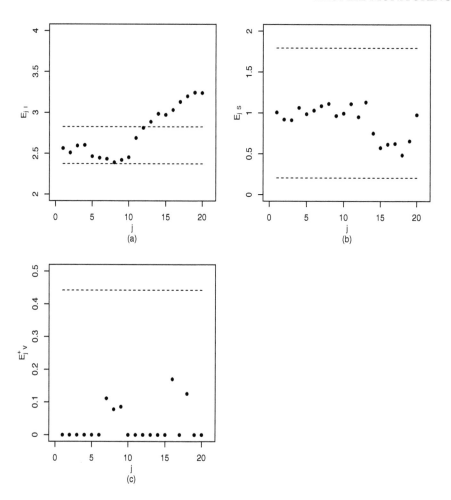

Figure 10.2 *(a) EWMA chart (10.9)–(10.10) for monitoring the estimated intercepts $\{\widehat{a}_j, j = 1, 2, \ldots, 20\}$ shown in Figure 10.1(a). (b) EWMA chart (10.11)–(10.12) for monitoring the estimated slopes $\{\widehat{b}_j, j = 1, 2, \ldots, 20\}$ shown in Figure 10.1(b). (c) EWMA chart (10.13)–(10.14) for monitoring the estimated error variances $\{\widehat{\sigma}_j^2, j = 1, 2, \ldots, 20\}$ shown in Figure 10.1(c). In these charts, $\lambda_I, \lambda_S,$ and λ_V are all chosen to be 0.2, $\rho_I, \rho_S,$ and ρ_V are chosen to be 3.0156, 3.0109, and 1.3723, respectively, so that the ARL_0 values of the three charts are all around 584.5 and the joint monitoring scheme has the ARL_0 value of about 200. The dashed horizontal lines in the three plots denote the upper or lower control limits.*

of each other, for each j. In such cases, the IC intercept $a_0^* = a_0 + b_0 \bar{x}$ of the centered model can be estimated by

$$\widehat{a}_0^* = \widehat{a}_0 + \widehat{b}_0 \bar{x} = \frac{1}{m} \sum_{j=1}^{m} \bar{y}_j. \tag{10.16}$$

Because of the independency among $\widehat{a}_j^*, \widehat{b}_j$, and $\widehat{\sigma}_j^2$, we can monitor the three sequences $\{\widehat{a}_j^*, j = 1, 2, \ldots, m\}$, $\{\widehat{b}_j, j = 1, 2, \ldots, m\}$, and $\{\widehat{\sigma}_j^2, j = 1, 2, \ldots, m\}$ separately.

To monitor the estimated intercepts $\{\widehat{a}_j^*, j = 1, 2, \ldots, m\}$, it is easy to check that when the process is IC, we have

$$\frac{\widehat{a}_j^* - \widehat{a}_0^*}{\sqrt{\frac{m-1}{mn} \widehat{\sigma}_0^2}} \sim t_{(n-2)m}.$$

Therefore, for a given false alarm rate $\alpha \in (0, 1)$, it is natural to define the control limits of the Shewhart chart for monitoring $\{\widehat{a}_j^*, j = 1, 2, \ldots, m\}$ to be

$$
\begin{aligned}
U &= \widehat{a}_0^* + t_{1-\alpha/2,(n-2)m} \widehat{\sigma}_0 \sqrt{\frac{m-1}{mn}} \\
C &= \widehat{a}_0^* \\
L &= \widehat{a}_0^* - t_{1-\alpha/2,(n-2)m} \widehat{\sigma}_0 \sqrt{\frac{m-1}{mn}}
\end{aligned}
\tag{10.17}
$$

where $t_{1-\alpha/2,(n-2)m}$ is the $(1 - \alpha/2)$-th quantile of the $t_{(n-2)m}$ distribution. A Shewhart chart for monitoring the estimated slopes $\{\widehat{b}_j, j = 1, 2, \ldots, m\}$ can be constructed in a similar way, with the control limits

$$
\begin{aligned}
U &= \widehat{b}_0 + t_{1-\alpha/2,(n-2)m} \widehat{\sigma}_0 \sqrt{\left(\frac{m-1}{m}\right) \Big/ sxx} \\
C &= \widehat{b}_0 \\
L &= \widehat{b}_0 - t_{1-\alpha/2,(n-2)m} \widehat{\sigma}_0 \sqrt{\left(\frac{m-1}{m}\right) \Big/ sxx}.
\end{aligned}
\tag{10.18}
$$

To monitor the estimated error variances $\{\widehat{\sigma}_j^2, j = 1, 2, \ldots, m\}$, Mahmoud and Woodall (2004) suggested using the charting statistic

$$F_j = \frac{\widehat{\sigma}_j^2}{\frac{1}{m-1} \sum_{l=1, l \neq j}^m \widehat{\sigma}_l^2},
\tag{10.19}$$

which has the IC distribution of $F_{(n-2),(m-1)(n-2)}$. The corresponding Shewhart chart has the control limits

$$
\begin{aligned}
U &= F_{(1-\alpha/2),(n-2),(m-1)(n-2)} \\
C &= 1 \\
L &= F_{\alpha/2,(n-2),(m-1)(n-2)},
\end{aligned}
\tag{10.20}
$$

where $F_{(1-\alpha/2),(n-2),(m-1)(n-2)}$ and $F_{\alpha/2,(n-2),(m-1)(n-2)}$ are the $(1 - \alpha/2)$-th and $\alpha/2$-th quantiles of the $F_{(n-2),(m-1)(n-2)}$ distribution.

Example 10.2 *In this example, we consider the calibration problem that was origi-*
nally discussed by Mestek et al. (1994) and was subsequently analyzed by Mahmoud
and Woodall (2004). In this application, we are mainly concerned about the stability
of the calibration curve in the photometric determination of Fe^{3+} with sulfosalicylic
acid (denoted as x). In each profile, 5 volumes of 0, 1, 2, 3, and 4 mL of 50 $\mu g/mL$
Fe^{3+} solution are diluted with water to 25 mL, denoted by 5 levels of x: 0, 1, 2, 3,
and 4. Then, for each volume, 2.5 mL of a 20% solution of sulfosalicylic acid and
1.5 mL of a concentrated solution of ammonia are added to the diluted solution, and
the absorbance of the solution (denoted as y) at 420 nm is measured on a Spekol
11 using 1-cm cells. Each volume is repeated twice, resulting in 10 observations in
each profile. Table 10.2 presents 22 observed profiles. We then apply the Shewhart
charts (10.17), (10.18), and (10.19)–(10.20) to this dataset to monitor the estimated
intercepts $\{\hat{a}_j^*, j = 1, 2, \ldots, m\}$, the estimated slopes $\{\hat{b}_j, j = 1, 2, \ldots, m\}$, and the
estimated error variances $\{\hat{\sigma}_j^2, j = 1, 2, \ldots, m\}$. In all three charts, α is chosen to
be 0.005, making the ARL_0 value of each individual chart 200. The three charts are
shown in Figure 10.3(a),(b),(d), and the estimated error variances are shown in Fig-
ure 10.3(c). From the plots, it can be seen that the estimated slopes and the estimated
error variances are quite stable, but the estimated intercepts seem very unstable.

Table 10.2 *This table presents 22 observed profiles in the calibration problem discussed by*
Mestek et al. (1994) and Mahmoud and Woodall (2004). In each profile, five levels of x are
considered, and two replicated observations of y are collected at each level of x.

j	y_{1j}, y_{2j}	y_{3j}, y_{4j}	y_{5j}, y_{6j}	y_{7j}, y_{8j}	y_{9j}, y_{10j}
1	1, 3	104, 104	206, 206	307, 308	409, 412
2	4, 2	104, 103	206, 204	308, 307	412, 413
3	3, 2	105, 104	207, 207	311, 309	414, 411
4	4, 2	104, 104	206, 207	308, 312	411, 413
5	9, 8	92, 95	195, 197	296, 299	397, 400
6	3, 3	107, 105	209, 207	311, 308	412, 410
7	3, 2	104, 105	207, 208	311, 308	414, 410
8	2, 2	105, 104	208, 208	310, 309	412, 412
9	6, 7	95, 94	196, 197	297, 300	401, 401
10	2, 4	104, 105	206, 207	311, 310	413, 412
11	1, 2	103, 104	205, 206	309, 307	412, 411
12	7, 7	94, 96	198, 199	298, 301	404, 402
13	5, 7	105, 107	210, 208	313, 315	415, 415
14	3, 2	106, 104	208, 207	311, 308	411, 414
15	8, 6	94, 95	196, 199	299, 302	400, 404
16	4, 6	104, 106	207, 210	311, 310	415, 413
17	2, 4	105, 106	206, 208	308, 310	410, 413
18	2, 0	104, 103	206, 206	309, 308	414, 409
19	0, 1	101, 102	203, 206	305, 307	409, 411
20	1, 4	104, 106	206, 208	311, 309	410, 414
21	9, 10	92, 92	194, 194	298, 297	400, 398
22	8, 8	95, 95	195, 199	298, 301	401, 403

In the above discussion, within-profile observations are assumed to be independent. In practice, they can be correlated due to serial and/or spatial correlation. Jensen

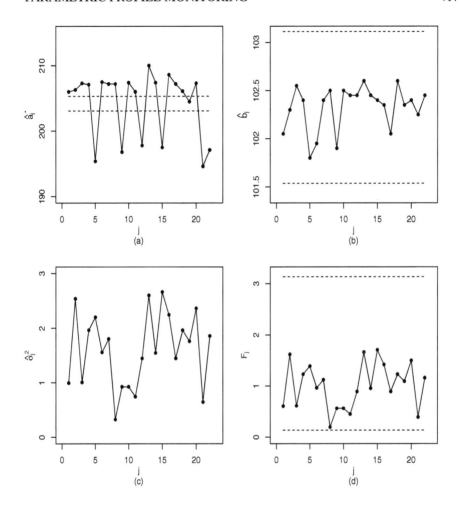

Figure 10.3 *(a) Shewhart chart (10.17) for monitoring the estimated intercepts $\{\hat{a}_j^*, j = 1, 2, \ldots, m\}$ in the calibration problem discussed in Example 10.2. (b) Shewhart chart (10.18) for monitoring the estimated slopes $\{\hat{b}_j, j = 1, 2, \ldots, m\}$. (c) Estimated error variances $\{\hat{\sigma}_j^2, j = 1, 2, \ldots, m\}$. (d) Shewhart chart (10.19)–(10.20) for monitoring the estimated error variances $\{\hat{\sigma}_j^2, j = 1, 2, \ldots, 20\}$. The dashed horizontal lines in plots (a), (b), and (d) denote the upper and lower control limits.*

et al. (2008) discussed the LPM problem in cases when the within-profile observations were correlated. Mahmoud et al. (2007) and Zou et al. (2006) discussed the LPM problem under the framework of change-point detection. Zou et al. (2007a) generalized the LPM problem from cases with a single predictor to cases with multiple predictors, and proposed an MEWMA chart for handling the generalized LPM problem. A self-starting control chart was proposed by Zou et al. (2007b) for phase II LPM in cases when the IC parameters are unknown.

10.2.2 Nonlinear profile monitoring

In the previous subsection, it is assumed that the IC profile is linear (cf., model (10.1)). In some applications, however, the linearity assumption may not be valid. For instance, the profile describing the relationship between the total stamping force (y) and the crank angle (x) in a manufacturing process of sheet-metal stamping is often nonlinear (cf. Ding et al., 2006; Jin and Shi, 1999). In dose-response studies, people often use the nonlinear regression model

$$y = a + \frac{d-a}{1+(x/c)^b} + \varepsilon \qquad (10.21)$$

to describe the relationship between the dose level (x) of a medicine and the corresponding response (y) of a patient, where $a, b, c,$ and d are four parameters, and ε is the random error. Model (10.21) describes a specific nonlinear profile. More generally, a nonlinear profile can be described by the model

$$y_i = f(x_i, \theta) + \varepsilon_i, \qquad \text{for } i = 1, 2, \ldots, n, \qquad (10.22)$$

where $\{(x_i, y_i), i = 1, 2, \ldots, n\}$ are n observations of (x, y), f is a *known* nonlinear function, θ is a q-dimensional parameter vector, and $\{\varepsilon_i, i = 1, 2, \ldots, n\}$ are i.i.d. random errors with the common distribution $N(0, \sigma^2)$. The nonlinear profile monitoring (NLPM) problem is for handling cases when the IC profile of a production process can be described by the nonlinear regression model (10.22).

Let us first discuss the phase II NLPM problem, in which the IC value of θ is assumed known to be θ_0. Assume that $\{(x_i, y_{ij}), i = 1, 2, \ldots, n\}$ is the j-th observed profile, for $j = 1, 2, \ldots,$ and that it follows the model (10.22). Namely, we have

$$y_{ij} = f(x_i, \theta) + \varepsilon_{ij}, \qquad \text{for } i = 1, 2, \ldots, n, \ j = 1, 2, \ldots, \qquad (10.23)$$

where $\{\varepsilon_{ij}, i = 1, 2, \ldots, n, j = 1, 2, \ldots\}$ are i.i.d. random errors with the common distribution $N(0, \sigma^2)$. For the j-th profile, the value of θ can be estimated from the observations $\{(x_i, y_{ij}), i = 1, 2, \ldots, n\}$ using the nonlinear least squares estimation described as follows. Let $\theta^{(0)}$ be an initial estimator of θ, and $G^{(0)} = (\partial f(x_i, \theta)/\partial \theta_l|_{\theta=\theta^{(0)}}, i = 1, 2, \ldots, n, \ l = 1, 2, \ldots, q)$ be an $n \times q$ matrix, where $\theta = (\theta_1, \theta_2, \ldots, \theta_q)'$. Then, for each x in the design space, by the Taylor's expansion, we have

$$f(x, \theta) \approx f(x, \theta^{(0)}) + \left(\frac{\partial f(x, \theta)}{\partial \theta} \bigg|_{\theta=\theta^{(0)}} \right)' \left(\theta - \theta^{(0)} \right).$$

So, the objective function of the least squares estimation is

$$\sum_{i=1}^{n} (y_{ij} - f(x_i, \theta))^2$$

$$\approx \sum_{i=1}^{n} \left[(y_{ij} - f(x_i, \theta^{(0)})) - \left(\frac{\partial f(x_i, \theta)}{\partial \theta} \bigg|_{\theta=\theta^{(0)}} \right)' \left(\theta - \theta^{(0)} \right) \right]^2.$$

Then, from the least squares estimation (i.e., $\min_{\theta} \sum_{i=1}^{n} (y_{ij} - f(x_i, \theta))^2$), the updated estimator of θ is

$$\theta^{(1)} = \theta^{(0)} + \left(G^{(0)'} G^{(0)}\right)^{-1} G^{(0)'} y_j^{(0)},$$

where $y_j^{(0)} = (y_{1j}, y_{2j}, \ldots, y_{nj})' - (f(x_1, \theta^{(0)}), f(x_2, \theta^{(0)}), \ldots, f(x_n, \theta^{(0)}))'$. The above algorithm can be made iterative, and at the s-th iteration the updated estimator of θ is

$$\theta^{(s)} = \theta^{(s-1)} + \left(G^{(s-1)'} G^{(s-1)}\right)^{-1} G^{(s-1)'} y_j^{(s-1)}, \qquad \text{for } s \geq 1, \qquad (10.24)$$

where $G^{(s-1)} = (\partial f(x_i, \theta)/\partial \theta_l|_{\theta = \theta^{(s-1)}}, i = 1, 2, \ldots, n, l = 1, 2, \ldots, q)$ is a $n \times q$ matrix, and $y_j^{(s-1)} = (y_{1j}, y_{2j}, \ldots, y_{nj})' - (f(x_1, \theta^{(s-1)}), f(x_2, \theta^{(s-1)}), \ldots, f(x_n, \theta^{(s-1)}))'$. Under some regularity conditions, the iterative algorithm (10.24) will converge when the number of iterations s increases. The limit is the final estimator of θ, denoted as $\widehat{\theta}_j$. When the sample size n is large, it has been shown (cf., Seber and Wild, 2003) that the distribution of this estimator is close to the normal distribution

$$N_q \left(\theta_j, \sigma^2 \left(G_j' G_j\right)^{-1}\right), \qquad (10.25)$$

where θ_j is the true parameter vector of the j-th profile, and

$$G_j = \left(\frac{\partial f(x_1, \theta)}{\partial \theta}, \frac{\partial f(x_2, \theta)}{\partial \theta}, \ldots, \frac{\partial f(x_n, \theta)}{\partial \theta}\right)'\bigg|_{\theta = \theta_j}$$

is an $n \times q$ matrix. The algorithm (10.24) is the well-known Gauss-Newton algorithm. It has several variants. See Bates and Watts (2007) for a systematic discussion on nonlinear least squares estimation. As a side note, the estimator $\widehat{\theta}_j$ can be computed by the R-function nls().

Based on the result (10.25), it is natural to consider the Shewhart chart with the decision rule that gives a signal at time j when

$$T_j^2 = \frac{1}{\sigma^2} \left(\widehat{\theta}_j - \theta_0\right)' \left(G_j' G_j\right) \left(\widehat{\theta}_j - \theta_0\right) > \chi_{1-\alpha,q}^2, \qquad (10.26)$$

where $\alpha \in (0, 1)$ is a given significance level, and $\chi_{1-\alpha,q}^2$ is the $(1-\alpha)$-th quantile of the χ_q^2 distribution. It is also possible to construct CUSUM, EWMA, and CPD charts. For instance, we can consider the EWMA charting statistic

$$E_j = \lambda \left(\widehat{\theta}_j - \theta_0\right) + (1 - \lambda) E_{j-1}, \qquad \text{for } j \geq 1, \qquad (10.27)$$

where $E_0 = 0$, and $\lambda \in (0, 1]$ is a weighting parameter. Then, the chart gives a signal at time j if

$$W_j^2 = E_j' \Sigma_{E_j}^{-1} E_j > \rho, \qquad (10.28)$$

where $\rho > 0$ is a control limit chosen to achieve a given ARL_0 value, and $\Sigma_{\mathbf{E}_j}$ is the covariance matrix of $\mathbf{E}_j$ which can be computed recursively by the formula

$$
\begin{aligned}
\Sigma_{\mathbf{E}_j} &= \lambda^2 \Sigma_{\widehat{\theta}_j} + (1-\lambda)^2 \Sigma_{\mathbf{E}_{j-1}} \\
&\approx \lambda^2 \sigma^2 \left(G_j' G_j\right)^{-1} + (1-\lambda)^2 \Sigma_{\mathbf{E}_{j-1}}, \quad \text{for } j \geq 1, \tag{10.29}
\end{aligned}
$$

in which $\Sigma_{\mathbf{E}_0} = 0$. The computation involved in computing $\Sigma_{\mathbf{E}_j}$ can be greatly reduced by using (10.29).

Example 10.3 *In a dose-response application, assume that the relationship between the response y and the dose level x follows the following 2-parameter nonlinear regression model:*

$$
y = \frac{d}{1+x^b} + \varepsilon, \qquad \text{for } x \in [0,1], \tag{10.30}
$$

where b and d are two parameters, and ε is the random error with the distribution $N(0, 0.25^2)$. It is further assumed that the IC values of b and d are 1 and 2, respectively. Table 10.3 presents the first 20 observed profiles obtained during phase II SPC, in each of which the design points are fixed at 0.1, 0.25, 0.4, 0.55, 0.7, 0.85, and 1.0. In this example, it is assumed that the within-profile and between-profile observations are independent of each other. We then use the R function nls() to compute the values of $\{(\widehat{b}_j, \widehat{d}_j), j = 1, 2, \ldots, 20\}$. The computed values are shown in Figure 10.4(a)–(b). From the plots, it seems that both b and d shift around $j = 11$, with b shifted upward and d shifted downward. Next, we use the Shewhart chart (10.26) and the EWMA chart (10.27)–(10.28) to monitor the estimated values $\{(\widehat{b}_j, \widehat{d}_j), j = 1, 2, \ldots, 20\}$. In both charts, we use $\chi^2_{0.995,2} = 10.597$ as the control limits. The charts are shown in Figure 10.4(c)–(d). It can be seen that the Shewhart chart gives its first signal at the 11th time point, and the EWMA chart gives its first signal at the 15th time point.

Next, we discuss the phase I NLPM problem, in which the IC value of the parameter vector θ is unknown. Assume that there are m observed profiles $\{(x_i, y_{ij}), i = 1, 2, \ldots, n, j = 1, 2, \ldots, m\}$, and each observed profile follows the model (10.23). Then, by the nonlinear least squares estimation discussed above (cf., (10.24)), we can compute the estimators of θ, denoted as $\{\widehat{\theta}_j, j = 1, 2, \ldots, m\}$, from the m observed profiles. One natural idea to monitor these estimators of θ is to use the Shewhart charting statistic

$$
\widetilde{T}_j^2 = \left(\widehat{\theta}_j - \overline{\widehat{\theta}}\right)' S_{\widehat{\theta}}^{-1} \left(\widehat{\theta}_j - \overline{\widehat{\theta}}\right), \tag{10.31}
$$

where $\overline{\widehat{\theta}}$ and $S_{\widehat{\theta}}$ are the sample mean and sample covariance matrix of $\{\widehat{\theta}_j, j = 1, 2, \ldots, m\}$. As discussed in Subsection 7.2.2, the IC distribution of $\frac{m}{(m-1)^2}\widetilde{T}_j^2$ is $Beta(q/2, (m-q-1)/2)$. So, the chart gives a signal at time j if

$$
\widetilde{T}_j^2 > \frac{(m-1)^2}{m} Beta_{1-\alpha}(q/2, (m-q-1)/2), \tag{10.32}
$$

where $\alpha \in (0, 1)$ is a given significance level, and $Beta_{1-\alpha}(q/2, (m-q-1)/2)$ is the $(1-\alpha)$-th quantile of the $Beta(q/2, (m-q-1)/2)$ distribution.

Table 10.3 *The first 20 observed profiles of a production process obtained during phase II SPC. The design points of all profiles are the same, with values 0.1, 0.25, 0.4, 0.55, 0.7, 0.85, and 1.0.*

j	y_{1j}	y_{2j}	y_{3j}	y_{4j}	y_{5j}	y_{6j}	y_{7j}
1	1.768	1.613	1.421	1.379	1.188	1.113	0.942
2	1.890	1.517	1.393	1.299	1.186	1.061	1.074
3	1.831	1.597	1.390	1.341	1.085	1.312	0.956
4	1.895	1.626	1.506	1.209	1.133	1.009	1.023
5	1.702	1.625	1.419	1.466	1.163	1.070	0.931
6	1.796	1.618	1.470	1.397	1.273	1.071	1.140
7	1.641	1.662	1.376	1.423	1.140	1.213	1.004
8	1.630	1.555	1.255	1.308	1.366	0.854	1.098
9	1.678	1.782	1.567	1.206	1.150	1.074	0.962
10	2.076	1.613	1.357	1.354	1.197	1.074	0.991
11	1.035	0.835	0.746	0.933	0.465	0.582	0.391
12	1.017	1.042	0.655	0.857	0.666	0.446	0.307
13	1.061	0.925	0.884	0.849	0.844	0.570	0.444
14	1.133	0.852	0.746	0.715	0.916	0.497	0.541
15	0.872	0.824	0.829	0.904	0.624	0.665	0.354
16	0.950	0.864	0.825	0.892	0.660	0.598	0.525
17	0.929	0.798	0.829	0.781	0.773	0.555	0.470
18	1.152	0.864	0.904	0.709	0.713	0.426	0.448
19	0.962	1.042	0.815	0.798	0.629	0.496	0.569
20	0.944	1.076	0.906	0.753	0.717	0.577	0.546

Williams et al. (2007b) discussed two alternative Shewhart charts for solving the phase I NLPM problem. The first one uses the charting statistic

$$\widetilde{T}_{j,D}^2 = \left(\widehat{\theta}_j - \overline{\overline{\theta}} \right)' S_D^{-1} \left(\widehat{\theta}_j - \overline{\overline{\theta}} \right),$$

where

$$S_D = \frac{1}{2(m-1)} \sum_{j=1}^{m-1} \left(\widehat{\theta}_{j+1} - \widehat{\theta}_j \right) \left(\widehat{\theta}_{j+1} - \widehat{\theta}_j \right)'.$$

The matrix S_D provides a point estimator of the covariance matrix of $\widehat{\theta}$. It was first discussed by Hawkins and Merriam (1974). Several papers, including Holmes and Mergen (1993), Sullivan and Woodall (1996), and Vargas (2003), studied the properties of the charting statistic $\widetilde{T}_{j,D}^2$. They found that $\widetilde{T}_{j,D}^2$ was effective in detecting both step and ramp shifts, and it was invariant to the full-rank linear transformations on the original observations. Williams et al. (2006) provided a formula for determining the control limit of the Shewhart chart using $\widetilde{T}_{j,D}^2$. Because this formula is quite complicated, it is omitted here. The second alternative Shewhart chart discussed by Williams et al. (2007b) uses the so-called minimum volume ellipsoid (MVE) estimator of the covariance matrix of $\widehat{\theta}$. That chart was found ineffective in detecting step shifts in θ, but effective in detecting outliers.

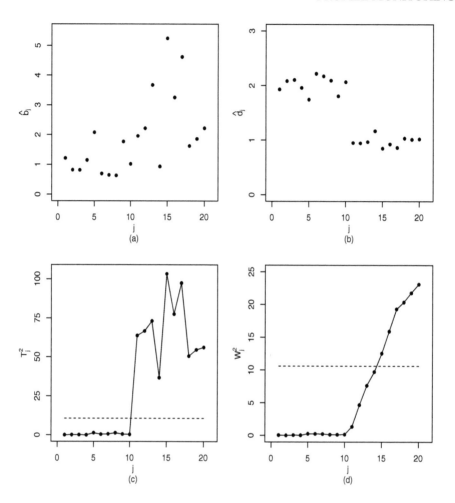

Figure 10.4 *(a)* $\{\hat{b}_j, j = 1, 2, \ldots, 20\}$. *(b)* $\{\hat{d}_j, j = 1, 2, \ldots, 20\}$. *(c) Shewhart chart (10.26) with control limit* $\chi^2_{0.995,2} = 10.597$. *(d) EWMA chart (10.27)–(10.28) with control limit* $\chi^2_{0.995,2} = 10.597$. *In plots (c) and (d), the dashed horizontal lines denote the control limits.*

Example 10.3 (continued) *For the 20 observed profiles presented in Table 10.3, let us now assume that they are collected during phase I NLPM, and each profile still follows the model (10.30) with the IC values of the parameters b and d unknown. For each profile, the estimates of b and d are obtained by the nonlinear least squares estimation, as was done in Example 10.3, and their values are shown in Figure 10.4(a)–(b). Then, the Shewhart chart (10.31)–(10.32) is applied to these estimated values. In the chart, α is chosen to be 0.005, and the corresponding control limit is $\frac{19^2}{20} Beta_{0.995}(1, 8.5) = 8.372$. The Shewhart chart is shown in Figure 10.5. From the plot, it can be seen that no signal is given by this chart.*

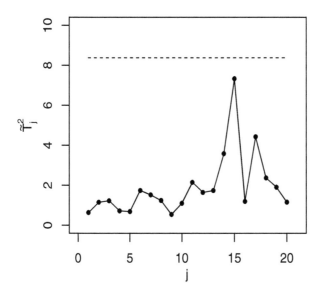

Figure 10.5 *Shewhart chart (10.31)–(10.32) with $\alpha = 0.005$ when it is applied to the data presented in Table 10.3. The horizontal dashed line denotes the control limit.*

In the literature, there are some other proposed methods for handling the NLPM problem. For instance, Chicken et al. (1998) proposed describing a nonlinear profile by a linear combination of a finite number of wavelet basis functions, and then profile monitoring could be accomplished by monitoring the estimated wavelet coefficients. Fan et al. (2012) proposed a similar idea by describing a nonlinear profile by a linear combination of some sine functions. Jensen and Birch (2009) proposed using a nonlinear mixed-effects model for describing nonlinear profiles, which can accommodate possible correlation among within-profile observations.

10.3   Nonparametric Profile Monitoring

The LPM and NLPM problems discussed in the previous section are under the assumption that observed profiles can be described by a linear regression model or a parametric nonlinear regression model. In a specific application, if this assumption is invalid and an LPM or NLPM control chart is still used, then the results could be unreliable or even misleading because the deviation of the true profile model from the assumed profile model is also a shift and its impact on the performance of the control chart is difficult to eliminate. To overcome this limitation of the LPM and NLPM control charts, some nonparametric profile monitoring (NPPM) charts have been proposed in the literature. For instance, Qiu and Zou (2010), and Zou et al. (2008, 2009b) proposed several NPPM charts under various assumptions on IC parameters, design points, and the correlation structure among within-profile observations. Wei et al.

(2012) proposed an NPPM Shewhart chart using quantile regression (cf., Koenker, 2005). In this section, we describe the NPPM chart proposed by Qiu et al. (2010), which does not require restrictive model assumptions. Our description is divided into two parts. Estimation of the IC profile model based on mixed effects modeling is discussed in Subsection 10.3.1. A phase II EWMA chart using the estimated IC profile model is discussed in Subsection 10.3.2.

10.3.1 Nonparametric mixed-effects modeling

The phase II NPPM chart described in the next subsection does not require IC parameters to be known. Instead, we estimate the related IC parameters from an IC dataset, using the nonparametric mixed-effects (NME) modeling. In the literature, the mixed-effects modeling is often used in longitudinal data analysis (e.g., Laird and Ware, 1982; Diggle et al., 1994). It has become a major tool for accommodating possible correlation among observed data. NME modeling for analyzing longitudinal data has been discussed by several authors, including Shi et al. (1996) and Rice and Wu (2001). We follow this framework for modeling possible correlation among within-profile observations of an IC dataset. It should be noted that, in the literature on mixed-effects modeling, profiles are also called "clusters" or "subjects." To simplify the presentation, we choose to discuss cases with a single predictor here; this discussion can be generalized easily to cases with multiple predictors. In the IC dataset, assume that there are m profiles and the j-th profile has n_j observations, for $j = 1, 2, \ldots, m$. Then, the NME model can be written as

$$y_{ij} = g(x_{ij}) + f_j(x_{ij}) + \varepsilon_{ij}, \qquad \text{for } i = 1, 2, \ldots, n_j, \ j = 1, 2, \ldots, m, \qquad (10.33)$$

where g is the population profile function (i.e., the fixed-effects term), f_j is the random-effects term describing the deviation of the j-th individual profile from g, $\{(x_{ij}, y_{ij}), i = 1, 2, \ldots, n_j\}$ is the sample collected for the j-th profile, and $\{\varepsilon_{ij}, i = 1, 2, \ldots, n_j\}$ are i.i.d. random errors with mean 0 and variance σ^2. In model (10.33), it is routinely assumed that the random-effects term f_j and the random errors ε_{ij} are independent of each other, and f_j is a realization of a mean 0 process with a common covariance function

$$\gamma(x_1, x_2) = E[f_j(x_1) f_j(x_2)].$$

Without loss of generality, we further assume that $x_{ij} \in [0, 1]$, for all i and j.

Model (10.33) is fairly flexible. It includes many common correlation structures as special cases. For instance, if $f_j(x_{ij}) = \xi_j$ and ξ_j is a mean 0 random variable, then within-profile correlation would have the compound symmetry structure. If $\text{Cor}(f_j(x_1), f_j(x_2)) = \rho(|x_1 - x_2|; \nu)$, for some correlation function ρ and a coefficient ν, then the correlation structure includes the nonhomogeneous Ornstein-Uhlenbeck process and the Gaussian correlation model (cf., Zhang et al., 1998) as special cases. When the design points are equally spaced and unchanged among different profiles, this model can also be used for describing the autoregressive correlation structure. Because of its flexibility, model (10.33) requires a relatively large set of IC profiles for model estimation and calibration, compared to its parametric counterparts. Thanks to fast progress in sensor and information technology, automatic data acquisition is becoming increasingly common in industry. Consequently, a large amount of IC data is often available, and model (10.33) allows us to make use of such data without imposing a parametric form on the profile model.

Next, we discuss estimation of g, γ, and σ^2 in model (10.33) when the related production process is IC. In the literature, there are some existing discussions about statistical analysis of correlated data under various settings and assumptions, including those in Altman (1990), Fan and Zhang (2000), Hart (1991), Hoover et al. (1998), Lin and Carroll (2000), Wang (1998), Zhang et al. (1998), and many others. Wu and Zhang (2002) proposed a method for estimating model (10.33) by combining linear mixed-effects (LME) modeling and local linear kernel smoothing (cf., Subsection 2.8.5). They demonstrated that their estimator of g, which was referred to as local LME (LLME) estimator, was often more efficient than some alternative estimators in terms of the mean squared error (cf., Subsection 2.7.1). Furthermore, by their approach, it is fairly easy to obtain consistent estimators of γ and σ^2, which is important for constructing a phase II control chart in the current NPPM problem. For these reasons, Qiu et al. (2010) adopted Wu and Zhang's method, which is briefly described below.

For a given point $s \in [0, 1]$, LLME estimators of $g(s)$ and $f_j(s)$ are obtained by minimizing the following penalized, negative-log, local linear kernel likelihood function:

$$\sum_{j=1}^{m} \left\{ \frac{1}{\sigma^2} \sum_{i=1}^{n_j} \left[y_{ij} - \mathbf{z}'_{ij}(\mathbf{a} + \mathbf{b}_j) \right]^2 K_h\left(x_{ij} - s \right) + \mathbf{b}'_j D^{-1} \mathbf{b}_j + \log |D| + n_j \log \left(\sigma^2 \right) \right\},$$

(10.34)

where $K_h(x) = K(x/h)/h$, K is a symmetric density kernel function, h is a bandwidth, $\mathbf{z}'_{ij} = (1, x_{ij} - s)$, $\mathbf{a}$ is a deterministic two-dimensional coefficient vector, and $\mathbf{b}_j$ is a two-dimensional vector of the random effects with mean $\mathbf{0}$ and covariance matrix D. Minimization of (10.34) with respect to $\mathbf{a}$ and $\mathbf{b}_j$ can be accomplished by the 4-step iterative procedure below.

Step 1 Set the initial values for D and σ^2, denoted as $D_{(0)}$ and $\sigma^2_{(0)}$.

Step 2 At the l-th iteration, for $l \geq 0$, compute estimators of $\mathbf{a}$ and $\mathbf{b}_j$ by solving the so-called mixed-model equation (cf., Davidian and Giltinan, 1995; Wu and Zhang, 2002), and the resulting estimators are

$$\widehat{\mathbf{a}}^{(l)} = \left(\sum_{j=1}^{m} Z'_j \Sigma_j Z_j \right)^{-1} \sum_{j=1}^{m} Z'_j \Sigma_j \mathbf{y}_j \qquad (10.35)$$

$$\widehat{\mathbf{b}}_j^{(l)} = \left(Z'_j K_j Z_j + \sigma^2_{(l)} D^{-1}_{(l)} \right)^{-1} Z'_j K_j \left(\mathbf{y}_j - Z_j \widehat{\mathbf{a}}^{(l)} \right), \qquad (10.36)$$

where $Z_j = (\mathbf{z}_{1j}, \mathbf{z}_{2j}, \ldots, \mathbf{z}_{n_j j})'$, $\mathbf{y}_j = (y_{1j}, y_{2j}, \ldots, y_{n_j j})'$, $\Sigma_j = (Z_j D_{(l)} Z'_j + \sigma^2_{(l)} K_j^{-1})^{-1}$, and $K_j = \text{diag}\{K_h(x_{1j} - s), K_h(x_{2j} - s), \ldots, K_h(x_{n_j j} - s)\}$.

Step 3 Based on $\widehat{\mathbf{a}}^{(l)}$ and $\widehat{\mathbf{b}}_j^{(l)}$, update the estimators of D and σ^2 by

$$D_{(l+1)} = \frac{1}{m} \sum_{j=1}^{m} \widehat{\mathbf{b}}_j^{(l)} \left(\widehat{\mathbf{b}}_j^{(l)} \right)' \qquad (10.37)$$

$$\sigma_{(l+1)}^2 = \frac{1}{m} \sum_{j=1}^{m} \frac{1}{n_j} \left[\mathbf{y}_j - Z_j \left(\widehat{\mathbf{a}}^{(l)} + \widehat{\mathbf{b}}_j^{(l)} \right) \right]' \left[\mathbf{y}_j - Z_j \left(\widehat{\mathbf{a}}^{(l)} + \widehat{\mathbf{b}}_j^{(l)} \right) \right]. \quad (10.38)$$

Step 4 Repeat Steps 2–3 until the following condition is satisfied:

$$\|D_{(l)} - D_{(l-1)}\|_1 / \|D_{(l-1)}\|_1 \leq \tau,$$

where τ is a pre-specified small positive number (e.g., $\tau = 10^{-4}$), and $\|A\|_1$ denotes the sum of absolute values of all elements of a matrix A. Then, the algorithm stops at the l-th iteration.

Note that, in Step 4, we use the relative difference of the successive estimators of D in the convergence criterion. In fact, other estimators can also be used for this purpose. We use D here because our simulation shows that it gives good results in various cases. As a side note, nonconvergence of the above iterative procedure can occasionally happen, although we find that the frequency of nonconvergence is negligible in all our simulation studies, except certain extreme cases such as the ones when m or n_j's are too small. To reduce the frequency of nonconvergence, it is suggested in the literature to use good initial values for D and σ^2. A simple but effective method is to set $D_{(0)}$ to be the identity matrix and

$$\sigma_{(0)}^2 = \frac{1}{m} \sum_{j=1}^{m} \left\{ \frac{1}{n_j} \sum_{i=1}^{n_j} \left[y_{ij} - \widehat{g}^{(P)}(x_{ij}) \right]^2 \right\},$$

where $\widehat{g}^{(P)}(x_{ij})$ is the standard local linear kernel estimator constructed from the pooled data (cf., Hoover et al., 1998).

After obtaining solutions to $\mathbf{a}$ and $\mathbf{b}_j$, denoted as $\widehat{\mathbf{a}}(s)$ and $\widehat{\mathbf{b}}_j(s)$, of the minimization problem using the above 4-step algorithm, we can define

$$\widehat{g}(s) = \mathbf{e}_1' \widehat{\mathbf{a}}(s), \quad \widehat{f}_j(s) = \mathbf{e}_1' \widehat{\mathbf{b}}_j(s), \quad \widehat{\gamma}(s_1, s_2) = \frac{1}{m} \sum_{j=1}^{m} \widehat{f}_j(s_1) \widehat{f}_j(s_2), \qquad (10.39)$$

where $s, s_1, s_2 \in [0, 1]$, and $\mathbf{e}_1 = (1, 0)'$. Note that the variance estimator from the above iterative procedure depends on s. Since σ^2 is a population parameter that does not depend on s, we suggest estimating it by

$$\widehat{\sigma}^2 = \frac{1}{m} \sum_{j=1}^{m} \left\{ \frac{1}{n_j} \sum_{i=1}^{n_j} \left[y_{ij} - \widehat{g}(x_{ij}) - \widehat{f}_j(x_{ij}) \right]^2 \right\}. \qquad (10.40)$$

Example 10.4 *Qiu et al. (2010) considered a dataset obtained from a manufacturing process of aluminium electrolytic capacitors (AECs). This process transforms* *raw materials, such as anode aluminum foil, cathode aluminum foil, guiding pin, electrolyte sheet, plastic cover, aluminum shell, and plastic tube, into AECs that are appropriate for use in low leakage circuits and are well adapted to a wide range of environmental temperatures. The whole manufacturing process consists of a sequence of operations, including clenching, rolling, soaking, assembly, cleaning, aging, and classifying. Before packing, a careful quality monitoring step is required by sampling from a batch of products. Regarding the quality of AECs, the most important characteristic is the dissipation factor (DF), which can be automatically measured by an electronic device. However, it is known that DF measurements would change significantly with environmental temperature, and there is a specific requirement about the adaptability of AECs to the temperature. To monitor the adaptability, engineers put a sampled AEC in a container. Then, the container's temperature is controlled, and the temperature is supposed to stably increase from $-26^o F$ to $78^o F$. In this process, measurements of DF and the actual temperature inside the container are taken at 53 equally spaced time points. The actual temperature inside the container is reported by a temperature sensor. So, for each sampled AEC, a set of 53 observations of the pair (temperature, DF), which corresponds to (x, y) in model (10.33), are obtained for monitoring the adaptability of the AEC to the temperature. Figure 10.6 shows three AEC profiles along with an NME estimator of the IC profile function (see related discussion below). It should be noted that the actual temperature inside a container would fluctuate around its nominal level at each observation time. Therefore, the actual temperature readings of different containers at a given observation time are all different, although the differences are usually small.*

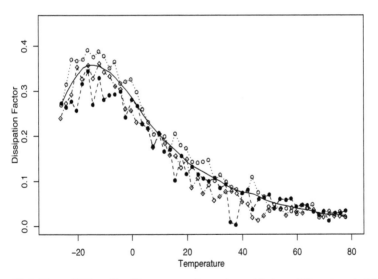

Figure 10.6 *Three AEC profiles (lines connecting points with three different symbols) and the NME estimator (solid curve) of the IC profile function.*

The entire AEC dataset contains 144 profiles, and each profile has $n = 53$ observations. We use the first 96 profiles as the IC data to estimate the IC profile function and the remaining ones as phase II observed profiles. We first estimate model (10.33) from the IC data by the iterative procedure (10.35)–(10.38), using the suggested ini-

tial values of D and σ^2 given above, the Epanechnikov kernel function, and the CV bandwidth selection procedure suggested by Wu and Zhang (2002) (cf., Subsections 2.8.4 and 2.8.5 for related descriptions). The resulting IC profile estimator $\widehat{g}$ is the one displayed in Figure 10.6 by the solid curve. From (10.39) and (10.40), we can also compute the estimated correlation of two within-profile observations of the response variable y at any two points s_1 and s_2 in the design interval

$$\widehat{\rho}(s_1,s_2) = \widehat{\gamma}(s_1,s_2)/[\widehat{v}(s_1)\widehat{v}(s_2)],$$

where $\widehat{\gamma}(s_1,s_2)$ is defined in (10.39), $\widehat{v}^2(s) = \widehat{\gamma}(s,s) + \widehat{\sigma}^2$ is the estimated variance of y at s, and $\widehat{\sigma}^2$ is defined in (10.40). Let $x_i^* = 2(i-1) - 26$, for $i = 1,2,\ldots,53$, be 53 equally spaced points in the design interval $[-26,78]$, which denote the nominal temperature levels used when taking DF measurements of the sampled AECs. The estimated correlations $\widehat{\rho}(x_i^*,x_{i+1}^*)$, $\widehat{\rho}(x_i^*,x_{i+3}^*)$, $\widehat{\rho}(x_1^*,x_i^*)$, and $\widehat{\rho}(x_i^*,x_{53}^*)$, for $i = 1,2,\ldots,53$, are shown in Figure 10.7(a). From the plot, we can see that correlation within AEC profiles is substantial; thus, it should not be ignored. Figure 10.7(b) shows the estimated standard deviation $\widehat{v}(x_i^*)$ of the response variable y at x_i^*, for $i = 1,2,\ldots,53$, from which heteroscedasticity of the response variable y at different positions of x is clearly seen. In addition, we can obtain an estimate of the error standard deviation σ to equal 0.016, by formula (10.40), which is much smaller than $\widehat{v}(x_i^*)$, especially when $i \in [12,45]$. This result implies that the random-effects term in model (10.33) describes a substantial amount of random variation in the data.

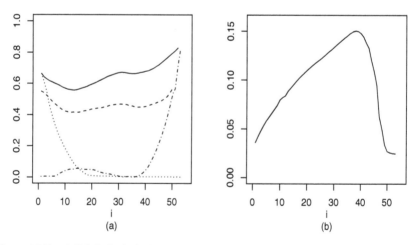

Figure 10.7 *(a) Solid, dashed, dotted, and dash-dotted curves represent estimated within-profile correlations $\widehat{\rho}(x_i^*,x_{i+1}^*)$, $\widehat{\rho}(x_i^*,x_{i+3}^*)$, $\widehat{\rho}(x_1^*,x_i^*)$, and $\widehat{\rho}(x_i^*,x_{53}^*)$, for $i = 1,2,\ldots,53$, where $\{x_i^*, i = 1,2,\ldots,53\}$ are 53 equally spaced points in the design interval $[-26,78]$. (b) Estimated standard deviation $\widehat{v}(x_i^*)$ of the response variable y at x_i^*, for $i = 1,2,\ldots,53$.*

10.3.2 Phase II nonparametric profile monitoring

In this part, we describe the phase II NPPM chart proposed by Qiu et al. (2010) in general cases when within-profile data might be correlated and the design points within and between profiles are arbitrary. Profile monitoring in such cases is challenging mainly due to the following two reasons. First, because the within-profile data might be correlated, estimation of the profile function g involves a considerable amount of computation if the NME modeling approach described in the previous subsection is also used in phase II NPPM. However, a good on-line control chart should maintain a reasonable efficiency while being effective in detecting profile shifts. Second, in cases when the design points $\mathbf{x}_j = (x_{ij}, x_{2j}, \ldots, x_{n_j j})'$ are unchanged from one profile to another, one natural profile monitoring method is to first average observed responses y_{ij}'s across different profiles and then detect potential profile shifts using a generalized likelihood ratio test statistic (cf., Fan et al., 2001). This idea cannot be applied to the current problem directly because the response is observed at different design points in different profiles. One immediate alternative is to estimate g from individual profile data at a given set of points in $[0,1]$. But the resulting estimators would be inefficient since they are constructed from individual profile data instead of from all observed data.

To overcome the above difficulties, at a given point $s \in [0,1]$, let us consider the following local weighted negative-log likelihood:

$$WL(a,b;s,\lambda,t) = \sum_{j=1}^{t} \sum_{i=1}^{n_j} [y_{ij} - a - b(x_{ij} - s)]^2 K_h(x_{ij} - s)(1-\lambda)^{t-j}/v^2(x_{ij}),$$

where t denotes the current time point during phase II profile monitoring, $\lambda \in (0,1]$ is a weighting parameter, and $v^2(x) = \gamma(x,x) + \sigma^2$ is the variance function of the response. Note that $WL(a,b;s,\lambda,t)$ combines the exponential weighting scheme used in EWMA at different time points through the term $(1-\lambda)^{t-j}$ and the local linear kernel smoothing procedure (cf., Subsection 2.8.5). At the same time, it takes into account the heteroscedasticity of observations by using $v^2(x_{ij})$. Then, the local linear kernel estimator of $g(s)$, defined as the solution to a of the minimization problem $\min_{a,b} WL(a,b;s,\lambda,t)$, has the expression

$$\widehat{g}_{t,h,\lambda}(s) = \sum_{j=1}^{t} \sum_{i=1}^{n_j} U_{ij}^{(t,h,\lambda)}(s) y_{ij} \Big/ \sum_{j=1}^{t} \sum_{i=1}^{n_j} U_{ij}^{(t,h,\lambda)}(s), \tag{10.41}$$

where

$$U_{ij}^{(t,h,\lambda)}(s) = \frac{(1-\lambda)^{t-j} K_h(x_{ij} - s)}{v^2(x_{ij})} \left[m_2^{(t,h,\lambda)}(s) - (x_{ij} - s) m_1^{(t,h,\lambda)}(s) \right],$$

$$m_l^{(t,h,\lambda)}(s) = \sum_{j=1}^{t} (1-\lambda)^{t-j} \sum_{i=1}^{n_j} (x_{ij} - s)^l K_h(x_{ij} - s)/v^2(x_{ij}), \tag{10.42}$$

for $l = 0,1,2$. Note that $m_0^{(t,h,\lambda)}(s)$ is not used in (10.42), but it will be used later.

From (10.41) and (10.42), we can see that $\widehat{g}_{t,h,\lambda}(s)$ makes use of all the available observations up to the current time point t, and different profiles are weighted as in a conventional EWMA chart (i.e., more recent profiles get more weights and the weights change exponentially over time). When $\lambda = 0$ (i.e., all profiles receive equal weight), the resulting estimator is similar to the local linear generalized estimating equations (GEE) estimator considered in Lin and Carroll (2000). The GEE estimator can accommodate within-profile correlation without specifying the correlation structure (it uses the so-called independent working correlation matrix). Under certain mild conditions, Lin and Carroll showed that it was asymptotically the best estimator. Although Wu and Zhang (2002) demonstrated that their LLME estimator performed better in certain cases, especially when within-profile correlation was strong, this latter estimator involves a considerable amount of computation, and may not be feasible for phase II profile monitoring, which is an on-line sequential procedure. As a comparison, the estimator (10.41) has an explicit formula, and the related computation is relatively fast.

Following the convention of phase II SPC, we assume that the IC profile function, denoted as g_0, and the variance function $v^2(\cdot)$ are both known. In practice, they need to be estimated from an IC dataset, as described in Subsection 10.3.1. Let $\xi_{ij} = y_{ij} - g_0(x_{ij})$, for all i and j, and $\widehat{\xi}_{t,h,\lambda}(s)$ be the estimator defined in (10.41) after y_{ij} are replaced by ξ_{ij}. Then, the IC distribution of $\widehat{\xi}_{t,h,\lambda}(s)$ does not depend on g_0, and the original testing problem with $H_0 : g = g_0$ versus $H_1 : g \neq g_0$, which is associated with the profile monitoring problem, is changed to the one with $H_0 : g = 0$ versus $H_1 : g \neq 0$. Consequently, the IC distribution of the proposed control chart defined below and all quantities related to this distribution (e.g., the control limit ρ) do not depend on g_0 either, which should simplify the design and implementation of our proposed control chart.

When the process is IC, $|\widehat{\xi}_{t,h,\lambda}(s)|$ should be small. So, a natural statistic that can be used for SPC is

$$T_{t,h,\lambda} = c_{0,t,\lambda} \int \frac{[\widehat{\xi}_{t,h,\lambda}(s)]^2}{v^2(s)} \Gamma_1(s) ds,$$

where

$$c_{t_0,t_1,\lambda} = a^2_{t_0,t_1,\lambda} / b_{t_0,t_1,\lambda},$$

$$a_{t_0,t_1,\lambda} = \sum_{j=t_0+1}^{t_1} (1-\lambda)^{t_1-j} n_j,$$

$$b_{t_0,t_1,\lambda} = \sum_{j=t_0+1}^{t_1} (1-\lambda)^{2(t_1-j)} n_j,$$

and Γ_1 is some pre-specified density function. In the expression of $T_{t,h,\lambda}$, quantities $c_{0,t,\lambda}$ and $v(\cdot)$ are used for unifying its asymptotic variance. In practice, we suggest using the following discrete version:

$$T_{t,h,\lambda} \approx \frac{c_{0,t,\lambda}}{n_0} \sum_{k=1}^{n_0} \frac{\left[\widehat{\xi}_{t,h,\lambda}(s_k)\right]^2}{v^2(s_k)}, \qquad (10.43)$$

where $\{s_k, k = 1, \ldots, n_0\}$ are some i.i.d. random numbers generated from Γ_1. Then, the chart gives a signal if

$$T_{t,h,\lambda} > \rho,$$

where $\rho > 0$ is a control limit chosen to achieve a given ARL_0 level. Hereafter, this chart is referred to as the nonparametric mixed-effects profile monitoring (NMEPM) chart.

In phase II SPC, it is a convention that the IC distribution of the process observations is assumed known. Then, the control limit ρ can be searched for by simulation based on this distribution. In practice, the IC distribution is often unknown. Instead, we usually have a quite large IC dataset. In such cases, ρ can be searched for by a resampling algorithm, briefly described below. In each simulation run, we resample the IC dataset by randomly choosing a sequence of profiles with replacement. The sequence of profiles is sequentially chosen until a signal of shift is triggered by the chart NMEPM. Then, an estimated ARL_0 value is computed based on B simulation runs, and ρ is searched for by matching the estimated ARL_0 value to the nominal value. In all our numerical examples presented below, we choose $B = 10,000$.

For on-line process monitoring, which generally handles a large number of profiles, fast implementation is important and some computational issues deserve our careful examination. For the chart NMEPM, computing the test statistic $T_{t,h,\lambda}$ by formulas (10.41)–(10.43) requires a considerable amount of computing time and a substantial amount of storage space as well to save all past profile observations. Next, we provide updating formulas for computing $T_{t,h,\lambda}$, which can greatly simplify the computation and reduce the storage requirement. Let

$$\widetilde{m}_l^{(t,h)}(s) = \sum_{i=1}^{n_t}(x_{it}-s)^l K_h(x_{it}-s)/v^2(x_{it}), \quad \text{for } l=0,1,2,$$

$$\widetilde{q}_l^{(t,h)}(s) = \sum_{i=1}^{n_t}(x_{it}-s)^l K_h(x_{it}-s)y_{it}/v^2(x_{it}), \quad \text{for } l=0,1.$$

Then, $m_l^{(t,h,\lambda)}(s)$ in (10.42) can be recursively updated by

$$m_l^{(t,h,\lambda)}(s) = (1-\lambda)m_l^{(t-1,h,\lambda)}(s) + \widetilde{m}_l^{(t,h)}(s), \quad \text{for } l=0,1,2,$$

where $m_l^{(0,h,\lambda)}(s) = 0$. Let $q_l^{(t,h,\lambda)}(s)$, for $l=0,1$, be two working functions defined by the recursive formula

$$q_l^{(t,h,\lambda)}(s) = (1-\lambda)q_l^{(t-1,h,\lambda)}(s) + \widetilde{q}_l^{(t,h)}(s),$$

where $q_l^{(0,h,\lambda)}(s) = 0$. Then, we have

$$\begin{aligned}
\widehat{g}_{t,h,\lambda}(s) &= \Big[(1-\lambda)^2 M^{(t-1,h,\lambda)}\widehat{g}_{t-1,h,\lambda} + \Big(\widetilde{q}_0^{(t,h)}m_2^{(t,h,\lambda)} - \widetilde{q}_1^{(t,h)}m_1^{(t,h,\lambda)}\Big) \\
&\quad + (1-\lambda)\Big(q_0^{(t-1,h,\lambda)}\widetilde{m}_2^{(t,h)} - q_1^{(t-1,h,\lambda)}\widetilde{m}_1^{(t,h)}\Big)\Big] \Big/ M^{(t,h,\lambda)}, \quad (10.44)
\end{aligned}$$

where $M^{(t,h,\lambda)}(s) = m_2^{(t,h,\lambda)}(s)m_0^{(t,h,\lambda)}(s) - [m_0^{(t,h,\lambda)}(s)]^2$. On the right-hand side of the above equation, dependence on s in each function is not made explicit in notation for simplicity, which should not cause any confusion.

By using the above updating formulas, implementation of the NMEPM chart can be briefly summarized as follows. At time point t, we first compute quantities $\widetilde{m}_l^{(t,h)}(s)$, for $l = 0, 1, 2$, and $\widehat{q}_l^{(t,h)}(s)$, for $l = 0, 1$, at n_0 pre-determined s locations $\{s_k, k = 1, \ldots, n_0\}$ (cf., (10.43)). Then, $m_l^{(t,h,\lambda)}(s_k)$, for $l = 0, 1, 2$, and $q_l^{(t,h,\lambda)}(s_k)$, for $l = 0, 1$, are updated by the above formulas. Finally, $\widehat{g}_{t,h,\lambda}(s)$ is computed from (10.44), and the test statistic $T_{t,h,\lambda}$ is computed from $\widehat{g}_{t,h,\lambda}(s)$, after y_{ij} is replaced by ξ_{ij}.

Qiu et al. (2010) provided some practical guidelines for choosing some parameters involved in the chart NMEPM, which are summarized below. (i) To attain desirable IC distributional properties, the IC data should be large enough that $n_j \geq 20$, for each j, and $m \geq 500$. (ii) In estimating the NME model (10.33) by the iterative procedure described in Subsection 10.3.1, we can use a data-driven bandwidth selection technique, such as the CV procedure by Wu and Zhang (2002). In phase II profile monitoring, we suggest using the following empirical formula to choose the bandwidth:

$$
h_E = \begin{cases} c_1 n^{-1/5} \left(\sum_{i=1}^n (x_i - \bar{x})^2 / n \right)^{1/2} & \text{for balanced design cases} \\ c_2 [\tilde{n}(2 - \lambda)/\lambda]^{-1/5} \sqrt{\text{Var}(x)} & \text{for random design cases,} \end{cases} \tag{10.45}
$$

where $\bar{x} = \frac{1}{n} \sum_{i=1}^n x_i$ is the mean of the n design points in the balanced design cases (i.e., in cases when all profiles have the same design points), $\tilde{n}$ and $\text{Var}(x)$ are the average number of design points and the variance of design points within a profile, respectively, in the random design cases, which can be estimated from the IC data, and c_1 and c_2 are two constants. Empirically, c_1 and c_2 can be chosen as any values in the interval $[1.0, 2.0]$. (iii) We suggest choosing $\lambda \in [0.02, 0.1]$ if h_E in (10.45) is used in phase II SPC. (iv) Based on our numerical experience, selection of $\{s_k, k = 1, 2, \ldots, n_0\}$ does not much affect the performance of the NMEPM chart, as long as n_0 is not too small and s_k's cover all the key parts of g_0 (e.g., peaks/valleys or oscillating regions) well. Our numerical examples show that results would hardly change when $n_0 \geq 40$.

Example 10.4 (continued) *For the AEC data discussed in Example 10.4 of the previous subsection, we construct the NMEPM chart here for phase II monitoring of the last 48 profiles in the data, using the estimated IC parameters discussed in Example 10.4. In designing the NMEPM chart, the IC ARL is fixed at 200, and λ is chosen to be 0.1. The bandwidth h used in (10.41) and (10.42) is chosen to be $h = 1.5[n(2 - \lambda)/\lambda]^{-1/5} \sqrt{\text{Var}(x)}$. For simplicity, we choose $n_0 = n = 53$ and $\{s_k, k = 1, 2, \ldots, n_0\}$ to be equally spaced in the design interval $[-26, 78]$ of the predictor "temperature". The control limit ρ is computed to be 18.24 by simulation. Values of the charting statistic $T_{t,h,\lambda}$, for $t = 97, 98, \ldots, 144$, are shown in Figure 10.8 along with the control limit by the solid curve and solid horizontal line, respectively. As a comparison, we also compute the nonparametric fixed-effects profile monitoring (NFEPM) chart, which is the same as the NMEPM chart except that the random-effects term $f_j(x_{ij})$ is removed from the model (10.33). Note that the NFEPM* *chart can be regarded as a generalization of the NPPM chart by Zou et al. (2008); the latter assumes that design points in different profiles are deterministic and unchanged from one profile to another while the former can handle arbitrary designs. The NFEPM chart, using the same λ and h as those in the NMEPM chart, is also presented in Figure 10.8 along with its control limit 34.52, by the dashed curve and the dashed horizontal line. From the plot, it can be seen that the NMEPM chart gives a signal of profile shift around the 112th time point, and remains above the control limit for several profiles until the 120th profile. This result confirms a marked step-change that seems to have occurred around the 108th profile. The process seems to have been adjusted around the 119th profile; thus, the NMEPM charting statistic goes back below its control limit afterward. As a comparison, the NFEPM chart does not give a signal until the 118th profile.*

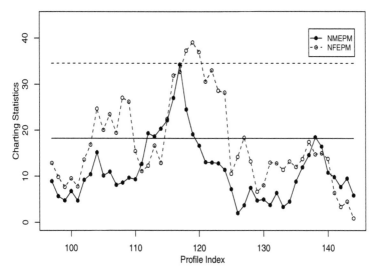

Figure 10.8 *NMEPM and NFEPM control charts for monitoring the AEC process. The solid and dashed horizontal lines indicate their control limits, respectively.*

10.4   Some Discussions

Profile monitoring is a relatively new research topic. But, it is in rapid development because of its broad applications. In this chapter, we briefly describe some existing parametric and nonparametric profile monitoring methods. Regarding the two types of methods, parametric methods would be more efficient than nonparametric methods if the related parametric model assumptions are valid. But proper justification of such parametric model assumptions is currently lacking in the profile monitoring literature. In cases when parametric models are inappropriate or unavailable for describing the observed profile data, nonparametric methods should be used.

Our description about profile monitoring in this chapter is far from complete. There are new profile monitoring methods appearing in academic journals on a daily basis. For instance, Yeh et al. (2009) discussed the profile monitoring problem with

a binary response. Yu et al. (2012) recently proposed a nonparametric profile monitoring method based on outlier detection in analyzing functional data. Paynabar et al.
(2013) proposed a nonlinear profile monitoring method for monitoring a sequence of
multichannel profiles (i.e., the response variable y is multivariate). In several recent
years, statistical monitoring of spatial profile data has attracted much attention from
statisticians. One important application of this research topic concerns the monitoring of image data. See, for instance, Megahed et al. (2011b, 2012). Another important application concerns the monitoring of environmental data (cf., Anderson and
Thompson, 2004; Morrison, 2008). Qiu and Xiang (2014, 2013) recently proposed
univariate and multivariate dynamic screening systems (DySS) for identifying irregular longitudinal patterns. The so-called dynamic screening (DS) problem handled
by these DySS methods is related to the profile monitoring problem discussed in this
chapter.

   There are many open research problems in the area of profile monitoring. For
instance, in most existing profile monitoring charts, observations of different profiles are routinely assumed to be independent. In practice, they can be autocorrelated
over time or spatially correlated, as in the conventional SPC problem. In previous
chapters, we have demonstrated several times that results from control charts that are
based on the assumption of independent observations would be unreliable or even
misleading in cases when the observations are actually correlated (cf., e.g., Example
4.5 in Subsection 4.2.2, and Example 5.3 in Subsection 5.2.2). It therefore requires
much future research to handle the profile monitoring problem in cases when both
between-profile observations and within-profile observations are correlated. Furthermore, almost all practical issues related to the conventional control charts are shared
by profile monitoring charts, including proper estimation of the IC values of certain
process parameters used in phase II process monitoring (e.g., g_0 and σ^2 used in the
chart (10.43)) and determination of some procedure parameters (e.g., λ used in the
chart (10.43)) when designing a phase II control chart. Therefore, self-starting and
adaptive control charts are also relevant in profile monitoring, which have not been
discussed sufficiently yet in the literature. When control charts are used for monitoring spatial data (e.g., images), the spatial data structures (e.g., step and roof/valley
edges in images) should be well accommodated in process monitoring. For instance,
when monitoring a sequence of images, it is critical to align them properly beforehand, which is related to the research area of image registration (cf., Qiu and Xing,
2013a,b; Xing and Qiu, 2011). Otherwise, some statistics summarizing the difference between an observed image and an IC image may not be meaningful. It therefore requires much future research to properly accommodate spatial data structures
in the related control charts. For basic image processing methods, read Gonzalez and
Woods (2007), Pratt (2007), Qiu (2005), and the references cited therein.

10.5   Exercises

10.1 For the charting statistic T_j^2 defined in (10.6), verify that its IC distribution is
      χ_2^2.

10.2 For the data presented in Table 10.1, assume that the IC error variance is known to be 0.5^2. Construct a phase II Shewhart chart to monitor the estimated error variances $\{\widehat{\sigma}_j^2, j = 1, 2, \ldots\}$, and apply this chart to the data in Table 10.1.

10.3 For the centered linear regression model (10.8), verify that

(i) $a_j^* = a_j + b_j \bar{x}$, for each j, where a_j and b_j are regression coefficients in model (10.1),

(ii) under some regularity conditions, the IC distribution of $\widehat{a}_j^*$ is $N(a_0 + b_0 \bar{x}, \sigma_0^2/n)$, where $\widehat{a}_j^*$ is the LS estimator of a_j^*, a_0, b_0, and σ_0^2 are IC values of a_j, b_j, and σ^2,

(iii) under some regularity conditions, $\widehat{a}_j^*$ and $\widehat{b}_j$ are independent of each other.

10.4 Reproduce the results in Figure 10.2.

10.5 Verify that (10.16) is valid for defining the LS estimator of the IC intercept a_0^* discussed immediately below (10.15).

10.6 For constructing the phase I Shewhart chart (10.17), verify that the IC distribution of the statistic $(\widehat{a}_j^* - \widehat{a}_0^*)/\sqrt{\frac{m-1}{mn}\widehat{\sigma}_0^2}$ is $t_{(n-2)m}$. Also, verify that the IC distribution of F_j defined in (10.19) is $F_{(n-2),(m-1)(n-2)}$.

10.7 Redo Example 10.2 using the first 20 observed profiles presented in Table 10.2.

10.8 The data presented in Table 10.4 are the first 20 observed profiles of a production process obtained during a phase II process monitoring. The design points of all profiles are the same, with values 0.1, 0.25, 0.4, 0.55, 0.7, 0.85, and 1.0. As in Example 10.3, the 2-parameter nonlinear regression model (10.30) is believed to be appropriate to describe these profiles. Use the phase II Shewhart chart (10.26) and the phase II EWMA chart (10.27)–(10.28) to monitor the estimated values $\{(\widehat{b}_j, \widehat{d}_j), j = 1, 2, \ldots, 20\}$. In both charts, use $\chi^2_{0.995,2} = 10.597$ as the control limits.

10.9 For the calibration data presented in Table 10.2, assume that the following parametric model is appropriate to describe the j-th observed profile:

$$y_{ij} = a_j + b_j x_i^{c_j} + \varepsilon_{ij}, \qquad i = 1, 2, \ldots, n, j = 1, 2, \ldots, m,$$

where a_j, b_j, and c_j are parameters. Apply the Shewhart chart (10.31)–(10.32) with $\alpha = 0.005$ to this dataset for phase I profile monitoring.

10.10 In model (10.33), for each j and $i_1, i_2 = 1, 2, \ldots, n_j$, verify that

$$\mathrm{Cov}(y_{i_1 j}, y_{i_2 j}) = \begin{cases} \gamma(x_{i_1 j}, x_{i_1 j}) + \sigma^2, & \text{when } i_1 = i_2 \\ \gamma(x_{i_1 j}, x_{i_2 j}), & \text{when } i_1 \neq i_2. \end{cases}$$

Table 10.4 *The first 20 observed profiles of a production process obtained during phase II SPC. The design points of all profiles are the same, with values 0.1, 0.25, 0.4, 0.55, 0.7, 0.85, and 1.0.*

j	y_{1j}	y_{2j}	y_{3j}	y_{4j}	y_{5j}	y_{6j}	y_{7j}
1	1.768	1.613	1.421	1.379	1.188	1.113	0.942
2	1.890	1.517	1.393	1.299	1.186	1.061	1.074
3	1.831	1.597	1.390	1.341	1.085	1.312	0.956
4	1.895	1.626	1.506	1.209	1.133	1.009	1.023
5	1.702	1.625	1.419	1.466	1.163	1.070	0.931
6	1.796	1.618	1.470	1.397	1.273	1.071	1.140
7	1.641	1.662	1.376	1.423	1.140	1.213	1.004
8	1.630	1.555	1.255	1.308	1.366	0.854	1.098
9	1.678	1.782	1.567	1.206	1.150	1.074	0.962
10	2.076	1.613	1.357	1.354	1.197	1.074	0.991
11	1.530	1.305	1.177	1.316	0.801	0.872	0.641
12	1.512	1.513	1.086	1.241	1.002	0.736	0.557
13	1.556	1.396	1.315	1.233	1.179	0.860	0.694
14	1.628	1.322	1.177	1.099	1.251	0.788	0.791
15	1.367	1.294	1.260	1.288	0.960	0.955	0.604
16	1.445	1.334	1.256	1.276	0.996	0.888	0.775
17	1.424	1.269	1.260	1.164	1.109	0.845	0.720
18	1.647	1.334	1.336	1.093	1.048	0.716	0.698
19	1.457	1.513	1.246	1.181	0.965	0.786	0.819
20	1.439	1.547	1.337	1.137	1.052	0.867	0.796

10.11 Derive formulas (10.41) and (10.42) from the minimization problem $\min_{a,b} WL(a, b; s, \lambda, t)$, where $WL(a,b; s, \lambda, t)$ is defined in the second paragraph of Subsection 10.3.2.

Appendix A

R Functions for SPC

R is a freely available language and environment for statistical computing and graphics. It can be downloaded from the Comprehensive R Archive Network (CRAN) web page of the address http://cran.r-project.org/. During the past 15 years, R has become a standard and popular statistical software package that is widely used in academia and industries. Most figures in this book are generated in R. Computation involved in many examples and tables in the book is also accomplished by using computer codes in R. In this appendix, we make a brief introduction of some basic functions in R and some R packages that are developed specifically for SPC analysis. We also make a list of all R source codes that are written by the author and used in the book. These codes are posted on the author's home page for free download.

A.1 Basic R Functions

To enter the R software, double click the R icon if you are using a computer running a windows system, or type the following UNIX/LINUX command after the command prompt "%":

    % R

Note that you need to hit the **Return** or **Enter** key to execute each UNIX/LINUX or R command after typing the command. After the above command is executed, you will enter the R software and get some description about it, followed by the R command prompt

    >

You can type any legitimate R commands after the prompt ">". Here are some examples of some basic algebraic manipulations.

    > 1+2
    [1] 3
    > 4*2/2
    [1] 4
    > 4**2
    [1] 16
    > 4^2   # 4^2 is the same as 4**2
    [1] 16

Note that "#" is a sign for comments. R will ignore all material written after this sign.
Here are several examples about how to enter data from a keyboard.

```
> tmp <- c(1,1,2,2,3,3)   # create a numerical vector (1,1,2,2,
>                         # 3,3) and assign it to the variable
>                         # named "tmp."

> tmp =  c(1,1,2,2,3,3)   # same as the previous command.

> c(1,1,2,2,3,3)          # show the vector on screen.
[1] 1 1 2 2 3 3

> tmp <- c(T,F,F,F)       # create a logical vector (T,F,F,F),
>                         # where T denotes "true" and F
>                         # denotes "false."

> tmp <- c(''HOT'',''HOT'',''COLD'') # create a character
>                                    # variable.

> length(tmp)            # check the length of the vector
[1] 3                    # "tmp."

> mode(tmp)              # check the mode of the vector "tmp."
[1] ''character''

> rep(5,3)               # create a 3-dimensional vector by
[1] 5 5 5                # repeating 5 three times.

> rep(c(1,2,3),2)        # repeat the vector (1,2,3) 2 times.
[1] 1 2 3 1 2 3

> rep(c(1,2),c(3,4))     # create a vector by repeating 1
[1] 1 1 1 2 2 2 2        # three times and then 2 four times.

> 1:10                   # create a vector consisting of
[1] 1 2 3 4 5 6 7 8 9 10 # integers from 1 to 10.

> seq(2,4,0.5)           # create a vector consisting of
[1] 2 2.5 3 3.5 4        # numbers from 2 to 4 with step 0.5.

> tmp _ scan()           # Entering data in the subsequent
1: 2 2 3                 # lines from keyboard.
4: 4 9 6
7:
> tmp
[1] 2 2 3 4 9 6

> help(seq)              # get some online help about the
>                        # command "seq."

> q()                    # exit R
```

Sometimes, we need to enter data into R from an external file, or write some results to an external file. The following R commands are for this purpose.

```
> x <- scan("filename")    # Read numerical data from the file
>                          # "filename" into a vector and assign
>                          # the vector to the variable x.
> x <- read.table("filename") # Read data in table format
>                          # from the external file "filename."
> write(x,"filename")      # Write the data assigned to x to
>                          # the external file "filename."
> write.table(x,"filename")  # Write the data in table format
>                          # to the external file "filename."
```

Here are some R commands about the binomial distribution. The use of R commands for other parametric distributions is similar.

```
> dbinom(3,10,0.3)         # compute P(X=3) when X has binomial
[1] .2668279               # distribution with n=10 and pi=0.3.

> pbinom(3,10,0.3)         # compute P(X<=3) when X has binomial
[1] .6496107               # distribution with n=10 and pi=0.3.

> qbinom(0.4,10,0.3)       # Find the smallest x so that P(X<=x)
[1] .6496107               # >= 0.4 (x is the 0.4 quantile).

> rbinom(3,10,0.3)         # Generate 3 random numbers from the
[1] 6 3 2                  # binomial(10,0.3) distribution.

> x <- seq(0,10)             # These commands compute the mean
> dist <- dbinom(x,10,0.3) # of the binomial(10,0.3) distri-
> sum(x*dist)                # bution. The command "sum"
[1] 3                        # computes the summation of all
                             # elements of a vector.

> sum(x**2*dist)-3^2       # Computes the variance of the
                           # binomial(10,0.3) distribution.
```

Here is an example to make a plot including lines of $l(x) = (1-x)^{10}$ and $l(x) = 210x^6(1-x)^4$ on the computer screen.

```
> X11()                    # Open a window on computer screen.
> x <- seq(0,100)/100      # Create a sequence for x.
> y1 <- (1-x)^10           # Compute function values for line 1.
> y2 <- 210*x^6*(1-x)^4    # Compute function values for line 2.
> plot(x,y1,type=''l'',lty=1,xlab=''pi'',ylab=''Likelihood'')
> lines(x,y2,lty=2)
> dev.off()                # Close the plot window.
```

If you want to put the plot in a postscript file, instead of showing it in a window, then you can replace the first and last commands above by the following two commands, respectively.

```
> postscript(file=''likelihood.ps'',width=4in,height=4in,
  horizontal=F)
> graphics.off()
```

Again, you can use the command "help(postscript)" to read more about the R function postscript().

A.2   R **Packages for SPC**

There are several R packages written specifically for SPC. In Chapters 4 and 5, we have used one of them (i.e., the package spc). This package along with several others are briefly described below.

cpm **(Ross, 2012)** The package cpm can be used for phase II SPC of univariate processes under the change point detection (CPD) framework (cf., Chapter 6). It contains R functions for parametric monitoring of processes with normal and Bernoulli distributions. It also provides R functions for executing certain nonparametric CPD charts.

mnspc **(Bezener and Qiu, 2011)** The package mnspc accomplishes the MLL-CUSUM chart (9.38)–(9.39) discussed in Subsection 9.3.2. A function is provided to categorize components of multivariate response vectors. Tools for setting up a CUSUM chart for the transformed data are included. The CUSUM chart can also be applied to cases when some (or all) of the multivariate response components are binary-categorical.

msqc **(Santos-Fernandez, 2012)** The package msqc provides R functions for executing certain conventional multivariate control charts, such as the multivariate Shewhart charts based on Hotelling's T^2 statistic, MCUSUM, and MEWMA charts. It also includes some tools for assessing multivariate normality.

qcc **(Scrucca, 2010)** The package qcc contains functions for making various Shewhart charts, and the conventional CUSUM and EWMA charts.

spc **(Knoth, 2011)** The package spc can be used to compute the zero-state and steady-state IC and OC ARL values of the one- and two-sided EWMA, CUSUM, and Shiryaev-Roberts charts for monitoring the mean or variance of a normally distributed univariate production process. This package can also be used for determining the control limits of the above-mentioned control charts when their ARL_0 values and other parameters are given, and the optimal allowance constant k of a CUSUM chart or the optimal weighting parameter λ of an EWMA chart when the ARL_0 value and a target are given.

A.3   **List of R Functions Used in the Book**

This part gives a list of all R functions written by the author for this book. The R functions are organized by chapters. A brief description about the main purpose of each function is also given.

Chapter 2

fig21.r R-code for making Figure 2.1.

fig22.r R-code for making Figure 2.2.

fig23.r R-code for making Figure 2.3.

fig24.r R-code for making Figure 2.4.

fig26.r R-code for making Figure 2.6.

fig27.r R-code for making Figure 2.7.

fig28.r R-code for making Figure 2.8.

fig29.r R-code for making Figure 2.9.

fig210.r R-code for making Figure 2.10.

fig211.r R-code for making Figure 2.11.

fig212.r R-code for making Figure 2.12.

Chapter 3

fig31.r R-code for making Figure 3.1. It reads data from the file "example31.summary."

fig32.r R-code for making Figure 3.2.

fig33.r R-code for making Figure 3.3.

fig34.r R-code for making Figure 3.4.

fig35.r R-code for making Figure 3.5. It reads data from the file "example31.summary."

fig36.r R-code for making Figure 3.6. It reads data from the file "example33.dat."

fig37.r R-code for making Figure 3.7.

fig38.r R-code for making Figure 3.8.

fig39.r R-code for making Figure 3.9.

fig310.r R-code for making Figure 3.10. It reads data from "example37.dat."

fig311.r R-code for making Figure 3.11.

table33.r R-code for making Table 3.3

Chapter 4

ex415.r R-code for generating the data file "ex415.dat" used in Exercise 4.15.

example45ARL0.r R-code for computing the ARL_0 values of the upward CUSUM chart when the data are auto-correlated and follow the AR(1) model. It is used in Example 4.5 and Table 4.3.

example45ARL1.r R-code for computing the ARL_1 values of the upward CUSUM chart when the data are auto-correlated and follow the AR(1) model. It is used in Example 4.5 and Table 4.3.

example46.r R-code for computing the residuals of the soil data used in Example 4.6.

example412.r R-code for making Figure 4.12. It also generates the data file "example412.dat" used in Example 4.12.

fig41.r R-code for making Figure 4.1. It reads data from "example41.dat."

fig42.r R-code for making Figure 4.2. It reads data from "example41.dat."

fig43.r R-code for making Figure 4.3. It reads data from "example41.dat."

fig44.r R-code for making Figure 4.4. It reads data from "example41.dat."

fig45.r R-code for making Figure 4.5.

fig46.r R-code for making Figure 4.6.

fig47.r R-code for making Figure 4.7. It reads data from "soil.dat"

fig48.r R-code for making Figure 4.8. It also generates the data file "example48.dat" used in Example 4.8.

fig49.r R-code for making Figure 4.9. It reads data from "example48.dat."

fig410.r R-code for making Figure 4.10. It also generates the data file "example49.dat" used in Example 4.9.

fig411.r R-code for making Figure 4.11.

table44.r R-code for computing the ARL_0 values of the upward CUSUM chart when the data are auto-correlated and follow the AR(1) model and when they are grouped into batches with batch size m. It is used in Table 4.4.

Chapter 5

example53ARL0.r R-code for computing the ARL_0 values of the EWMA chart when the data are auto-correlated. It is used in Example 5.3 and Table 5.2.

example53ARL1.r R-code for computing the ARL_1 values of the EWMA chart when the data are auto-correlated. It is used in Example 5.3 and Table 5.2.

example57RhoU.r R-code for searching ρ_U value used in Example 5.7.

example59.r R-code for generating data file "example59.dat" and making Figure 5.13 in Example 5.9.

fig51.r R-code for making Figure 5.1.

fig52.r R-code for making Figure 5.2. It reads data from "example51.dat."

fig53.r R-code for making Figure 5.3.

fig54.r R-code for making Figure 5.4.

fig55.r R-code for making Figure 5.5.

fig55ARL0chisq.r R-code for computing the actual ARL_0 values of the EWMA chart when the actual process distribution is chi-square. It is used for making Figure 5.5.

fig55ARL0normal.r R-code for computing the actual ARL_0 values of the EWMA chart when the actual process distribution is normal. It is used for making Figure 5.5.

fig55ARL0t.r R-code for computing the actual ARL_0 values of the EWMA chart when the actual process distribution is t. It is used for making Figure 5.5.

fig56.r R-code for making Figure 5.6. It also generates the data file "example54.dat" used in Example 5.4.

fig57.r R-code for making Figure 5.7.

fig58.r R-code for making Figure 5.8. It also generates the data file "example55.dat" used in Example 5.5.

fig59.r R-code for making Figure 5.9. It reads data from "example55.dat."

fig510.r R-code for making Figure 5.10. It also generates the data file "example56.dat" used in Example 5.6.

fig511TwoSideRho.r R-code for computing the actual ARL_0 value of a 2-sided EWMA chart for detecting process variance shifts. It is used in Example 5.6 and Figure 5.11.

fig511.r R-code for making Figure 5.11. It also generates the data file "example56.dat" used in Example 5.6.

fig512.r R-code for making Figure 5.12. It also generates the data file "example57.dat" used in Example 5.7.

fig514.r R-code for making Figure 5.14.

fig515.r R-code for making Figure 5.15. It also generates the data file "example510.dat" used in Example 5.10.

table53.r R-code for computing numbers in Table 5.3. It needs to use the R-package spc which should be installed beforehand.

table54RhoU.r R-code for computing ρ_U values in Table 5.4.

table54RhoL.r R-code for computing ρ_L values in Table 5.4.

table55ARL0TwoSided.r R-code for computing ARL_0 values of the two-sided EWMA chart considered in Table 5.5.

table55ARL0Upward.r R-code for computing ARL_0 values of the upward EWMA chart considered in Table 5.5.

table55ARL1TwoSided.r R-code for computing ARL_1 values of the two-sided EWMA chart considered in Table 5.5.

table55ARL1Upward.r R-code for computing ARL_1 values of the upward EWMA chart considered in Table 5.5.

Chapter 6

example61.r R-code for Example 6.1. It writes the related data to the file "example61.dat."

example62.r R-code for Example 6.2. It writes the related data to the file "example62.dat." It also creates "fig61.ps."

example64.r R-code for Example 6.4. It writes the related data to the file "example64.dat." It also creates "fig63.ps."

example65.r R-code for Example 6.5. It writes the related data to the file "example65.dat." It also creates "fig65.ps."

fig62.r R-code for making Figure 6.2.

fig66.r R-code for making Figure 6.6.

Chapter 7

ex722.r R-code for generating data used in Exercise 7.22.

example71.r R-code for Example 7.1. It writes the related data to the file "example71.dat." It also makes "fig73.ps."

example71a.r R-code for Example 7.1(continued). It also makes "fig74.ps."

example72.r R-code for Example 7.2. It writes the related data to the file "example72.dat." It also makes "fig75.ps."

example73.r R-code for Example 7.3. It writes the related data to the file "example73.dat." It also makes "fig76.ps."

example73Searchh1.r R-code for searching for the control limit h of the chart (7.25).

example73Searchh2.r R-code for searching for the control limit h of the chart (7.26)–(7.27).

example74.r R-code for Example 7.4. It writes the related data to the file "example74a.dat," "example74b.dat," and "example74c.dat." It also makes "fig77.ps."

example74Searchh1.r R-code for searching for the control limit h of the chart (7.32)–(7.33).

example74Searchh2.r R-code for searching for the control limit h of the chart (7.30)–(7.31).

example75.r R-code for Example 7.5. It writes the related data to the file "example75.dat." It also makes "fig78.ps."

example76.r R-code for Example 7.6. It writes the related data to the file "example76.dat." It also makes "fig79.ps."

example76Searchh.r R-code for searching for the control limit h of the chart (7.45)–(7.46).

example77.r R-code for Example 7.7. It writes the related data to the file "example77.dat." It also makes "fig710.ps."

example78.r R-code for Example 7.8. It writes the related data to the file "example78.dat." It also makes "fig711.ps."

fig71.r R-code for making Figure 7.1.

fig72.r R-code for making Figure 7.2.

Chapter 8

ex85.r R-code for generating data used in Exercise 8.5.

ex86.r R-code for generating data used in Exercise 8.6.

ex810.r R-code for generating data used in Exercise 8.10.

ex815.r R-code for generating data used in Exercise 8.15.

ex816.r R-code for generating data used in Exercise 8.16.

example81.r R-code for Example 8.1. It writes the related data to the file "example81.dat." It also makes "fig82.ps."

example82.r R-code for Example 8.2. It writes the related data to the file "example82.dat." It also makes "fig83.ps."

example83.r R-code for Example 8.3. It writes the related data to the file "example83.dat." It also makes "fig84.ps."

example84.r R-code for Example 8.4. It writes the related data to the file "example84.dat." It also makes "fig85.ps."

example85.r R-code for Example 8.5. It writes the related data to the file "example85.dat." It also makes "fig86.ps."

example86.r R-code for Example 8.6. It writes the related data to the file "example86.dat." It also makes "fig88.ps."

example87.r R-code for Example 8.7. It writes the related data to the file "example87.dat." It also makes "fig89.ps."

example88.r R-code for Example 8.8. It writes the related data to the file "example88.dat." It also makes "fig810.ps."

fig87.r R-code for making Figure 8.7.

Chapter 9

ex93.r R-code for generating data used in Exercise 9.3.

ex97.r R-code for generating data used in Exercise 9.7.

ex99.r R-code for generating data used in Exercise 9.9.

ex913.r R-code for generating data used in Exercise 9.13.

example91.r R-code for Example 9.1. It writes the related data to the file "example91.dat." It also makes "fig93.ps."

example92.r R-code for Example 9.2. It writes the related data to the file "example92.dat." It also makes "fig94.ps."

example93.r R-code for Example 9.3. It writes the related data to the file "example93.dat." It also makes "fig95.ps."

fig91.r R-code for making Figure 9.1.

fig92.r R-code for making Figure 9.2.

fig96.r R-code for making Figure 9.6. It reads data from "soilResidual.dat."

fig97.r R-code for making Figure 9.7. It reads data from "soil.dat."

fig98.r R-code for making Figure 9.8. It reads data from files "fig98crit1.dat" and "fig98crit2.dat."

fig99.r R-code for making Figure 9.9.

Chapter 10

ex107.r R-code for generating data used in Exercise 10.7.

example101.r R-code for Example 10.1. It writes the related data to the file "example101.dat." It also makes "fig101.ps."

example101a.r R-code for Example 10.1 (continued). It also makes "fig102.ps."

example102.r R-code for Example 10.2. It writes the related data to the file "example102.dat." It also makes "fig103.ps."

example103.r R-code for Example 10.3. It writes the related data to the file "example103.dat." It also makes "fig104.ps."

example103a.r R-code for Example 10.3 (continued). It also makes "fig105.ps."

Appendix B

Datasets Used in the Book

This appendix gives a list of all datasets used in the book, along with some brief descriptions. They are posted on the author's home page for free download.

ex35.dat Data used in Exercise 3.5.

ex39.dat Data used in Exercise 3.9.

ex318.dat Data used in Exercise 3.18.

ex415.dat Data used in Exercise 4.15.

ex421.dat Data used in Exercise 4.21.

ex722.dat Data used in Exercise 7.22.

ex85.dat Data used in Exercise 8.5.

ex86.dat Data used in Exercise 8.6.

ex810.dat Data used in Exercise 8.10.

ex815.dat Data used in Exercise 8.15.

ex816.dat Data used in Exercise 8.16.

ex93.dat Data used in Exercise 9.3.

ex97.dat Data used in Exercise 9.7.

ex99.dat Data used in Exercise 9.9.

ex913.dat Data used in Exercise 9.13.

ex107.dat Data used in Exercise 10.7.

example31.dat Original data used in Example 3.1. Note that we only use the first 120 data points of this file in the example.

example31.summary Summary statistics included in Table 3.2. Note that we only use the first 24 lines of this file in Table 3.2.

example33.dat Original data used in Example 3.3.

example37.dat Original data used in Example 3.7.

example41.dat Original data used in Example 4.1.

example48.dat Original data used in Example 4.8.

example49.dat Original data used in Example 4.9.

example412.dat Original data used in Example 4.12.

example51.dat Original data used in Example 5.1.

example54.dat Original data used in Example 5.4.

example55.dat Original data used in Example 5.5.

example56.dat Original data used in Example 5.6.

example57.dat Data used in Example 5.7. The first four columns are the sample means in four different cases, and the remaining four columns are the sample standard deviations.

example59.dat The columns in this dataset are n, X_n, $\overline{X}_n$, s_n, Z_n, E_n, and $E_{n,SS}$ used in Example 5.9.

example510.dat Original data used in Example 5.10.

example61.dat The three columns in the data are X_i, $\widetilde{S}_i^2$, and $\widetilde{\widetilde{S}}_i^2$ presented in Table 6.1 of Example 6.1.

example62.dat The three columns in the data are i, X_i, $X^2(2,h) + Q(h,30)$, and $X^2(1,h) + Q(h,20)$ shown in Figure 6.1 of Example 6.2.

example64.dat The five columns in the data are X_n, W_n, S_n^2, $T_{max,n}$, and h_n presented in Table 6.3 of Example 6.4.

example65.dat The five columns in the data denote X_n, W_n, S_n^2, $B_{max,n}$, and h_n presented in Table 6.6 of Example 6.5.

example71.dat The five columns in the data are X_1, X_2, X_3 presented in Table 7.1, and $\overline{X}$ and $T_{1,i}^2$ used in Example 7.1.

example72.dat The first three columns in the data are X_1, X_2, X_3 presented in Figure 7.5(a)(c)(e), and the last column is $T_{0,n}^2$ used in Example 7.2.

example73.dat The first three columns in the data are X_1, X_2, X_3 presented in Figure 7.6(a)(b)(c), the fourth column is C_n defined in (7.25), and the fifth column is $\widetilde{C}_n$ defined in (7.26). This data is used in Example 7.3.

example74a.dat Data in case (i) considered in Example 7.4. The first three columns contain observations of X_1, X_2, X_3, the remaining three columns contain values of T_n^2, C_n defined in (7.32), and $\widetilde{C}_n$ defined in (7.31).

example74b.dat Data in case (ii) considered in Example 7.4. The first three columns contain observations of X_1, X_2, X_3, the remaining three columns contain values of T_n^2, C_n defined in (7.32), and $\widetilde{C}_n$ defined in (7.31).

example74c.dat Data in case (iii) considered in Example 7.4. The first three columns contain observations of X_1, X_2, X_3, the remaining three columns contain values of T_n^2, C_n defined in (7.32), and $\widetilde{C}_n$ defined in (7.31).

example75.dat The first three columns in the data are X_1, X_2, X_3, the fourth–sixth columns are E_n defined in (7.39), and the last column is V_n^2 defined in (7.40). This data is used in Example 7.5.

example76.dat The first three columns in the data are X_1, X_2, X_3, the fourth–sixth columns are u_n defined in (7.44), the seventh–ninth columns are $E_{n,SS}$ defined in

(7.45), and the last column is $V^2_{n,SS}$ defined in (7.46). This data is used in Example 7.6.

example77.dat The first three columns in the data are X_1, X_2, X_3, the fourth–sixth columns are $\mathbf{Y}_n$ defined in (7.50), and the last column is C_n defined in (7.52). This data is used in Example 7.7.

example78.dat The first three columns in the data are X_1, X_2, X_3, the fourth–sixth columns are $\overline{\mathbf{X}}_{0,r}$, and the last two columns are $T^2_{max,n}$ and h_n defined in (7.54) and (7.56). This data is used in Example 7.8.

example81.dat The first ten columns are the batch data with batch size $m = 10$, and the last column is Ψ_i defined in (8.1). This data is used in Example 8.1.

example82.dat The first five columns are the batch data with batch size $m = 5$, and the last column is the median $X_{i(3)}$ shown in Figure 8.3(b). This data is used in Example 8.2.

example83.dat The first six columns are the batch data with batch size $m = 6$, and the last two columns are Ψ_n and C_n^+ defined in (8.7) and (8.8). This data is used in Example 8.3.

example84.dat The first five columns are the batch data with batch size $m = 5$, the sixth column is W_n, and the last two columns are C_n^+ and C_n^- defined in (8.10) and (8.12). This data is used in Example 8.4.

example85.dat The first ten columns are the batch data with batch size $m = 10$, the eleventh column is Ψ_n, and the last column is E_n defined in (8.17). This data is used in Example 8.5.

example86.dat The first column is the reference sample, the second column contains the phase II data, the third column is $F_{n,\lambda}(X_n)$, the fourth column is $\widehat{F}_0(X_n)$, the fifth column is $\widetilde{G}_n$, and the last column is $E_{n,SS}$. This data is used in Example 8.6.

example87.dat The first and the second columns are X_n and $T_{max,n}$ defined in (8.28). This data is used in Example 8.7.

example88.dat The first column is the data X_n, and the last two columns are B_n and $C_{n,P}$ defined in (8.34). This data is used in Example 8.8.

example91.dat Original data used in Example 9.1.

example92.dat The first two columns are the original data, and the last three columns are $r(X_n; F_{0M})$, $S_n(F_{0M})$ and $S_n^*(F_{0M})$ defined in (9.19). This data is used in Example 9.2.

example93.dat The first two columns are the original data, and the last column is $r(X_{ni}; F_{0M})$ used in (9.22). This data is used in Example 9.3.

example101.dat The profile data presented in Table 10.1 of Example 10.1.

example102.dat The profile data presented in Table 10.2 of Example 10.2.

example103.dat The profile data presented in Table 10.3 of Example 10.3.

fig87.dat The first five columns are the batch data with batch size $m = 5$, the sixth column is W_n, and the last column is E_n defined in (8.20). This data is used in Figure 8.7.

fig98crit1.dat The first column is the case id, and the second column is C_{n1} used in Figure 9.8(a).

fig98crit2.dat The first column is the case id, and the second column is C_{n15} used in Figure 9.8(b).

fig910.dat The first column is the case id, and the second column is C_n used in Figure 9.10.

soil.dat This is the aluminum smelter dataset. It is used in Example 2.5, Figure 2.12, Example 4.6, Figure 4.7, Figure 9.7.

soilResidual.dat This dataset contains the residuals of the soil.dat after auto-correlation is removed by a time series model. It is used in Figure 9.7.

Bibliography

B.M. Adams and I.T. Tseng. Robustness of forecast-based monitoring schemes. *Journal of Quality Technology*, 30:328–339, 1998.

A. Agresti. *Categorical data analysis*. John Wiley & Sons, New York, NY, USA, 2nd edition, 2002.

A. Agresti and B.A. Coull. Approximate is better than 'exact' for interval estimation of binomial proportions. *The American Statistician*, 52:119–126, 1998.

W. Albers. Control charts for health care monitoring under overdispersion. *Metrika*, 74:67–83, 2011.

W. Albers and W.C.M. Kallenberg. Empirical nonparametric control charts: estimation effects and corrections. *Journal of Applied Statistics*, 31:345–360, 2004.

W. Albers and W.C.M. Kallenberg. CUMIN charts. *Metrika*, 70:111–130, 2009.

W. Albers, W.C.M. Kallenberg, and S. Nurdiati. Data driven choice of control charts. *Journal of Statistical Planning and Inference*, 136:909–941, 2006.

D. Allen. The relationship between variable selection and data augmentation and a method for prediction. *Technometrics*, 16:125–127, 1974.

J.A. Alloway, Jr. and M. Raghavachari. Control chart based on the Hodges-Lehmann estimator. *Journal of Quality Technology*, 23:336–347, 1991.

F.B. Alt. Multivariate quality control. *The Encyclopedia of Statistical Sciences (S. Kotz, N.L. Johnson, and C.R. Read, eds.)*, pages 110–122, 1985.

N.S. Altman. Kernel smoothing of data with correlated errors. *Journal of the American Statistical Association*, 85:749–759, 1990.

L.C. Alwan. Effects of autocorrelation on control chart performance. *Communications in Statistics - Theory and Methods*, 21:1025–1049, 1992.

L.C. Alwan and H.V. Roberts. Time-series modeling for statistical process control. *Journal of Business and Economic Statistics*, 6:87–95, 1988.

R. Amin and A.J. Searcy. A nonparametric exponentially weighted moving average control scheme. *Communications in Statistics - Simulation and Computation*, 20:1049–1072, 1991.

R. Amin and O. Widmaier. Sign control charts with variable sampling intervals. *Communications in Statistics - Theory and Methods*, 28:1961–1985, 1999.

R. Amin, M.R. Reynolds, Jr., and S.T. Bakir. Nonparametric quality control charts based on the sign statistic. *Communications in Statistics - Theory and Methods*, 24:1597–1623, 1995.

M.J. Anderson and A.A. Thompson. Multivariate control charts for ecological and environmental monitoring. *Ecological Applications*, 14:1921–1935, 2004.

T.W. Anderson. *An Introduction to Multivariate Statistical Analysis*. John Wiley & Sons, New York, NY, USA, 3rd edition, 2003.

F. Aparisi, J. Jabaloyes, and A. Carrion. Statistical properties of the $|S|$ multivariate control chart. *Communications in Statistics - Theory and Methods*, 28:2671–2686, 1999.

F. Aparisi, J. Jabaloyes, and A. Carrion. Generalized variance chart design with adaptive sample sizes. *Communications in Statistics - Simulation and Computation*, 30:931–948, 2001.

D.W. Apley and H.C. Lee. Design of exponentially weighted moving average control charts for autocorrelated processes with model uncertainty. *Technometrics*, 45:187–198, 2003.

D.W. Apley and J. Shi. The GLRT for statistical process control of autocorrelated processes. *IIE Transactions*, 31:1123–1134, 1999.

R.B. Ash. *Real Analysis and Probability*. Academic Press, San Diego, CA, USA, 1972.

S.T. Bakir. A distribution-free Shewhart quality control chart based on signed-ranks. *Quality Engineering*, 16:613–623, 2004.

S.T. Bakir. Distribution-free quality control charts based on signed-rank-like statistics. *Communications in Statistics - Theory and Methods*, 35:743–757, 2006.

S.T. Bakir and M.R. Reynolds, Jr. A nonparametric procedure for process control based on within group ranking. *Technometrics*, 21:175–183, 1979.

M.S. Bartlett. Property of sufficiency and statistical tests. *Proceedings of the Royal Society of London (Series A)*, 160:268–282, 1937.

M.S. Bartlett. Approximate confidence intervals: III. a bias correction. *Biometrika*, 42:201–204, 1955.

M.S. Bartlett and D.G. Kendall. The statistical analysis of variance-heterogeneity and the logarithmic transformation. *Supplement to the Journal of the Royal Society of London*, 8:128–138, 1946.

D.M. Bates and D.G. Watts. *Nonlinear Regression Analysis and Its Applications*. John Wiley & Sons, New York, NY, USA, 2007.

R.E. Bellman and S.E. Dreyfus. *Applied Dynamic Programming*. Princeton University Press, Princeton, NJ, USA, 1962.

S. Bersimis, S. Psarakis, and J. Panaretos. Multivariate statistical process control charts: an overview. *Quality and Reliability Engineering International*, 23:517–543, 2007.

M. Bezener and P. Qiu. mnspc: multivariate nonparametric statistical process control. R *package version 1.01*, pages http://CRAN.r–project.org/package=mnspc, 2011.

P.K. Bhattacharya. Some aspects of change-point analysis. In *Change-Point Prob-*

lems (E. Carlstein, H.G. Müller, and D. Siegmund, eds.), pages 28–56, Hayward, CA, USA, 1994. IMS Monograph.

P.J. Bickel and B. Li. Regularization in statistics. *Test*, 15:271–344, 2006.

A.F. Bissell. CUSUM techniques for quality control (with discussions). *Applied Statistics*, 18:1–30, 1969.

A.F. Bissell. The performance of control charts and Cusums under linear trend. *Applied Statistics*, 33:145–151, 1984a.

A.F. Bissell. Estimation of linear trend from a CUSUM chart or tabulation. *Applied Statistics*, 33:152–157, 1984b.

G. Black, J. Smith, and S. Wells. The impact of Weibull data and autocorrelation on the performance of the Shewhart and exponentially weighted moving average control charts. *International Journal of Industrial Engineering Computations*, 2:575–582, 2011.

J.M. Boone and S. Chakraborti. Two simple Shewhart-type multivariate nonparametric control charts. *Applied Stochastic Models in Business and Industry*, 28:130–140, 2012.

C.M. Borror, C.W. Champ, and S.E. Ridgon. Poisson EWMA control charts. *Journal of Quality Technology*, 30:352–361, 1998.

C.M. Borror, D.C. Montgomery, and G.C. Runger. Robustness of the EWMA control chart to non-normality. *Journal of Quality Technology*, 31:309–316, 1999.

P.D. Bourke. The geometric CUSUM chart with sampling inspection for monitoring fraction defective. *Journal of Applied Statistics*, 28:951–972, 2001.

G.E.P. Box and G.C. Tiao. A change in level of non-stationary time series. *Biometrika*, 52:181–192, 1965.

G.E.P. Box, G.M. Jenkins, and G.C. Reinsel. *Time Series Analysis: Forecasting and Control*. Springer, New York, NY, USA, 4th edition, 2008.

L. Breiman. Heuristics of instability and stabilization in model selection. *The Annals of Statistics*, 24:2350–2383, 1996.

L. Breiman, J.H. Friedman, R.A. Olshen, and C.J. Stone. *Classification and Regression Trees*. Wadsworth, Belmont, CA, USA, 1984.

P.J. Brockwell and R.A. Davis. *Time Series: Theory and Methods*. Springer, New York, NY, USA, 2nd edition, 2009.

B.E. Brodsky and B.S. Darkhovsky. *Nonparametric Methods in Change-Point Problems*. Kluwer Academic Publishers, AA Dordrecht, The Netherlands, 1993.

D. Brook and D.A. Evans. An approach to the probability distribution of CUSUM run length. *Biometrika*, 59:539–549, 1972.

L.D. Brown, T.T. Cai, and A. DasGupta. Interval estimation for a binomial proportion. *Statistical Science*, 16:101–133, 2001.

G. Capizzi and G. Masarotto. An adaptive exponentially weighted moving average control chart. *Technometrics*, 45:199–207, 2003.

G. Capizzi and G. Masarotto. A least angle regression control chart for multidimensional data. *Technometrics*, 53:285–296, 2011.

G. Casella and R.L. Berger. *Statistical Inference*. Duxbury Press, Pacific Grove, CA, USA, 2nd edition, 2002.

P. Castagliola and P.E. Maravelakis. A CUSUM control chart for monitoring the variance when parameters are estimated. *Journal of Statistical Planning and Inference*, 141:1463–1478, 2011.

S. Chakraborti and S. Eryilmaz. A nonparametric Shewhart-type signed-rank control chart based on runs. *Communications in Statistics - Simulation and Computation*, 36:335–356, 2007.

S. Chakraborti and P. van der Laan. Precedence tests and confidence bounds for complete data: an overview and some results. *Journal of the Royal Statistical Society (Series D) - The Statistician*, 45:351–369, 1996.

S. Chakraborti and P. van der Laan. An overview of precedence tests for censored data. *Biometrical Journal*, 39:99–116, 1997.

S. Chakraborti and P. van der Laan. Precedence probability and prediction intervals. *Journal of the Royal Statistical Society (Series D) - The Statistician*, 49:219–228, 2000.

S. Chakraborti, P. van der Laan, and S.T. Bakir. Nonparametric control charts: an overview and some results. *Journal of Quality Technology*, 33:304–315, 2001.

S. Chakraborti, P. van der Laan, and M.A. van de Wiel. A class of distribution-free control charts. *Journal of the Royal Statistical Society (Series C) - Applied Statistics*, 53:443–462, 2004.

S. Chakraborti, S. Eryilmaz, and S.W. Human. A phase II nonparametric control chart based on precedence statistics with runs-type signaling rules. *Computational Statistics and Data Analysis*, 53:1054–1065, 2009.

I.M. Chakravarti, R.G. Laha, and J. Roy. *Handbook of Methods of Applied Statistics*. John Wiley & Sons, New York, NY, USA, 1967.

C.W. Champ and W.H. Woodall. Exact results for Shewhart control charts with supplementary runs rules. *Technometrics*, 29:393–399, 1987.

S. Chatterjee and P. Qiu. Distribution-free cumulative sum control charts using bootstrap-based control limits. *Annals of Applied Statistics*, 3:349–369, 2009.

J. Chen and A.K. Gupta. Testing and locating variance changepoints with application to stock prices. *Journal of the American Statistical Association*, 92:739–747, 1997.

J. Chen and A.K. Gupta. *Parametric Statistical Change Point Analysis*. Birkhäuser Boston, Boston, MA, USA, 2000.

S. Chib. Estimation and comparison of multiple change-points models. *Journal of Economics*, 86:221–241, 1998.

E. Chicken, J.J. Pignatiello, Jr., and J.R. Simpson. Statistical process monitoring of nonlinear profiles using wavelets. *Journal of Quality Technology*, 41:198–212, 1998.

H. Choi, H. Ombao, and B. Ray. Sequential change-point detection methods for nonstationary time series. *Technometrics*, 50:40–52, 2008.

Y.M. Chou, A.M. Polansky, and R.L. Mason. Transforming non-normal data to normality in statistical process control. *Journal of Quality Technology*, 30: 133–141, 1998.

K.L. Chung. *A Course In Probability Theory*. Academic Press, San Diego, CA, USA, 3rd edition, 2001.

W.G. Cochran. *Sampling Techniques: Probability and Mathematical Statistics*. John Wiley & Sons, New York, NY, USA, 2nd edition, 1977.

A.F.B. Costa. Joint $\overline{X}$ and R charts with variable parameters. *IIE Transactions*, 30: 505–514, 1998.

R.B. Crosier. Multivariate generalizations of cumulative sum quality-control schemes. *Technometrics*, 30:291–303, 1988.

S.V. Crowder. Average run lengths of exponentially weighted moving average control charts. *Journal of Quality Technology*, 19:161–164, 1987a.

S.V. Crowder. A simple method for studying run length distributions of exponentially weighted moving average control charts. *Technometrics*, 29:401–407, 1987b.

S.V. Crowder. Design of exponentially weighted moving average schemes. *Journal of Quality Technology*, 21:155–162, 1989.

S.V. Crowder and M. Hamilton. Average run lengths of EWMA control charts for monitoring a process standard deviation. *Journal of Quality Technology*, 24: 44–50, 1992.

M. Davidian and D.M. Giltinan. *Nonlinear Models for Repeated Measurement Data*. Chapman and Hall, London, UK, 1995.

R.B. Davis and W.H. Woodall. Performance of the control chart trend rule under linear shift. *Journal of Quality Technology*, 20:260–262, 1988.

C. Derman and S.M. Ross. *Statistical Aspects of Quality Control*. Academic Press, San Diego, CA, USA, 1997.

J.L. Devore. *Probability and Statistics for Engineering and the Sciences*. Duxbury Press, Pacific Grove, CA, USA, 8th edition, 2011.

P.J. Diggle, K.Y. Liang, and S.L. Zeger. *Analysis of Longitudinal Data*. Oxford University Press, New York, USA, 1994.

Y. Ding, L. Zeng, and S. Zhou. Phase I analysis for monitoring nonlinear profiles in manufacturing processes. *Journal of Quality Technology*, 38:199–216, 2006.

W.J. Dixon and F.J. Massey. *Introduction to Statistical Analysis*. McGraw-Hill Book Company, New York, NY, USA, 3rd edition, 1969.

R. Domangue and S.C. Patch. Some omnibus exponentially weighted moving average statistical process monitoring schemes. *Technometrics*, 33:299–313, 1991.

J.L. Doob. *Stochastic Processes*. John Wiley & Sons, New York, NY, USA, 1953.

F.D.J. Dunstan, A.B.J. Nix, and J.F. Reynolds. *Statistical Tables*. R.N.D. Publications, Cardiff, United Kingdom, 1979.

M.L. Eaton. *Multivariate Statistics*. John Wiley & Sons, New York, NY, USA, 1983.

M.L. Eaton. *Multivariate Statistics: A Vector Space Approach*. Lecture Notes–Monograph Series, Volume 53, Institute of Mathematical Statistics, Beachwood, OH, USA, 2007.

B. Efron. Bootstrap methods: another look at the jackknife. *The Annals of Statistics*, 7:1–26, 1979.

B. Efron. Least angle regression. *The Annals of Statistics*, 32:407–489, 2004.

B. Efron and R. Tibshirani. *An Introduction to the Bootstrap*. Chapman & Hall/CRC, Boca Raton, FL, USA, 1993.

J.R. English, S.C. Lee, T.W. Martin, and C. Tilmon. Detecting changes in autoregressive processes with X and EWMA charts. *IIE Transactions*, 32:1103–1113, 2000.

W.D. Ewan. When and how to use CUSUM charts. *Technometrics*, 5:1–22, 1963.

J. Fan and R. Li. Variable selection via nonconcave penalized likelihood and its oracle properties. *Journal of the American Statistical Association*, 96:1348–1360, 2001.

J. Fan and J. Zhang. Two-step estimation of functional linear models with applications to longitudinal data. *Journal of the Royal Statistical Society (Series B)*, 62:303–322, 2000.

J. Fan, C. Zhang, and J. Zhang. Generalized likelihood ratio statistics and Wilks phenomenon. *The Annals of Statistics*, 29:153–193, 2001.

S.K.S. Fan, Y.J. Chang, and N. Aidara. Nonlinear profile monitoring of reflow process data based on the sum of sine functions. *Quality and Reliability Engineering International*, page doi: 10.1002/qre.1425, 2012.

K.T. Fang, S. Kotz, and K.W. Ng. *Symmetric Multivariate and Related Distributions*. Chapman and Hall, New York, NY, USA, 1990.

P. Fearnhead. Exact and efficient bayesian inference for multiple changepoint problems. *Statistics and Computing*, 16:203–213, 2006.

W.H. Fellner. AS 258: average run length for cumulative sum scheme. *Applied Statistics*, 39:402–412, 1990.

T.S. Ferguson. *Mathematical Statistics: A Decision Theoretical Approach*. Academic Press, Inc., New York, NY, USA, 1967.

C. Fuchs and R.S. Kenett. *Multivariate Quality Control*. Marcel Dekker, New York, NY, USA, 1998.

F.F. Gan. Monitoring Poisson observations using modified exponentially weighted moving average control charts. *Communications in Statistics: Simulation and Computation*, 19:103–124, 1990a.

F.F. Gan. Monitoring observations generated from a binomial distribution using modified exponentially weighted moving average control chart. *Journal of Statistical Computation and Simulation*, 37:45–60, 1990b.

F.F. Gan. An optimal design of CUSUM quality control charts. *Journal of Quality Technology*, 23:279–286, 1991a.

F.F. Gan. EWMA control chart under linear drift. *Journal of Statistical Computation and Simulation*, 38:181–200, 1991b.

F.F. Gan. Exact run length distributions for one-sided exponential CUSUM schemes. *Statistica Sinica*, 2:197–312, 1992a.

F.F. Gan. CUSUM control charts under linear drift. *The Statistician*, 41:71–84, 1992b.

F.F. Gan. The run length distribution of a cumulative sum control chart. *Journal of Quality Technology*, 25:205–215, 1993a.

F.F. Gan. An optimal design of CUSUM control charts for binomial counts. *Journal of Applied Statistics*, 20:445–460, 1993b.

F.F. Gan. Joint monitoring of process mean and variance using exponentially weighted moving average control charts. *Technometrics*, 37:446–453, 1995.

D.A. Garvin. Competing in the eight dimensions of quality. *Harvard Business Review*, 87:101–109, 1987.

E.I. George and D.P. Foster. The risk inflation criterion for multiple regression. *The Annals of Statistics*, 22:1947–1975, 1994.

J.D. Gibbons and S. Chakraborti. *Nonparametric statistical inference.* Marcel Dekker, Inc., New York, NY, USA, 4th edition, 2003.

I. Gijbels, A. Lambert, and P. Qiu. Jump-preserving regression and smoothing using local linear fitting: a compromise. *Annals of the Institute of Statistical Mathematics*, 59:235–272, 2007.

A.L. Goel and S.M. Wu. Determination of A.R.L. and a contour nomogram for CUSUM charts to control normal mean. *Technometrics*, 13:221–230, 1971.

E. Gombay. Sequential change-point detection and estimation. *Sequential Analysis*, 22:203–222, 2003.

E. Gombay. Change detection in autoregressive time series. *Journal of Multivariate Analysis*, 99:451–464, 2008.

R.C. Gonzalez and R.E. Woods. *Digital Image Processing.* Prentice Hall, Upper Saddle River, NJ, USA, 3rd edition, 2007.

M.A. Graham, S.W. Human, and S. Chakraborti. A Phase I nonparametric Shewhart-type control chart based on the median. *Journal of Applied Statistics*, 37:1795–1813, 2010.

M.A. Graham, S. Chakraborti, and S.W. Human. A nonparametric exponentially weighted moving average signed-rank chart for monitoring location. *Computational Statistics and Data Analysis*, 55:2490–2503, 2011.

O.A. Grigg, D.J. Spiegelhalter, and H.E. Jones. Local and marginal control charts applied to methicillin resistant Staphylococcus aureus bacteraemia reports in UK acute National Health Service trusts. *Journal of the Royal Statistical Society (Series A)*, 172:49–66, 2009.

F. Gustafsson. *Adaptive Filtering and Change Detection*. John Wiley & Sons, New York, NY, USA, 2000.

S.J. Haberman. Log-linear models and frequency tables with small expected cell counts. *The Annals of Statistics*, 5:1148–1169, 1977.

P. Hackl and J. Ledolter. A control chart based on ranks. *Journal of Quality Technology*, 23:117–124, 1991.

P. Hackl and J. Ledolter. A new nonparametric quality control technique. *Communications in Statistics-Simulation and Computation*, 21:423–443, 1992.

D. Han and F. Tsung. A generalized EWMA control chart and its comparison with the optimal EWMA, CUSUM and GLR schemes. *The Annals of Statistics*, 32:316–339, 2004.

D. Han, F. Tsung, X. Hu, and K. Wang. CUSUM and EWMA multi-charts for detecting a range of mean shifts. *Statistica Sinica*, 17:1139–1164, 2007.

Western Electric Handbook. *Statistical Quality Control Handbook*. Western Electric Corporation, Indianapolis, IN, USA, 1956.

W. Härdle, P. Hall, and J.S. Marron. Regression smoothing parameters that are not far from their optimal. *Journal of the American Statistical Association*, 87:227–233, 1992.

J.D. Hart. Kernel regression estimation with time series errors. *Journal of the Royal Statistical Society (Series B)*, 53:173–187, 1991.

J.D. Hart. *Nonparametric Smoothing and Lack-of-Fit Tests*. Springer-Verlag, New York, NY, USA, 1997.

D.M. Hawkins. On the distribution and power of a test for a single outlier. *South African Statistical Journal*, 3:9–15, 1969.

D.M. Hawkins. Testing a sequence of observations for a shift in location. *Journal of the American Statistical Association*, 72:180–186, 1977.

D.M. Hawkins. A CUSUM for a scale parameter. *Journal of Quality Technology*, 13:228–231, 1981.

D.M. Hawkins. Self-starting cusums for location and scale. *The Statistician*, 36:299–315, 1987.

D.M. Hawkins. Multivariate quality control based on regression-adjusted variables. *Technometrics*, 33:61–75, 1991.

D.M. Hawkins. Evaluation of the average run length for cumulative sum charts for an arbitrary data distribution. *Communications in Statistics (Series B)*, 21:1001–1020, 1992a.

D.M. Hawkins. A fast accurate approximation of average run lengths of cusum control charts. *Journal of Quality Technology*, 24:37–42, 1992b.

D.M. Hawkins. Cumulative sum control charting: an underutilized SPC tool. *Quality Engineering*, 5:463–477, 1993a.

D.M. Hawkins. Regression adjustment for variables in multivariate quality control. *Journal of Quality Technology*, 25:170–182, 1993b.

D.M. Hawkins. Fitting multiple change-point models to data. *Computational Statistics & Data Analysis*, 37:323–341, 2001.

D.M. Hawkins and Q. Deng. A nonparametric change-point control chart. *Journal of Quality Technology*, 42:165–173, 2010.

D.M. Hawkins and E.M. Maboudou-Tchao. Self-starting multivariate exponentially weighted moving average control charting. *Technometrics*, 49:199–209, 2007.

D.M. Hawkins and E.M. Maboudou-Tchao. Multivariate exponentially weighted moving covariance matrix. *Technometrics*, 50:155–166, 2008.

D.M. Hawkins and D.F. Merriam. Zonation of multivariate sequences of digitized geologic data. *Mathematical Geology*, 6:263–269, 1974.

D.M. Hawkins and D.H. Olwell. *Cumulative Sum Charts and Charting for Quality Improvement*. Springer-Verlag, New York, NY, USA, 1998.

D.M. Hawkins and K.D. Zamba. A change-point model for a shift in variance. *Journal of Quality Technology*, 37:21–31, 2005a.

D.M. Hawkins and K.D. Zamba. Statistical process control for shifts in mean or variance using a change-point formulation. *Technometrics*, 47:164–173, 2005b.

D.M. Hawkins, P. Qiu, and C.W. Kang. The changepoint model for statistical process control. *Journal of Quality Technology*, 35:355–366, 2003.

D.M. Hawkins, S. Choi, and S. Lee. A general multivariate exponentially weighted moving-average control chart. *Journal of Quality Technology*, 39:118–125, 2007.

J.D. Healy. A note on multivariate CUSUM procedures. *Technometrics*, 29:409–412, 1987.

R. Henderson. Change-point problem with correlated observations, with an application in material accountancy. *Technometrics*, 28:381–389, 1986.

T. Herberts and U. Jensen. Optimal detection of a change point in a Poisson process for different observation schemes. *Scandinavian Journal of Statistics*, 31:347–366, 2004.

T.P. Hettmansperger. Multivariate location tests. In *Encyclopedia of Statistical Sciences*. John Wiley & Sons, New York, NY, USA, 2006.

D.V. Hinkley. Inference about the change-point in a sequence of random variables. *Biometrika*, 57:1–17, 1970.

P.G. Hoel, S.C. Port, and C.J. Stone. *Introduction to Stochastic Processes*. Waveland Press, Inc., Prospect Heights, Illinois, USA, 1972.

M. Hollander and D.A. Wolfe. *Nonparametric statistical methods*. John Wiley & Sons, New York, NY, USA, 2nd edition, 1999.

D.S. Holmes and A.E. Mergen. Improving the performance of the T^2 control chart. *Quality Engineering*, 5:619–625, 1993.

D.R. Hoover, J.A. Rice, C.O. Wu, and L.P. Yang. Nonparametric smoothing estimates of time-varying coefficient models with longitudinal data. *Biometrika*, 85:809–822, 1998.

J.L. Horowitz and V.G. Spokoiny. An adaptive, rate-optimal test of a parametric mean-regression model against a nonparametric alternative. *Econometrica*, 69: 599–631, 2001.

H. Hotelling. The generalization of student's ratio. *Annals of Mathematical Statistics*, 2:360–378, 1931.

H. Hotelling. Multivariate quality control. *Techniques of Statistical Analysis (C. Eisenhart, M. Hastay, and W.A. Wallis, eds.)*, pages 111–184, 1947.

S.W. Human, P. Kritzinger, and S. Chakraborti. Robustness of the EWMA control chart for individual observations. *Journal of Applied Statistics*, pages 2071–2087, 2011.

L. Huwang, A. Yeh, and C.W. Wu. Monitoring multivariate process variability for individual observations. *Journal of Quality Technology*, 39:258–278, 2007.

L. Huwang, Y.T. Wang, A. Yeh, and Z.J. Chen. On the exponentially weighted moving variance. *Naval Research Logistics*, 56:659–668, 2009.

L. Huwang, C.J. Huang, and Y.H. Wang. New EWMA control charts for monitoring process dispersion. *Computational Statistics and Data Analysis*, 54: 2328–2342, 2010.

J.M. Irvine. *Changes in regime in regression models*. Ph.D. Dissertation, Yale University, New Haven, Connecticut, 1982.

J.E. Jackson. Quality control methods for several related variables. *Technometrics*, 1:359–377, 1959.

J.E. Jackson. Principal components and factor analysis: part I – principal components. *Journal of Quality Technology*, 12:201–213, 1980.

J.E. Jackson and G.S. Mudholkar. Control procedures for residuals associated with principal component analysis. *Technometrics*, 21:341–349, 1979.

W.A. Jensen and J.B. Birch. Profile monitoring via nonlinear mixed models. *Journal of Quality Technology*, 41:18–34, 2009.

W.A. Jensen, L.A. Jones-Farmer, C.W. Champ, and W.H. Woodall. Effects of parameter estimation on control chart properties: a literature review. *Journal of Quality Technology*, 38:349–364, 2006.

W.A. Jensen, J.B. Birch, and W.H. Woodall. Monitoring correlation within linear profiles using mixed models. *Journal of Quality Technology*, 40:167–183, 2008.

W. Jiang. Multivariate control charts for monitoring autocorrelated processes. *Jour-*

nal of Quality Technology, 36:367–379, 2004.

W. Jiang, K.L. Tsui, and W. Woodall. A new SPC monitoring method: the ARMA chart. *Technometrics*, 42:399–410, 2000.

J. Jin and J. Shi. Feature-preserving data compression of stamping tonnage information using wavelets. *Technometrics*, 41:327–339, 1999.

N.L. Johnson and F.C. Leone. Cumulative sum control charts - mathematical principles applied to their construction and use (Parts I, II, III). *Industrial Quality Control*, 19:15–36, 1962.

N.L. Johnson, A.W. Kemp, and S. Kotz. *Univariate Discrete Distributions*. John Wiley & Sons, New York, NY, USA, 2nd edition, 1992.

N.L. Johnson, S. Kotz, and N. Balakrishnan. *Continuous Univariate Distributions (Volume 1)*. John Wiley & Sons, New York, NY, USA, 2nd edition, 1994.

N.L. Johnson, S. Kotz, and N. Balakrishnan. *Continuous Univariate Distributions (Volume 2)*. John Wiley & Sons, New York, NY, USA, 2nd edition, 1995.

R.A. Johnson and M. Bagshaw. The effect of serial correlation on the performance of CUSUM tests. *Technometrics*, 16:103–112, 1974.

R.A. Johnson and D.W. Wichern. *Applied Multivariate Statistical Analysis*. Pearson Education, Inc., Upper Saddle River, NJ, USA, 6th edition, 2007.

L.A. Jones, C.W. Champ, and S.E. Rigdon. The performance of exponentially weighted moving average charts with estimated parameters. *Technometrics*, 43:156–167, 2001.

L.A. Jones, C.W. Champ, and S.E. Rigdon. The run length distribution of the CUSUM with estimated parameters. *Industrial Quality Control*, 36:95–108, 2004.

M. Jones, S. Marron, and S. Sheather. A brief survey of bandwidth selection for density estimation. *Journal of the American Statistical Association*, 91:401–407, 1996.

L.A. Jones-Farmer, V. Jordan, and C.W. Champ. Distribution-free phase I control charts for subgroup location. *Journal of Quality Technology*, 41:304–317, 2009.

J. Joo and P. Qiu. Jump detection in a regression curve and its derivative. *Technometrics*, 51:289–305, 2009.

L. Kang and S.L. Albin. On-line monitoring when the process yields a linear profile. *Journal of Quality Technology*, 32:418–426, 2000.

M.G. Kendall and A. Stuart. *The Advanced Theory of Statistics, Volume 2*. Hafner Publishing Company, New York, NY, USA, 1961.

J.F. Kenney and E.S. Keeping. *Mathematics of Statistics, Part Two*. Van Nostrand Company Inc., Princeton, NJ, USA, 2nd edition, 1951.

M.B. Khoo and S.H. Quah. Multivariate control chart for process dispersion based on individual observations. *Quality Engineering*, 15:639–642, 2003.

H.J. Kim. Change-point detection for correlated observations. *Statistica Sinica*, 6: 275–287, 1996.

K. Kim, M.A. Mahmoud, and W.H. Woodall. On the monitoring of linear profiles. *Journal of Quality Technology*, 35:317–328, 2003.

S.H. Kim, C. Alexopoulos, K.L. Tsui, and J.R. Wilson. A distribution-free tabular CUSUM chart for autocorrelated data. *IIE Transactions*, 39:317–330, 2007.

M. Klein. Two alternatives to the Shewhart $\overline{X}$ control chart. *Journal of Quality Technology*, 32:427–431, 2000.

S. Knoth. Accurate ARL calculation for EWMA control charts monitoring normal mean and variance simultaneously. *Sequential Analysis*, 26:251–263, 2007.

S. Knoth. spc: statistical process control. *R package version 0.4.0*, pages http://CRAN.r–project.org/package=spc, 2011.

K. Koehler. Goodness-of-fit tests for loglinear models in sparse contingency tables. *Journal of the American Statistical Association*, 81:483–493, 1986.

K. Koehler and K. Larntz. An empirical investigation of goodness-of-fit statistics for sparse multinomials. *Journal of the American Statistical Association*, 75: 336–344, 1980.

R. Koenker. *Quantile Regression*. Cambridge University Press, Cambridge, UK, 2005.

P. Kokoszka and R. Leipus. Change-point in the mean of dependent observations. *Statistics and Probability Letters*, 40:385–393, 1998.

A.N. Kolmogorov. Sulla determinazione empirica di una legge di distribuzione. *Giornio Instituto Italia Attuari*, 4:83–91, 1933.

S. Kotz and N.L. Johnson. Process capability indices: a review 1992–2000 (with discussion). *Journal of Quality Technology*, 34:2–19, 2002.

S. Kotz and C.R. Lovelace. *Process Capability Indices in Theory and Practice*. Hodder Arnold Publication, London, UK, 1998.

H. Kramer and W. Schmid. EWMA charts for multivariate time series. *Sequential Analysis*, 16:131–154, 1997.

P.H. Kvam and B. Vidakovic. *Nonparametric Statistics with Applications to Science and Engineering*. John Wiley & Sons, New York, NY, USA, 2007.

T.L. Lai. Sequential change-point detection in quality control and dynamical systems (with discussions). *Journal of the Royal Statistical Society (Series B)*, 57: 613–658, 1995.

T.L. Lai. Sequential analysis: some classical problems and new challenges (with discussions). *Statistica Sinica*, 11:303–408, 2001.

N.M. Laird and J.H. Ware. Random effects models for longitudinal data. *Biometrics*, 38:963–974, 1982.

M. Lavielle. Using penalized contrasts for the change-points problems. *Signal Processing*, 85:1501–1510, 2005.

M. Lavielle and C. Ludena. The multiple change-points problem for the spectral distribution. *Bernoulli*, 6:845–869, 2000.

M. Lavielle and E. Moulines. Least-squares estimation of an unknown number of shifts in a time series. *Journal of Time Series Analysis*, 21:33–59, 2000.

D.N. Lawley. A general method for approximation to the distribution of likelihood ratio criteria. *Biometrika*, 43:295–303, 1956.

E. Lebarbier. Detecting multiple change-points in the mean of a Gaussian process by model selection. *Signal Processing*, 85:717–736, 2005.

E.L. Lehmann and G. Casella. *Theory of Point Estimation*. Springer-Verlag, New York, NY, USA, 2nd edition, 1998.

E.L. Lehmann and J.P. Romano. *Testing Statistical Hypotheses*. Springer-Verlag, New York, NY, USA, 3rd edition, 2005.

S.M. Lesch and D.R. Jeske. Some suggestions for teaching about normal approximation to Poisson and binomial distribution functions. *The American Statistician*, 63:274–277, 2009.

W. Levinson, D.S. Holmes, and A.E. Mergen. Variation charts for multivariate processes. *Quality Engineering*, 14:539–545, 2002.

W.A. Levinson. *Statistical Process Control for Real-World Applications*. CRC Press, Boca Raton, FL, USA, 2010.

B. Li, K. Wang, and A.B. Yeh. Monitoring covariance matrix via penalized likelihood estimation. *IIE Transactions*, 45:132–146, 2013a.

S.Y. Li, L.C. Tang, and S.H. Ng. Nonparametric CUSUM and EWMA control charts for detecting mean shifts. *Journal of Quality Technology*, 42:209–226, 2010.

Z. Li and P. Qiu. Statistical process control using dynamic sampling. *Technometrics*, accepted, 2013.

Z. Li, P. Qiu, S. Chatterjee, and Z. Wang. Using *p*-values to design statistical process control charts. *Statistical Papers*, 54:523–539, 2013b.

X. Lin and R.J. Carroll. Nonparametric function estimation for clustered data when the predictor is measured without/with error. *Journal of the American Statistical Association*, 95:520–534, 2000.

S. Ling. Testing for change points in time series models and limiting theorems for NED sequences. *The Annals of Statistics*, 35:1213–1237, 2007.

R.Y. Liu. On a notion of data depth based on random simplices. *Annals of Statistics*, 18:405–414, 1990.

R.Y. Liu. Control charts for multivariate processes. *Journal of the American Statistical Association*, 90:1380–1387, 1995.

R.Y. Liu and K. Singh. A quality index based on data depth and multivariate rank tests. *Journal of the American Statistical Association*, 88:257–260, 1993.

R.Y. Liu, J.M. Parelius, and K. Singh. Multivariate analysis by data depth: descriptive statistics, graphics and inference (with discussions). *Annals of Statistics*, 27:783–858, 1999.

R.Y. Liu, K. Singh, and J.H. Teng. DDMA-charts: nonparametric multivariate moving average control charts based on data depth. *Allgemeines Statisches Archiv*, 88:235–258, 2004.

W. Liu. *Some Charting Methodologies in MSPC (Ph.D. Thesis)*. School of Statistics, University of Minnesota, Minneapolis, MN, USA, 2010.

C.R. Loader. Change point estimation using nonparametric regression. *The Annals of Statistics*, 24:1667–1678, 1996.

C.R. Loader. Bandwidth selection: classical or plug-in? *The Annals of Statistics*, 27:415–438, 1999.

G. Lorden. Procedures for reacting to a change in distribution. *Annals of Mathematical Statistics*, 42:1897–1908, 1971.

C.A. Lowry and D.C. Montgomery. A review of multivariate control charts. *IIE Transactions*, 27:800–810, 1995.

C.A. Lowry, W.H. Woodall, C.W. Champ, and S.E. Rigdon. A multivariate exponentially weighted moving average control chart. *Technometrics*, 34:46–53, 1992.

C.W. Lu and M.R. Reynolds, Jr. Control chart for monitoring the mean and variance of autocorrelated processes. *Journal of Quality Technology*, 31:259–274, 1999.

C.W. Lu and M.R. Reynolds, Jr. Control charts for monitoring an autocorrelated process. *Journal of Quality Technology*, 33:316–334, 2001.

J.M. Lucas. A modified V-mask control scheme. *Technometrics*, 15:833–847, 1973.

J.M. Lucas. Combined Shewhart-CUSUM quality control schemes. *Journal of Quality Technology*, 14:51–59, 1982.

J.M. Lucas and R.B. Crosier. Fast initial response for CUSUM quality control schemes: give your CUSUM a head start. *Technometrics*, 24:199–205, 1982a.

J.M. Lucas and R.B. Crosier. Robust CUSUM: a robust study for CUSUM quality control schemes. *Communications in Statistics-Theory and Methods*, 11:2669–2687, 1982b.

J.M. Lucas and M.S. Saccucci. Exponentially weighted moving average control schemes: properties and enhancements (with discussions). *Technometrics*, 32:1–29, 1990.

A. Luceño. A process capability ratio with reliable confidence intervals. *Communications in Statistics - Simulation and Computation*, 25:235–246, 1996.

A. Luceño and J. Puig-Pey. Evaluation of the run-length probability distribution for CUSUM charts: assessing chart performance. *Technometrics*, 42:411–416, 2000.

Y. Luo, Z. Li, and Z. Wang. Adaptive CUSUM control chart with variable sampling intervals. *Computational Statistics and Data Analysis*, 53:2693–2701, 2009.

J.F. MacGregor and T.J. Harris. The exponentially weighted moving variance. *Journal of Quality Technology*, 25:106–118, 1993.

M.A. Mahmoud and W.H. Woodall. Phase I analysis of linear profiles with calibration applications. *Technometrics*, 46:380–391, 2004.

M.A. Mahmoud, P.A. Parker, W.H. Woodall, and D.M. Hawkins. A change point method for linear profile data. *Quality and Reliability Engineering International*, 23:247–268, 2007.

C.L. Mallows. Some comments on C_p. *Technometrics*, 15:661–675, 1973.

P.E. Maravelakis and P. Castagliola. An EWMA chart for monitoring the process standard deviation when parameters are estimated. *Computational Statistics and Data Analysis*, 53:2653–2664, 2009.

R.L. Mason and J.C. Young. *Multivariate Statistical Process Control with Industrial Applications*. ASA/SIAM, Philadelphia, PA, USA, 2002.

R.L. Mason, Y.M. Chou, and J.C. Young. Applying Hotelling's T^2 statistic to batch processes. *Journal of Quality Technology*, 33:466–479, 2001.

D. McDonald. A CUSUM procedure based on sequential ranks. *Naval Research Logistics*, 37:627–646, 1990.

F.M. Megahed, J.L.K. Kensler, K. Bedair, and W.H. Woodall. A note on the ARL of two-sided Bernoulli-based CUSUM control charts. *Journal of Quality Technology*, 43:43–49, 2011a.

F.M. Megahed, W.H. Woodall, and J.A. Camelio. A review and perspective on control charting with image data. *Journal of Quality Technology*, 43:83–98, 2011b.

F.M. Megahed, L.J. Wells, J.A. Camelio, and W.H. Woodall. A spatiotemporal method for the monitoring of image data. *Quality and Reliability Engineering International*, 28:967–980, 2012.

Y. Mei. Sequential change-point detection when unknown parameters are present in the pre-change distribution. *Annals of Statistics*, 34:92–122, 2006.

O. Mestek, J. Pavlik, and M. Suchánek. Multivariate control charts: control charts for calibration curves. *Fresenius' Journal of Analytical Chemistry*, 350:344–351, 1994.

D.C. Montgomery. *Introduction to Statistical Quality Control*. John Wiley & Sons, New York, NY, USA, 6th edition, 2009.

D.C. Montgomery and C.M. Mastrangelo. Some statistical process control methods for autocorrelated data. *Journal of Quality Technology*, 23:179–204, 1991.

L.W. Morrison. The use of control charts to interpret environmental monitoring data. *Natural Areas Journal*, 28:66–73, 2008.

S. Mousavi and M.R. Reynolds, Jr. A CUSUM chart for monitoring a proportion with autocorrelated binary observations. *Journal of Quality Technology*, 41:401–414, 2009.

G.V. Moustakides. Optimal stopping times for detecting changes in distributions.

The Annals of Statistics, 14:1379–1387, 1986.

C.H. Müller. Robust estimators for estimating discontinuous functions. *Metrika*, 55:99–109, 2002.

H.G. Müller. Change-points in nonparametric regression analysis. *The Annals of Statistics*, 20:737–761, 1992.

E.A. Nadaraya. On estimating regression. *Theory of Probability and Its Applications*, 9:141–142, 1964.

J. Needham. *Science and Civilisation in China*. Cambridge University Press, New York, NY, USA, 1986.

L.S. Nelson. Tables for a precedence life test. *Technometrics*, 5:491–499, 1963.

L.S. Nelson. The Shewhart control chart - tests for special causes. *Journal of Quality Technology*, 16:237–239, 1984.

L.S. Nelson. Tests on early failures - the precedence life test. *Journal of Quality Technology*, 25:140–149, 1993.

R. Noorossana, A. Saghaei, and A. Amir. *Statistical Analysis of Profile Monitoring*. John Wiley & Sons, New York, NY, USA, 2011.

H. Oja. *Multivariate Nonparametric Methods with R*. Springer-Verlag, New York, NY, USA, 2010.

H. Oja and R. Randles. Multivariate nonparametric tests. *Statistical Science*, 19: 598–605, 2004.

D.A. Olteanu. *Cumulative Sum Control Charts for Censored Reliability Data, Ph.D. Thesis*. Department of Statistics, Virginia Polytechnic Institute and State University, Blacksburg, Virginia, USA, 2010.

E.S. Page. Continuous inspection scheme. *Biometrika*, 41:100–115, 1954.

E.S. Page. Cumulative sum control charts. *Technometrics*, 3:1–9, 1961.

X. Pan and J. Jarrett. Applying state space to SPC: monitoring multivariate time series. *Journal of Applied Statistics*, 31:397–418, 2004.

X. Pan and J. Jarrett. Using vector autoregressive residuals to monitor multivariate processes in the presence of serial correlation. *International Journal of Production Economics*, 106:204–216, 2007.

E.A. Pappanastos and B.M. Adams. Alternative designs of the Hodges-Lehmann control chart. *Journal of Quality Technology*, 28:213–223, 1996.

C. Park and M.R. Reynolds, Jr. Nonparametric procedures for monitoring a location parameter based on linear placement statistics. *Sequential Analysis*, 6:303–323, 1987.

E. Parzen. On estimation of a probability density function and mode. *Annals of Mathematical Statistics*, 33:1065–1076, 1962.

F. Pascual. EWMA Charts for the Weibull Shape Parameter. *Journal of Quality Technology*, 42:400–416, 2010.

K. Paynabar, J. Jin, and M. Pacella.   Analysis of multichannel nonlinear profiles using uncorrelated multilinear principal component analysis with applications in fault detection and diagnosis. *IIE Transactions*, page DOI:10.1080/0740817X.2013.770187, 2013.

W.L. Pearn, S. Kotz, and N.L. Johnson. Distributional and inferential properties of process capability indices. *Journal of Quality Technology*, 24:216–231, 1992.

K. Pearson. On the criterion that a given system of deviations from the probable in the case of a correlated system of variables is such that it can be reasonably supposed to have arisen from random sampling. *Philosophical Magazine*, 50: 157–175, 1900.

R. Peck and J. Devore. *Statistics: The Exploration and Analysis of Data*. Duxbury-Thomson Brooks/Cole, Pacific Grove, CA, USA, 7th edition, 2012.

M.B. Perry and J.J. Pignatiello, Jr. Estimation of the change point of the process fraction nonconforming in SPC applications. *International Journal of Reliability, Quality and Safety Engineering*, 12:95–110, 2005.

M.B. Perry and J.J. Pignatiello, Jr. A change point model for the location parameter of exponential family densities. *IIE Transactions*, 40:947–956, 2008.

M.B. Perry, J.J. Pignatiello, Jr., and J.R. Simpson. A magnitude-robust control chart for monitoring and estimating step changes in a Poisson rate parameter. *International Journal of Reliability, Quality and Safety Engineering*, 14:1–19, 2007.

M.B. Perry, G.R. Mercado, and J.J. Pignatiello, Jr. Phase II monitoring of covariance stationary autocorrelated processes. *Quality and Reliability Engineering International*, 27:35–45, 2011.

A.N. Pettitt. A non-parametric approach to the change-point problem. *Applied Statistics*, 28:126–135, 1979.

J.J. Pignatiello, Jr. and T.R. Samuel. Estimation of the change point of a normal process mean in SPC applications. *Journal of Quality Technology*, 33:82–95, 2001.

W.K. Pratt. *Digital Image Processing*. John Wiley & Sons, New York, NY, USA, 4th edition, 2007.

P. Qiu. Estimation of the number of jumps of the jump regression functions. *Communications in Statistics-Theory and Methods*, 23:2141–2155, 1994.

P. Qiu. Discontinuous regression surfaces fitting. *The Annals of Statistics*, 26: 2218–2245, 1998.

P. Qiu. A jump-preserving curve fitting procedure based on local piecewise-linear kernel estimation. *Journal of Nonparametric Statistics*, 15:437–453, 2003.

P. Qiu. The local piecewisely linear kernel smoothing procedure for fitting jump regression surfaces. *Technometrics*, 46:87–98, 2004.

P. Qiu. *Image Processing and Jump Regression Analysis*. John Wiley & Sons, New York, NY, USA, 2005.

P. Qiu. Distribution-free multivariate process control based on log-linear modeling.

IIE Transactions, 40:664–677, 2008.

P. Qiu and D.M. Hawkins. A rank based multivariate CUSUM procedure. *Technometrics*, 43:120–132, 2001.

P. Qiu and D.M. Hawkins. A nonparametric multivariate CUSUM procedure for detecting shifts in all directions. *Journal of the Royal Statistical Society (Series D) - The Statistician*, 52:151–164, 2003.

P. Qiu and Z. Li. On nonparametric statistical process control of univariate processes. *Technometrics*, 53:390–405, 2011a.

P. Qiu and Z. Li. Distribution-free monitoring of univariate processes. *Statistics and Probability Letters*, 81:1833–1840, 2011b.

P. Qiu and D. Xiang. Multivariate dynamic screening system for identifying irregular longitudinal patterns. *in review*, 2013.

P. Qiu and D. Xiang. Univariate dynamic screening system: an approach for identifying individuals with irregular longitudinal behavior. *Technometrics*, 56:in press, 2014.

P. Qiu and C. Xing. On nonparametric image registration. *Technometrics*, 55:174–188, 2013a.

P. Qiu and C. Xing. Feature based image registration using non-degenerate pixels. *Signal Processing*, 93:706–720, 2013b.

P. Qiu and B. Yandell. A local polynomial jump detection algorithm in nonparametric regression. *Technometrics*, 40:141–152, 1998.

P. Qiu and C. Zou. Control chart for monitoring nonparametric profiles with arbitrary design. *Statistica Sinica*, 20:1655–1682, 2010.

P. Qiu, C. Asano, and X. Li. Estimation of jump regression function. *Bulletin of Informatics and Cybernetics*, 24:197–212, 1991.

P. Qiu, C. Zou, and Z. Wang. Nonparametric profile monitoring by mixed effects modeling (with discussions). *Technometrics*, 52:265–293, 2010.

C.P. Quesenberry. *SPC Methods for Quality Improvement*. John Wiley & Sons, New York, NY, USA, 1997.

M.R. Reynolds, Jr. A sequential signed-rank test for symmetry. *Annals of Statistics*, 3:382–400, 1975.

M.R. Reynolds, Jr. and J.C. Arnold. EWMA control charts with variable sample sizes and variable sampling intervals. *IIE Transactions*, 33:511–530, 2001.

M.R. Reynolds, Jr. and Z.G. Stoumbos. A general approach to modeling CUSUM charts for a proportion. *IIE Transactions*, 32:515–535, 2000.

M.R. Reynolds, Jr. and Z.G. Stoumbos. Should exponentially weighted moving average and cumulative sum charts be used with Shewhart limits. *Technometrics*, 47:409–424, 2005.

M.R. Reynolds, Jr. and Z.G. Stoumbos. Comparisons of some exponentially weighted moving average control charts for monitoring the process mean and variance. *Technometrics*, 48:550–567, 2006.

M.R. Reynolds, Jr. and Z.G. Stoumbos. Comparisons of multivariate Shewhart and MEWMA control charts for monitoring the mean vector and covariance matrix. *Journal of Quality Technology*, 40:381–393, 2008.

M.R. Reynolds, Jr., R.W. Amin, and J.C. Arnold. CUSUM charts with variable sampling intervals. *Technometrics*, 32:371–384, 1990.

J.A. Rice and C.O. Wu. Nonparametric mixed effects models for unequally sampled noisy curves. *Biometrics*, 57:253–259, 2001.

Y. Ritov. Decision theoretic optimality of the CUSUM procedure. *The Annals of Statistics*, 18:1464–1469, 1990.

S.V. Roberts. Control chart tests based on geometric moving averages. *Technometrics*, 1:239–250, 1959.

D.M. Rocke. Robust control charts. *Technometrics*, 31:173–184, 1989.

R.N. Rodriguez. Recent developments in process capability analysis. *Journal of Quality Technology*, 24:176–187, 1992.

M. Rosenblatt. Remarks on some nonparametric estimates of a density function. *Annals of Mathematical Statistics*, 27:832–837, 1956.

G.J. Ross. cpm: sequential parametric and nonparametric change detection. *R package version 1.0*, pages http://CRAN.r–project.org/package=cpm, 2012.

G.J. Ross and N.M. Adams. Two nonparametric control charts for detecting arbitrary distribution changes. *Journal of Quality Technology*, 44:102–116, 2012.

G.J. Ross, D.K. Tasoulis, and N.M. Adams. Nonparametric monitoring of data streams for changes in location and scale. *Technometrics*, 53:379–389, 2011.

T.D. Ross. Accurate confidence intervals for binomial proportion and Poisson rate estimation. *Computers in Biology and Medicine*, 33:509–531, 2003.

G.C. Runger and T.R. Willemain. Model-based and model-free control of autocorrelated processes. *Journal of Quality Technology*, 27:283–292, 1995.

G.C. Runger and T.R. Willemain. Batch means control charts for autocorrelated data. *IIE Transactions*, 28:483–487, 1996.

D. Ruppert, S.J. Sheather, and M.P. Wand. An efficient bandwidth selector for local least squares regression. *Journal of the American Statistical Association*, 90:1257–1270, 1995.

T.P. Ryan. *Statistical Methods for Quality Improvement*. John Wiley & Sons, New York, NY, USA, 2nd edition, 2000.

J.H. Ryu, H. Wan, and S. Kim. Optimal design of a CUSUM chart for a mean shift of unknown size. *Journal of Quality Technology*, 42:1–16, 1995.

T.R. Samuel and J.J. Pignatiello, Jr. Estimation of the change point of a normal process mean in SPC applications. *Journal of Quality Technology*, 33:82–95, 2001.

E. Santos-Fernandez. msqc: multivariate statistical quality control. *R package ver-*

sion 1.0.0, pages http://CRAN.r–project.org/package=msqc, 2012.

D.W. Scott. *Multivariate Density Estimation*. John Wiley & Sons, New York, NY, USA, 1992.

L. Scrucca. qcc: quality control charts. *R package version 2.0.1*, pages http://CRAN.r–project.org/package=qcc, 2010.

G.A.F. Seber and C.J. Wild. *Nonlinear Regression*. John Wiley & Sons, New York, NY, USA, 2003.

X. Shao and X. Zhang. Testing for change points in time series. *Journal of the American Statistical Association*, 105:1228–1240, 2010.

S.S. Shapiro and M.B. Wilk. An analysis of variance test for normality (complete samples). *Biometrika*, 52:591–611, 1965.

W.A. Shewhart. *Economic control of quality of manufactured product*. D. Van Nostrand Company, New York, NY, USA, 1931.

M. Shi, Weiss R.E., and J.M.G. Taylor. An analysis of paediatric CD4 counts for acquired immune deficiency syndrome using flexible random curves. *Applied Statistics*, 45:151–163, 1996.

L. Shu and W. Jiang. A Markov chain model for the adaptive CUSUM control chart. *Journal of Quality Technology*, 38:135–147, 2006.

L. Shu and W. Jiang. A new EWMA chart for monitoring process dispersion. *Journal of Quality Technology*, 40:319–331, 2008.

L. Shu, W. Jiang, and K.L. Tsui. A weighted CUSUM chart for detecting patterned mean shifts. *Journal of Quality Technology*, 40:194–213, 2008.

D. Siegmund. *Sequential Analysis Tests and Confidence Intervals*. Springer-Verlag, New York, USA, 1985.

G. Simons. An improved statement of optimality for sequential probability ratio tests. *The Annals of Statistics*, 4:1240–1243, 1976.

N.V. Smirnov. On the estimation of the discrepancy between empirical curves of distribution for two independent samples. *Bulletin Moscow University*, 2:3–16, 1939.

A.F.M. Smith. A Bayesian approach to inference about a change-point in a sequence of random variables. *Biometrika*, 62:407–416, 1975.

R.S. Sparks. CUSUM charts for signalling varying location shifts. *Journal of Quality Technology*, 32:157–171, 2000.

R.S. Sparks, T. Keighley, and D. Muscatello. Optimal exponentially weighted moving average (EWMA) plans for detecting seasonal epidemics when faced with non-homogeneous negative binomial counts. *Journal of Applied Statistics*, 38: 2165–2181, 2011.

F. Spiring, B. Leung, S. Cheng, and A. Yeung. A bibliography of process capability papers. *Quality and Reliability Engineering International*, 19:445–460, 2003.

V. Spokoiny. Multiscale local change point detection with applications to value-at-risk. *Annals of Statistics*, 37:1405–1436, 2009.

S.H. Steiner. Grouped data exponentially weighted moving average control charts. *Applied Statistics*, 47:203–216, 1998.

S.H. Steiner and R.J. MacKay. Monitoring processes with highly censored data. *Journal of Quality Technology*, 32:199–208, 2000.

S.H. Steiner, P.L. Geyer, and G.O. Wesolowsky. Grouped data-sequential probability ratio tests and cumulative sum control charts. *Technometrics*, 38:230–237, 1996.

Z.G. Stoumbos and L.A. Jones. On the properties and design of individuals control charts based on simplicial depth. *Nonlinear Studies*, 7:147–178, 2000.

Z.G. Stoumbos and J.H. Sullivan. Robustness to non-normality of the multivariate EWMA control chart. *Journal of Quality Technology*, 34:260–276, 2002.

F.S. Stover and R.V. Brill. Statistical quality control applied to ion chromatography calibrations. *Journal of Chromatography*, A804:37–43, 1998.

Y. Su, L. Shu, and K.L. Tsui. Adaptive EWMA procedures for monitoring processes subject to linear drifts. *Computational Statistics and Data Analysis*, 55:2819–2829, 2011.

J.H. Sullivan and L.A. Jones. A self-starting control chart for multivariate individual observations. *Technometrics*, 44:24–33, 2002.

J.H. Sullivan and W.H. Woodall. A comparison of multivariate control charts for individual observations. *Journal of Quality Technology*, 28:398–408, 1996.

R. Sun and F. Tsung. A kernel-distance-based multivariate control chart using support vector methods. *International Journal of Production Research*, 41:2975–2989, 2003.

P.F. Tang and N.S. Barnett. Dispersion control for multivariate processes. *The Australian Journal of Statistics*, 38:235–251, 1996a.

P.F. Tang and N.S. Barnett. Dispersion control for multivariate processes - some comparisons. *The Australian Journal of Statistics*, 38:253–273, 1996b.

M.C. Testik, G.C. Runger, and C.M. Borror. Robustness properties of multivariate EWMA control charts. *Quality and Reliability Engineering International*, 19:31–38, 2003.

R.J. Tibshirani. Regression shrinkage and selection via the LASSO. *Journal of the Royal Statistical Society (Series B)*, 58:267–288, 1996.

D.H. Timmer, J. Pignatiello, and M. Longnecker. The development and evaluation of CUSUM-based control charts for an AR(1) process. *IIE Transactions*, 30:525–534, 1998.

E. Topalidou and S. Psarakis. Review of multinomial and multiattribute quality control charts. *Quality and Reliability Engineering International*, 25:773–804, 2009.

N.D. Tracy, J.C. Young, and R.L. Mason. Multivariate control charts for individual observations. *Journal of Quality Technology*, 24:88–95, 1992.

S.T. Tseng, F. Tsung, and P.Y. Liu. Variable EWMA run-to-run controller for drifted processes. *IIE Transactions*, 39:291–301, 2007.

S.T. Tseng, B.Y. Jou, and C.H. Liao. Adaptive variable EWMA controller for drifted processes. *IIE Transactions*, 42:247–259, 2010.

C.S. Van Dobben de Bruyn. *Cumulative Sum Tests: Theory and Practice*. Lubrecht & Cramer Ltd, London, UK, 1968.

L.N. Vanbrackle and M.R. Reynolds, Jr. EWMA and cusum control charts in the presence of correlation. *Communications in Statistics - Simulation and Computation*, 26:979–1008, 1997.

L.C. Vance. Average run lengths of cumulative sum control charts for controlling normal means. *Journal of Quality Technology*, 18:189–193, 1986.

J.A. Vargas. Robust estimation in multivariate control charts for individual observations. *Journal of Quality Technology*, 35:367–376, 2003.

H.M. Wadsworth, K.S. Stephens, and A.B. Godfrey. *Modern Methods for Quality Control and Improvement*. John Wiley & Sons, New York, NY, USA, 2nd edition, 2002.

A. Wald. Sequential tests of statistical hypotheses. *Annals of Mathematical Statistics*, 16:117–186, 1945.

A. Wald and J. Wolfowitz. Optimum character of the sequential probability ratio test. *Annals of Mathematical Statistics*, 19:326–339, 1948.

K.H. Waldmann. Bounds for the distribution of the run length of one-sided and two-sided CUSUM quality control schemes. *Technometrics*, 28:61–67, 1986.

E. Walker and S.P. Wright. Comparing curves using additive models. *Journal of Quality Technology*, 34:118–129, 2002.

M.P. Wand and M.C. Jones. *Kernel Smoothing*. Chapman & Hall/CRC, Boca Raton, FL, USA, 3rd edition, 1995.

K. Wang and W. Jiang. High-dimensional process monitoring and fault isolation via variable selection. *Journal of Quality Technology*, 41:247–258, 2009.

K. Wang and F. Tsung. Using profile monitoring techniques for a data-rich environment with hugh sample sizes. *Quality and Reliability Engineering International*, 21:677–688, 2005.

Y. Wang. Smoothing spline models with correlated random errors. *Journal of the American Statistical Association*, 93:341–348, 1998.

D.G. Wardell, H. Moskowitz, and R.D. Plante. Run length distributions of special-cause control charts for correlated processes. *Technometrics*, 36:3–17, 1994.

G.S. Watson. Smooth regression analysis. *Sankhya (Series A)*, 26:359–372, 1964.

Y. Wei, Z. Zhao, and D.K.J. Lin. Profile control charts based on nonparametric L_1 regression methods. *Annals of Applied Statistics*, 6:409–427, 2012.

S. Weisberg. *Applied Linear Regression*. John Wiley & Sons, New York, NY, USA, 3rd edition, 2005.

C.H. Weiß. EWMA monitoring of correlated processes of Poisson counts. *Quality Technology & Quantitative Management*, 6:137–153, 2009.

F. Wilcoxon, S.K. Katti, and R.A. Wilcox. Critical values and probability levels for the Wilcoxon rank sum test and the Wilcoxon signed rank test. In *Selected Tables in Mathematical Statistics*, volume I, pages 171–259. American Mathematical Society, Providence, RI, USA, 1972.

S.S. Wilks. The large-sample distribution of the likelihood ratio for testing composite hypotheses. *The Annals of Mathematical Statistics*, 9:60–62, 1938.

T.R. Willemain and G.C. Runger. Designing control charts using an empirical reference distribution. *Journal of Quality Technology*, 28:31–38, 1996.

J.D. Williams, W.H. Woodall, J.B. Birch, and J.H. Sullivan. On the distribution of T^2 statistics based on successive differences. *Journal of Quality Technology*, 38:217–229, 2006.

J.D. Williams, J.B. Birch, W.H. Woodall, and N.M. Ferry. Statistical monitoring of heteroscedastic dose-response profiles from high-throughput screening. *Journal of Agriculture, Biological, and Environmental Statistics*, 12:216–235, 2007a.

J.D. Williams, W.H. Woodall, and J.B. Birch. Statistical monitoring of nonlinear product and process quality profiles. *Quality and Reliability Engineering International*, 23:925–941, 2007b.

W.H. Woodall. The distribution of the run length of one-sided CUSUM procedures for continuous random variables. *Technometrics*, 25:295–301, 1983.

W.H. Woodall. Current research on profile monitoring. *Producáo*, 17:420–425, 2007.

W.H. Woodall and B.M. Adams. The statistical design of CUSUM charts. *Quality Engineering*, 5:559–570, 1993.

W.H. Woodall and M.M. Ncube. Multivariate CUSUM quality-control procedures. *Technometrics*, 27:285–292, 1985.

W.H. Woodall, D.J. Spitzner, D.C. Montgomery, and S. Gupta. Using control charts to monitor process and product profiles. *Journal of Quality Technology*, 36: 309–320, 2004.

K.J. Worsley. On the likelihood ratio test for a shift in location of normal populations. *Journal of the American Statistical Association*, 74:365–367, 1979.

K.J. Worsley. An improved Bonferroni inequality and applications. *Biometrika*, 69: 297–302, 1982.

K.J. Worsley. The power of the likelihood ratio and cumulative sum tests for a change in a binomial probability. *Biometrika*, 70:455–464, 1983.

K.J. Worsley. Confidence regions and test for a change-point in a sequence of exponential family random variables. *Biometrika*, 73:91–104, 1986.

H. Wu and J. Zhang. Local polynomial mixed-effects models for longitudinal data. *Journal of the American Statistical Association*, 97:883–897, 2002.

J.S. Wu and C.K. Chu. Kernel type estimators of jump points and values of a regression function. *The Annals of Statistics*, 21:1545–1566, 1993.

W.B. Wu and Z. Zhao. Inference of trends in time series. *Journal of the Royal Statistical Society (Series B)*, 69:391–410, 2007.

Z. Wu, S. Zhang, and P. Wang. A CUSUM scheme with variable sample sizes and sampling intervals for monitoring the process mean and variance. *Quality and Reliability Engineering International*, 23:157–170, 2007.

Z. Wu, J. Jiao, and Y. Liu. A binomial CUSUM chart for detecting large shifts in fraction nonconforming. *Journal of Applied Statistics*, 35:1267–1276, 2008a.

Z. Wu, M. Yang, W. Jiang, and M.B.C. Khoo. Optimization designs of the combined Shewhart-CUSUM control charts. *Computational Statistics and Data Analysis*, 53:496–506, 2008b.

M. Xie, T.N. Goh, and X.S. Lu. A comparative study of CCC and CUSUM charts. *Quality and Reliability Engineering International*, 14:339–345, 1998.

M. Xie, T.N. Goh, and V. Kuralmani. *Statistical Models and Control Chart for High Quality Processes*. Kluwer, Boston, MA, USA, 2002.

C. Xing and P. Qiu. Intensity based image registration by nonparametric local smoothing. *IEEE Transactions on Pattern Analysis and Machine Intelligence*, 33:2081–2092, 2011.

Y. Yang. Can the strengths of AIC and BIC be shared?–a conflict between model identification and regression estimation. *Biometrika*, 92:937–950, 2005.

Y.C. Yao. Approximating the distribution of the maximum likelihood estimate of the change-point in a sequence of independent random variables. *The Annals of Statistics*, 15:1321–1328, 1987.

Y.C. Yao and S.T. Au. Least-squares estimation of a step function. *Sankhya (Series A)*, 51:370–381, 1989.

E. Yashchin. Analysis of CUSUM and other Markov-type control schemes by using empirical distributions. *Technometrics*, 34:54–63, 1992.

E. Yashchin. Performance of CUSUM control schemes for serially correlated observations. *Technometrics*, 35:37–52, 1993a.

E. Yashchin. Statistical control schemes: methods, applications, and generalizations. *International Statistical Review*, 61:41–66, 1993b.

A.B. Yeh and S. Bhattacharya. A robust process capability index. *Communications in Statistics - Simulation and Computation*, 27:565–589, 1998.

A.B. Yeh and H. Chen. A non-parametric multivariate process capability index. *International Journal of Modeling and Simulation*, 21:218–224, 2001.

A.B. Yeh and D.K.J. Lin. A new variables control chart for simultaneously monitoring multivariate process mean and variability. *International Journal of Reliability, Quality and Safety Engineering*, 9:41–59, 2002.

A.B. Yeh, D.K.J. Lin, H. Zhou, and C. Venkataramani. A multivariate exponentially weighted moving average control chart for monitoring process variability.

Journal of Applied Statistics, 30:507–536, 2003.

A.B. Yeh, L. Huwang, and Y.F. Wu. A likelihood ratio based EWMA control chart for monitoring variability of multivariate normal processes. *IIE Transactions*, 36:865–879, 2004a.

A.B. Yeh, D.K.J. Lin, and C. Venkataramani. Unified CUSUM charts for monitoring process mean and variability. *Quality Technology and Quantitative Management*, 1:65–86, 2004b.

A.B. Yeh, L. Huwang, and Y.F. Wu. A multivariate EWMA control chart for monitoring process variability with individual observations. *IIE Transactions*, 37: 1023–1035, 2005.

A.B. Yeh, D.K.J. Lin, and R.N. McGrath. Multivariate control charts for monitoring covariance matrix: a review. *Quality Technology and Quantitative Management*, 3:415–436, 2006.

A.B. Yeh, L. Huwang, and Y.M. Lee. Profile monitoring for binary response. *IIE Transactions*, 41:931–941, 2009.

A.B. Yeh, L. Huwang, R.N. McGrath, and Z. Zhang. On monitoring process variance with individual observations. *Quality and Reliability Engineering International*, 26:631–641, 2010.

S.A. Yourstone and W.J. Zimmer. Non-normality and the design of control charts for averages. *Decision Sciences*, 23:1099–1113, 1992.

G. Yu, C. Zou, and Z. Wang. Outlier detection in functional observations with applications to profile monitoring. *Technometrics*, 54:308–318, 2012.

B.J. Yum and K.W. Kim. A bibliography of the literature on process capability indices: 2000–2009. *Quality and Reliability Engineering International*, 27: 251–268, 2011.

K.D. Zamba and D.M. Hawkins. A multivariate change-point model for statistical process control. *Technometrics*, 48:539–549, 2006.

K.D. Zamba and D.M. Hawkins. A multivariate change-point model for change in mean vector and/or covariance structure. *Journal of Quality Technology*, 41: 285–303, 2009.

D. Zhang, X. Lin, J. Raz, and M. Sowers. Semiparametric stochastic mixed models for longitudinal data. *Journal of the American Statistical Association*, 93:710–719, 1998.

J. Zhang. Powerful goodness-of-fit tests based on likelihood ratio. *Journal of the Royal Statistical Society (Series B)*, 64:281–294, 2002.

L. Zhang and G. Chen. EWMA charts for monitoring the mean of censored Weibull lifetimes. *Journal of Quality Technology*, 36:321–328, 2004.

N.F. Zhang. Detection capability of residual chart for autocorrelated data. *Journal of Applied Statistics*, 24:475–492, 1997.

N.F. Zhang. A statistical control chart for stationary process data. *Technometrics*, 40:24–38, 1998.

N.F. Zhang, G.A. Stenback, and D.M. Wardrop. Interval estimation of process capability index C_{pk}. *Communications in Statistics - Theory and Methods*, 19: 4455–4470, 1990.

P. Zhao and B. Yu. On model selection consistency of lasso. *Journal of Machine Learning Research*, 7:2541–2563, 2006.

Y. Zhao, F. Tsung, and Z. Wang. Dual CUSUM control schemes for detecting a range of mean shifts. *IIE Transactions*, 37:1047–1057, 2005.

C. Zhou, C. Zou, Y. Zhang, and Z. Wang. Nonparametric control chart based on change-point model. *Statistical Papers*, 50:13–28, 2009.

S. Zhou and J. Jin. Automatic feature selection for unsupervised clustering of cycle-based signals in manufacturing processes. *IIE Transactions*, 37:569–584, 2005.

C. Zou and P. Qiu. Multivariate statistical process control using LASSO. *Journal of the American Statistical Association*, 104:1586–1596, 2009.

C. Zou and F. Tsung. Likelihood ratio-based distribution-free EWMA control charts. *Journal of Quality Technology*, 42:1–23, 2010.

C. Zou and F. Tsung. A multivariate sign EWMA control chart. *Technometrics*, 53: 84–97, 2011.

C. Zou, Y. Zhang, and Z. Wang. Control chart based on change-point model for monitoring linear profiles. *IIE Transactions*, 38:1093–1103, 2006.

C. Zou, F. Tsung, and Z. Wang. Monitoring general linear profiles using multi-variate exponentially weighted moving average schemes. *Technometrics*, 49: 395–408, 2007a.

C. Zou, C. Zhou, Z. Wang, and F. Tsung. A self-starting control chart for linear profiles. *Journal of Quality Technology*, 39:364–375, 2007b.

C. Zou, F. Tsung, and Z. Wang. Monitoring profiles based on nonparametric regres-sion methods. *Technometrics*, 50:512–526, 2008.

C. Zou, Y. Liu, and Z. Wang. Comparisons of control schemes for monitoring the mean of processes subject to drifts. *Metrika*, 70:141–163, 2009a.

C. Zou, P. Qiu, and D.M. Hawkins. Nonparametric control chart for monitoring profiles using change point formulation and adaptive smoothing. *Statistica Sinica*, 19:1337–1357, 2009b.

C. Zou, W. Jiang, and F. Tsung. A LASSO-based SPC diagnostic framework for multivariate statistical process control. *Technometrics*, 53:297–309, 2011.

C. Zou, Z. Wang, and F. Tsung. A spatial rank-based multivariate EWMA control chart. *Naval Research Logistics*, 59:91–110, 2012.

H. Zou. The adaptive lasso and its oracle properties. *Journal of the American Statistical Association*, 101:1418–1429, 2006.

H. Zou and R. Li. One-step sparse estimates in nonconcave penalized likelihood models (with discussions). *The Annals of Statistics*, 36:1509–1533, 2008.

H. Zou, T. Hastie, and R. Tibshirani. On the 'Degrees of Freedom' of Lasso. *The Annals of Statistics*, 35:2173–2192, 2007c.

Index

For Product Safety Concerns and Information please contact our EU
representative GPSR@taylorandfrancis.com
Taylor & Francis Verlag GmbH, Kaufingerstraße 24, 80331 München, Germany